Characterization of Protein Therapeutics using Mass Spectrometry

Guodong Chen
Editor

Characterization of Protein Therapeutics using Mass Spectrometry

Springer

Editor
Guodong Chen
Bioanalytical and Discovery Analytical Sciences
Bristol-Myers Squibb
Princeton, NJ
USA

ISBN 978-1-4419-7861-5 ISBN 978-1-4419-7862-2 (eBook)
DOI 10.1007/978-1-4419-7862-2
Springer New York Heidelberg Dordrecht London

Library of Congress Control Number: 2013938173

Printed on acid-free paper

Springer is part of Springer Science+Business Media (www.springer.com)

Preface

Since the introduction of the first recombinant DNA-derived protein insulin in the 1980s, protein therapeutics market has shown a steady growth. Their high efficacy, safety, and ability to treat life-threatening diseases such as cancer, inflammation, and genetic disorders have revolutionized modern medicine. This is primarily due to advances in recombinant DNA technology that have provided the means to produce therapeutic proteins. However, there are significant challenges in characterizing protein therapeutics, including heterogeneity associated with post-translational modifications, protein conformational dynamics upon modifications, and the complexity in analysis due to the presence of biological matrices. Mass spectrometry (MS) is one of the most highly utilized analytical techniques in the characterization of protein therapeutics because of its unique analytical sensitivity, selectivity, and specificity. Advances in ionization methods including electrospray ionization and matrix-assisted laser desorption ionization, the improvement of MS instrumentation, and the growth in the data processing have greatly contributed to wide applications of MS in biopharmaceutical research and development.

I am delighted to bring together the work of contributors from academia and industry in highlighting current analytical approaches and industry practices for the characterization of protein therapeutics using MS. The book begins with an overview on protein MS (Chap. 1), followed by descriptions of quantitative analysis of therapeutic peptides and proteins in biological matrices (Chaps. 2, 3). Structural characterization of protein therapeutics is discussed in Chaps. 4–6 with the focus on modifications of protein therapeutics from discovery to development. A unique class of protein therapeutics, antibody–drug conjugates, is also described in detail (Chap. 7). The remaining chapters (Chaps. 8–10) cover emerging MS techniques for the characterization of protein therapeutics, including hydrogen/deuterium exchange MS, fast photochemical oxidation of proteins, and ion mobility MS.

I would like to acknowledge the special efforts and patience of all the authors, who have made significant contributions to this book.

Guodong Chen

Contents

Contributors

Fotini N. Bazoti The Goulandris Natural History Museum, Kifissia, Greece

Yulia Benitex Bristol-Myers Squibb, Wallingford, CT, USA

Jacob Bongers Bristol-Myers Squibb, East Syracuse, NY, USA

Asish Chakraborty Waters Corporation, Milford, MA, USA

Erin E. Chambers Waters Corporation, Milford, MA, USA

Guodong Chen Bristol-Myers Squibb, Princeton, NJ, USA

Hao Chen Ohio University, Athens, OH, USA

Weibin Chen Waters Corporation, Milford, MA, USA

Ismael Cotte-Rodriguez Procter & Gamble, Loveland, OH , USA

Randall Dere Genentech, South San Francisco, CA, USA

Dieter M. Drexler Bristol-Myers Squibb, Wallingford, CT, USA

Jinmei Fu Bristol-Myers Squibb, East Syracuse, NY, USA

Yunping Huang Bristol-Myers Squibb, Hopewell, NJ, USA

Lisa M. Jones Indiana University-Purdue University, Indianapolis, IN, USA

Surinder Kaur Genentech, South San Francisco, CA, USA

Luna Liu Genentech, South San Francisco, CA, USA

Peiran Liu Bristol-Myers Squibb, Hopewell, NJ, USA

Richard Ludwig Bristol-Myers Squibb, Hopewell, NJ, USA

John Mehl Bristol-Myers Squibb, Princeton, NJ, USA

Zhixin Miao Ohio University, Athens, OH, USA

Jingjie Mo Bristol-Myers Squibb, Princeton, NJ, USA

Timothy V. Olah Bristol-Myers Squibb, Princeton, NJ, USA

Asoka Ranasinghe Bristol-Myers Squibb, Princeton, NJ, USA

Reb Russell Bristol-Myers Squibb, Bloomsbury, NJ, USA

Ola Saad Genentech, South San Francisco, CA, USA

Eric Shields Bristol-Myers Squibb, Bloomsbury, NJ, USA

Tim Slattery Genentech, South San Francisco, CA, USA

Bogdan Sleczka Bristol-Myers Squibb, Princeton, NJ, USA

James Smalley Bristol-Myers Squibb, Hopewell, NJ, USA

Hangtian Song Bristol-Myers Squibb, East Syracuse, NY, USA

Li Tao Bristol-Myers Squibb, Hopewell, NJ, USA

Anthony Tsarbopoulos University of Athens Medical School, Athens, Greece

Adrienne A. Tymiak Bristol-Myers Squibb, Princeton, NJ, USA

Hui Wei Bristol-Myers Squibb, Princeton, NJ, USA

Richard L. Wong Bristol-Myers Squibb, Hopewell, NJ, USA

Steven Wu Bristol-Myers Squibb, Princeton, NJ, USA

Wei Wu Bristol-Myers Squibb, Bloomsbury, NJ, USA

Baomin Xin Bristol-Myers Squibb, Hopewell, NJ, USA

Keyang Xu Genentech, South San Francisco, CA, USA

Hongwei Zhang Bristol-Myers Squibb, Hopewell, NJ, USA

Yun Zhang Ohio University, Athens, OH, USA

Yongxin Zhu Bristol-Myers Squibb, Princeton, NJ, USA

Chapter 1
Introduction to Protein Mass Spectrometry

Ismael Cotte-Rodriguez, Zhixin Miao, Yun Zhang and Hao Chen

1.1 Introduction

Proteins fulfill a plethora of biochemical functions within every living organism, and mass spectrometry (MS) has become one of the most powerful and popular modern physical–chemical methods to study the complexities of proteins. In particular, the invention of matrix-assisted laser desorption/ionization (MALDI) [1] and electrospray ionization (ESI) technologies [2, 3] allows one to measure protein molecular weights and sequences, and to probe conformations and post-translational modifications of proteins. In addition, the mass range of species amenable for MS analysis has increased, enabling the transfer of ionized non-covalent species with masses well over one million (e.g., 1.5 MDa 24-Mer flavoprotein vanillyl-alcohol oxidase (VAO) from Penicillium simplicissimum [4] into the gas phase). These advances moved MS into the range of intact protein oligomers and functional machineries.

This chapter serves as an introduction to protein MS. As it is a broad topic with a vast literature coverage [5–17], we first introduce spray- and laser-based protein ionization techniques used for protein ionization, beginning with an introduction on the historical development of protein ionization methods, followed by the description of several methods including their principles, strengths, and analytical applications. In addition, we also survey various ion activation methods used for protein/peptide structure analysis (viz. tandem mass spectrometry), including collision-, photon-, surface-, and electron-based ion dissociation strategies.

I. Cotte-Rodriguez
Procter & Gamble, Cincinnati, OH, USA

Z. Miao · Y. Zhang · H. Chen (✉)
Department of Chemistry and Biochemistry, Ohio University, 391 Clippinger Laboratory 100 University Terrace, 45701 Athens, OH, USA
e-mail: chenh2@ohio.edu

G. Chen (ed.), *Characterization of Protein Therapeutics using Mass Spectrometry*,
DOI: 10.1007/978-1-4419-7862-2_1,

1.2 History of the Development of Protein Mass Spectrometry

MS originates from nineteenth century physics regarding the nature of matter. The first known mass spectrometer was built by Sir J. J. Thomson in the early 1900s to study and measure the *m/z* values of the "corpuscles" that make up "positive rays" [18], a type of radiation initially observed by German physicist Eugen Goldstein. Following the seminal work of Thomson, MS underwent countless improvements in instrumentation, ionization methods, and applications. The classical ionization method, electron ionization (EI), was devised by Dempster and improved later by Bleakney [19] and Nier [20] and became a widely used standard for ionization of gases and volatile organic molecules. This ionization technique requires extensive derivatization for non-volatile molecules and evaporation of the analyte to the ion source, and it involves numerous fragmentation and rearrangement reactions.

Peptide applications of MS were begun in the late 1950s by Biemann [21] and McLafferty [22]. The first methods that allowed analysis of non-derivatized peptides were field desorption (FD) and chemical ionization (CI) developed in the 1960s [23, 24]. Ionization by CI is achieved by interaction of its volatile molecules with reagent ions. CI allows ionization without a significant degree of ion fragmentation but still requires gas-phase samples. FD was reported by Beckey in 1969 [25], in which electron tunneling triggered by a very high electric field resulted in ionization of gaseous analyte molecules.

It was plasma desorption (PD) [26] and fast atom bombardment (FAB) [27] that opened the way to protein analysis. PD ionization, invented by Macfarlane in 1976 [28], a breakthrough in the analysis of solid samples, involves ionization of materials in the solid state by bombardment with ions or neutral atoms generated by nuclear fission of californium isotope ^{252}Cf. In 1982, Sundqvist obtained the first spectrum of the protein insulin (Fig. 1.1), using bombardment with a beam of 90 MeV $^{127}I^{20+}$ ions from a tandem accelerator [26]. Later, FAB, involving the focusing of a beam of neutral atoms or molecules on a sample contained in a liquid matrix, was implemented for the ionization of proteins up to 24 kDa [29]. In 1983, Blakely and Vestal [30] introduced thermospray ionization (TSI) to produce ions from an aqueous solution sprayed directly into a mass spectrometer. Thermospray is a form of atmospheric pressure ionization in MS, transferring ions from the liquid phase to the gas phase for analysis. It was particularly useful in coupling liquid chromatography with MS [31].

The breakthrough for large molecule laser desorption/ionization analysis came in 1987 when Tanaka combined 30 nm cobalt particles in glycerol with a 337 nm nitrogen laser for ionization which showed singly charged protein molecular ions up to about 35 kDa [32]. During that time [33], MALDI [1], first reported in 1985 by Hillenkamp, Karas, and their colleagues, emerged as the culmination of a long series of experiments using desorption ionization (DI). MALDI is a soft ionization technique for the analysis of biomolecules and large organic molecules and has gained wide success in protein analysis, particularly when coupled with time-of-flight (TOF) instruments [34, 35].

Another breakthrough occurred in 1984 when Fenn and co-workers used electrospray to ionize biomolecules [2]; the first ESI analyses of biopolymers

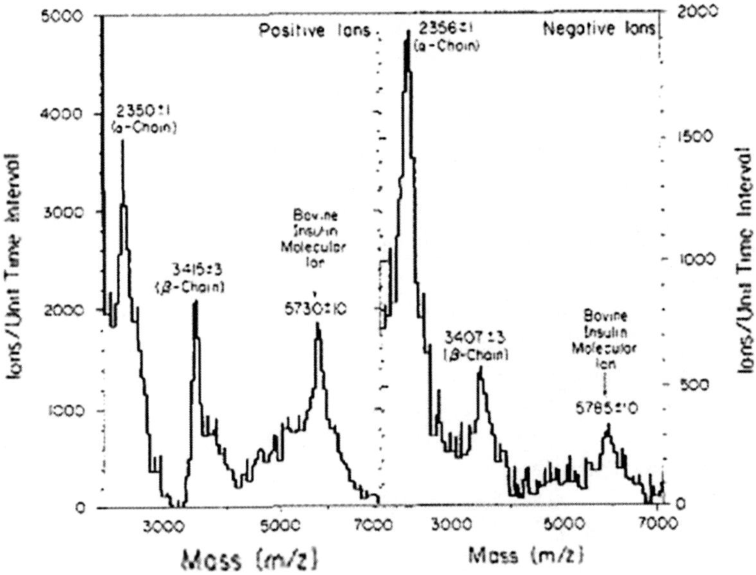

Fig. 1.1 ^{127}I-PDMS spectra of bovine insulin recorded over a 1.5 h period with a 90 MeV ^{127}I (+20) beam current [26]. Reproduced with permission from ACS, copyright 1982

including proteins were published in 1989 [3]. MALDI and ESI have conquered protein MS since their invention in 1980s, and they have triggered an explosion in applications of MS for protein studies [36].

The advent of ambient ionization methods [37, 38] is a recent advancement in the field, in which sample analysis can be conducted with little or no sample preparation. Desorption electrospray ionization (DESI) [39] and direct analysis in real time (DART) [40] are the first two representative methods of this family. There have been about 30 ambient ionization methods for MS reported [37], such as electrospray-assisted laser desorption/ionization (ELDI) [41], laser ablation electrospray ionization mass spectrometry (LAESI) [42], laser desorption electrospray ionization (LDESI) [43], laser-induced acoustic desorption electrospray ionization (LIAD-ESI) [44], matrix-assisted laser desorption electrospray ionization (MALDESI) [45], and radio frequency acoustic desorption and ionization (RADIO) [46], which have been used for protein ionization.

1.3 Ionization Methods

1.3.1 Electrospray Ionization

The principle of ESI was first described by Dole in 1968 [47] and coupled to MS in 1984 by Yamashita and Fenn [2]. ESI usually generates intact, multiply charged ions, generally in the form $[M+nH]^{n+}$ in both the positive-ion mode (e.g., protonated) and negative-ion mode (e.g., deprotonated). In ESI-MS, "naked" ions form via progressive solvent evaporation from charged droplets of a liquid sample, sprayed in the presence of a strong electrical field. The formation of gaseous analyte ions by electrospray involves three steps: formation of charged droplets, shrinkage of the droplets owing to solvent evaporation, and transfer of ions into the gas phase. Although the macroscopic aspects of electrospray are generally well understood, the mechanisms for the final generation of desolvated (or nearly desolvated) ions from a charged droplet are not yet fully resolved. Two models describe this process. The charged residue model (CRM), conceived by Dole et al. [47], invokes successive cycles of solvent evaporation and coulombic fission at the Rayleigh limit until a droplet containing a single residual analyte ion remains. Complete evaporation of the solvent comprising this droplet eventually yields a "naked" analyte ion, the charged residue. The ion evaporation model (IEM) proposed by Iribarne and Thomson [48] is based on the transition-state theory. Prior to complete desolvation of the droplet, sufficiently strong repulsions between the charged analyte ions and the other charges in the droplet overcomes solvation forces, resulting in the analyte ion ejected (field desorbed) from the droplet surface into the gas phase [49].

With the advent of ESI, it became possible to study protein conformations. Different from traditional methods to investigate protein conformations such as circular dichroism (CD), NMR, and X-ray, ESI-MS offers several advantages for this purpose. First, ESI-MS is sensitive, requiring fmol and amol amounts of protein samples [12, 50, 51]. Second, ESI analysis makes use of a protein solution, which is important because most of biology and much of separations take place in solution. In traditional ESI experiments, organic compounds are often used as co-solvents; however, the use of highly organic solvents is no longer mandatory. This has led to the birth of an emerging field in biomolecular MS, termed native ESI-MS [49, 52–54]; the objective of this field is the analysis of intact proteins and protein complexes under near-physiological conditions achieved by using neutral volatile buffer salts like ammonium acetate for protein sample preparation. The third is that gas-phase, multiply charged ions are generated from the protein sample [3]. This point plays a central role in protein studies, given that the charge-state distributions (CSDs) observable in protein ESI mass spectra are affected by the conformations that the protein held in the solution at the moment of its transfer into the gas phase [12, 55]. Typically, when a protein is in the folded structure, a narrow CSD in low charge states is observed, whereas the CSD is broadened and shifted to high charge states after unfolding, probably because the unfolded protein has a greater capacity to accommodate charges on its surface [50, 56, 57]. Therefore, information about the conformational states of the protein can often be extracted based on the structural interpretation of CSDs in ESI-MS, upon controlling other experimental conditions [12]. Another MS-based approach to protein conformation study is to monitor protein hydrogen/deuterium exchange reactions, which are conformation sensitive; thus, the exchange level determined by MS can be related to protein conformational structures [58–64], and this subject is covered in an article by Kerfoot and Gross in this volume.

In 1994, Wilm and Mann introduced an important variant of conventional ESI, termed nanoelectrospray (nESI) [65]. While this technique uses the same fundamental sequence of charged-droplet generation followed by solvent evaporation, Coulombic fission events, and finally ion formation, it is distinguished from regular ESI in several ways. First, nESI is typically performed using glass or quartz capillaries that are pulled to a fine tip (~1 μm inner diameter) and given a metallic (usually gold) coating to hold the electric potential; these are used instead of the metallic capillary used for conventional ESI. Approximately 1–3 μL of sample is injected into the glass capillary and electrosprayed at flow rates in the range of ~1 nL/min to several tens of nL/min [66, 67]. The spray is driven primarily by the approximately 0.5–1.5 kV potential applied to the capillary, although it is often necessary to provide an auxiliary backing gas pressure to the sample to initiate and/or to maintain a steady stream of the solution through the tip [49]. Second, in comparison with conventional ESI, a smaller initial droplet size in nESI leads to less non-specific aggregation (both protein–protein and protein–salt) and its gentler

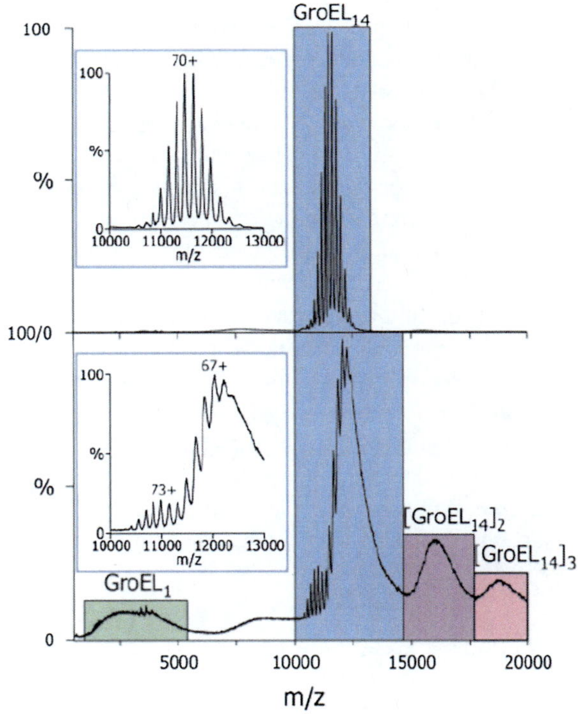

Fig. 1.2 Conventional and nanoelectrospray MS of a protein complex. MS of the GroEL complex ionized by means of ESI (*lower*) and nESI (*upper*). Solution conditions were 200 mM ammonium acetate, pH 6.9, and a protein concentration of 2 μM tetradecamer. The nESI spectrum displays a series of peaks around 11,500 *m/z* which correspond to the 800 kDa tetradecamer. Conventional ESI of the same solution results in poorly resolved "humps" centered on 12,500, 16,000, and 18,500 *m/z*. These are assigned to the tetradecamer, a dimer of tetradecamers, and a trimer of tetradecamers, respectively. There is also a signal at low *m/z* which corresponds to the GroEL monomer [49]. Reproduced with permission from ACS, copyright 2007

interface conditions, while still allowing adequate desolvation, causing less dissociation and disruption of oligomeric and higher-order structures (Fig. 1.2 shows the contrast between nESI and ESI for the ionization of a GroEL complex). The benefits of nESI analysis include high ionization efficiency and well-resolved protein complex ion peaks with narrow charge-state distributions, reduced non-specific adduct formation, and high salt tolerance.

1.3.2 Matrix-Assisted Laser Desorption/Ionization

Investigations of wavelength influence on ultraviolet laser desorption [68] led to invention of ultraviolet laser matrix-assisted laser desorption/ionization (UV-MALDI) between 1984 and 1986 (summarized in a 1987 paper) [69]. In 1988, Karas and Hillenkamp reported ultraviolet laser desorption (UVLD) of bioorganic compounds in the mass range above 10 kDa [1]. As a soft DI method, MALDI handles thermolabile, non-volatile organic compounds, especially those with high molecular weight, and can be successfully used for the analysis of proteins, peptides, glycoproteins, oligosaccharides, and oligonucleotides. Its operation is relatively straightforward although matrix preparation requires experience and perhaps some artistry.

MALDI is based on the bombardment of sample molecules with laser light to bring about sample ionization. It requires a specific matrix consisting of small organic compounds (e.g., nicotinic acid) that exhibit a strong resonance absorption at the laser wavelength used. The sample is premixed and diluted with the highly absorbing matrix and allowed to dry on a sample target. A range of compounds are suitable as matrices: sinapinic acid is a common one for protein analysis, while *alpha*-cyano-4-hydroxycinnamic acid is often used for peptide analysis (structures shown in Scheme 1.1.). These acids serve well as a matrix for MALDI due to their ability to absorb laser radiation and also to donate protons (H^+) to the analyte of interest. Upon laser irradiation, energy is absorbed by the matrix in a localized region of the surface. As a result of this local, rapid heating, the analyte acquires translational energy without being internally excited.

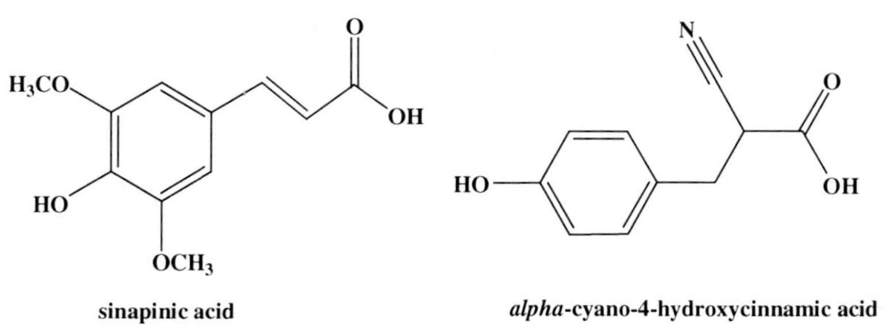

Scheme 1.1 The structures of two common MALDI matrices

The translational excitation occurs by the expansion of the vaporized matrix in MALDI and accounts for the release of the analyte molecules from the surface of the condensed-phase sample into vacuum. The analyte ions leave the surface with significant kinetic energies, entrained in a microsupersonic molecular beam of expanding matrix vapor. The analyte may be pre-charged (e.g., exist as a salt), and the intact analyte ion may simply be transferred as an ion from the solid to the vapor state upon laser irradiation of the matrix. Alternatively, a neutral analyte may be ionized through ion/molecule reactions (e.g., proton transfer reaction) occurring in the energized selvedge [70].

MALDI has remarkable efficiency in producing intact molecular ions (often $[M+H]^+$, $[M+Na]^+$) of large biological compounds. An even more remarkable characteristic of MALDI is its extraordinary sensitivity. Total amounts of sample loaded onto the target surface are often in the picomole to femtomole range, and much of this sample is not used. The method has reasonable tolerance to buffers and other additives and gives predominantly singly charged ions for large biomolecules [35]. TOF mass analyzers are ideal to use with this ionization technique because they are compatible with high-mass ions and pulsed ion production [34, 35]. TOF analyzers separate ions according to their *m/z* ratios by measuring the time it takes for ions, accelerated to the same kinetic energy, to travel through a field-free region known as the flight or drift tube. The heavier ions are slower than the lighter ones.

MALDI Applications

An important application of MALDI is chemical imaging, using a technique called matrix-assisted laser desorption/ionization imaging mass spectrometry (MALDI-IMS) [16]. Imaging combines parallel, high-throughput molecular analysis with location-specific information for the characterization of protein distributions directly from thin sections of intact biological tissue [71, 72] and offers complementary information to two-dimensional (2D) gel electrophoresis and to shotgun proteomics for investigating proteomic differences. For example, MALDI-IMS strategy can be applied for plant protein analysis. The epidermis of whole leaves is sprayed directly with a matrix, or the tissue is cut into small sections (10–15 μm) and coated with a matrix on glass slides. During analysis, the matrix heavily absorbs the laser energy to be vaporized, and the analyte embedded in the matrix is carried in the fast vaporization process. The molecules pick up a charge and travel down the TOF tube and are analyzed on basis of their *m/z* ratios. The data is acquired across a section in the form of a raster of predefined resolution, and then, the mass spectrum of each raster point is recorded. After that, the ion intensity and distribution for each *m/z* value are collected. Herein, there are two options for identification of potential biomarkers. The slide sometimes is coated with enzyme for on-slide protein digestion, and the generated peptides are analyzed by MS/MS. Usually, the tissue region of interest needs to be excised and separated in order to isolate the *m/z* ion of interest, which can be further digested with enzyme and analyzed using LC–MS/MS [73]. MALDI-IMS has become a powerful technique capable of identifying and localizing biological compounds directly on tissue surfaces.

1.3.3 Ambient Ionization Methods

In most applications, MS required moderate to extensive sample preparation followed by introduction of the sample into the high-vacuum conditions prior to analysis, limiting in situ analysis and increasing the possibility of contamination during sample handling. These drawbacks are overcome with the introduction of DESI and DART, which can be viewed as ambient ionization methods. In these methods, samples are examined in the open environment (natural or in the laboratory), and typically, no sample preparation is required, allowing for in situ analysis while preserving all attributes associated with MS analysis. These approaches open a new era in MS.

After the first reported applications using DESI and DART [39, 40], a whole new family of ambient methods and variants emerged. DESI variants such as reactive DESI (reactions accompanying desorption), non-proximate detection DESI (transport of sample ions at long distances), geometry-independent DESI, transmission-mode DESI, liquid sample DESI and continuous flow-extractive desorption electrospray ionization (CF-EDESI) were soon introduced either to increase selectivity and sensitivity for trace analysis or to facilitate direct sample analysis [74–82]. Another ionization method termed desorption atmospheric pressure chemical ionization (DAPCI) was also developed to study ionization mechanisms in explosive compounds [76]. Other established ambient ionization methods include electrospray-assisted laser desorption/ionization (ELDI) [83], MALDESI [84], extractive electrospray ionization (EESI) [85], atmospheric solids analysis probe (ASAP) [86], jet desorption ionization (JeDI) [87], desorption sonic spray ionization (DeSSI) [88], field-induced droplet ionization (FIDI) [89], desorption atmospheric pressure photoionization (DAPPI) [90], plasma-assisted desorption ionization (PADI) [91], dielectric barrier discharge ionization (DBDI) [92], liquid microjunction surface sampling probe method (LMJ-SSP) [93], atmospheric pressure thermal desorption ionization (APTDI) [94], surface sampling probe (SSP) [95], fused-droplet electrospray ionization (FD-ESI) [96], helium atmospheric pressure glow discharge ionization (HAPGDI) [97], neutral desorption extractive electrospray ionization (ND-EESI) [98], LAESI [99], low-temperature plasma (LTP) ionization [100], and laser spray ionization (LSI) [101]. Although, these methods can be used for ambient analysis, protein or peptide analysis has been achieved by a few owing to the ionization process involved (i.e., the amount of internal energy deposited into a protein). In the following, we focus on the analysis of proteins and peptides by DESI.

Desorption Electrospray Ionization

DESI allows one to record spectra of condensed-phase samples (pure, mixtures, or tissue) under ambient conditions, making the samples accessible during analysis for manipulation by ordinary physical or chemical means [76, 102–105]. Analysis of small and large molecules, very short analysis time (high-throughput), high selectivity (reactive DESI and MS/MS), and sensitivity are other attributes of this method. The DESI method is based on directing a pneumatically assisted electrospray onto a surface (e.g., paper, metal, plastic, glass, and biological tissue), from which small organics and large biomolecules are picked up, ionized, and delivered as desolvated ions into the mass spectrometer. Ions are generated by the interaction of charged microdroplets or gas-phase ions derived from the electrospray with neutral molecules of analyte present on the surface [39, 76]. DESI is a soft ionization method and shows ESI-like spectra of proteins, primarily attributed to some common features of the ionization process that produces low-energy, intact molecular ions through fast collisional cooling under atmospheric conditions [104]. The method can be used for many types of compounds (polar/non-polar and low/high molecular weight) in forensics and homeland security (e.g., explosives, chemical warfare agents, and bacteria) [76, 78, 79, 105–107], biomedical (e.g., tissue imaging, proteomics, lipidomics, and pathology) [26, 108–111], pharmaceutical/industrial (e.g., drug analysis, pharmacokinetics, polymers, process monitoring, metabolomics, environmental analysis) [112–117], and other fields. Many of these applications can be implemented with various mass spectrometers including triple quadrupoles [118], linear ion traps [119], Orbitrap [120], quadrupole time-of-flight (QTOF) [121], ion mobility/TOF and ion mobility/QTOF hybrids [121], Q-traps [80], Fourier transform ion cyclotron resonance (FT-ICR) instruments [122], and miniature ion trap mass spectrometers [123].

DESI Ionization Source

In a typical DESI setup (Fig. 1.3a), the source consists of a solvent nebulizer made of deactivated fused silica capillary, similar to the one used in electrosonic spray ionization (ESSI) [124]. Nitrogen (N_2) is used as the nebulizing gas at a linear velocity of approximately 350 m/s. The solvent (typically mixtures of methanol, water, and small amount of acetic acid) is sprayed under the influence of an applied high voltage (typically in the range of 3 to 6 kV). The gas jet composed of electrosprayed aqueous microdroplets and free gas-phase ions is directed onto the analyte on a surface at various incident angles (usually from as low as 25° up to 80° depending on the analyte) to the normal. The resulting droplets, ions, and neutrals are collected at a shallow angle from the surface. The ions are then transferred as a result of electrostatic and pneumatic forces to a mass spectrometer equipped with an atmospheric pressure interface. The source is typically mounted on an xyz moving stage, allowing it to be positioned at any chosen point with respect to the sample. The moving stage also has a tangent arm drive miniature stage that allows precise angular adjustment from 0 to 90° (Fig. 1.3a).

DESI Ionization Mechanisms

Droplet pickup has been suggested as the primary ionization mechanism in DESI although there is evidence for chemical sputtering (reactive ion/surface collisions) and gas-phase ionization processes (e.g., charge transfer, ion–molecule reactions, volatilization/desorption of neutrals followed by ionization) [39, 76, 103, 125, 126]. According to the droplet pickup mechanism, the surface is pre-wetted by initial droplets (velocities in excess of 100 m/s and diameters of less than 10 μm), forming a solvent layer that helps surface analytes to become dissolved. These dissolved

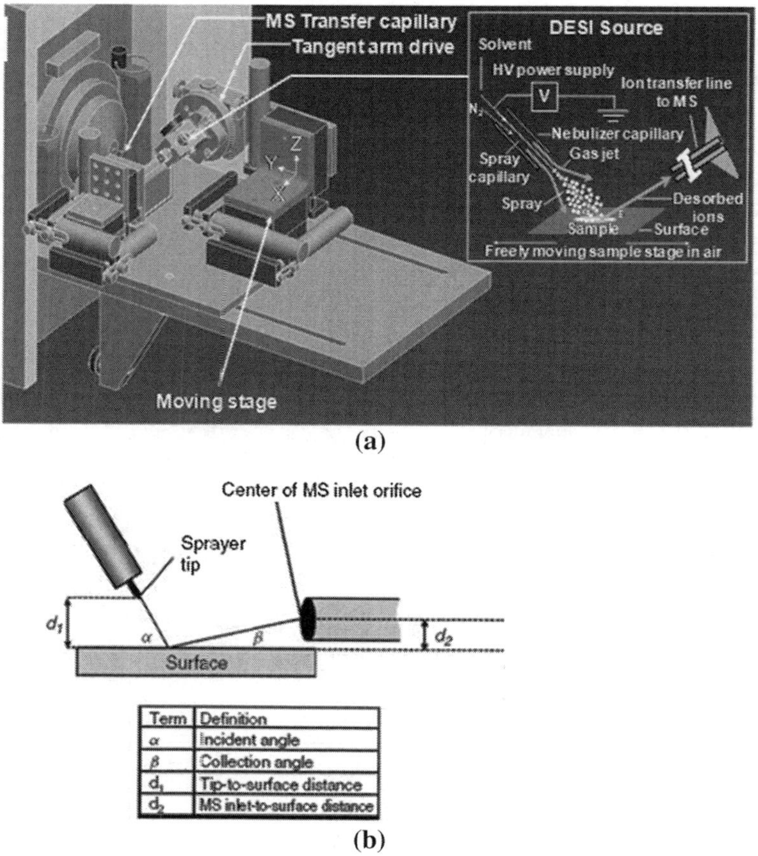

Fig. 1.3 **a** DESI source and moving stage used to position the source; an early prototype of the OmniSpray source of Prosolia, Inc. The source is fitted with an ion transfer capillary [76]. Reproduced with permission from ACS, copyright 2005. **b** Definitions of terms used in conjunction with DESI [103]. Reproduced with permission from John Wiley and Sons, copyright 2005

analytes are picked up by later arriving droplets that are impacting the surface, creating secondary droplets containing the dissolved analytes. Gas-phase ions are then formed from these secondary droplets by ESI-like mechanisms [103, 125, 126]. The resulting gas-phase ions have internal energy values similar to those in ESI and ESSI [127]. The formation of cold ions gives DESI its soft ionization character that affords ESI-like spectra, especially for proteins and polypeptides.

DESI Analytical Performance

Signal intensity in DESI spectra depends on incident angle (β), collection angle (α), tip-to-surface distance (d_1), MS inlet-to-surface distance (d_2), and other geometric parameters, as defined in Fig. 1.3b. Nebulization gas velocity, spray solvent flow rate, and spray potential also affect performance. The type of surface analyzed (its texture and electrical conductivity) is also a factor that affects the ionization process. The limits of detection (LOD) are in the low picogram to femtogram range for small molecules and some biopolymers [39, 128]. The dynamic range is 5 orders of magnitude, and relative standard deviations (RSD) of 5 % for quantitation (lower if using an internal standard) can be achieved [103]. For imaging applications, spatial resolution approaching 40 μm can be obtained [129]. Accuracies in the range of ±7 % relative errors are possible [39, 130].

DESI for Protein Analysis

Since the first reported applications of DESI for protein and peptide analysis [39, 103], various research groups implemented applications ranging from solid sample analysis (from surfaces) to direct analysis of liquid samples or liquid films [80, 81, 131–134]. A feature of liquid DESI is that it can desorb large proteins and protein complexes directly from solution (Fig. 1.4) [135]. For example, the

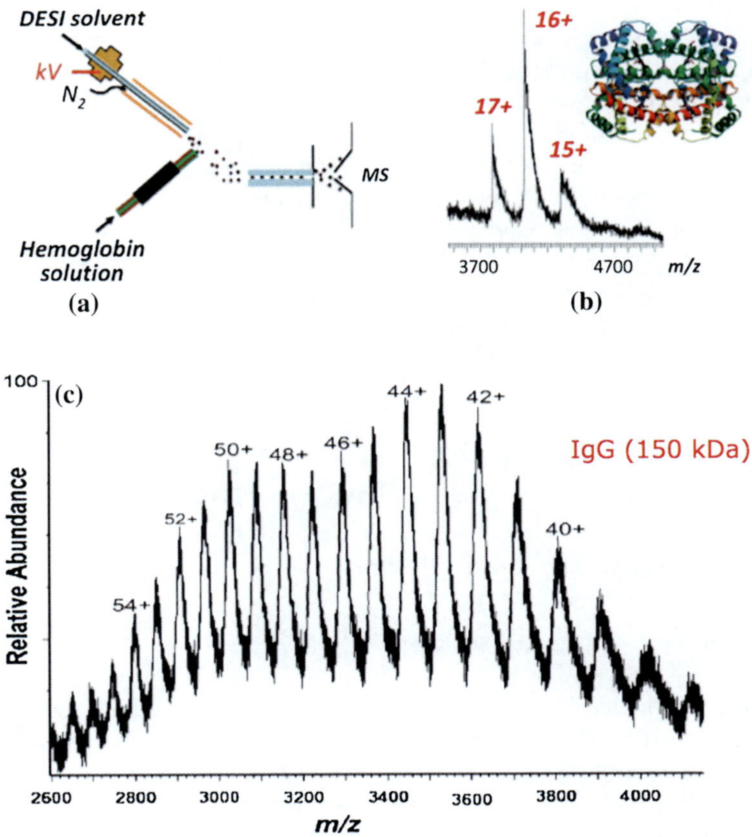

Fig. 1.4 Liquid sample DESI **a** apparatus; **b** mass spectrum of human hemoglobin (50 μM) in 20 mM NH_4OAc; **c** IgG (6 μM) in H_2O/ACN/FA (90:10:0.1 by volume) [135]. Reproduced with permission from ACS, copyright 2011

ionization of intact hemoglobin tetramer (Fig. 1.4b) and IgG (150 kDa) (Fig. 1.4c) was demonstrated by liquid sample DESI. It appears that large proteins and protein complexes are relatively easier to be desorbed and ionized from solution than from dried samples on surface, probably due to less protein aggregation in solution than in the solid form [135]. High mass resolving power can be obtained in protein and peptide identification by coupling DESI with Fourier transform ion cyclotron resonance mass spectrometry (FT-ICR-MS) [122]. Other applications of peptide analysis can be envisioned for the direct identification of tryptic digests; examples are cytochrome c and myoglobin deposited on high-performance thin-layer chromatographic (HPTL) plates. After separation on the HPTL plates, the resulting bands are exposed to the DESI sprayer for peptide identification.

There are other ionization methods which have been introduced with specific advantages such as high sensitivity in terms of protein analysis. For example, surface-enhanced laser desorption/ionization (SELDI) [136], a prominent form of laser desorption/ionization (LDI) MS, can be readily used to analyze the major and minor protein components in heterogeneous samples. This ionization method, used for analysis of macromolecules, efficiently facilitates the investigation of biological molecules on probe and simplifies sample purification and extraction steps in contrast to conventional LDI and MALDI. Also, SELDI is a rapid and highly reproducible method with high sensitivity for trace protein (<fmol/mL using chemical arrays) analysis, offering a significant advancement in protein analysis by MS. Furthermore, other methods nanostructure-initiator mass spectrometry (NIMS) [137, 138], sonic spray ionization (SSI) [139], ESSI [124], fused-droplet electrospray ionization (FD-ESI) [140, 141], ELDI [142], EESI [143], inlet ionization [144, 145], etc. are widely used for protein analysis, generating multiply charged ESI-like spectra. The details for these different methods, including apparatus, mechanism, and applications, can be find in our previous book chapter [146].

1.4 Ion Activation and Tandem Mass Spectrometry

1.4.1 Collisional Activation Methods

1.4.1.1 Collision-Induced Dissociation

Dissociation of molecular ions in the gas phase can be traced back to the first mass spectra recorded by English physicist Sir J. J. Thomson with his parabola mass spectrograph at the Cavendish Laboratories, leading to what is known today as collision-induced dissociation (CID) or also known as collision-activated dissociation (CAD) [147, 148]. Consequently, evolution of CID as a tool for ion structural elucidation can be traced directly to the work of Jennings [149] and McLafferty [150]. CID is one of the most popular activation methods used today to detect, identify, and obtain structural information from small and large molecules such as peptides and proteins. Complex mixture analysis and biopolymer sequencing (e.g., peptides and proteins) are areas where this technique has proved extremely useful [151]. Even though unimolecular dissociation theory [152, 153] is important to describe the behavior of an ionized molecule after it has been activated, it is beyond the scope of this chapter and we will focus on the specifics of the CID process and its applications toward protein and peptide ion activation.

In tandem MS analysis, the precursor ions are isolated, accelerated to higher kinetic energies, and allowed to collide with a neutral target gas (usually nitrogen, helium, or argon). As a result of these inelastic collisions, part of the translational energy is converted into internal energy of the ion, leading to subsequent decomposition (i.e., CID) [154–157]. Fragment ions produced during CID of the precursor are recorded, giving a fingerprint spectrum of the parent ion [155]. The CID process is assumed to occur by a two-step process, where activation of the precursor ion and its dissociation are separated in time (activation time is typically orders of magnitude faster than dissociation) as shown in Eqs. 1.1 and 1.2 below, where Q is the change in kinetic energy of the system, m_p^+ is the precursor ion, N is the target gas, and $m_p^{+\cdot}$ and N′ are the collision partners in their post-collision state [157]. This process is followed by unimolecular dissociation (Eq. 1.2), where m_a^+ and m_b are the products of the unimolecular dissociation of the activated ion ($m_p^{+\cdot}$) [157]. Collision yield depends on the activated precursor ion decomposition probability according to the quasi-equilibrium or RRKM theory, and fundamental aspects of both theories are detailed in the literature [152, 153]. Other dissociation mechanisms such as stripping, where the ion is torn away during the collision event (i.e., collision and dissociation cannot be separated), might occur under specific conditions.

$$Q + m_p^+ + N \rightarrow m_p^{+\cdot} + N' \tag{1.1}$$

$$m_p^{+\cdot} \rightarrow m_a^+ + m_b \tag{1.2}$$

Fragmentation of the precursor ion will occur if the collision energy is high enough to excite the ion beyond its dissociation threshold. In order to simplify and make easier to visualize the dynamics of the kinetic to internal energy conversion process between a mobile specie (the ion) and a static target (the collision gas), a simpler center-of-mass (com) framework has been adapted instead of the laboratory reference frame, since the "com" momentum is always zero [157, 158]. The total energy available for kinetic to internal energy conversion is the relative kinetic energy of the collision partners. This center-of-mass collision energy, E_{COM}, is a fraction of the laboratory kinetic energy, E_{Lab}, if the velocity of the neutral is ignored [151]. As described in Eq. 1.3, E_{COM} depends on the masses of the collision partners, where m is the mass of the neutral collision gas and M is the mass of the ion to be activated.

$$E_{COM} = \frac{m}{m + M} E_{LAB} \tag{1.3}$$

E_{COM} increases with the mass of the target (m) and decreases as a function of $1/M$. This means that large precursor ions have less internal energy deposited to undergo fragmentation during the collision process. This is very important for protein and peptide activation since typical collision partners are much lighter, often depositing energies orders of magnitude smaller, as calculated using Eq. 1.3. To maximize E_{COM} and minimize ion–molecule reactions and charge transfer, relatively heavy atomic targets (e.g., Ar, Xe) with high ionization potentials have been used in the electron volt E_{Lab} collisional activation [157]. Lighter targets such as helium are used in the kiloelectron volt E_{Lab} collisional activation of small- to moderate-sized precursor ions to minimize scattering and charge exchange. Besides the parent ion internal energy distribution after collision, there are other factors that dictate the appearance of the resulting CID spectra, and these include the time frame of the experiment, instrumental discrimination effects, and partitioning of energy within the ion [157].

The overall CID process can be achieved via fast (high-energy collision) or slow activation (low-energy collision). As the word implies, fast activation methods are those in which the energy input occurs rapidly relative to unimolecular dissociation and typically involves a single high-energy collision, higher than 1,000 eV of laboratory collision energy [151]. Slow activation occurs via multiple small collisions (1–100 eV) with long intervals between individual collisions [151, 158]. If deactivation processes such as cooling collisions or photon emission occur between activation events, the process is considered a "very slow" or "slow heating" method. At high energy, the ion excitation is mostly electronic [159], while at low energy, the ion excitation energy is mostly vibrational [160], coinciding with a bond's vibrational period. Although the energy deposited per collision in low-energy collisions is slightly lower than in high energy, collision yields are extremely high as compared to the total energy available. This is due to the multiple collisions allowed by the gas pressures typically employed and length of the collision cell (QqQ), or as the case for ion traps, time allotted for CID. Slow activation is mostly performed in collision cells such as the ones found in triple quadrupole (QqQ) instruments or in instruments that use multipole collision regions at pressures ranging from few millitorr up to a torr [161], while very slow activation or slow heating methods are performed in trapping devices such as quadrupole ion traps (e.g., in approximately 1 millitorr He bath gas), linear trapping quadrupoles (Q-CAD), and sustained off-resonance irradiation (SORI) in Fourier transform ion cyclotron resonance (FT-ICR) [162–165]. In contrast, fast activation is mostly performed in beam-type instruments such sectors and TOF [166–168].

CID of Peptide and Protein Ions

Currently, one of the most active and challenging areas for CID applications is to address the fragmentation of large peptide and protein ions. Besides providing critical information for protein and peptide sequencing, CID has been used to understand dissociation mechanisms and to obtain threshold energies from these systems [169]. As discussed earlier, there are some factors that influence the CID behavior of peptide and protein ions. For large ions, a typical challenge is the efficiency of the CID process itself, that is, the amount of energy that can be deposited and redistributed across the many degrees of freedom, which will eventually induce dissociation of such large systems. For peptides ions, typical backbone cleavage during CID occurs at the peptide amide bond (weakest bond) to produce **b**-type (amino-terminal retains the charge) and **y**-type (carboxyl-terminal retains the charge) ions (Schemes 1.2 and 1.3) [170]. These **b/y** ion pairs dominate the low-energy CID spectra, in some instances including ions resulting from small molecule losses from the sequence fragment ions such as ammonia, water, or carbon monoxide (the loss of CO from **b** ions produces **a** type ions) [151, 170, 171]. Peptide ion dissociation pathways have been studied extensively [172–178] and have been rationalized by the "mobile proton model" [175, 177, 178]. In this model, fragmentation is initiated by the transfer of a proton or protons (triggered by collisional activation) intramolecularly to cleavage sites among backbone protonation sites (Scheme 1.4). Proton transfer is facilitated by the proton affinity of the heteroatoms, for example, oxygen or nitrogen of the amide bond to be cleaved. Under these mobile proton conditions (number of ionizing protons > number of strong basic sites), charge-directed fragmentation pathways become energetically available at several sites, with enhanced cleavage at the N-terminal side of proline (known as proline effect) [170, 179, 180]. Enhanced cleavage is also noted C-terminal to a protonated histidine residue. This could be attributed to its ability to transfer a proton to the backbone, a process

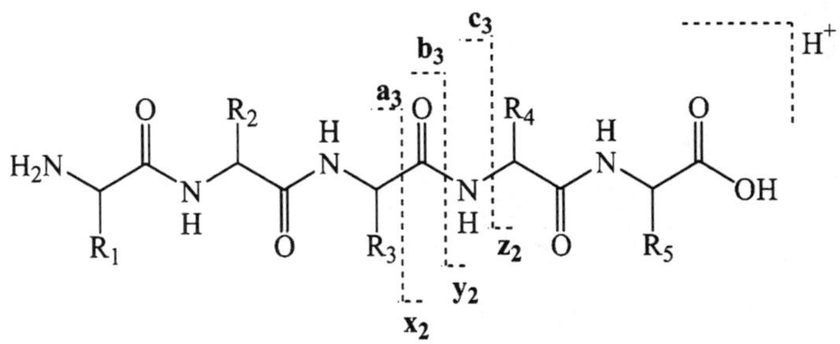

Scheme 1.2 Nomenclature of common ion types [181]. Reproduced with permission from John Wiley and Sons, copyright 1984

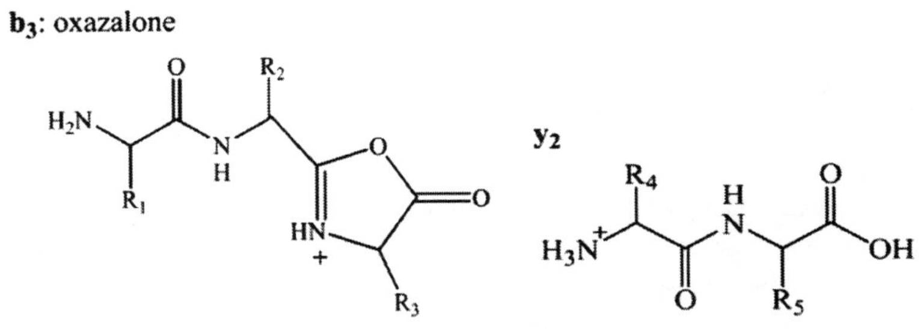

Scheme 1.3 Representative structures of **b** ions and **y** ions [170]. Reproduced with permission from Elsevier, copyright 2005

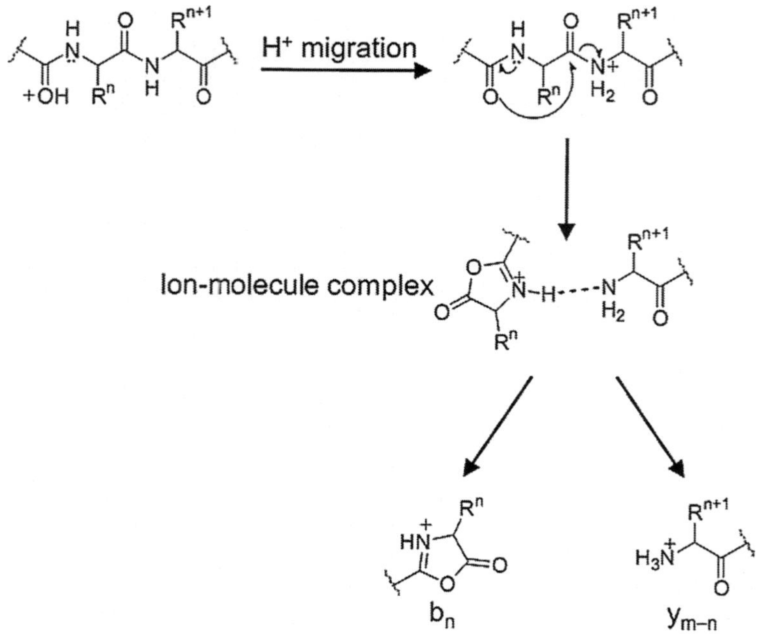

Scheme 1.4 Amide bond cleavage typically involves proton migration as an initial step, followed by neighboring group–assisted bond cleavage between the *n*th and $(n + 1)$th residue [179]. Reproduced with permission from John Wiley and Sons, copyright 2008

that triggers formation of a resonance-stabilized cyclic *b* ion. When all protons are bound to basic residues such as arginine (number of ionizing protons ≤ number of strong basic sites), cleavage often occurs selectively at the C-terminus of aspartic or glutamic acid via a charge-remote mechanism (proton derived from the acidic side chain) [170]. Peptides with basic amino acids have higher energy thresholds for fragmentation as compared to those with less basic amino acid groups. The basic sites sequester the protons, requiring more energy in order to facilitate fragmentation [169].

Even though **b** and **y** ions are the most useful ion types for sequencing, there are other ion types (e.g., **c**, **x**, **z**, **a**, **d**, **w**, **v**, ammonium ions) used for spectral identification and database searches, but these are typically observed at higher collision energies (e.g., high-energy CID) [169–171]. A marked difference between low- and high-energy CID is the abundant dissociation of amino acid side chains forming **d**-, **w**-, and **v**-type ions (Scheme 1.5). Side-chain cleavages are useful for distinction of isomeric and isobaric amino acids in peptide sequencing [170]. Another characteristic of the high-energy CID process is the abundance of ammonium ions in the low mass range region [151, 171]. Ammonium ions are used as sequence qualifier, that is, good indicators of the presence or absence of a particular amino acid in the peptide sequence. Figure 1.5 shows the CID mass spectra for ACTH (decapeptide) collected at high (a) and low (b) collision energies, showing the characteristic ion types formed in each case.

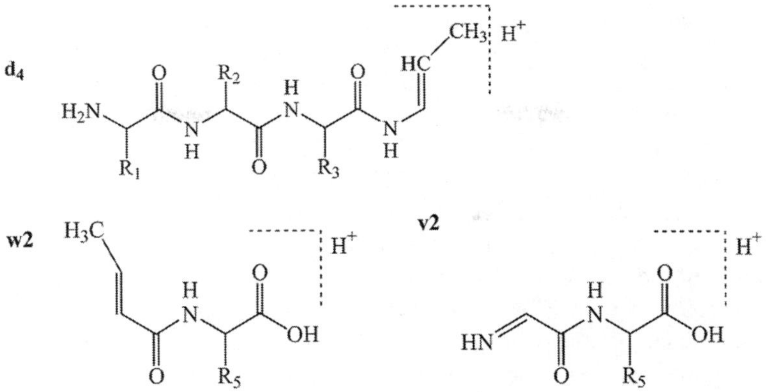

Scheme 1.5 Structure of fragments involving side-chain cleavage (**d**, **v**, and **w** ions), with Val as residue 4 [170]. Reproduced with permission from Elsevier, copyright 2005

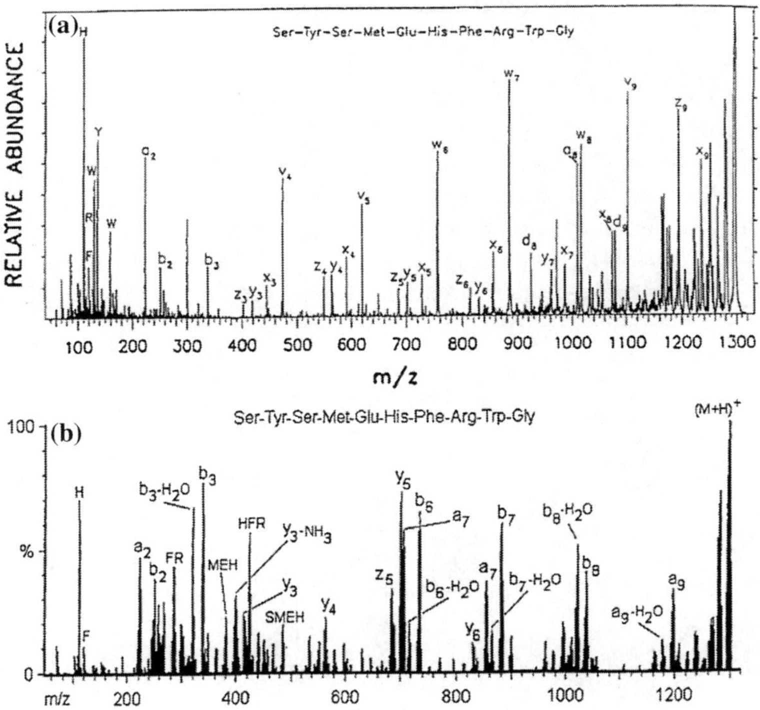

Fig. 1.5 Tandem CID mass spectra of ACTH (1–10), at 7 keV (**a**) and 45 eV (**b**) collision energies (high- and low-energy spectra, respectively). Both precursor ions are singly charged and were generated by Cs bombardment of the sample dissolved in glycerol (FAB). The high-energy MS/MS spectrum was obtained with a four-sector magnetic deflection mass spectrometer, using helium as the collision gas; only the ^{12}C isotope of the precursor ion was selected, and the product ion spectrum was acquired at unit mass resolution. For the low-energy MS/MS spectrum, a triple quadrupole instrument was used, with argon as the collision gas; the entire precursor ion isotope cluster was selected, and the product ion spectrum was acquired at less than unit mass resolution [171]. Reproduced with permission from John Wiley and Sons, copyright 1995

Even though CID has been widely implemented in bottom-up, top-down, and shotgun proteomics, the technique has shown several limitations, especially in the analysis of post-translational modifications (PTMs). Many PMTs are fairly labile, that is, weaker than the peptide amide bond. When ions containing PTMs are collisionally activated, PTMs are the first to dissociate (at expense of sequence fragments), making the localization of the PTMs a challenge [182]. Other effects such as sequence scrambling have been observed, which can produce misleading ions, that is, making accurate peptide and protein sequencing a more challenging task [183, 184]. In the field of structural biology, CID has shown limitations in determining subunit topology for some quaternary non-covalent protein assemblies, due to significant protein subunit unfolding during activation (multicollision CID), yielding highly charge-asymmetric monomers, thus limiting the amount of relevant substructure information for such non-covalent complexes [185–187].

1.4.1.2 Surface-Induced Dissociation

Surface-induced dissociation (SID) was pioneered in the mid-1970s by R. Graham Cooks at Purdue University [188–190]. Other groups have studied SID, including Wysocki [191–200], McLafferty [201, 202], Futrell [203–207], and Laskin [206–215]. The SID process is analogous to CID, except that the neutral gas used as target in a typical CID experiment is replaced by a surface, as shown in Fig. 1.6 (ion–surface collision event) [189, 198]. The initial translational energy of the precursor ion is converted into internal energy upon collision with the surface, causing its activation and subsequent dissociation. The main motivation to develop SID was

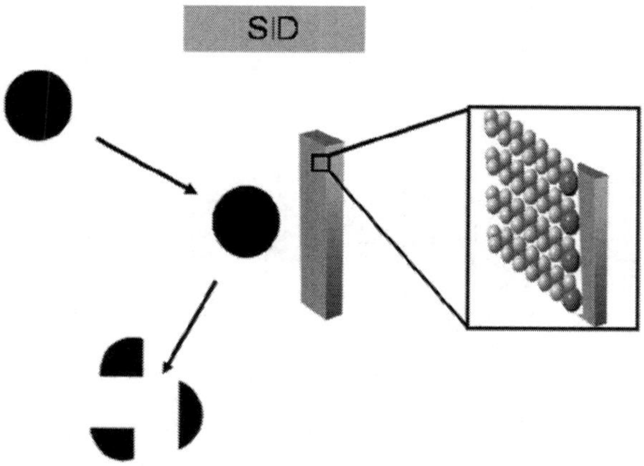

Fig. 1.6 Diagram of surface-induced dissociation where surface collision event deposits energy into the precursor ion. Collisions can generate fragment ions, neutralized precursor molecules, sputtered surface atoms, and ion–surface reaction products. *Inset* shows an all-*trans* configuration alkanethiolate (e.g., $CH_3(CH_2)_{11}SAu$) self-assembled monolayer on gold [198]. Reproduced with permission from Springer, copyright 2008

the fact that energy transfer in CID is limited by the energy available in the center-of-mass reference frame (E_{COM}) [158], which depends on the mass of the target gas as described in Eq. 1.3, where m is the mass of the neutral collision gas, M is the mass of the ion to be activated, and E_{LAB} is the laboratory collision energy.

Since the mass of the target is increased (assuming entire surface as a collision partner) in a typical SID experiment, as compared to collisions with individual surface molecules in the gas phase (CID), E_{COM} becomes larger and energy transfer into the projectile ion can be improved. Although conversion of translational energy into internal energy should be more efficient in SID, there are cases where the mass of terminal groups on the target surface influences the amount of energy transfer [216]. SID reactions are prevalent in the hyperthermal collision energy regime (i.e., 1–100 eV), reaching energies of the order of or greater than chemical bond energies, allowing bond cleavages and rearrangement reactions [158, 217]. Upon impact with the target surface, projectile ions can scatter in an elastic, inelastic, or in a chemically reactive way [218]. Reactive scattering includes charge-changing collisions and ion–molecule association reactions (new bonds formed). Out of those, inelastic and reactive scatterings are the most common processes observed in SID, based on the typical SID collision energies implemented [158]. Generally, the SID activation mechanism has been rationalized as a two-step process where the projectile ion hits the surface inelastically (interaction time of ion with surface $\approx 10^{-14}$ s) forming an exited ion ($m_1^{+\cdot}$) with enough internal energy, which then undergoes delayed gas-phase dissociation (typical maximum rate constants for unimolecular dissociations are in the order of 10^{10} to 10^{12} s^{-1}) after it leaves the surface, as described in Eq. 1.4. This mechanism has been suggested as a plausible explanation to account for similarities found in CID and SID product ion spectra for some systems [158, 218]. Although fragmentation of organic ions by SID and CID shows similar sets of fragment ions in some systems, their abundances vary considerably [218]. Another mechanism has been proposed, and this involves decomposition of the projectile ion at the surface (at high collision energies reaching 30 eV), resulting in fragment ions with the same kinetic energy [215].

$$m_1^{+} \rightarrow m_1^{+\cdot} \rightarrow m_2^{+} + m_3 \tag{1.4}$$

In SID, energy is deposited in a large and fast deposition step. This helps to minimize rearrangement product formation (low-energy competitive reactions) and allows access to higher-energy pathways such as side-chain cleavage, as compared to a slow heating CID process where ions are gradually activated [170, 213, 219]. SID also allows obtaining narrow internal energy distributions and provides excellent control of energy deposition into the projectile ion. This allows deposition of small and large amounts of internal energy, providing access to a variety of fragmentation pathways dependent on the collision energy used (see Fig. 1.7) [198, 218, 220]. Besides obvious advantages such as activation of high-mass ions (peptides and proteins) with high dissociation thresholds, other advantages of this technique over CID are the absence of collision gas (i.e., eliminating the gas load from the instrument) and 100 % interaction efficiency since the surface is put directly in the path of the ion beam [158, 198]. Also, it has been demonstrated that lower kinetic energies can be used for SID, since the target is substantially more massive as compared to typical CID lower-mass targets

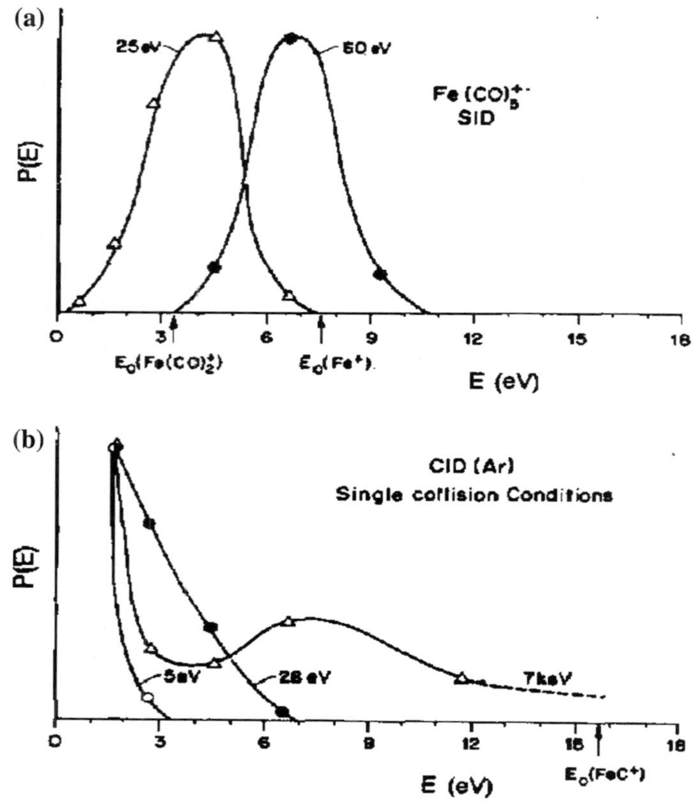

Fig. 1.7 Comparison of internal energy distributions in $Fe(CO)_5{}^{+\cdot}$ upon collision (**a**) at a solid surface and (**b**) with Ar under single-collision conditions [218]. Reproduced with permission from Elsevier, copyright 1990

(Fig. 1.8). The SID technique has provided reproducible spectra, including good signal-to-noise ratios, and has shown its potential for isomer distinction, based on the narrow range of internal energies deposited [221]. Some disadvantages of the technique are poor ion-optical quality of the emerging ion beam and the nonlinear collision geometry, differing from linear arrangement typically used in MS/MS instrument employing CID [218].

Many surfaces and mass analyzers have been used for SID applications. Surface studied includes metals, graphite, diamond, and functionalized alkanethiolate self-assembled monolayer (SAM) films (e.g., hydrocarbon and fluorocarbon). Surface selection is very important since this can affect the dissociation efficiency by adding other competing processes. Examples include neutralization, chemical sputtering from the surface, and ion surface complex reactions [215, 218, 222]. A strategy to minimize neutralization of the incident ion beam has been the utilization of SAM surfaces, specifically with high ionization potentials (e.g., fluorinated SAMs) to avoid charge transfer mechanisms between projectile ions

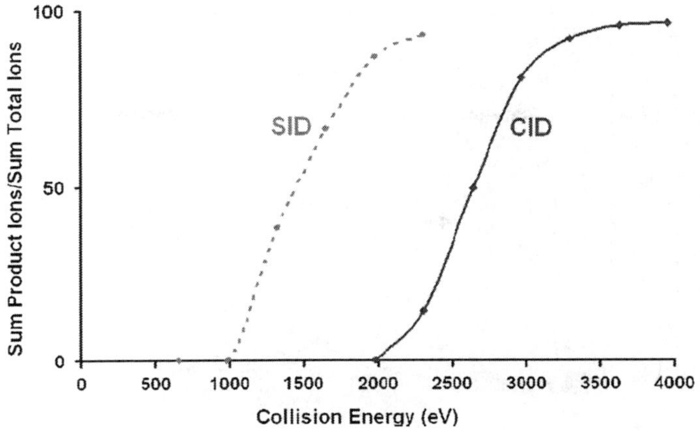

Fig. 1.8 Fragmentation efficiency curves for dodecamers of *Ta*HSP16.9 dissociated via SID or CID. Much lower laboratory collision energy is needed to fragment *Ta*HSP16.9 when a surface is used as the collision partner [196]. Reproduced with permission from Springer, copyright 2008

and surface [223]. Some mass analyzers reported for SID applications include dual time-of-flight (TOF) ion mobility (IM) SID [224], matrix-assisted laser desorption/ionization (MALDI)-IM-SID-TOF [225, 226], sector-TOF [227], dual quadrupole [227], quadrupole TOF [228], Fourier transform ion cyclotron resonance (FT-ICR) [229], and magnetic sector–electric sector–quadrupole mass filter (BEEQ) [230]. While most SID applications are focused on protein and peptide analysis, which is the focus of this chapter, this technique has been used for small molecule analysis [190, 231–238] such as fullerenes [239, 240] and metal clusters [241], most of which involved surface composition analysis, characterization of reactions between surface adsorbates and projectile ions, and projectile ion structure determinations.

Structural characterization of proteins and peptides via SID has been extensively explored, taking advantage of the well-defined internal energy distributions deposited and the ability to overcome energy thresholds in large biological species, which allows their activation and broad sequence coverage, not always provided by methods such as CID [201, 202, 223]. Cooks et al. showed the first applications toward peptide characterization (tetrapeptide Met-Arg-Phe-Ala) [242]. Later, applications involved the use of SID in FTMS, which showed extensive fragmentation of oligopeptides at 48 eV collision energies, making sequence-specific dissociations possible for large ions (>3,000 *m*/*z*, usually not possible via CID for large oligopeptides, Fig. 1.9) [243]. SID has been used to refine the "mobile proton" model (proton transferred to cleavage site by internal motions in the peptide, i.e., among backbone protonation sites) and to understand unusual peptide fragmentation pathways, providing critical knowledge on how peptide/protein fragment helping protein identification via a more complete peptide sequence coverage [199].

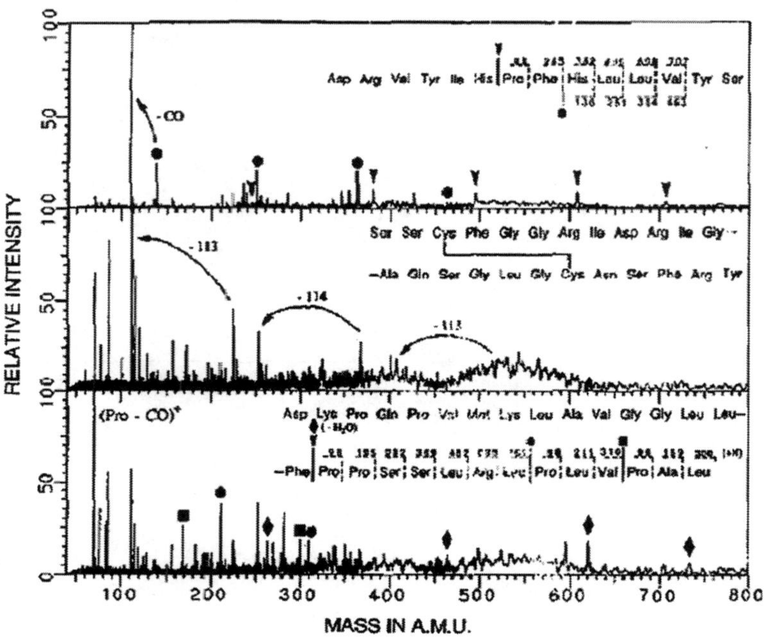

Fig. 1.9 48 eV SID spectra of renin substrate tetradecapeptide (MW 1758, 100 scans), atriopeptin III (MW 2549, 200 scans), and RSBP (MW 3054, 200 scans) (*top* to *bottom*) [243]. Reproduced with permission from Springer, copyright 1990

Since early applications involving characterization of small peptides [244–246], SID has found an integral role in the characterization of larger systems such as proteins [202]. Figure 1.10 shows the SID spectrum for carbonic anhydrase $(M+23H)^{23+}$, a 29 kDa protein which primarily yields fragments via amide bond cleavages (y/b) [202]. SID has been successfully used to fragment protein complexes over 200 kD [196]. Recently, SID has played a complementary role in structural biology, that is, characterization of macromolecular assemblies (e.g., large protein complexes) [185, 186, 196, 228]. An example of such application is shown in Fig. 1.11 for cytochrome c dimmers [196]. By using SID over a broad range of collision energies, it leads to the formation of 5+ and 6+ monomers from the 11+ charge state of cytochrome c, that is, the expected charge states if 11 charges are split between two equivalent subunits. When the same experiment is performed using CID, charge distributions of 8+ and 3+ monomers are achieved, suggesting an unfolding mechanism where one monomer gets enriched. This shows the ability of the SID process to reduce unfolding (as a consequence of the one-step activation of SID, which provides high-energy deposition within a short time frame, allowing subunit dissociation on a time scale faster than that of protein unfolding) during activation of the protein complex, allowing access to dissociation pathways that lead to charge-symmetric dissociation, that is, obtaining critical substructure information [196]. Charge-symmetric product ions have been obtained via SID for several

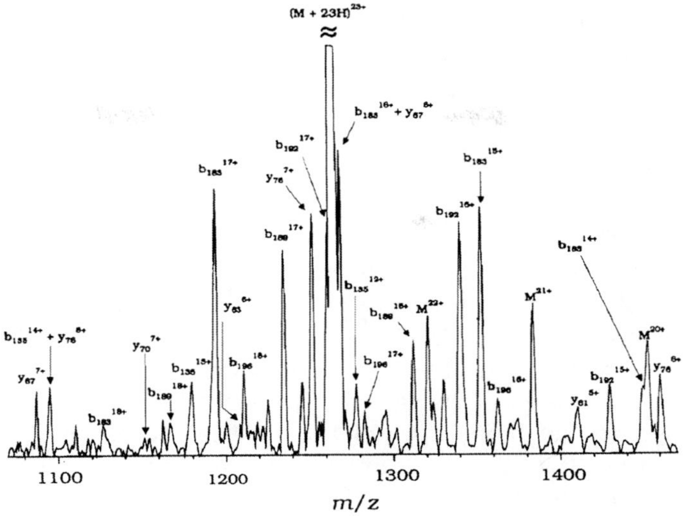

Fig. 1.10 SID spectrum of carbonic anhydrase $(M+23H)^{23+}$ using $V_{Surf} = -52$ V, $V_{excrec} = +0$ V, and $V_{ft} = +52$ V for 37 μs; sum of eight scans [202]. Reproduced with permission from ACS, copyright 1995

Fig. 1.11 Energy-resolved SID of the 11+ dimer of cytochrome c. The dimer (D11+) dissociates into monomers with an approximately equal partitioning of the precursor ion charge, regardless of the SID collision energy. Monomer product ions are labeled with an M, followed by their respective charge state [196]. Reproduced with permission from ASMS, copyright 2008

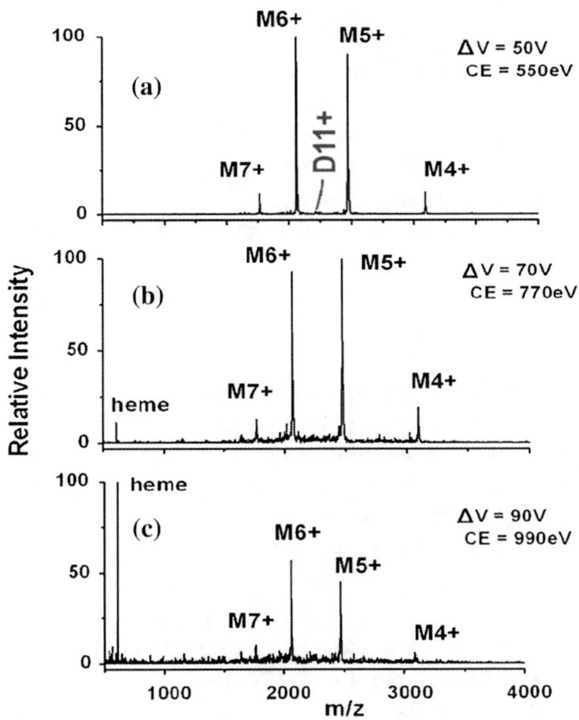

protein complexes including dimmers, tetramers, pentamers, dodecamers, and hexameric complex containing distinct types of subunits [186, 196, 247, 248].

1.4.2 Photodissociation

The exponential growth in applications of MS in biological fields has stimulated the exploration of new ion activation/dissociation methods capable of meeting the versatile needs for determining sequences, modifications, interactions, and conformations of biopolymers [249, 250]. Most agree that CID dominates the field, but recent inroads in electron- and photon-based activation methods have cemented their role as outstanding alternatives [249]. Photodissociation (PD) generally entails using a laser to irradiate ions with UV, visible, or IR photons, thus resulting in internal energy deposition. The activation process can be extremely rapid and efficient, as well as having the potential for high total energy deposition [250]. PD offers several compelling advantages over other fragmentation techniques. First, PD allows better control of energy. In other words, the specificity/selectivity of photon/chromophore interactions determines the location and amount of energy introduced into a molecule [249, 251]. Second, ultrafast energy deposition allows high-throughput analysis. Even the low-energy PD such as infrared multiphoton dissociation (IRMPD) takes place in millisecond time scale [252, 253] while for UVPD occurs in the nanosecond to microsecond range [251]. Third, PD only needs simple implementation of "trapping" mass spectrometers by drilling a hole in the trap and adding a quartz window [251]. Fourth, ion-scattering effects can be alleviated by using photons instead of collision [249]. Both precursor ions and product ions may undergo photoactivation and dissociation, leading to extensive fragment ions that may provide a more specific structural fingerprint or may be useful for database search.

However, the various advantages of PD do not come without any drawbacks: the expense of a laser is an unavoidable upfront cost and the concerns about laser safety [249].

1.4.2.1 Implementation of Photodissociation

In PD experiments, ions and photons should be located in the same place at the same time to enable absorption of light by the ions to trigger bond cleavage. This requirement needs "trapping" instruments, in which ions can be stored and accumulated for some period of time. Linear ion trap and ion cyclotron resonance mass spectrometers are well suited for PD owing to the instrument geometries which facilitate the introduction of the laser pulse [251]. Besides the addition of a quartz window, access holes in ion traps are required. These holes are typically drilled through the ring electrode of the ion trap (Fig. 1.12). An alternative setup for laser introduction is through a hole in the end-cap electrode which has also been proved to be an effective option.

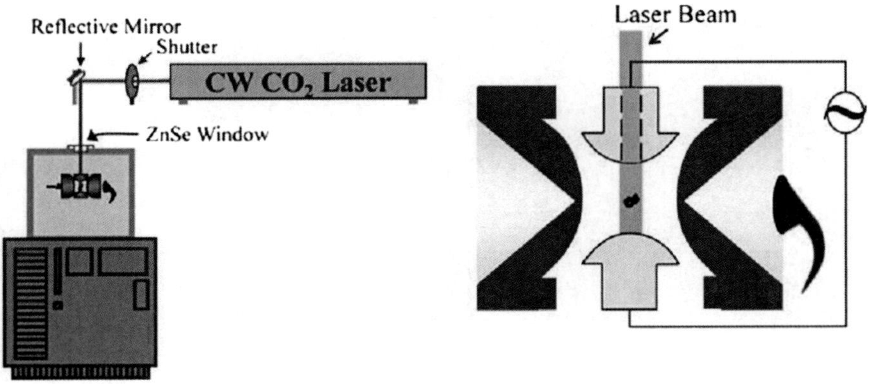

Fig. 1.12 Photodissociation in a quadrupole ion trap [250]. Reproduced with permission from John Wiley and Sons, copyright 2009

1.4.2.2 Infrared Multiple Photon Dissociation

Infrared multiple photon dissociation (IRMPD), like traditional CID, is classified as a "slow heating" method due to the relative low energy of photons (~0.1 eV for a typical CO_2 laser), allowing multiple photons to be absorbed to promote dissociation [250, 254]. The photons used in IRMPD may originate from a laser with wavelengths in the infrared range, or from the blackbody irradiation, which is called blackbody infrared radiative dissociation (BIRD) [255]. CO_2 laser is the most popular choice for IRMPD since it is widely available and affordable. Also, the wavelength of CO_2 (10.6 μm) can be absorbed by a large amount of organic molecules [250].

Generally, IRMPD offers several advantages over CID. First, the non-resonant nature of IRMPD allows the simultaneous activation and dissociation of both precursor ions and the primary fragment ions, resulting in a richer array of fragment ions. Also, since IRMPD is not a collision-based activation process, the loss of ions or collision scattering can be minimized compared to CID. In addition, the energy deposition in IRMPD is independent of the rf voltage, so it displays the collection of full MS/MS spectra due to the alleviation of the low-mass cutoff associated with traditional CID in QIT instrument [250]. IRMPD has proved to be an effective activation/dissociation method for analytes from drug to natural products, peptides, nucleic acids, DNA/drug complexes, etc. [250, 256].

Figure 1.13 shows the performance of IRMPD versus CID in terms of the dissociation of multiply charged peptide ions, such as melittin (MW 2845.7 Da). The low-mass cutoff feature of IRMPD is clearly demonstrated in which a higher sequence coverage is obtained.

Low energy of IR photons requires multiple photons to be absorbed for effective activation of large molecules. The energy deposition needs time to accumulate which causes the low IRMPD efficiency for large molecules [250].

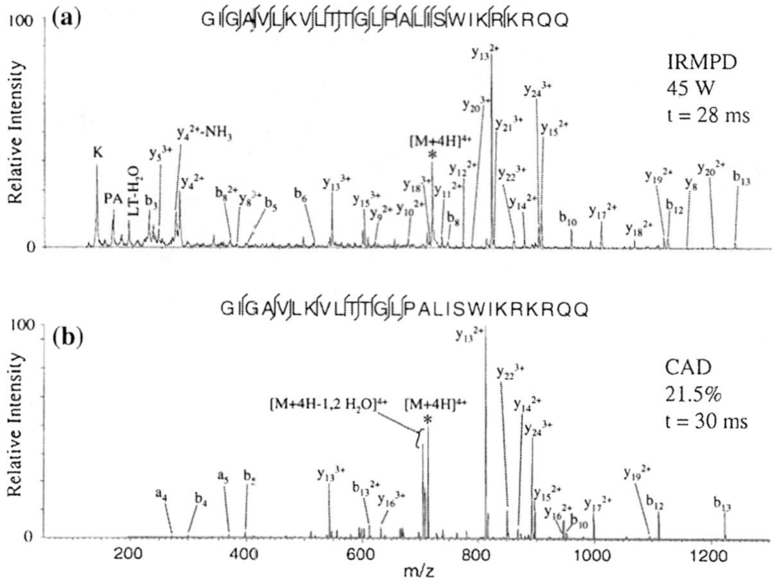

Fig. 1.13 **a** IRMPD and **b** CAD mass spectra of the $[M+4H]^{4+}$ melittin precursor ion. The precursor ions are indicated by asterisks [257]. Reproduced with permission from Elsevier, copyright 2004

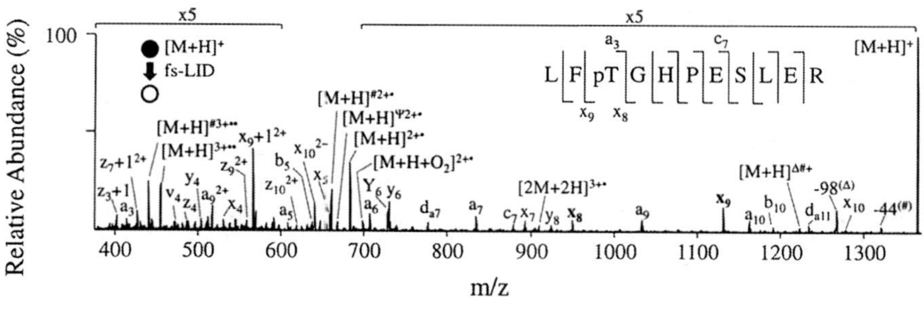

Fig. 1.14 fs-LID mass spectrum of the singly phosphorylated peptide LFpT-GHPESLER [260]. Reproduced with permission from Elsevier, copyright 2010

Several approaches have been developed to overcome this problem, which includes the use of pulsed-valve introduction of helium, thermally assisted IRMPD, and attachment of IR chromophores to analytes of interests [258, 259]. The later one is the most commonly used. For example, phosphate and phosphonate groups are the outstanding IR absorption groups.

Recently, an alternative PD technique termed "femtosecond laser-induced ionization/dissociation" (fs-LID) has been introduced [260, 261]. In this technique, a pack of ~1×10^5 near-infrared photons (~800 nm, 1.55 eV/photon) which has relative low energies is used for energy deposition and dissociation in ultrashort time scale (<35 fs). The fs-LID is characterized as a radical-directed dissociation in which the losses or rearrangements of the labile functional groups such as phosphate are minimized (Fig. 1.14). This feature allows fs-LID to be used to characterize the phosphopeptide precursor ions as well as to locate the phosphate groups. Also, in fs-LID, the precursor ions with low charge states undergo the most efficient fragmentation due to the low ionization potential compared to ions with higher charge states [260]. The mechanism involved in fs-LID is probably via a tunneling ionization mechanism. The high electric field caused by the packet of photons is capable of warping the pseudo-potential energy surface of the ions to allow an electron to escape efficiently. Then, the resulting radical intermediate ($[M+H]^{+} \rightarrow [M+H]^{2+\cdot}$) undergoes relative non-selective dissociation to form abundant peptide sequence ions (*a*, *b*, *c*, *x*, *y*, and *z* ions) [260].

1.4.2.3 Ultraviolet

Compared to IRMPD, photons used in UV dissociation have higher energy, which enables single-photon fragmentation [262]. These beams of higher energy expand the utility of PD methods for MS [262]. UVPD has some advantages over other dissociation methods. First, UVPD occurs on a time scale in the nanosecond to microsecond range, which is several orders of magnitude faster than IRMPD [251]. This is due to the UV lasers exciting the peptide electronically instead of through vibrational energies [251]. Second, energy deposition into the molecules is controllable due to the specificity of photon/chromophore interactions. Typical UV wavelengths are commonly used for dissociation, as listed in Table 1.1.

Absorption of UV photons leads to higher internal energy deposition than IRMPD. However, absorption of UV photons is a more selective process, that is, to say that only the ions with suitable UV chromophores can absorb. Recently, there have been significant applications of UVPD on the analysis of nucleic acids, peptides, and oligosaccharides [263–273]. In the case of molecules in the absence of natural chromophores, derivation with the high photoabsorptivity functional groups allows the molecules to absorb the specific wavelength. For example, peptides do not contain natural chromophores at 355 nm, i.e. cannot be dissociated upon absorbing photons at 355 nm. By attaching a molecule, 7-amino-4-methyl coumarin-3-acetic acid succinimidyl ester (AMCA), containing a UV

Table 1.1 Relevant energies for UV lasers [251]. Reproduced with permission from John Wiley and Sons, copyright 2009

Laser	λ (nm)	*E* (kj mol^{-1})	*E* [eV]
F_2 excimer	157	762	7.9
ArF excimer	193	620	6.4
Nd:YAG (4th harmonic)	266	450	4.7
Nd:YAG (3rd harmonic)	355	337	3.5

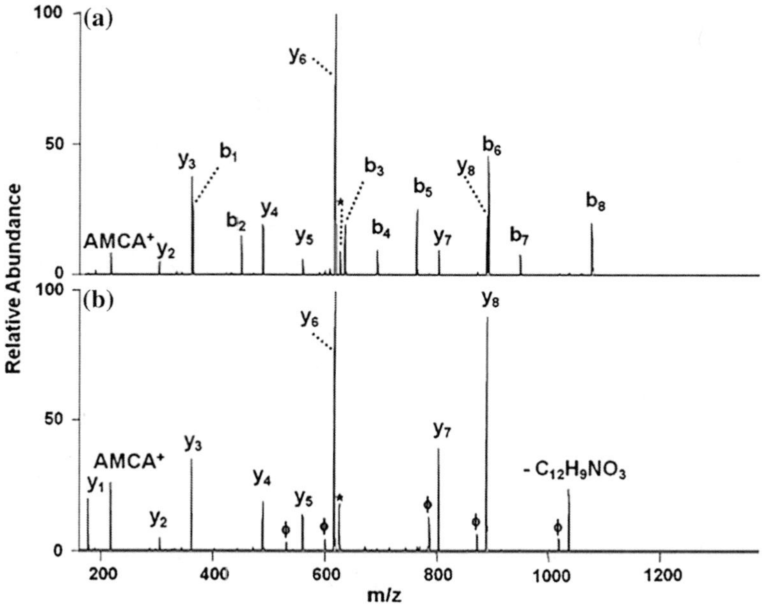

Fig. 1.15 ESI-MS/MS spectra of the doubly protonated N-terminally modified AMCA-modified peptide FSWGAEGQR by (**a**) CID (0.50 V) and (**b**) UVPD (15 pulses at 10 Hz) [249]. Reproduced with permission from Elsevier, copyright 2011

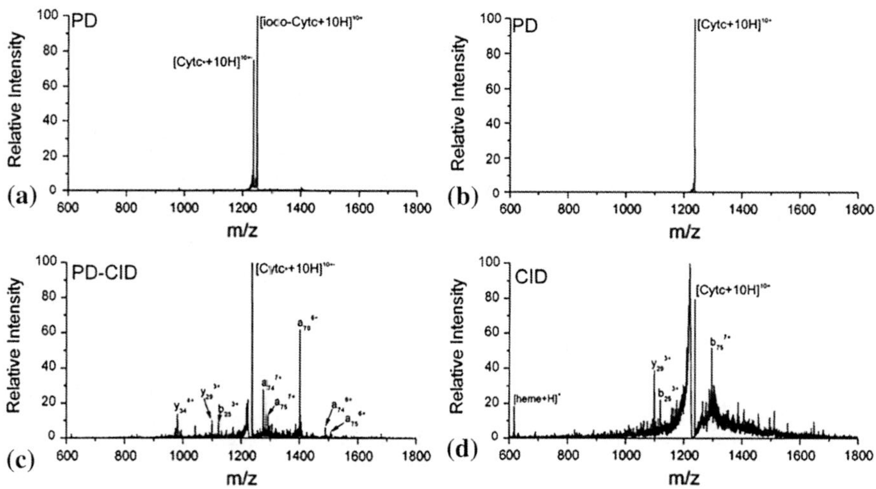

Fig. 1.16 **a** Photodissociation spectrum for the +10 charge state of iodo-cytc. **b** Photoexcitation of the unmodified protein. **c** The radical protein generated in (**a**) is fragmented by CID. **d** CID spectrum of the unmodified protein [273]. Reproduced with permission from ACS, copyright 2008

chromophore at this wavelength, this problem can be solved [249, 274]. A complete series of y ions down to y_1 were observed, which helped to sequence the peptide (shown in Fig. 1.15). The lack of b ions in UVPD is caused by the high photoabsorptivity for the incorporation of the modifications, in which the b ions undergo rapid and effective consecutive fragmentation, finally leading to disappearance from the spectra [249].

Intact protein ions can also be efficiently and selectively cleaved at tyrosine or histidine residues in gas phase upon the UV irradiation (266 nm) after iodination modification [273]. Modification of tyrosine to iodo-tyrosine followed by UVPD of the carbon–iodine bond generated a radial site specifically at the modified residue. The subsequent dissociation of the protein is largely dominated by radical-directed reactions, such as the backbone fragmentation at the modified tyrosine site. This method is very useful in terms of reducing the database searching time by several orders of magnitudes [273].

Figure 1.16 shows an example for the iodine-modified cytochrome c dissociation. PD of the I-modified Cytoc gives rise to the radical protein ion by losing I• with excellent yield compared to the unmodified protein with no PD (Fig. 1.16a, b). Radical protein ions generated from PD for the modified protein undergo further CID, providing further fragmentation (Fig. 1.16c). Each of these radical-directed fragments is produced within four amino acids of the modified tyrosine residue. In comparison, few fragments are resolved from the bulk of non-selective cleavages for the unmodified proteins (Fig. 1.16d). This demonstrates that photoactivation of iodo-tyrosine is an effective method for generating a radical site selectively, even when attached to an intact protein [273].

1.4.3 Electron-Induced Dissociation

1.4.3.1 Electron Capture Dissociation

Gas-phase ion–electron and ion–ion reactions are getting popularity for the activation of peptide ions in tandem mass spectrometry. Electron capture dissociation (ECD) [275] is a gas-phase ion fragmentation technique for structural elucidation in FT-ICR-MS [275, 276], which has been established for top-down sequencing. The first ECD-type mass spectra were observed in UVPD experiments, in which the protein samples were trapped in an FT-MS cell and irradiated with 193 nm laser pulses, followed by a charge reduction effect producing *c*, *z* fragments [277]. After that, the UV laser was replaced by a standard EI source (filament-based electron gun) and the ECD technique was born. ECD was first demonstrated by Roman Zubarev and Neil Kelleher in Fred McLafferty's laboratory in 1998 [275], and since then, more and more groups are adopting this approach.

ECD spectra of multiply charged protein ions usually produce the *c*- and *z*-type ions by cleavage of the N–C_α bond [275], which has some unique attributes

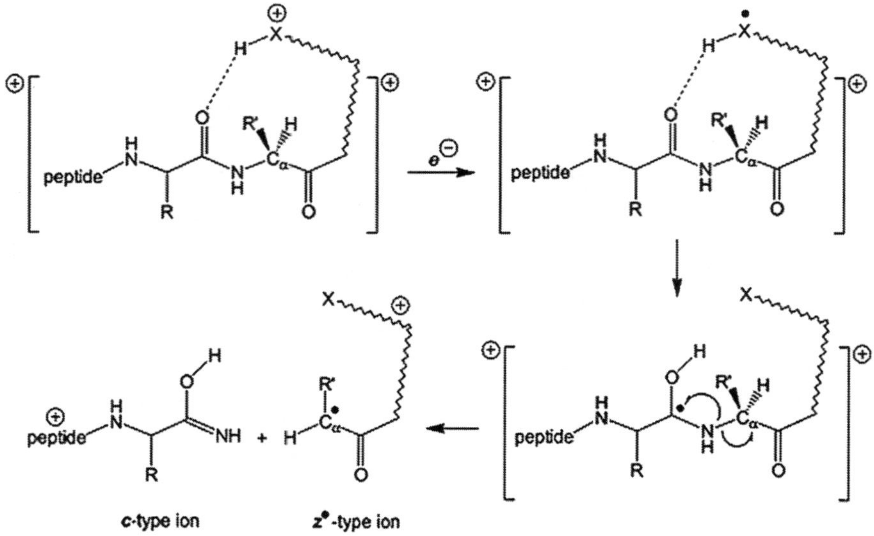

Scheme 1.6 Cornell mechanism [285]. Reproduced with permission from ACS, copyright 2006

including the preferential cleavage of disulfide bonds in gaseous multiply protonated proteins by low-energy electrons due to the high S–S affinity for H• atoms [278], location of PTMs [279–282], preservation of the non-covalent bonds [283], and extensive backbone cleavage [284]. However, the mechanism of ECD is still under debate, and both the Cornell mechanism and the Utah-Washington mechanism (UW) have been proposed [285]. The first one was originally formulated by the Cornell group [278, 286]. An electron is captured at a charge state, followed by hydrogen transfer to a proximate amide carbonyl forming an aminoketyl intermediate, and the resulting fragments are the residues of the peptide N-terminus and C-terminus, denoted as c and z ions, respectively [285] (Scheme 1.6).

The second one has been proposed independently by the group of Simons [287] and Tureček [288]. The UW mechanism considers that the electron capture was directly attached to amide π^* orbitals, which generates an anion-radical superbase, with high proton affinity. It is possible to abstract an H• from a protonated site to generate fragile aminoketyl radicals, thus triggering backbone dissociation via facile N–C_α bond cleavage to generate c and z ions [285] (Scheme 1.7).

However, there are still some challenges for this ECD method. For example, sometimes it is difficult to separate the fragments cleaved by ECD due to the intramolecular interactions between the residues in large proteins. In addition, the non-covalent intramolecular bonding surviving ECD is an obstacle to obtain abundant backbone fragmentation in the sequencing of protein larger than 20 kDa [289]. To solve this problem, activated ion electron capture dissociation (AI-ECD) was introduced [290]. Protein ions are first preheated by collisions with inert gas molecules or atoms or by irradiation with infrared laser light, and then, ECD is applied. This preheating of protein ions can cause unfolding of the protein and break the

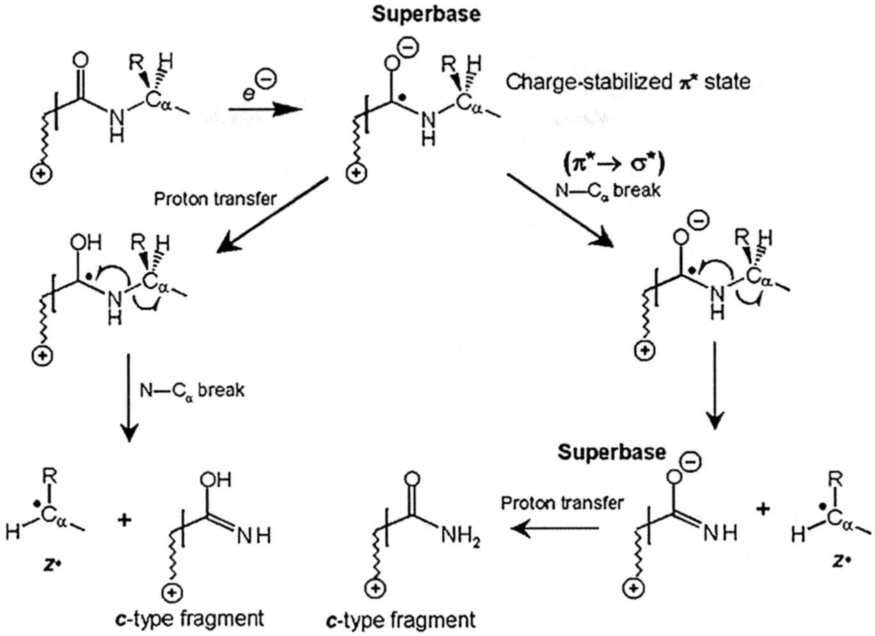

Scheme 1.7 Utah-Washington mechanism [285]. Reproduced with permission from ACS, copyright 2006

intramolecular interactions. A protein solution was sprayed into the MS entrance capillary from the right side and then heated for ion desolvation and folding retardation. The ions were reaccelerated by V_{pre} through the pre-skimmer collision region of short mean-free path to cleave weak non-covalent bonds. Ions were accelerated by V_{post} through the post-skimmer region to cleave the strong backbone covalent bonds in order to get more sequence information [291]. Thus, this approach extended top-down mass spectrometry to proteins with masses greater than 200 kDa.

FT-MS is the most effective device for protein sequence analysis by ECD, even if it is costly and requires more expertise than other mass spectrometers. The high mass resolving power of FT-MS instrumentation [292] affords resolution of overlapping fragment ions to obtain a more confident assignment. Bruker [293] and IonSpec [294] produce commercial FT-MS instruments that are used for ECD experiments without hardware modification. Also, a more efficient low-energy source, based on an indirectly heated dispenser cathode, can reduce the ECD experiment time from seconds to milliseconds, which improved the ion dissociation efficiency [283]. Thus, most top-down analyses have been performed with FT-MS instruments.

With recent technological developments, ECD has found a lot of applications in biomacromolecular research, including the characterization of intact proteins [279–282], nucleic acids, characterization of protein conformation [295–299], location of protein–ligand binding sites [300–302], and protein–protein assembly

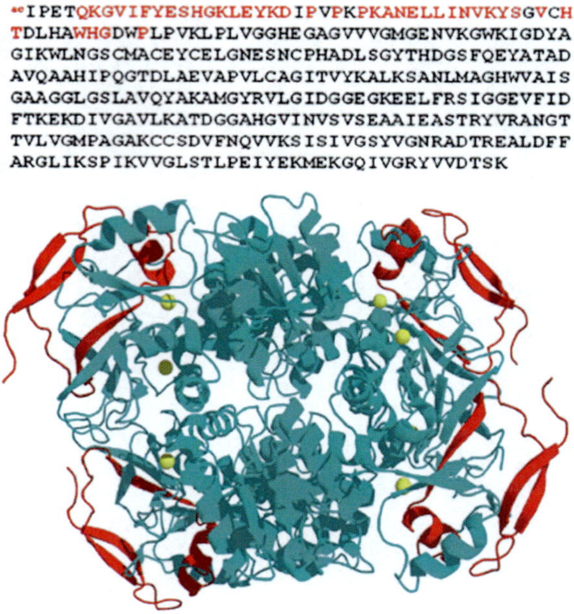

Fig. 1.17 ADH sequence showing the sites of ECD cleavages (*top*) and crystal structure of the yeast ADH tetramer (2HCY in PDB) with the N-terminal 55 residues that were sequenced, highlighted in *red* (*bottom*) [303]. Reproduced with permission from Springer, copyright 2010

characterization [289, 303, 304]. For example, Gross group introduced ECD for the sequence analysis of intact yeast alcohol dehydrogenase (ADH) tetramer with 147 kDa using native electrospray MS [303, 304] (shown in Fig. 1.17). This novel approach to study protein assemblies that combines ECD, native ESI, and FT-ICR-MS can sequence the flexible regions of the assembly subunits of ADH, concanavalin A, and photosynthetic Fenna–Matthews–Olson antenna protein complex by ECD or AI-ECD. In addition, non-covalent metal-binding sites for the concanavalin A assembly can also be determined. Furthermore, the regions undergoing fragmentation initiated by CID are consistent with the B-factor from X-ray crystallography of that protein, which is a measure of the extent that an atom can move from its coordinated position as a function of temperature or crystal imperfections. The approach can provide sequence information, non-covalent metal-binding sites, assembly stoichiometry, structural insights that pinpoint flexible regions [304], opening up a new top-down approach to characterize macromolecular assemblies [289].

In addition, negative ion electron capture dissociation (niECD) [305] was recently developed to allow the localization of PTMs and de novo sequencing for acidic peptides, which have an improved ionization efficiency in the negative-ion mode compared to the positive-ion mode. The peptide anions can capture 3.5–6.5 eV electrons, generating radical species with increased charges and yielding peptide backbone bond fragmentation analogous to that observed in cation ECD/ETD (including the preservation of PTMs and higher sequence coverage compared to CAD, Table 1.2) [305]. Note that backbone N–C_α bond cleavages to yield *c*/*z* ions are indicated with red lines, and backbone amide bond cleavages to

Table 1.2 Comparison of niECD and CAD for phosphopeptide anions [305]

z	m/z	niECD	CAD
-1	1236.541	H-KRSpYEEHIP-OH	H-KRSpYEEHIP-OH
-1	1568.634	H-RRREEEpSEEEAA-OH	H-RRREEEpSEEEAA-OH
-1	1540.632	H-TSTEPQpYQPGENL-NH_2	H-TSTEPQpYQPGENL-NH_2
-1	1658.780 (α-casein 121-134)	H-VPQLEIVPNpSAEER-OH	H-VPQLEIVPNpSAEER-OH
-2	1029.403 (β-casein 48-63)	H-FQpSEEQQQTEDELQDK-OH	H-FQpSEEQQQTEDELQDK-OH

yield *b*/*y* ions are indicated with green lines. Dashed lines indicate accompanying phosphate loss. Lack of indicated fragments in CAD is due to extensive neutral losses (e.g., HPO_3, H_3PO_4, and H_2O).

1.4.3.2 Electron Transfer Dissociation

In addition to the prevalent ECD approaches, electron transfer dissociation (ETD) is also one of the electron-induced fragmentation methods for fragmenting protein/peptide ions. ETD was first invented by the Donald Hunt group [306, 307], which shares many similarities with ECD, such as the preferential and extensive N–C_α bond cleavage, the preservation of labile modifications, and the ability to differentiate certain isomeric amino acid residues by secondary and radical-induced rearrangements [308–310]. Furthermore, the mechanism of ETD is still in debate as for ECD. ETD method dissociates ions utilizing ion/ion chemistry [307, 311–314], in which an electron from a radical anion is transferred to a protonated peptide, inducing the fragmentation of the peptide backbone via the cleavage of the N–C_α bond (Scheme 1.8). Similar to ECD, ECD also produces the *c*- and *z*-type ions [308].

The differences between ETD and ECD are that ETD uses a RF quadrupole ion trapping device instead of an FT-ICR-MS for ion trapping and detection [307]. The advantage is that RF ion trap mass spectrometers are of low cost, of low maintenance, and widely accessible compared to FT-ICR-MS [308]. In addition, for ETD, some energy is consumed to overcome the electron affinity of the anion reagent, and the collisional cooling is provided by a higher-pressure ion trap. Thus, ETD is considered to be a "colder" fragmentation than ECD, which can preserve PTMs, even sulfations, the most labile of PTMs [315].

Although ETD has advanced performance, the efficiency of top-down ETD fragmentation is reduced by non-covalent interactions, which are necessary to be ruptured before the product ions are detected individually for ETD. This can be

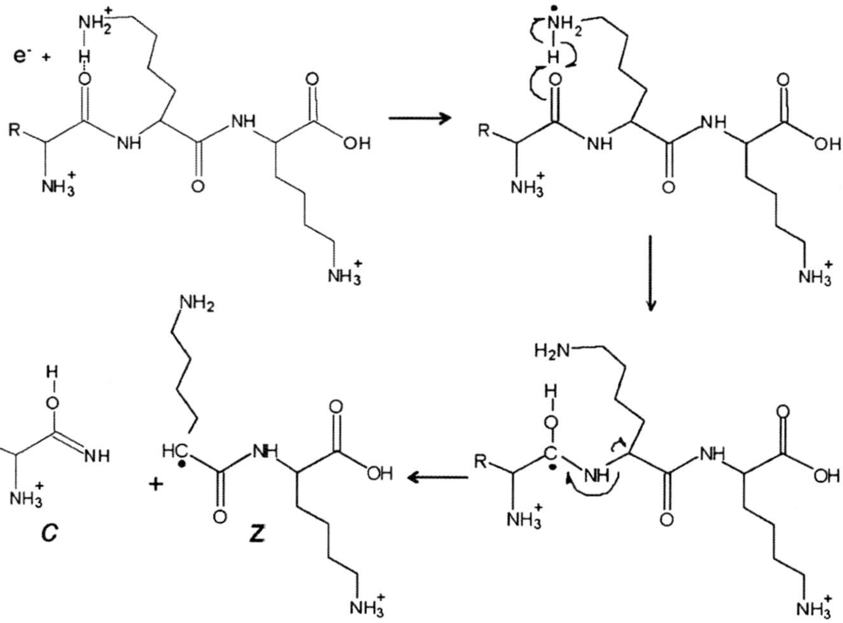

Scheme 1.8 ETD fragmentation scheme [307]. Reproduced with permission from PNAS, copyright 2004

achieved by post-ETD collisional activation (ETcaD) [316]. ETcaD is a supplemental collisional activation (CAD) method targeting the non-dissociated (intact) electron transfer (ET) product species ($[M+2H]^{+\cdot}$) to improve the ETD efficiency for the doubly charged peptide ions [316]. In addition, ETD efficiency can be enhanced by modifying the peptides with fixed charge tags [317].

ETD-based top-down MS has been used in-depth characterization of PTMs in large peptides, small- and medium-sized proteins, and non-covalent protein complexes [41, 307, 310, 318, 319]. To date, it has been applied for the structural analysis of intact 150 kDa monoclonal antibodies and immunoglobulins G (IgGs) [320]. In addition, negative electron transfer dissociation (NETD) is compatible with fragmenting peptide and protein ions, with backbone cleavage at the C–C_α bond, resulting in *a*- and *x*-type product ions (shown in Fig. 1.18) [311, 321].

1.4.3.3 Electron Detachment Dissociation

Electron detachment dissociation (EDD) is another the electron-induced dissociation method, used for fragmenting anionic species [322]. This technique is mainly applied to the structural analysis of peptides [323], carbohydrates [324], oligonucleotides [325–328], and proteins [329]. EDD was first introduced by Zubarev and co-workers in 2001 [322]. In EDD, irradiation of multiply deprotonated ions

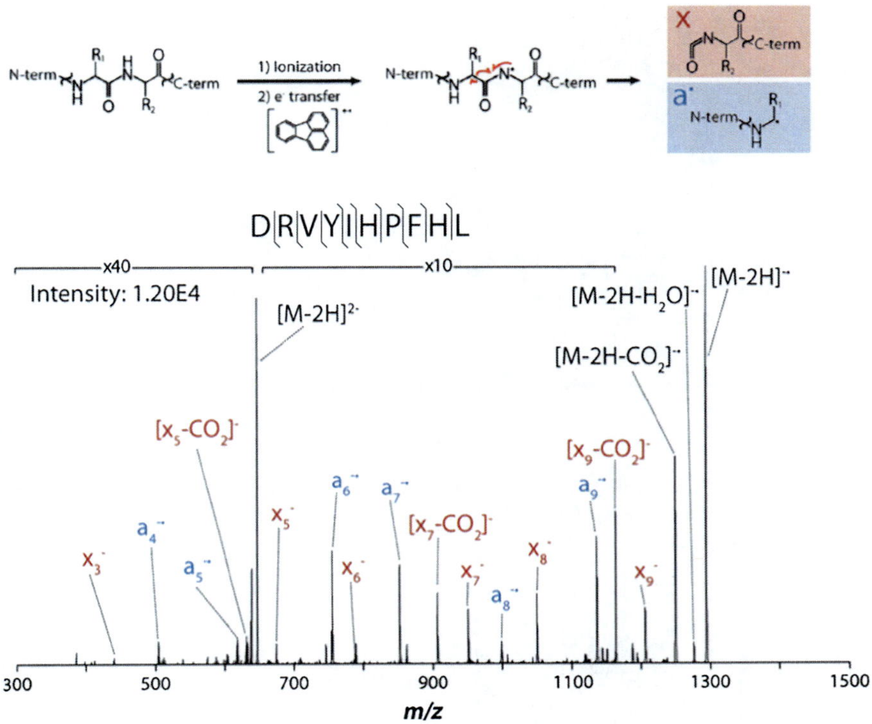

Fig. 1.18 Generalized reaction scheme by which peptide anions are oxidized by fluoranthene radical cations to produce *a*- and *x*-type product ions (*above*) [311]. *Below* is an example NETD MS/MS mass spectrum resulting from the fragmentation of the standard peptide human angiotensin I ions [321]. Reproduced with permission from ASMS, copyright 2005 and ACS, copyright 2012

with >10 eV electrons results in electron detachment to form the radical ions $[M{-}nH]^{(n-1)-\cdot}$ which can further undergo backbone dissociation, leading to the formation of *a*-, *c*-, and *z*-type ions [322]. The first demonstration of EDD for the analysis of the large proteins was with ubiquitin [329], whereas EDD can be applied to the analysis of RNAs and 34 nt RNA dissociated by EDD [330]. Figure 1.19 shows the EDD spectrum of di-anions of the sulfated peptide caerulein obtained at 21 eV, and the most prominent losses from $[M{-}2H]^{-\cdot}$ are $-CO_2$ and $-SO_3$, along with the backbone cleavages, generating the *a*-, *c*-, and *z*-type ions. In addition, most fragment ions preserved the sulfate group [322].

1.4.3.4 Electron Ionization Dissociation

Electron ionization dissociation (EID) can be applied to singly charged precursor ions (as compared to ECD and EDD). The mechanism of EID is probably attributed to irradiation of trapped cations $[M{+}nH]^{n+}$ by high-energy electrons (at least

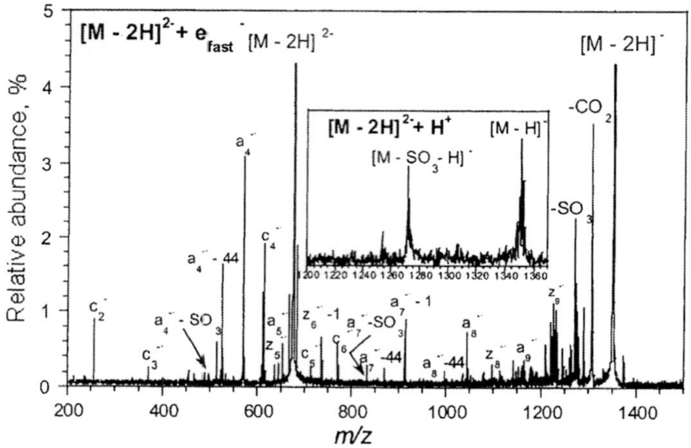

Fig. 1.19 EDD spectrum of di-anions of the sulfated peptide caerulein. The *inset* shows the results of the capture of the hydrogen ions by the di-anions [322]. Reproduced with permission from Elsevier, copyright 2001

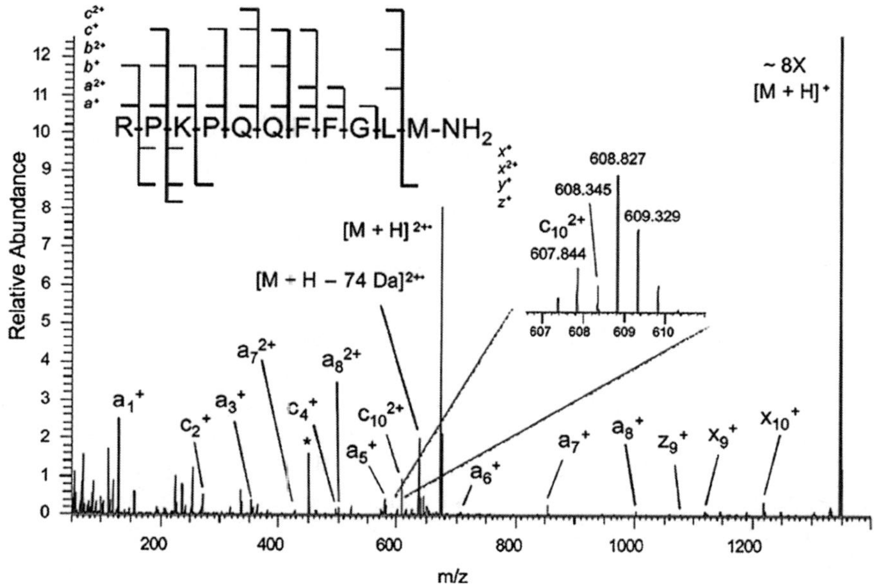

Fig. 1.20 EID spectrum of singly protonated substance P. (*inset*) Isotopic distribution of c_{10}^{2+} ions [331]. Reproduced with permission from ACS, copyright 2009

10 eV higher than the ionization threshold of the cations), which causes simultaneous ionization and electronic excitation of $[M+nH]^{n+}$, generating a radical $[M+nH]^{(n+1)+\cdot}$ that subsequently undergoes fragmentation and gives both side-chain losses and backbone fragmentation. EID often leads to N–C_α bond cleavage,

giving rise to *c*- and *z*-type ions. The cleavage of C–C bonds is also observed, generating preferentially *a*- and *x*-type ions [331]. Thus, the EID spectra are very complicated, in which ions are generated by both radical pathways and ergodic processes. EID can be used in bottom-up proteomics for structural analysis. It can also be adopted in the top-down analysis for probing the folded gas-phase protein conformations [331]. Figure 1.20 shows the EID spectrum of singly charged ions of substance P, showing backbone cleavages for the generation of *a*-, *x*-, *c*-, and *z*-type ions [331].

1.4.4 Atmospheric Pressure Ion Dissociation Methods

1.4.4.1 Atmospheric Pressure Thermal Dissociation

Atmospheric pressure thermal dissociation (APTD) [332] is a new dissociation method used to fragment protein/peptide ions, which is performed under atmospheric pressure outside of the mass spectrometer. Typically, protein/peptide ions generated from an ESSI [124] source, a variant of ESI, go through a heated coiled metal tube undergoing thermal fragmentation assisted by the high pressure of N_2 gas (175 psi as optimized), which results in extensive fragmentation. Complementary structure information can be obtained by utilizing neutral re-ionization (NRI), subsequently after APTD using the online corona discharge, making it an attractive method to elucidate the primary sequence of proteins/peptides [332]. Figure 1.21a shows the apparatus for the methodology of APTD with subsequent NRI. In this APTD-NRI, a pair of metal plates, serving as an "ion switch," was used to deflect all the fragment ions from the gas stream and to leave the neutral species for re-ionization by corona discharge. Temperatures ranging from 230–380 °C were typically used to heat the coiled metal tube.

An important feature of APTD is that some PTMs such as phosphorylation and sulfonation in polypeptides could survive the dissociation process. Figure 1.21b shows the APTD of phosphopeptide DHTGFLpTEpYVATR at 290 °C, producing many *b* and *y* ions with preservation of labile phosphate groups (e.g., the ions b_7^-, b_8^-, y_5^-, y_7^-, y_9^-, y_{10}^-, and y_{12}^-). These fragments were not observed in CID (the inset of Fig. 1.21b shows that few fragment ions are formed upon CID of the singly charged deprotonated phosphopeptide ion $[M-H]^-$) [332]. Also, the capability of online re-ionization of the neutral fragments following APTD was successfully achieved, as exemplified by the case of angiotensin II in both positive and negative ion modes (Fig. 1.21c, d) [136]. The formation of a series of positively charged y ions (y_2^+ to y_7^+ in Fig. 1.21c) as well as a series of negatively charged *y* ions (y_2^- to y_7^- in Fig. 1.21d) points to the presence of the corresponding neutrals (i.e., deprotonated forms of y ions) emerging from the hot coiled tube. With this additional information, high fragmentation coverage was achieved [332]. Therefore, APTD coupling with subsequent NRI provides rich sequence information for proteins/peptides.

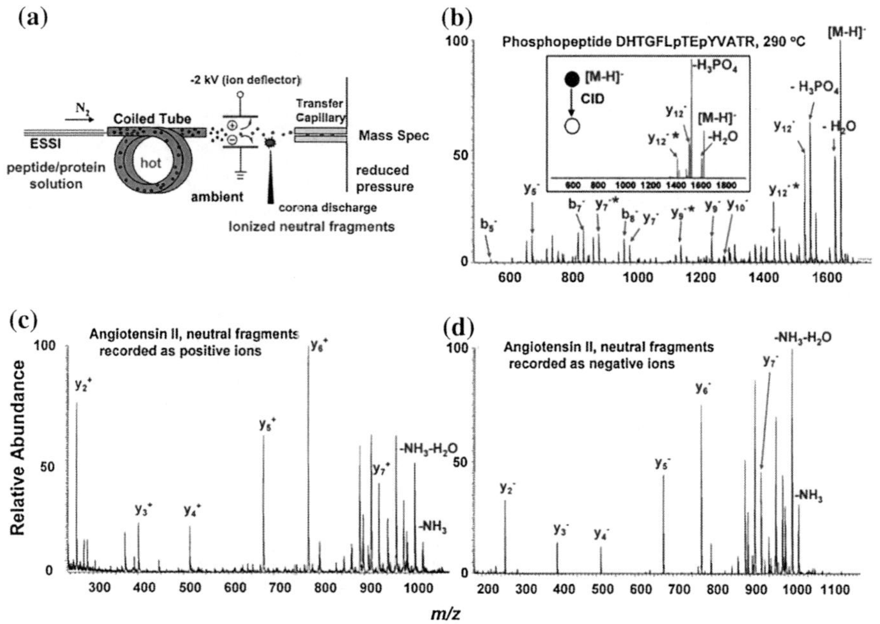

Fig. 1.21 **a** Experimental apparatus for recording mass spectrum of ionized neutral fragments after thermal dissociation of peptide/protein ions. **b** Mass spectrum of phosphopeptide DHTGFLpTEpYVATR at 290 °C (*y* ions with the loss of one phosphate group is labeled with an asterisk); mass spectra recorded by ionizing the neutral fragment generated from angiotensin II by thermal dissociation showing **c** positive and **d** negative ionization modes [332]. Reproduced with permission from ACS, copyright 2007

1.4.4.2 Low-Temperature Plasma

Low-temperature plasma (LTP) ionization was first developed by R. Graham Cooks and Zheng Ouyang et al. in 2008, for ambient DI. Later, the observation of peptide fragmentation in nano-ESI after pre-treatment of the nano-ESI emitter with a helium LTP at atmospheric pressure was reported by Xia et al. [333]. Figure 1.22a shows the experimental apparatus including a T-shaped glass tube with a nano-ESI emitter inside. Helium discharge gas is introduced from the bottom into the glass tube, and the LTP is produced by using a dielectric barrier discharge (DBD). In addition, an alternating current is applied to the two electrodes. For this experiment, the LTP was first turned on for 2 min without operating the nano-ESI emitter, and then, a high voltage was applied to the nano-ESI emitter containing sample solutions, while the plasma was off. This new fragmentation method was applied to more than 30 peptides with different sequences and sizes (5–26 amino acids). The results show that *a*-, *b*-, and *y*-type fragment ions were generated via the cleavage of amide bonds. This method was also applied to phosphopeptides, and it shows the preservation of the modification. Figure 1.22b shows the nano-ESI spectrum for a phosphopeptide after plasma

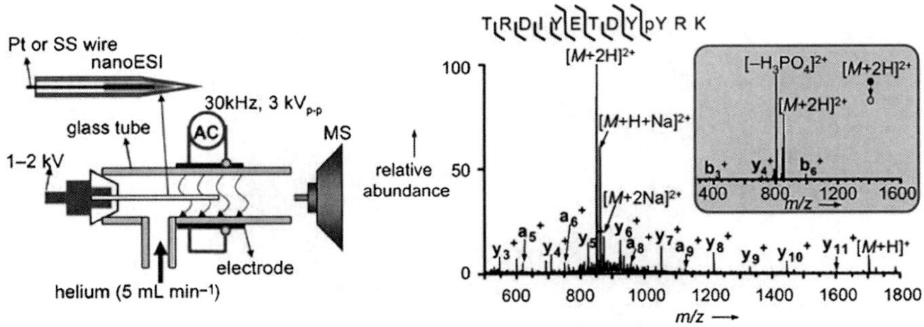

Fig. 1.22 **a** The experimental setup. **b** Mass spectrum derived from positive nano-ESI of a phosphopeptide TRDIYETDYpYRK. The *inset* shows data acquired from conventional ion trap CID of the $[M+2H]^{2+}$ peptide ions [333]. Reproduced with permission from John Wiley and Sons, copyright 2008

exposure. Compared to the conventional ion trap CID spectrum shown in the inset of Fig. 1.22b, the extensive backbone fragmentation including a series of *y* ions and several *a* ions were observed with little loss in phosphorylation [333]. This method demonstrates that ESI as a soft ionization technique can also generate peptide fragmentation.

1.4.4.3 Atmospheric Pressure Electron Capture Dissociation

When atmospheric pressure photoionization was used to ionize hydrophobic peptides, in-source dissociation probably resulting from ECD processes was observed. Based on this phenomenon, atmospheric pressure electron capture dissociation (AP-ECD) was developed. In this method, photoelectrons were generated within a commercial PhotoSpray atmospheric pressure photoionization (APPI) source, which was used to induce ECD of multiply charged peptide ions produced by an upstream heated nebulizer device, leading to *b*-, *y*-, and *c*-type ions. The strength of this method is that this technique does not need the specialized instruments normally used, such as FT-ICR-MS for ECD. Robb's group introduced a AP-ECD ion source where the conventional nanospray ionization emitters are coupled with the source block and photoionization lamp of a PhotoSpray APPI source (shown in Fig. 1.23) [334]. The multiply charged peptide ions generated in the spray chamber by the enclosed nanospray source are carried into the downstream source block by the nebulizer gas, where they interact with the photoelectrons generated by the photoionization lamp. When the lamp is turned off, the source operates as a normal nanospray source; when the lamp is turned on, photoelectrons are produced and subsequently captured by the peptide ions in the atmospheric pressure reaction/transport region, leading to ECD. Ions exit the source and are delivered through the vacuum interface for mass analysis and detection [334]. This new source shows the capability of providing qualitative analysis for low-femtomole sample. The coupling of AP-ECD

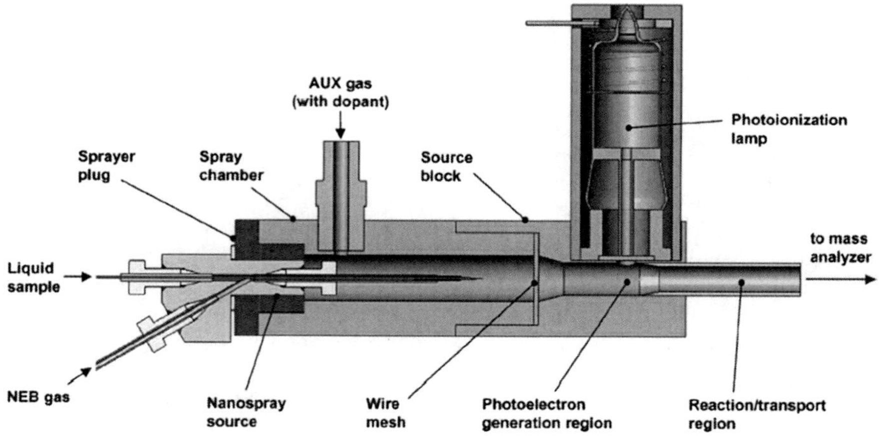

Fig. 1.23 Cross-sectional drawing of the AP-ECD source [334]. Reproduced with permission from ACS, copyright 2012

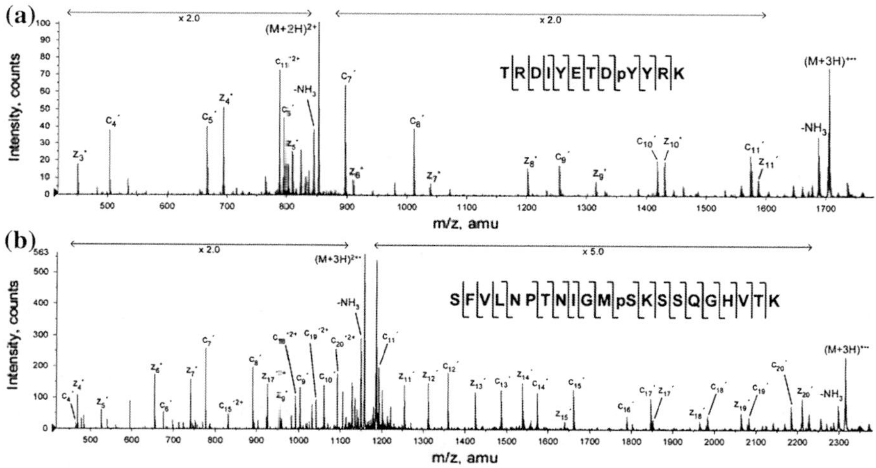

Fig. 1.24 AP-ECD spectra from 100 fmol each of two phosphopeptides [334]. Reproduced with permission from ACS, copyright 2012

source with capillary LC (LC–AP-ECD-MS) was also developed, which was applied to the analysis of substance P, a tryptic digest of bovine serum albumin (BSA), and a phosphopeptide mixture [334]. All samples at the femtomole levels were analyzed, and high-quality ECD spectra were obtained. This study also indicated that LC–AP-ECD-MS is suitable and useful for the structural analysis of peptides and protein digests. Figure 1.24 shows the AP-ECD spectra for two phosphorylated peptides separated by LC, showing extensive fragments [334]. This example demonstrated the utility of LC–AP-ECD-MS for peptides with labile modifications.

1.4.5 Other Fragmentation Methods

1.4.5.1 Metastable Ion Dissociation

Since 2005, several research groups proposed the use of metastable atoms to achieve precursor ion dissociation [335–340]. This type of dissociation method, termed as metastable atom dissociation (MAD), has been introduced in RF ion trap instruments. The precursors of interest that are stored in a RF ion trap are irradiated by a beam of particles produced by a FAB gun or a glow discharge (Fig. 1.25a [340]). A high voltage (up to 10 kV) was applied on the anode of the FAB gun, through which the noble gas flows (Fig. 1.25a). A gas discharge occurs when the high voltage bias is applied to the anode and surrounding cathode. With the generated electrostatic field, electrons flow along the symmetric axis of the anode to ionize the neutral gas. These formed ions are accelerated and subsequently converted into neutrals via charge exchange reactions with gas atoms. The beam of fast neutrals and few remaining ions thus formed exits the gun through the 0.5 mm orifice. Along with neutrals and ions in the electronic ground state, the beam contains a fraction of particles in electronically excited metastable states [340]. Since the particle beam is neutral, there is no problem inherent to the charge capacity limitations of ion traps [338]. The interaction of the isolated precursor

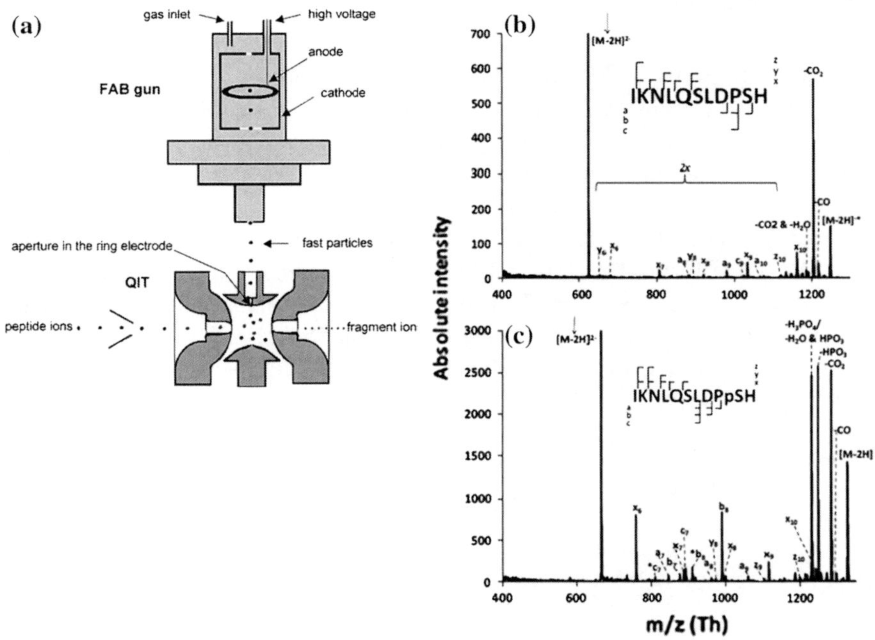

Fig. 1.25 **a** Scheme of the instrument for MAD (FAB gun and QIT) [340]. **b** MAD of non-phosphorylated IKNLQSLDPSH [336]. **c** MAD of phosphorylated IKNLQSLDPpSH [336]. Reproduced with permission from Wiley, copyright 2005 and Elsevier, copyright 2011

ions with a high or low kinetic energy beam of noble gas metastable atoms produces extensive peptide backbone cleavages in the form of *a*-, *b*-, *c*-, *x*-, *y*-, and *z*-type ions, as well as preserving PTMs (Fig. 1.25b, c) [336].

It was shown that peptide ion dissociation occurs following two competing mechanisms in MAD, penning ionization (PI) and charge reduction [335–337].

$$\left[\mathrm{P}+\mathrm{nH}^{+}\right]^{\mathrm{n}+} + \mathrm{M}^{*} \rightarrow \left[\mathrm{P} + \mathrm{nH}^{+}\right]^{(\mathrm{n}+1)+\cdot} + \mathrm{M} + \mathrm{e}^{-} \tag{1.5}$$

M* is the metastable atom. This PI process (Eq. 1.5) generates the radical cation which is more reactive than the even-electron cation. Then, the radical cations undergo some rearrangement reaction to provide fragments [334, 336]. The second mechanism of charge reduction is similar to ECD and ETD, which occurs when an electron is transferred to the polycation of interests to reduce charge, which is still in debate [336]. However, both of the mechanisms agree that an electron transfer process is involved to generate a radical bioion that can rearrange, propagate along the peptide backbone, and subsequently fragment [336]. MAD is able to analyze the same peptides as ECD techniques as well as singly charged peptides which are difficult to be fragmented using ECD/ETD methods. MAD can also be used for the study of phosphorylated peptides [341, 342].

1.5 Conclusions

In summary, this chapter introduces several classical protein ionization techniques including their working principles, strengths, and analytical applications. In addition, various ion activation methods used for protein/peptide structure analysis (viz. tandem mass spectrometry) are covered to illustrate the versatilities of protein MS. As this field is very dynamic and rapidly developing, it is not possible to cover all the methods for protein analysis. The authors apologize for omissions. As the performance of current MS ionization and dissociation technologies, although highly effective, cannot meet all real-world demands in biochemistry, structural biology, and biopharmaceutical development, we foresee that these methods will undergo further development; the only limitation seems to be our imagination [36].

Acknowledgements This work was supported by the NSF (CHE-0911160) and NSF Career (CHE-1149367) funds.

References

1. Karas M, Hillenkamp F (1988) Laser desorption ionization of proteins with molecular masses exceeding 10000 daltons. Anal Chem 60:2299–2301
2. Yamashita M, Fenn JB (1984) Electrospray ion source. Another variation on the free-jet theme. J Phys Chem 88:4451–4459

3. Fenn JB, Mann M, Meng CK et al (1989) Electrospray ionization for mass spectrometry of large biomolecules. Science 246:64–71
4. Van Berkel WJH, Van Den Heuvel RHH, Versluis C et al (2000) Detection of intact megaDalton protein assemblies of vanillyl-alcohol oxidase by mass spectrometry. Protein Sci 9:435–439
5. Cole RB (ed) (1997) Electrospray ionization mass spectrometry: fundamentals, instrumentation, and applications. Wiley-Interscience, New York
6. Gross JH (ed) (2004) Mass spectrometry: a textbook. Springer, New York
7. Fenn JB (2000) Mass spectrometric implications of high-pressure ion sources. Int J Mass Spectrom 200:459–478
8. Winston RL, Fitzgerald MC (1997) Mass spectrometry as a readout of protein structure and function. Mass Spectrom Rev 16:165–179
9. Loo JA (2000) Electrospray ionization mass spectrometry: a technology for studying noncovalent macromolecular complexes. Int J Mass Spectrom 200:175–186
10. Loo JA (1997) Studying noncovalent protein complexes by electrospray ionization mass spectrometry. Mass Spectrom Rev 16:1–23
11. Kaltashov IA, Zhang M, Eyles SJ et al (2006) Investigation of structure, dynamics and function of metalloproteins with electrospray ionization mass spectrometry. Anal Bioanal Chem 386:472–481
12. Grandori R (2003) Electrospray-ionization mass spectrometry for protein conformational studies. Curr Org Chem 7:1589–1603
13. Calvo E, Camafeita E, Fernando Díaz J et al (2008) Mass spectrometry for studying the interaction between small molecules and proteins. Curr Proteom 5:20–34
14. Dalluge JJ (2002) Matrix-assisted laser desorption ionization-mass spectrometry (MALDI-MS). Anal Bioanal Chem 372:18–19
15. Harvey DJ (2003) Matrix-assisted laser desorption/ionization mass spectrometry of carbohydrates and glycoconjugates. Int J Mass Spectrom 226:1–35
16. Reyzer ML, Caprioli RM (2007) MALDI-MS-Based imaging of small molecules and proteins in tissues. Curr Opin Chem Biol 11:29–35
17. Najam-ul-Haq M, Szabó RZ, Vallant R et al (2007) Role of carbon nano-materials in the analysis of biological materials by laser desorption/ionization-mass spectrometry. J Biochem Biophys Methods 70:319–328
18. Thomson JJ (1913) Rays of positive electricity and their application to chemical analysis. Longmans, London
19. Bleakney W (1929) A new method of positive ray analysis and its application to the measurement of ionization potentials in mercury vapor. Phys Rev 34:157–160
20. Nier AO (1947) A mass spectrometer for isotope and gas analysis. Rev Sci Instrum 18:398
21. Biemann K, Gapp F, Seibl J (1959) Application of mass spectrometry to structure problems. I. amino acid sequence in peptides. J Am Chem Soc 81:2274–2275
22. Senn M, McLafferty FW (1966) Automatic amino acid sequence determination in peptides. Biochem Biophys Res Commun 23:381–385
23. Munson MSB, Field FH (1966) Chemical ionization mass spectrometry. I. general introduction. J Am Chem Soc 88:2621
24. Harrison AG (1992) Chemical ionization mass spectrometry. CRC Press, Boca Raton
25. Beckey HD (1969) Field ionization mass spectrometry. Research/Development 20:26
26. Hakansson P, Kamensky I, Sundqvist B et al (1982) ^{127}I-Plasma desorption mass spectrometry of insulin. J Am Chem Soc 104:2948–2949
27. Morris HR, Panico M, Taylor GW (1983) FAB-mapping of recombinant-DNA protein products. Biochem Biophys Res Commun 117:299–305
28. Macfarlane RD, Torgerson DF (1976) Californium-252 plasma desorption mass spectroscopy. Science 191:920–925
29. Barber M, Green BN (1987) The analysis of small proteins in the molecular weight range 10–24 kDa by magnetic sector mass spectrometry. Rapid Commun Mass Spectrom 1:80–83

30. Blakley CR, Vestal ML (1983) Thermospray interface for liquid chromatography/mass spectrometry. Anal Chem 55:750–754
31. Koropchak JA, Veber M, Browner RF (1992) Thermospray sample introduction to atomic spectrometry. Crit Rev Anal Chem 23:113–141
32. Tanaka K, Waki H, Ido Y et al (1988) Protein and polymer analyses up to m/z 100 000 by laser ionization time-of flight mass spectrometry. Rapid Commun Mass Spectrom 2:151–153
33. Wikimedia Foundation I (2012) http://en.wikipedia.org/wiki/History_of_mass_spectrometry. 29 Feb 2012
34. Ingendoh A, Karas M, Hillenkamp F et al (1994) Factors affecting the mass resolution in matrix-assisted laser desorption-ionization mass spectrometry. Int J Mass Spectrom Ion Process 131:345–354
35. Vestal ML, Juhasz P, Martin SA (1995) Delayed extraction matrix-assisted laser desorption time-of-flight mass spectrometry. Rapid Comm Mass Spectrom 9:1044–1050
36. Roepstoref P (2002) Mass spectrometry of peptides and proteins a personal historical view. In: Silberrying J, Ekman R (eds) Mass spectrometry of peptides and proteins a personal historical view. Wiley, New York
37. Harris GA, Galhena AS, Fernandez FM (2011) Ambient sampling/ionization mass spectrometry: applications and current trends. Anal Chem 83:4508–4538
38. Takats Z, Wiseman JM, Cooks RG (2005) Ambient mass spectrometry using desorption electrospray ionization (DESI): instrumentation, mechanisms and applications in forensics, chemistry, and biology. J Mass Spectrom 40:1261–1275
39. Takats Z, Wiseman JM, Gologan B et al (2004) Mass spectrometry sampling under ambient conditions with desorption electrospray ionization. Science 306:471–473
40. Cody RB, Laramee JA, Durst HD (2005) Versatile new ion source for the analysis of materials in open air under ambient conditions. Anal Chem 77:2297–2302
41. Huang M, Jhang S, Cheng C et al (2010) Effects of matrix, electrospray solution, and laser light on the desorption and ionization mechanisms in electrospray-assisted laser desorption ionization mass spectrometry. Analyst 135:759–766
42. Peng IX, Loo RO, Margalith E et al (2010) Electrospray-assisted laser desorption ionization mass spectrometry (ELDI-MS) with an infrared laser for characterizing peptides and proteins. Analyst 135:767–772
43. Sampson JS, Murray KK, Muddimana DC (2009) Intact and top-down characterization of biomolecules and direct analysis using infrared matrix-assisted laser desorption electrospray ionization coupled to FT-ICR mass spectrometry. J Am Soc Mass Spectrom 20:667–673
44. Cheng S-C, Cheng T-L, Chang H-C et al (2009) Using laser-induced acoustic desorption/electrospray ionization mass spectrometry to characterize small organic and large biological compounds in the solid state and in solution under ambient conditions. Anal Chem 81:868–874
45. Dixon RB, Muddiman DC (2010) Study of the ionization mechanism in hybrid laser based desorption techniques. Analyst 135:880–882
46. Dixon RB, Sampson JS, Muddiman DC (2009) Generation of multiply charged peptides and proteins by radio frequency acoustic desorption and ionization for mass spectrometric detection. J Am Soc Mass Spectrom 20:597–600
47. Dole M, Mack LL, Hines RL (1968) Molecular Beams of Macroions. J Chem Phys 49:2240–2249
48. Iribarne JV, Thomson BA (1976) On the evaporation of small ions from charged droplets. J Chem Phys 64:2287–2294
49. Benesch JLP, Ruotolo BT, Simmons DA et al (2007) Protein complexes in the gas phase: technology for structural genomics and proteomics. Chem Rev 107:3544–3567
50. Kaltashov IA, Abzalimov RR (2008) Do ionic charges in ESI MS provide useful information on macromolecular structure? J Am Soc Mass Spectrom 19:1239–1246
51. Przybylski M, Glocker MO (1996) Electrospray mass spectrometry of biomacromolecular complexes with non-covalent interactions-new analytical perspectives for supramolecular chemistry and molecular recognition processes. Angew Chem Int Ed 35:806–826

52. van den Heuvel RHH, Heck AJR (2004) Native protein mass spectrometry: from intact oligomers to functional machineries. Curr Opin Chem Biol 8:519–526
53. Kaddis CS, Loo JA (2007) Native protein MS and ion mobility: large flying proteins with ESI. Anal Chem 79:1778–1784
54. Benesch JLP, Robinson CV (2006) Mass spectrometry of macromolecular assemblies: preservation and dissociation. Curr Opin Struct Biol 16:245–251
55. Chowdhurry SK, Katta V, Chait BT (1990) Probing conformational changes in proteins by mass spectrometry. J Am Chem Soc 112:9012–9013
56. Nemirovskiy O, Giblin DE, Gross ML (1999) Electrospray ionization mass spectrometry and hydrogen/deuterium exchange for probing the interaction of calmodulin with calcium. J Am Soc Mass Spectrom 10:711–718
57. Simmons DA, Konermann L (2002) Characterization of transient protein folding intermediates during myoglobin reconstitution by time-resolved electrospray mass spectrometry with on-line isotopic pulse labeling. Biochemistry 41:1906–1914
58. Winger BE, Light-Wahl KJ, Rockwood AL et al (1992) Probing qualitative conformation differences of multiply protonated gas-phase proteins via H/D isotopic exchange with D_2O. J Am Chem Soc 114:5898–5900
59. Campbell S, Rodgers MT, Marzluff EM et al (1995) Deuterium exchange reactions as a probe of biomolecule structure. Fundamental studies of gas phase H/D exchange reactions of protonated glycine oligomers with D_2O, CD_3OD, CD_3CO_2D, and ND_3. J Am Chem Soc 117:12840–12854
60. Engen JR (2009) Analysis of protein conformation and dynamics by hydrogen/deuterium exchange MS. Anal Chem 81:7870–7875
61. Zhang Z, Smith DL (1996) Thermal-induced unfolding domains in aldolase identified by amide hydrogen exchange and mass spectrometry. Protein Sci 5:1282–1289
62. Zhu M, Rempel DL, Du Z et al (2003) Quantification of protein-ligand interactions by mass spectrometry, titration, and H/D exchange: PLIMSTEX. J Am Chem Soc 125:5252–5253
63. Zhu M, Rempel DL, Zhao J et al (2003) Probing Ca^{2+}-induced conformational changes in porcine calmodulin by H/D exchange and ESI-MS: effect of cations and ionic strength. Biochemistry 42:15388–15397
64. McLafferty FW, Guan Z, Haupts U et al (1998) Gaseous conformational structures of cytochrome c. J Am Chem Soc 120:4732–4740
65. Wilm MS, Mann M (1994) Electrospray and Taylor-Cone theory, Dole's beam of macromolecules at last? Int J Mass Spectrom Ion Proc 136:167–180
66. El-Faramawy YA, Siu KWM, Thomson BA (2005) Efficiency of nano-electrospray ionization. J Am Soc Mass Spectrom 16:1702–1707
67. Wilm M, Mann M (1996) Analytical properties of the nanoelectrospray ion source. Anal Chem 68:1–8
68. Karas M, Bachmann D, Hillenkamp F (1985) Influence of the wavelength in high irradiance ultraviolet laser desorption mass spectrometry of organic molecules. Anal Chem 57:2935–2939
69. Karas M, Bachmann D, Bahr U et al (1987) Matrix-assisted ultraviolet laser desorption of non-volatile compounds. Int J Mass Spectrom Ion Proc 78:53–68
70. Karas M, Krüger R (2003) Ion formation in MALDI: the cluster ionization mechanism. Chem Rev 103:427–439
71. Hardesty WM, Caprioli RM (2008) In situ molecular imaging of proteins in tissues using mass spectrometry. Anal Bioanal Chem 391:899–903
72. Cornett DS, Reyzer ML, Chaurand P et al (2007) MALDI imaging mass spectrometry: molecular snapshots of biochemical systems. Nat Methods 4:828–833
73. Grassl J, Taylor NL, Millar AH (2011) Matrix-assisted laser desorption/ionisation mass spectrometry imaging and its development for plant protein imaging. Plant Methods 7:21–32
74. Miao Z, Wu S, Chen H (2010) The study of protein conformation in solution via direct sampling by desorption electrospray ionization mass spectrometry. J Am Soc Mass Spectrom 21:1730–1736

75. Yang SH, Wijeratne AB, Li L, Edwards BL, Schug KA (2011) Manipulation of protein charge states through continuous flow-extractive desorption electrospray ionization: a new ambient ionization technique. Anal Chem 83:643–647
76. Cotte-Rodriguez I, Takats Z, Talaty N et al (2005) Desorption electrospray ionization of explosives on surfaces: sensitivity and selectivity enhancement by reactive desorption electrospray ionization. Anal Chem 77:6755–6764
77. Cotte-Rodriguez I, Mulligan CC, Cooks G (2007) Non-proximate detection of small and large molecules by desorption electrospray ionization and desorption atmospheric pressure chemical ionization mass spectrometry: instrumentation and applications in forensics, chemistry, and biology. Anal Chem 79:7069–7077
78. Cotte-Rodriguez I, Hernandez-Soto H, Chen H et al (2008) In situ trace detection of peroxide explosives by desorption electrospray ionization and desorption atmospheric pressure chemical ionization. Anal Chem 80:1512–1519
79. Cotte-Rodriguez I, Cooks RG (2006) Non-proximate detection of explosives and chemical warfare agent simulants by desorption electrospray ionization mass spectrometry. Chem Commun 28:2968–2970
80. Miao ZX, Chen H (2009) Direct analysis of liquid samples by desorption electrospray ionization-mass spectrometry (DESI-MS). J Am Soc Mass Spectrom 20:10–19
81. Chipuk JE, Brodbelt JS (2008) Transmission mode desorption electrospray ionization. J Am Soc Mass Spectrom 19:1612–1620
82. Venter A, Cooks RG (2007) Desorption electrospray ionization in a small pressure-tight enclosure. Anal Chem 79:6368–6403
83. Shiea J, Huang MZ, Hsu HJ et al (2005) Electrospray-assisted laser desorption/ionization mass spectrometry for direct ambient analysis of solids. Rapid Commun Mass Spectrom 19:3701–3704
84. Sampson JS, Hawkridge AM, Muddiman DC (2007) Direct characterization of intact polypeptides by matrix-assisted laser desorption electrospray ionization quadrupole fourier transform ion cyclotron resonance mass spectrometry. Rapid Commun Mass Spectrom 21:1150–1154
85. Chen H, Venter A, Cooks RG (2006) Extractive electrospray ionization for direct analysis of undiluted urine, milk and other complex mixtures without sample preparation. Chem Commun 19:2042–2044
86. McEwen CN, McKay RG, Larsen BS (2005) Analysis of solids, liquids, and biological tissues using solids probe introduction at atmospheric pressure on commercial LC/MS instruments. Anal Chem 77:7826–7831
87. Takáts Z, Czuczy N, Katona M et al (2006) Jet desorption ionization. In: Proceedings of the 54th ASMS conference on mass spectrometry, Seattle, 28 May–1 June 1
88. Haddad R, Sparrapan R, Eberlin MN (2006) Desorption sonic spray ionization for (high) voltage-free ambient mass spectrometry. Rapid Commun Mass Spectrom 20:2901–2905
89. Grimm RL, Beauchamp JL (2005) Dynamics of field-induced droplet ionization: time-resolved studies of distortion, jetting, and progeny formation from charged and neutral methanol droplets exposed to strong electric fields. J Phys Chem B 109:8244–8250
90. Haapala M, Pol J, Saarela V et al (2007) Desorption atmospheric pressure photoionization. Anal Chem 79:7867–7872
91. Ratcliffe LV, Rutten FJM, Barrett DA et al (2007) Surface analysis under ambient conditions using plasma-assisted desorption/ionization mass spectrometry. Anal Chem 79:6094–6101
92. Na N, Zhang C, Zhao M et al (2007) Direct detection of explosives on solid surfaces by mass spectrometry with an ambient ion source based on dielectric barrier discharge. J Mass Spectrom 42:1079–1085
93. Van Berkel GJ, Kertesz V, Koeplinger KA et al (2008) Liquid microjunction surface sampling probe electrospray mass spectrometry for detection of drugs and metabolites in thin tissue sections. J Mass Spectrom 43:500–508
94. Chen H, Ouyang Z, Cooks RG (2006) Thermal production and reactions of organic ions at atmospheric pressure. Angew Chem Int Ed 45:3656–3660

95. Ford MJ, Van Berkel GJ (2004) An improved thin-layer chromatography/mass spectrometry coupling using a surface sampling probe electrospray ion trap system. Rapid Communs in Mass Spectrom 18:1303–1309
96. Shieh IF, Lee CY, Shiea J (2005) Eliminating the interferences from TRIS Buffer and SDS in protein analysis by fused-droplet electrospray ionization mass spectrometry. J Proteome Res 4:606–612
97. Andrade FJ, Shelley JT, Wetzel WC et al (2008) Atmospheric pressure chemical ionization source. 1. Ionization of compounds in the gas phase. Anal Chem 80:2646–2653
98. Chen H, Yang S, Wortmann A et al (2007) Neutral desorption sampling of living objects for rapid analysis by extractive electrospray ionization mass spectrometry. Angew Chem Int Ed 46:7591–7594
99. Nemes P, Vertes A (2007) Laser ablation electrospray ionization for atmospheric pressure, in vivo, and imaging mass spectrometry. Anal Chem 79:8098–8106
100. Harper JD, Charipar NA, Mulligan CC et al (2008) Low-temperature plasma probe for ambient desorption ionization. Anal Chem 80:9097–9104
101. Murray KK, Russell DH (1994) Laser spray ionization for biological mass spectrometry. Am Lab 26:38–44
102. Venter A, Nefliu M, Cooks RG (2008) Ambient desorption ionization mass spectrometry. Trends Anal Chem 27:284–290
103. Takats Z, Wiseman JM, Cooks RG (2005) Ambient mass spectrometry using desorption electrospray ionization (DESI): instrumentation, mechanisms and applications in forensics, chemistry, and biology. J Mass Spectrom 40:1261–1275
104. Cooks RG, Ouyang Z, Takats Z et al (2006) Ambient mass spectrometry. Science 311:1566–1570
105. Cotte-Rodriguez I, Chen H, Cooks RG (2006) Rapid trace detection of triacetone triperoxide (TATP) by complexation reactions during desorption electrospray ionization. Chem Commun 953–955
106. Ifa DR, Manicke NE, Dill AL et al (2008) Latent fingerprint chemical imaging by mass spectrometry. Science 321:805
107. Meetani MA, Shin YS, Zhang S et al (2007) Desorption electrospray ionization mass spectrometry of intact bacteria. J Mass Spectrom 42:1186–1193
108. Wu C, Ifa DR, Manicke NE et al (2010) Molecular imaging of adrenal gland by desorption electrospray ionization mass spectrometry. Analyst 135:28–32
109. Dill AL, Ifa DR, Manicke NE et al (2009) Lipid profiles of canine invasive transitional cell carcinoma of the urinary bladder and adjacent normal tissue by desorption electrospray ionization imaging mass spectrometry. Anal Chem 81:8758–8764
110. Kertesz V, Van Berkel GJ, Vavrek M et al (2008) Comparison of drug distribution images from whole-body thin tissue sections obtained using desorption electrospray ionization tandem mass spectrometry and autoradiography. Anal Chem 80:5168–5177
111. Manicke NE, Wiseman JM, Ifa DR et al (2008) Desorption electrospray ionization (DESI) mass spectrometry and tandem mass spectrometry (MS/MS) of phospholipids and sphingolipids: ionization, adduct formation, and fragmentation. J Am Soc Mass Spectrom 19:531–543
112. Nefliu M, Venter A, Cooks RG (2006) Desorption electrospray ionization and electrosonic spray ionization for solid- and solution-phase analysis of industrial polymers. Chem Commun 888–890
113. Kennedy JH, Wiseman JM (2010) Evaluation and performance of desorption electrospray ionization using a triple quadrupole mass spectrometer for quantitation of pharmaceuticals in plasma. Rapid Commun Mass Spectrom 24:309–314
114. Harry EL, Reynolds JC, Bristow AWT et al (2009) Direct analysis of pharmaceutical formulations from non-bonded reversed-phase thin-layer chromatography plates by desorption electrospray ionisation ion mobility mass spectrometry. Rapid Commun Mass Spectrom 23:2597–2604
115. Luosujärvi L, Laakkonen UM, Kostiainen R et al (2009) Analysis of street market confiscated drugs by desorption atmospheric pressure photoionization and desorption electrospray ionization coupled with mass spectrometry. Rapid Commun Mass Spectrom 23:1401–1404

116. Soparawalla S, Salazar GA, Perry RH et al (2009) Pharmaceutical cleaning validation using non-proximate large-area desorption electrospray ionization mass spectrometry. Rapid Commun Mass Spectrom 23:131–137
117. Williams JP, Scrivens JH (2005) Rapid accurate mass desorption electrospray ionisation tandem mass spectrometry of pharmaceutical samples. Rapid Commun Mass Spectrom 19:3643–3650
118. Shin YS, Drolet B, Mayer R et al (2007) Desorption electrospray ionization-mass spectrometry of proteins. Anal Chem 79:3514–3518
119. Myung S, Wiseman JM, Valentine SJ et al (2006) Coupling desorption electrospray ionization with ion mobility/mass spectrometry for analysis of protein structure: evidence for desorption of folded and denatured states. J Phys Chem B 110:5045–5051
120. Huang M-Z, Hsu H-J, Lee J-Y et al (2006) Direct protein detection from biological media through electrospray-assisted laser desorption ionization/mass spectrometry. J Proteome Res 5:1107–1116
121. Weston DJ, Bateman R, Wilson ID et al (2005) Direct analysis of pharmaceutical drug formulations using ion mobility spectrometry/quadrupole-time-of-flight mass spectrometry combined with desorption electrospray ionization. Anal Chem 77:7572–7580
122. Bereman MS, Nyadong L, Fernandez FM et al (2006) Direct high-resolution peptide and protein analysis by desorption electrospray ionization fourier transform ion cyclotron resonance mass spectrometry. Rapid Commun Mass Spectrom 20:3409–3411
123. Patterson GE, Guymon AJ, Riter LS et al (2002) Miniature cylindrical ion trap mass spectrometer. Anal Chem 74:6145–6153
124. Takats Z, Wiseman JM, Gologan B et al (2004) Electrosonic spray ionization. A gentle technique for generating folded proteins and protein complexes in the gas phase and for studying ion—molecule reactions at atmospheric pressure. Anal Chem 76:4050–4058
125. Costa AB, Cooks RG (2007) Simulation of atmospheric transport and droplet-thin film collisions in desorption electrospray ionization. Chem Commun 3915–3917
126. Costa AB, Cooks GR (2008) Simulated splashes: elucidating the mechanism of desorption electrospray ionization mass spectrometry. Chem Phys Lett 464:1–8
127. Nefliu M, Smith JN, Venter A et al (2008) Internal energy distributions in desorption electrospray ionization (DESI). J Am Soc Mass Spectrom 19:420–427
128. Takats Z, Cotte-Rodriguez I, Talaty N et al (2005) Direct, trace level detection of explosives on ambient surfaces by desorption electrospray ionization mass spectrometry. Chem Commun 1950–1952
129. Kertesz V, Van Berkel GJ (2008) Improved imaging resolution in desorption electrospray ionization mass spectrometry. Rapid Commun Mass Spectrom 22:2639–2644
130. Ifa DR, Manicke NE, Rusine AL et al (2008) Quantitative analysis of small molecules by desorption electrospray ionization mass spectrometry from polytetrafluoroethylene surfaces. Rapid Commun Mass Spectrom 22:503–510
131. Mulligan CC, MacMillan DK, Noll RJ et al (2007) Fast analysis of high-energy compounds and agricultural chemicals in water with desorption electrospray ionization mass spectrometry. Rapid Commun Mass Spectrom 21:3729–3736
132. Miao Z, Chen H (2008) Analysis of continuous-flow liquid samples by desorption electrospray ionization-mass spectrometry (DESI-MS). In: Proceedings of the 56th annual American society for mass spectrometry conference on mass spectrometry, Denver, June 1–5
133. Ma X, Zhao M, Lin Z et al (2008) Versatile platform employing desorption electrospray ionization mass spectrometry for high-throughput analysis. Anal Chem 80:6131–6136
134. Chipuk JE, Brodbelt JS (2009) The influence of material and mesh characteristics on transmission mode desorption electrospray ionization. J Am Soc Mass Spectrom 20:584–592
135. Ferguson C, Benchaar S, Miao Z, Loo J, Chen H (2011) Direct ionization of large proteins and protein complexes by desorption electrospray ionization-mass spectrometry. Anal Chem 83:6468–6473
136. Hutchens TW, Yip TT (1993) New desorption strategies for the mass spectrometric analysis of macromolecules. Rapid Commun Mass Spectrom 7:576–580

137. Northen TR, Yanes O, Northen MT et al (2007) Clathrate nanostructures for mass spectrometry. Nature 449:1033–U1033
138. Woo HK, Northen TR, Yanes O et al (2008) Nanostructure-initiator mass spectrometry: a protocol for preparing and applying NIMS surfaces for high-sensitivity mass analysis. Nature Protocols 3:1341–1349
139. Hirabayashi A, Sakairi M, Koizumi H (1994) Sonic spray ionization method for atmospheric-pressure ionization mass-spectrometry. Anal Chem 66:4557–4559
140. Chang DY, Lee CC, Shiea J (2002) Detecting large biomolecules from high-salt solutions by fused-droplet electrospray ionization mass spectrometry. Anal Chem 74:2465–2469
141. Hong CM, Tsia FC, Shiea J (2000) A multiple channel electrospray source used to detect highly reactive ketenes from a flow pyrolyzer. Anal Chem 72:1175–1178
142. Huang M, Jhang S, Cheng C et al (2010) Effects of matrix, electrospray solution, and laser light on the desorption and ionization mechanisms in electrospray-assisted laser desorption ionization mass spectrometry. Analyst 135:759–766
143. Chen H, Yang S, Li M et al (2010) Sensitive detection of native proteins using extractive electrospray ionization mass spectrometry. Angew Chem Int Ed 49:3053–3056
144. Pagnotti VS, Chubatyi ND, McEwen CN (2011) Solvent assisted inlet ionization: an ultrasensitive new liquid introduction ionization method for mass spectrometry. Anal Chem 83:3981–3985
145. Pagnotti VS, Inutan ED, Marshall DD et al (2011) Inlet ionization: a new highly sensitive approach for liquid chromatography/mass spectrometry of small and large molecules. Anal Chem 83:7591–7594
146. Cotte-Rodriguez I, Zhang Y, Miao Z et al (2011) Ionization methods in protein mass spectrometry. In: Guodong C et al (eds) Protein mass spectrometry, John Wiley & Sons, pp 3–42
147. Cody RB, Freiser BS (1982) Collision-induced dissociation in a Fourier-transform mass spectrometer. Int J Mass Spectrom Ion Processes 41:199–204
148. Cooks RG (1995) Collision-induced dissociation: readings and commentary. J Mass Spectrom 30:1215–1221
149. Jennings KR (1968) Collision-induced decompositions of aromatic molecular ions. Int J Mass Spectrom Ion Phys 1:227–235
150. McLafferty FW, Bryce TA (1967) Metastable-ion characteristics. Characterization of isomeric molecules. Chem Commun 1215–1217
151. Wells JM, McLuckey SA (2005) Collision-induced dissociation (CID) of peptides and proteins. Methods Enzymol 402:148–185
152. Marcus RA (1952) Unimolecular dissociations and free-radical recombination reactions. J Chem Phys 20:359–364
153. Rosenstock HM, Wallenstein MB, Wahrhaftig AL et al (1952) Absolute rate theory for isolated systems and the mass spectra of polyatomic molecules. Proc Natl Acad Sci U S A 38:667–678
154. Cooks RG, Glish GL (1981) Mass-spectrometry. Chem Eng News 59:40–52
155. McLafferty FW (1983) Tandem mass spectrometry. Wiley, New York
156. McLafferty FW (1981) Tandem mass-spectrometry. Science 214:280–287
157. McLuckey SA (1992) Principles of collisional activation in analytical mass spectrometry. J Am Soc Mass Spectr 3:599–614
158. Sleno L, Volmer DA (2004) Ion activation methods for tandem mass spectrometry. J Mass Spectrom 39:1091–1112
159. Yamaoka H, Pham D, Durup J (1969) Energetics of the collision-induced dissociations C2H2+.far. C2H++H and C2H2+.far. H++C2H. J Chem Phys 51:3465–3476
160. Beynon JH, Boyd RK, Brenton AG (1986) Charge permutation reactions. Adv Mass Spectrom 10th:437–469
161. Dodonov A, Kozlovsky V, Loboda A et al (1997) A new technique for decomposition of selected ions in molecule ion reactor coupled with ortho-time-of-flight mass spectrometry. Rapid Commun Mass Sp 11:1649–1656
162. Laskin J, Futrell JH (2003) Collisional activation of peptide ions in FT-ICR mass spectrometry. Mass Spectrom Rev 22:158–181

163. Gauthier JW, Trautman TR, Jacobson DB (1991) Sustained off-resonance irradiation for collision-activated dissociation involving Fourier transform mass spectrometry. Collision-activated dissociation technique that emulates infrared multiphoton dissociation. Anal Chim Acta 246:211–225
164. Louris JN, Brodbelt-Lustig JS, Cooks RG et al (1990) Ion isolation and sequential stages of mass spectrometry in a quadrupole ion trap mass spectrometer. Int J Mass Spectrom Ion Processes 96:117–137
165. Yost RA, Enke CG, McGilvery DC et al (1979) High efficiency collision-induced dissociation in an rf-only quadrupole. Int J Mass Spectrom Ion Phys 30:127–136
166. Fabris D, Kelly M, Murphy C et al (1993) High-energy collision-induced dissociation of multiply charged polypeptides produced by electrospray. J Am Soc Mass Spectr 4:652–661
167. Medzihradszky KF, Burlingame AL (1994) The advantages and versatility of a high-energy collision-induced dissociation-based strategy for the sequence and structural determination of proteins. Methods (San Diego) 6:284–303
168. Medzihradszky KF, Campbell JM, Baldwin MA et al (2000) The characteristics of peptide collision-induced dissociation using a high-performance MALDI-TOF/TOF tandem mass spectrometer. Anal Chem 72:552–558
169. Shukla AK, Futrell JH (2000) Tandem mass spectrometry: dissociation of ions by collisional activation. J Mass Spectrom 35:1069–1090
170. Wysocki VH, Resing KA, Zhang Q et al (2005) Mass spectrometry of peptides and proteins. Methods (San Diego) 35:211–222
171. Papayannopoulos IA (1995) The interpretation of collision-induced dissociation tandem mass spectra of peptides. Mass Spectrom Rev 14:49–73
172. O'Hair RA (2000) The role of nucleophile–electrophile interactions in the unimolecular and bimolecular gas-phase ion chemistry of peptides and related systems. J Mass Spectrom 35:1377–1381
173. Schlosser A, Lehmann WD (2000) Five-membered ring formation in unimolecular reactions of peptides: a key structural element controlling low-energy collision-induced dissociation of peptides. J Mass Spectrom 35:1382–1390
174. Polce MJ, Ren D, Wesdemiotis C (2000) Dissociation of the peptide bond in protonated peptides. J Mass Spectrom 35:1391–1398
175. Wysocki VH, Tsaprailis G, Smith LL et al (2000) Mobile and localized protons: a framework for understanding peptide dissociation. J Mass Spectrom 35:1399–1406
176. Paizs B, Suhai S (2005) Fragmentation pathways of protonated peptides. Mass Spectrom Rev 24:508–548
177. Summerfield SG, Whiting A, Gaskell SJ (1997) Intra-ionic interactions in electrosprayed peptide ions. Int J Mass Spectrom Ion Processes 162:149–161
178. Dongre AR, Jones JL, Somogyi A et al (1996) Influence of peptide composition, gas-phase basicity, and chemical modification on fragmentation efficiency: evidence for the mobile proton model. J Am Chem Soc 118:8365–8374
179. Barlow CK, O'Hair RAJ (2008) Gas-phase peptide fragmentation: how understanding the fundamentals provides a springboard to developing new chemistry and novel proteomic tools. J Mass Spectrom 43:1301–1319
180. Breci LA, Tabb DL, Yates JR III et al (2003) Cleavage N-terminal to proline: analysis of a database of peptide tandem mass spectra. Anal Chem 75:1963–1971
181. Roepstorff P, Fohlman J (1984) Proposal for a common nomenclature for sequence ions in mass spectra of peptides. Biomed Mass Spec 11:601
182. Jones AW, Cooper HJ (2011) Dissociation techniques in mass spectrometry-based proteomics. Analyst (Cambridge) 136:3419–3429
183. Harrison AG, Young AB, Bleiholder C et al (2006) Scrambling of sequence information in collision-induced dissociation of peptides. J Am Chem Soc 128:10364–10365
184. Bleiholder C, Osburn S, Williams TD et al (2008) Sequence-scrambling fragmentation pathways of protonated peptides. J Am Chem Soc 130:17774–17789

185. Dodds ED, Blackwell AE, Jones CM et al (2011) Determinants of gas-phase disassembly behavior in homodimeric protein complexes with related yet divergent structures. Anal Chem (Washington) 83:3881–3889
186. Blackwell AE, Dodds ED, Bandarian V et al (2011) Revealing the quaternary structure of a heterogeneous noncovalent protein complex through surface-induced dissociation. Anal Chem (Washington) 83:2862–2865
187. Benesch JLP (2009) Collisional activation of protein complexes: picking up the pieces. J Am Soc Mass Spectrom 20:341–348
188. Cooks RG, Ast T, Beynon JH (1975) Anomalous metastable peaks. Int J Mass Spectrom Ion Phys 16:348–352
189. Cooks RG, Terwilliger DT, Ast T et al (1975) Surface modified mass spectrometry. J Am Chem Soc 97:1583–1585
190. Mabud MA, Dekrey MJ, Cooks RG (1985) Surface-induced dissociation of molecular ions. Int J Mass Spectrom Ion Processes 67:285–294
191. Wysocki VH (2002) Surface-induced dissociation for peptide/protein sequencing. In: Abstracts of papers, 224th ACS national meeting, Boston, 18–22 August 2002: ANYL-230
192. Wysocki VH, Ding JM, Jones JL et al (1992) Surface-induced dissociation in tandem quadrupole mass spectrometers: a comparison of three designs. J Am Soc Mass Spectrom 3:27–32
193. Wysocki VH, Galhena A, Gamage C (2005) Surface-induced dissociation in a Q-TOF mass spectrometer. In: Abstracts of papers, 229th ACS national meeting, San Diego, 13–17 March
194. Wysocki VH, Gamage C, Qi Z et al (2004) Integrating surface-induced dissociation into simple TOF mass spectrometers. In: Abstracts of papers, 227th ACS national meeting, Anaheim, 28 March–1 April
195. Wysocki VH, Jones CM, Galhena A et al (2008) Surface-induced dissociation of noncovalent protein–protein complexes. In: Abstracts of papers, 236th ACS national meeting, Philadelphia, 17–21 August
196. Wysocki VH, Jones CM, Galhena AS et al (2008) Surface-induced dissociation shows potential to be more informative than collision-induced dissociation for structural studies of large systems. J Am Soc Mass Spectrom 19:903–913
197. Wysocki VH, Jones JL, Dongre AR et al (1994) Surface-induced dissociation of peptides. Biol Mass Spectrom Present Future, [Proc Kyoto '92 Int Conf]:249–254
198. Wysocki VH, Joyce KE, Jones CM et al (2008) Surface-induced dissociation of small molecules, peptides, and non-covalent protein complexes. J Am Soc Mass Spectrom 19:190–208
199. Wysocki VH, Tsaprailis G, Smith LL et al (2000) Mobile and localized protons: a framework for understanding peptide dissociation. J Mass Spectrom 35:1399–1406
200. Dongre AR (1996) Surface-induced dissociation of polyatomic ions: structure elucidation, energetics and mechanisms for fragmentation of protonated peptides, 348
201. Williams ER, Henry KD, McLafferty FW et al (1990) Surface-induced dissociation of peptide ions in Fourier-transform mass spectrometry. J Am Soc Mass Spectrom 1:413–416
202. Chorush RA, Little DP, Beu SC et al (1995) Surface-induced dissociation of multiply-protonated proteins. Anal Chem 67:1042–1046
203. Futrell J, Laskin J, Shukla AK (2003) Kinetics and dynamics of surface-induced dissociation of complex ions. In: Abstracts of papers, 225th ACS national meeting, New Orleans, 23–27 March
204. Futrell JH, Zhong W, Nikolaev EN et al (1998) Tandem Fourier transform mass spectrometry studies of surface-induced dissociation of benzene monomer and dimer ions on a self-assembled fluoroalkanethiolate monolayer surface. Adv Mass Spectrom 14:65321–65391
205. Laskin J, Bailey TH, Futrell JH (2003) Shattering of peptide ions on self-assembled monolayer surfaces. J Am Chem Soc 125:1625–1632
206. Laskin J, Beck KM, Hache JJ et al (2004) Surface-induced dissociation of ions produced by matrix-assisted laser desorption/ionization in a Fourier transform ion cyclotron resonance mass spectrometer. Anal Chem 76:351–356

207. Laskin J, Denisov EV, Shukla AK et al (2002) Surface-induced dissociation in a Fourier transform ion cyclotron resonance mass spectrometer: instrument design and evaluation. Anal Chem 74:3255–3261
208. Laskin J (2004) Energetics and dynamics of peptide fragmentation from multiple-collision activation and surface-induced dissociation studies. Eur J Mass Spectrom 10:259–267
209. Laskin J (2006) Energetics and dynamics of fragmentation of protonated leucine enkephalin from time- and energy-resolved surface-induced dissociation studies. J Phys Chem A 110:8554–8562
210. Laskin J, Bailey TH, Futrell JH (2004) Fragmentation energetics for angiotensin II and its analogs from time- and energy-resolved surface-induced dissociation studies. Int J Mass Spectrom 234:89–99
211. Laskin J, Bailey TH, Futrell JH (2006) Mechanisms of peptide fragmentation from time- and energy-resolved surface-induced dissociation studies: dissociation of angiotensin analogs. Int J Mass Spectrom 249(250):462–472
212. Laskin J, Denisov E, Futrell J (2000) A comparative study of collision-induced and surface-induced dissociation. 1. Fragmentation of protonated dialanine. J Am Chem Soc 122:9703–9714
213. Laskin J, Denisov E, Futrell J (2001) Comparative study of collision-induced and surface-induced dissociation. 2. Fragmentation of small alanine-containing peptides in FT-ICR MS. J Phys Chem B 105:1895–1900
214. Laskin J, Denisov E, Futrell JH (2002) Fragmentation energetics of small peptides from multiple-collision activation and surface-induced dissociation in FT-ICR MS. Int J Mass Spectrom 219:189–201
215. Laskin J, Futrell JH (2003) Surface-induced dissociation of peptide ions: kinetics and dynamics. J Am Soc Mass Spectrom 14:1340–1347
216. Laskin J, Futrell JH (2003) Energy transfer in collisions of peptide ions with surfaces. J Chem Phys 119:3413–3420
217. Grill V, Shen J, Evans C et al (2001) Collisions of ions with surfaces at chemically relevant energies: instrumentation and phenomena. Rev Sci Instrum 72:3149–3179
218. Cooks RG, Ast T, Mabud MA (1990) Collisions of polyatomic ions with surfaces. Int J Mass Spectrom Ion Processes 100:209–265
219. Wysocki VH, Kenttamaa HI (1990) Collisional activation of distonic radical cations and their conventional isomers in quadrupole tandem mass spectrometry. J Am Chem Soc 112:5110–5116
220. DeKrey MJ, Kenttamaa HI, Wysocki VH et al (1986) Energy deposition in iron pentacarbonyl cation radical upon collision with a metal surface. Org Mass Spectrom 21:193–195
221. Schaaff TG, Qu Y, Farrell N et al (1998) Investigation of the trans effect in the fragmentation of dinuclear platinum complexes by electrospray ionization surface-induced dissociation tandem mass spectrometry. J Mass Spectrom 33:436–443
222. Franchetti V, Solka BH, Baitinger WE et al (1977) Soft landing of ions as a means of surface modification. Int J Mass Spectrom Ion Phys 23:29–35
223. Dongre AR, Somogyi A, Wysocki VH (1996) Surface-induced dissociation: an effective tool to probe structure, energetics and fragmentation mechanisms of protonated peptides. J Mass Spectrom 31:339–350
224. Sun W, May JC, Gillig KJ et al (2009) A dual time-of-flight apparatus for an ion mobility-surface-induced dissociation-mass spectrometer for high-throughput peptide sequencing. Int J Mass Spectrom 287:39–45
225. Sun W (2007) Development of a MALDI-ion mobility-surface-induced dissociation-time-of-flight mass spectrometer with novel collision source configurations for high throughput peptide sequencing. Texas A&M University, Texas, p 129
226. Stone EG (2003) Development of a MALDI-ion mobility-surface-induced dissociation-time-of- flight-mass spectrometer for the analysis of peptides and protein digests. Texas A&M University, Texas, p 169

227. Fernandez FM, Smith LL, Kuppannan K et al (2003) Peptide sequencing using a patchwork approach and surface-induced dissociation in sector-TOF and dual quadrupole mass spectrometers. J Am Soc Mass Spectrom 14:1387–1401
228. Galhena AS, Dagan S, Jones CM et al (2008) Surface-induced dissociation of peptides and protein complexes in a quadrupole/time-of-flight mass spectrometer. Anal Chem (Washington) 80:1425–1436
229. Fernandez FM, Wysocki VH, Futrell JH et al (2006) Protein identification via surface-induced dissociation in an FT-ICR mass spectrometer and a patchwork sequencing approach. J Am Soc Mass Spectrom 17:700–709
230. Winger BE, Laue HJ, Horning SR et al (1992) Hybrid BEEQ tandem mass spectrometer for the study of ion/surface collision processes. Rev Sci Instrum 63:5613–5625
231. Qayyum A, Herman Z, Tepnual T et al (2004) Surface-induced dissociation of polyatomic hydrocarbon projectile ions with different initial internal energy content. J Phys Chem A 108:1–8
232. Feketeova L, Grill V, Zappa F et al (2008) Charge exchange, surface-induced dissociation and reactions of doubly charged molecular ions SF_4^{2+} upon impact on a stainless steel surface: a comparison with surface-induced dissociation of singly charged SF_4^+ molecular ions. Int J Mass Spectrom 276:37–42
233. Feketeova L, Tepnual T, Grill V et al (2007) Surface-induced dissociation and reactions of cations and dications $C_7H_8^{+/2+}$, $C_7H_7^{+/2+}$ and $C_7H_6^{2+}$: dependence of mass spectra of product ions on incident energy of the projectiles. Int J Mass Spectrom 265:337–346
234. Jasik J, Roithova J, Zabka J et al (2006) Surface-induced dissociation and reactions of dications and cations: collisions of dications $C_7H_8^{2+}$, $C_7H_7^{2+}$, and $C_7H_6^{2+}$ and a comparison with the respective cations $C_7D_8^+$ and $C_7H_7^+$. Int J Mass Spectrom 249(250):162–170
235. Jasik J, Zabka J, Feketeova L et al (2005) Collisions of slow polyatomic ions with surfaces: dissociation and chemical reactions of $C_2H_2^+$, $C_2H_3^+$, $C_2H_4^+$, $C_2H_5^+$, and their deuterated variants $C_2D_2^+$ and $C_2D_4^+$ on room-temperature and heated carbon surfaces. J Phys Chem A 109:10208–10215
236. Jo S-C, Cooks RG (2003) Translational to vibrational energy conversion during surface-induced dissociation of n-butylbenzene molecular ions colliding at self-assembled monolayer surfaces. Eur J Mass Spectrom 9:237–244
237. Shukla AK, Futrell JH (2003) Surface-induced dissociation of acetone cations from self-assembled monolayer surface of fluorinated alkyl thiol on Au (1 1 1) substrate at low collision energies. Int J Mass Spectrom 228:563–576
238. Rakov VS, Denisov EV, Laskin J et al (2002) Surface-induced dissociation of the benzene molecular cation in Fourier transform ion cyclotron resonance mass spectrometry. J Phys Chem A 106:2781–2788
239. Zulauf A, Schmidt L, Jungclas H (2008) Grazing incidence surface-induced dissociation: molecules sliding along a surface. Anal Bioanal Chem 392:793–796
240. Biasioli F, Fiegele T, Mair C et al (2000) Surface-induced dissociation of singly and multiply charged fullerene ions. J Chem Phys 113:5053–5057
241. Waldschmidt B, Turra M, Schaefer R (2007) Surface-induced dissociation as a probe for the energetics and structure of lead clusters. Zeitschrift fuer Physikalische Chemie (Muenchen) 221:1569–1579
242. Cooks R, Amy J, Bier M et al (1989) New mass spectrometers. Adv Mass Spectrom 11A:33–52
243. Williams ER, Henry KD, McLafferty FW et al (1990) Surface-induced dissociation of peptide ions in Fourier-transform mass spectrometry. J Am Soc Mass Spectr 1:413–416
244. Cole RB, LeMeillour S, Tabet JC (1992) Surface-induced dissociation of protonated peptides: implications of initial kinetic energy spread. Anal Chem 64:365–371
245. McCormack AL, Somogyi A, Dongre AR et al (1993) Fragmentation of protonated peptides: surface-induced dissociation in conjunction with a quantum mechanical approach. Anal Chem 65:2859–2872

246. Jones JL, Dongre AR, Somogyi A et al (1994) Sequence dependence of peptide fragmentation efficiency curves determined by electrospray ionization/surface-induced dissociation mass spectrometry. J Am Chem Soc 116:8368–8369
247. Jones CM, Beardsley RL, Galhena AS et al (2006) Symmetrical gas-phase dissociation of noncovalent protein complexes via surface collisions. J Am Chem Soc 128:15044–15045
248. Beardsley RL, Jones CM, Galhena AS et al (2009) Noncovalent protein tetramers and pentamers with "n" charges yield monomers with n/4 and n/5 charges. Anal Chem (Washington) 81:1347–1356
249. Brodbelt JS (2011) Shedding light on the frontier of photodissociation. J Am Soc Mass Spectr 22:197–206
250. Brodbelt JS, Wilson JJ (2009) Infrared multiphoton dissociation in quadrupole ion traps. Mass Spec Rev 28:390–424
251. Ly T, Julian RR (2009) Ultraviolet photodissociation: developments towards applications for mass-spectrometry-based proteomics. Angew Chem Int Edit 48:7130–7137
252. Laskin J, Futrell JH (2005) Activation of large ions in FT-ICR mass spectrometry. Mass Spectrom Rev 24:135–167
253. Black DM, Payne AH, Glish GL (2006) Determination of cooling rates in a quadrupole ion trap. J Am Soc Mass Spectr 17:932–938
254. Palumbo AM, Smith SA, Kalcic CL et al (2011) Tandem mass spectrometry strategies for phosphoproteome analysis. Mass Spec Rev 30:600–625
255. Dunbar RC (2004) BIRD (blackbody infrared radiative dissociation): evolution, principles, and applications. Mass Spec Rev 23:127–158
256. Stephenson JL, Booth MM, Boue SM et al (1996) Analysis of biomolecules using electrospray ionization ion-trap mass spectrometry and laser photodissociation. Acs Sym Ser 619:512–564
257. Crowe MC, Brodbelt JS (2004) Infrared multiphoton dissociation (IRMPD) and collisionally activated dissociation of peptides in a quadrupole ion trap with selective IRMPD of phosphopeptides. J Am Soc Mass Spectr 15:1581–1592
258. Pikulski M, Wilson JJ, Aguilar A et al (2006) Amplification of infrared multiphoton dissociation efficiency in a quadruple ion trap using IR-active ligands. Anal Chem 78:8512–8517
259. Pikulski M, Hargrove A, Shabbir SH et al (2007) Sequencing and characterization of oligosaccharides using infrared multiphoton dissociation and boronic acid derivatization in a quadrupole ion trap. J Am Soc Mass Spectr 18:2094–2106
260. Smith SA, Kalcic CL, Safran KA et al (2010) Enhanced characterization of singly protonated phosphopeptide ions by femtosecond laser-induced ionization/dissociation tandem mass spectrometry (fs-LID-MS/MS). J Am Soc Mass Spectr 21:2031–2040
261. Kalcic CL, Gunaratne TC, Jonest AD et al (2009) Femtosecond laser-induced ionization/dissociation of protonated peptides. J Am Chem Soc 131:940–942
262. Reilly JP (2009) Ultraviolet photofragmentation of biomolecular ions. Mass Spectrom Rev 28:425–447
263. Thompson MS, Cui WD, Reilly JP (2004) Fragmentation of singly charged peptide ions by photodissociation at lambda = 157 nm. Angew Chem Int Edit 43:4791–4794
264. Cui WD, Thompson MS, Reilly JP (2005) Pathways of peptide ion fragmentation induced by vacuum ultraviolet light. J Am Soc Mass Spectr 16:1384–1398
265. Kim TY, Thompson MS, Reilly JP (2005) Peptide photodissociation at 157 nm in a linear ion trap mass spectrometer. Rapid Commun Mass Sp 19:1657–1665
266. Thompson MS, Cui WD, Reilly JP (2007) Factors that impact the vacuum ultraviolet photofragmentation of peptide ions. J Am Soc Mass Spectr 18:1439–1452
267. Kim TY, Reilly JP (2009) Time-resolved observation of product ions generated by 157 nm photodissociation of singly protonated phosphopeptides. J Am Soc Mass Spectr 20:2334–2341
268. Parthasarathi R, He Y, Reilly JP et al (2010) New insights into the vacuum UV photodissociation of peptides. J Am Chem Soc 132:1606–1610
269. Ly T, Julian RR (2010) Elucidating the tertiary structure of protein ions in vacuo with site specific photoinitiated radical reactions. J Am Chem Soc 132:8602–8609

270. Diedrich JK, Julian RR (2010) Site-Selective fragmentation of peptides and proteins at quinone-modified cysteine residues investigated by ESI-MS. Anal Chem 82:4006–4014
271. Sun QY, Yin S, Loo JA et al (2010) Radical directed dissociation for facile identification of iodotyrosine residues using electrospray ionization mass spectrometry. Anal Chem 82:3826–3833
272. Liu ZJ, Julian RR (2009) Deciphering the peptide iodination code: influence on subsequent gas-phase radical generation with photo dissociation ESI-MS. J Am Soc Mass Spectr 20:965–971
273. Ly T, Julian RR (2008) Residue-specific radical-directed dissociation of whole proteins in the gas phase. J Am Chem Soc 130:351–358
274. Wilson JJ, Brodbelt JS (2007) MS/MS simplification by 355 nm ultraviolet photodissociation of chromophore-derivatized peptides in 4–3 quadrupole ion trap. Anal Chem 79:7883–7892
275. Zubarev RA, Kelleher NL, McLafferty FW (1998) Electron capture dissociation of multiply charged protein cations. A nonergodic process. J Am Chem Soc 120:3265–3266
276. McLafferty F, Horn DM, Breuker K et al (2001) Electron capture dissociation of gaseous multiply charged ions by fourier-transform ion cyclotron resonance. J Am Soc Mass Spectrom 12:245–249
277. Guan Z, Kelleher NL, O'Connor PB et al (1996) 193 nm photodissociation of larger multiply-charged biomolecules. Int J Mass Spectrom Ion Processes 157(158):357–364
278. Zubarev RA, Kruger NA, Fridriksson EK et al (1999) Electron capture dissociation of gaseous multiply-charged proteins is favored at disulfide bonds and other sites of high hydrogen atom affinity. J Am Chem Soc 121:2857–2862
279. Kelleher NL, Zubarev RA, Bush K et al (1999) Localization of labile posttranslational modifications by electron capture dissociation: The case of γ-carboxyglutamic acid. Anal Chem 71:4250–4253
280. Shi SDH, Hemling ME, Carr SA (2001) Phosphopeptide/phosphoprotein mapping by electron capture dissociation mass spectrometry. Anal Chem 73:19–22
281. Mirgorodskaya E, Roepstorff P, Zubarev RA (1999) Localization of O-glycosylation sites in peptides by electron capture dissociation in a Fourier transform mass spectrometer. Anal Chem 71:4431–4436
282. Whitelegge JP, Zabrouskov V, Halgand F et al (2007) Protein-sequence polymorphisms and post-translational modifications in proteins from human saliva using top-down fourier-transform ion cyclotron resonance mass spectrometry. Int J Mass Spectrom 268:190–197
283. Haselmann KF, Budnik BA, Olsen JV et al (2001) Advantages of external accumulation for electron capture dissociation in Fourier-transform mass spectrometry. Anal Chem 73:2998–3005
284. Cooper HJ, Håkansson K, Marshall AG (2005) The role of electron capture dissociation in biomolecular analysis. Mass Spectrom Rev 24:201–222
285. Chen XH, Tureček F (2006) The arginine anomaly: Arginine radicals are poor hydrogen atom donors in electron transfer induced dissociations. J Am Chem Soc 128:12520–12530
286. Zubarev RA, Horn DM, Fridriksson EK et al (2000) Electron capture dissociation for structural characterization of multiply charged protein cations. Anal Chem 72:563–573
287. Sawicka A, Skurski P, Hudgins RR et al (2003) Model calculations relevant to disulfide bond cleavage via electron capture influenced by positively charged group. J Phys Chem B 107:13505–13511
288. Syrstad EA, Tureček F (2005) Toward a general mechanism of electron capture dissociation. J Am Soc Mass Spectrom 16:208–224
289. Cui W, Rohrs HW, Gross ML (2011) Top-down mass spectrometry: recent developments, applications, and perspectives. Analyst 136:3854–3864
290. Horn DM, Ge Y, McLafferty FW (2000) Activated ion electron capture dissociation for mass spectral sequencing of larger (42 kDa) proteins. Anal Chem 72:4778–4784
291. Han X, Jin M, Breuker K et al (2006) Extending top-down mass spectrometry to proteins with masses greater than 200 kilodaltons. Science 314:109–112

292. Henry KD, Quinn JP, McLafferty FW (1991) High-resolution electrospray mass spectra of large molecules. J Am Chem Soc 113:5447–5449
293. Axelsson J, Palmblad M, Håkansson K et al (1999) Electron capture dissociation of substance P using a commercially available Fourier transform ion cyclotron resonance mass spectrometer. Rapid Commun Mass Spectrom 13:474–477
294. Mirgorodskaya E, Roepstorff P, Zubarev RA (1999) Localization of O-glycosylation sites in peptides by electron capture dissociation in a Fourier transform mass spectrometer. Anal Chem 71:4431–4436
295. Breuker K, Oh H, Horn DM et al (2002) Detailed unfolding and folding of gaseous ubiquitin ions characterized by electron capture dissociation. J Am Chem Soc 124:6407–6420
296. Horn DM, Breuker K, Frank AJ et al (2001) Kinetic intermediates in the folding of gaseous protein ions characterized by electron capture dissociation mass spectrometry. J Am Chem Soc 123:9792–9799
297. Breuker K, McLafferty FW (2003) Native electron capture dissociation for the structural characterization of noncovalent interactions in native cytochrome c. Angre Chem Int Ed 42:4900–4904
298. Breuker K, McLafferty FW (2005) The thermal unfolding of native cytochrome c in the transition from solution to gas phase probed by native electron capture dissociation. Angre Chem Int Ed 44:4911–4914
299. Breuker K, Bruschweiler S, Tollinger M (2011) Electrostatic stabilization of a native protein structure in the gas phase. Angre Chem Int Ed 50:873–877
300. Xie Y, Zhang J, Yin S et al (2006) Top-down ESI-ECD-FT-ICR mass spectrometry localizes noncovalent protein-ligand binding sites. J Am Chem Soc 128:14432–14433
301. Yin S, Loo JA (2010) Elucidating the site of protein-ATP binding by top-down mass spectrometry. J Am Soc Mass Spectrom 21:899–907
302. Yin S, Loo JA (2011) Top-down mass spectrometry of supercharged native protein–ligand complexes. Int J Mass Spectrom 300:118–122
303. Zhang H, Cui W, Wen J et al (2010) Native electrospray and electron-capture dissociation in FTICR mass spectrometry provide top-down sequencing of a protein component in an intact protein assembly. J Am Soc Mass Spectrom 21:1966–1968
304. Zhang H, Cui W, Wen J et al (2011) Native electrospray and electron-capture dissociation FTICR mass spectrometry for top-down studies of protein assemblies. Anal Chem 83:5598–5606
305. Yoo HJ, Wang N, Zhuang S et al (2011) Negative-ion electron capture dissociation: radical-driven fragmentation of charge-increased gaseous peptide anions. J Am Chem Soc 133:16790–16793
306. Hunt DF, Coon JJ, Syka JEP et al (2009) Electron transfer dissociation for biopolymer sequence mass spectrometric analysis eds, US
307. Syka JEP, Coon JJ, Schroeder MJ et al (2004) Peptide and protein sequence analysis by electron transfer dissociation mass spectrometry. Proc Natl Acad Sci USA 101:9528–9533
308. Mikesh LM, Ueberheide B, Chi A et al (2006) The utility of ETD mass spectrometry in proteomic analysis. Biochim Biophys Acta 1764:1811–1822
309. O'Connor PB, Cournoyer JJ, Pitteri SJ et al (2006) Differentiation of aspartic and isoaspartic acids using electron transfer dissociation. J Am Soc Mass Spectrom 17:15–19
310. Wiesner J, Premsler T, Sickmann A (2008) Application of electron transfer dissociation (ETD) for the analysis of posttranslational modifications. Proteomics 8:4466–4483
311. Coon JJ, Shabanowitz J, Hunt DF et al (2005) Electron transfer dissociation of peptide anions. J Am Soc Mass Spectrom 16:880–882
312. Han H, Xia Y, McLuckey SA (2007) Ion trap collisional activation of c and z.bul. ions formed via gas-phase ion/ion electron-transfer dissociation. J Proteome Res 6:3062–3069
313. Han H, Pappin DJ, Ross PL et al (2008) Electron transfer dissociation of iTRAQ labeled peptide ions. J Proteome Res 7:3643–3648
314. Han H, Xia Y, Yang M et al (2008) Rapidly alternating transmission mode electron-transfer dissociation and collisional activation for the characterization of polypeptide ions. Anal Chem (Washington) 80:3492–3497

315. Lin C, O'Connor P (2012) Ion activation and mass analysis in protein mass spectrometry. In: Gross ML, Guodong C, Pramanik BN (eds) Ion activation and mass analysis in protein mass spectrometry. Wiley, Hoboken, pp 55–59
316. Swaney DL, McAlister GC, Wirtala M et al (2006) Supplemental activation method for high-efficiency electron-transfer dissociation of doubly protonated peptide precursors. Anal Chem 79:477–485
317. Vasicek L, Brodbelt JS (2009) Enhanced electron transfer dissociation through fixed charge derivatization of cysteines. Anal Chem 81:7876–7884
318. Pitteri SJ, Chrisman PA, Hogan JM et al (2005) Electron transfer ion/ion reactions in a three-dimensional quadrupole ion trap: reactions of doubly and triply protonated peptides with SO_2^{*}. Anal Chem 77:1831–1839
319. Gunawardena HP, Gorenstein L, Erickson DE et al (2007) Electron transfer dissociation of multiply protonated and fixed charge disulfide linked polypeptides. Int J Mass Spectrom 265:130–138
320. Tsybin YO, Fornelli L, Stoermer C et al (2011) Structural analysis of intact monoclonal antibodies by electron transfer dissociation mass spectrometry. Anal Chem 83:8919–8927
321. McAlister GC, Russell JD, Rumachik NG et al (2012) Analysis of the acidic proteome with negative electron-transfer dissociation mass spectrometry. Anal Chem 84:2875–2882
322. Budnik BA, Haselmann KF, Zubarev RA (2001) Electron detachment dissociation of peptide di-anions: an electron-hole recombination phenomenon. Chem Phys Lett 342:299–302
323. Kalli A, Håkansson K (2007) Preferential cleavage of S–S and C–S bonds in electron detachment dissociation and infrared multiphoton dissociation of disulfide-linked peptide anions. Int J Mass Spectrom 263:71–81
324. Wolff JJ, Laremore TN, Franklin E, Leach I et al (2009) Electron capture dissociation, electron detachment dissociation and infrared multiphoton dissociation of sucrose octasulfate. EurJ Mass Spectrom 15:275–281
325. Wolff JJ, Amster IJ, Chi L et al (2007) Electron detachment dissociation of glycosaminoglycan tetrasaccharides. J Am Soc Mass Spectrom 18:234–244
326. Wolff JJ, Chi L, Linhardt RJ et al (2007) Distinguishing glucuronic from iduronic acid in glycosaminoglycan tetrasaccharides by using electron detachment dissociation. Anal Chem 79:2015–2022
327. Wolff JJ, Laremore TN, Busch AM et al (2008) Electron detachment dissociation of dermatan sulfate oligosaccharides. J Am Soc Mass Spectrom 19:294–304
328. Yang J, Håkansson K (2008) Characterization and optimization of electron detachment dissociation Fourier transform ion cyclotron resonance mass spectrometry. Int J Mass Spectrom 276:144–148
329. Ganisl B, Valovka T, Ms Hartl et al (2011) Electron detachment dissociation for top-down mass spectrometry of acidic proteins. Chem Eur J 17:4460–4469
330. Taucher M, Breuker K (2010) Top-down mass spectrometry for sequencing of larger (up to 61 nt) RNA by CAD and EDD. J Am Soc Mass Spectrom 21:918–929
331. Fung YME, Adams CM, Zubarev RA (2009) Electron ionization dissociation of singly and multiply charged peptides. J Am Chem Soc 131:9977–9985
332. Chen H, Eberlin LS, Cooks RG (2007) Neutral fragment mass spectra via ambient thermal dissociation of peptide and protein ions. J Am Chem Soc 129:5880–5886
333. Xia Y, Ouyang Z, Cooks RG (2008) Peptide fragmentation assisted by surfaces treated with a low-temperature plasma in NanoESI. Angew Chem Int Edit 47:8646–8649
334. Robb DB, Rogalski JC, Kast J et al (2012) Liquid chromatography—atmospheric pressure electron capture dissociation mass spectrometry for the structural analysis of peptides and proteins. Anal Chem 84:4221–4226
335. Cook SL, Collin OL, Jackson GP (2009) Metastable atom-activated dissociation mass spectrometry: leucine/isoleucine differentiation and ring cleavage of proline residues. J Mass Spectrom 44:1211–1223
336. Cook SL, Jackson GP (2011) Characterization of tyrosine nitration and cysteine nitrosylation modifications by metastable atom-activation dissociation mass spectrometry. J Am Soc Mass Spectr 22:221–232

337. Berkout VD (2006) Fragmentation of protonated peptide ions via interaction with metastable atoms. Anal Chem 78:3055–3061
338. Berkout VD, Doroshenko VM (2008) Fragmentation of phosphorylated and singly charged peptide ions via interaction with metastable atoms. Int J Mass Spectrom 278:150–157
339. Berkout VD (2009) Fragmentation of singly protonated peptides via interaction with metastable rare gas atoms. Anal Chem 81:725–731
340. Misharin AS, Silivra OA, Kjeldsen F et al (2005) Dissociation of peptide ions by fast atom bombardment in a quadrupole ion trap. Rapid Commun Mass Sp 19:2163–2171
341. Berkout VD, Doroshenko VM (2008) Fragmentation of phosphorylated and singly charged peptide ions via interaction with metastable atoms. Int J Mass Spectrom 278:150–157
342. Berkout VD (2006) Fragmentation of protonated peptide ions via interaction with metastable atoms. Anal Chem 78:3055–3061

Chapter 2
Quantitative Analysis of Therapeutic and Endogenous Peptides using LC/MS/MS Methods

Erin E. Chambers

2.1 Introduction

Historically, new drug entities arose from combinatorial chemistry of small molecules. Today, a rapidly increasing number of new drugs, called biopharmaceuticals, are being developed which are based on biological molecules such as peptides, proteins, and oligonucleotides. For example, as of 2011, there were 60 peptide drugs available and at least another 400 in late-stage clinical trials. Overall, there are over 600 biologically based drugs in development, many of which are antibody based [1, 2]. In fact, according to a recent study through the Peptide Therapeutics Foundation, there are approximately 17 new peptide drug entities going into clinical trials each year. In addition, the study and identification of peptide biomarkers are critical areas of research. Advances in the various technologies applied for drug discovery and biomolecule characterization (i.e., recombinant DNA, fermentation, proteomics, genomics, and informatics) [3] have made it possible for drug manufacturers to successfully develop and characterize biopharmaceuticals. These types of compounds have been used to treat a variety of serious diseases such as diabetes, cancer, arthritis, and hemophilia. Perhaps, the oldest and best known peptide therapeutic is insulin. While originally dosed by injection, current research focuses on developing nasal or transdermal dose formulations [4]. Several peptide-based drug products, including goserelin, leuprolide, and octreotide, have even reached over a billion in sales. One additional example is the synthetic peptide desmopressin, a modified form of the human hormone arginine vasopressin. It is prescribed as a replacement for antidiuretic hormone (vasopressin) and is used to treat bed-wetting and diabetes insipidus. Desmopressin provides several treatment benefits over recombinant vasopressin. It degrades more slowly, enabling less frequent dosing, and it does not raise blood pressure unlike the unmodified peptide.

While peptide biomarkers are endogenous peptides whose relationship with a disease state has been identified, therapeutic peptides are typically modified or

E. E. Chambers (✉)
Waters Corporation, 34 Maple Street, Milford, 01757 MA, UK
e-mail: erin_chambers@waters.com

G. Chen (ed.), *Characterization of Protein Therapeutics using Mass Spectrometry*,
DOI: 10.1007/978-1-4419-7862-2_2,

synthetic versions of substances that already exist in the body and therefore, as drugs, tend to be better tolerated than many small molecule–based medicines. In addition, by their very nature, these drugs are often more specific, as they have been designed to mimic behavior of endogenous substances and are mapped to specific receptor proteins. Relative to small molecules, other benefits of peptide drugs include greater activity and potency, lower toxicity, improved molecular recognition, no accumulation in tissues and organs, and minimal drug–drug interactions. There are also drawbacks to peptide drugs and their development. Peptide drugs tend to have low oral bioavailability, often necessitating injection rather than preferred delivery methods such as oral dosing. In addition, their solubility is poor, they are rapidly cleared from the body and/or broken down by enzymatic activity, and finally, due to their hydrophilicity, it is difficult to get these drugs across biological membranes. Poor membrane transfer accounts for the fact that many peptide drugs are aimed at extracellular targets. A few companies, however, do focus on the development of peptide drugs for intracellular or "undruggable" targets. Some of these issues have been resolved through recent research efforts [5]. For example, PEGylation, liposomal encapsulation, and conjugation to small molecules, antibodies, or proteins have been shown to be effective means of improving the stability and in vivo half-life of peptide therapeutics [6].

Due to the increasing number of peptide therapeutics and biomarkers, there exists an immediate need for an efficient method development workflow for the bioanalysis of peptide therapeutics as well as a comprehensive understanding of their differences with respect to small molecules. Quantification of peptides is important not only for synthetic peptide drugs, but also for peptide biomarkers and for quantification of proteins based upon measurement of unique or signature peptides.

Naturally, the determination of pharmacokinetic (PK) parameters and metabolic fate is as critical during the drug discovery and development process for peptide therapeutics as for small molecules. These data are typically generated by liquid chromatography/tandem mass spectrometry (LC/MS/MS) analysis of in vitro (liver microsomes, $CaCo_2$ cells, etc.,) or in vivo (animal or human fluids or tissues) samples. In "small molecule" analysis, LC/MS/MS, specifically triple quadrupole MS, has become the technique of choice for these activities due to its unparalleled selectivity and sensitivity.

In contrast to small molecules, the gold standard for biomolecule quantitation has historically been ligand-binding assays (LBAs) such as enzyme-linked immunosorbent assays (ELISAs) and radioimmunoassays (RIAs). While these types of assays remain the primary and most widely accepted method of quantitating protein and antibody-based therapies, LC/MS/MS has begun to emerge as the technique of choice for quantitation of synthetic peptides and is increasingly being used to analyze endogenous peptide biomarkers [7]. While ligand-binding assays have high sensitivity, specificity, and rapid "plate reader" detection, they also have several shortcomings that are influencing the transition to LC/MS/MS methods. LBAs do not quite meet the demands of a high-throughput discovery setting where the specific biological reagents are not yet available. Not only are specific reagents necessary, but also the time required to develop these reagents may be on the order of several months. Furthermore, the reproducibility and reliability of quantitative results are highly dependent on reagent quality and batch-to-batch consistency. In addition, LBAs have limited linear dynamic range (often requiring extensive dilution of samples to accommodate the concentration range of a PK study), cross-reactivity (yielding erroneous or inaccurate results), and matrix interferences. LBAs also have trouble distinguishing between the drugs and metabolites or other closely related substances. Finally, an individual assay is required for each peptide of interest, limiting multiplexing ability.

LC/MS/MS is an attractive alternative to LBAs because it is characterized by short method development times, broad linear dynamic range, a higher degree of accuracy and precision (reflected in more stringent regulatory guidelines) [8], high specificity, and the ability to simultaneously quantify multiple peptide therapeutics within a single injection and/or method. In addition, LC/MS/MS is already widely used in most bioanalytical laboratories, thus making it accessible to those skilled with the technology.

Development of LC/MS/MS assays (including sample extraction prior to analysis) for peptides is not without its challenges. There are those challenges related specifically to the nature and handling of peptides, and there are others related to evolving regulatory guidelines and our growing understanding of the possible limitations of LC/MS/MS. For example, bioanalytical methods have historically relied on the selectivity of triple quadrupole mass spectrometry to generate acceptable data. Reliance on MS selectivity was so strong, in fact, that ballistic LC gradients and simple quick and dirty sample preparation techniques such as protein precipitation became commonplace. The evolution of regulatory guidelines since 2007 now requires that scientists develop more selective and reliable bioanalytical methods [9, 10]. Discussions relating to acceptable results from incurred sample reanalysis (ISR) not only impact assay reproducibility but also bring a new focus to analyte stability in various matrices (i.e., plasma, blood, urine). New guidelines for the acceptable variability of matrix effects will require the most selective bioanalytical method possible, placing renewed focus on the bioanalytical method as a whole. These guidelines force bioanalytical scientists to consider the chromatography, mass spectrometry, and extraction protocol with equal importance to provide accurate and reproducible results. From a handling and analysis standpoint, physiochemical properties of peptides differ from small molecules in many ways, and in some instances, conventional "small molecule" techniques and strategies may not be directly applicable to the analysis of peptides. One such difference is the existence of multiple charge states for peptides, which decreases MS sensitivity and limits the specific MS instrumentation that can be used. For instance, only triple quadrupole instruments with at least a 1,500 dalton mass range on the first resolving quadrupole can detect the triply charged state of certain larger therapeutic peptides (>3,000–4,000 MW). Careful attention must also be paid to the choice of precursor and fragment ions to avoid isobaric interferences and to improve specificity for the peptide of interest. Furthermore, the zwitterionic nature of peptides, their tendency to bind in a concentration-dependent manner to storage containers, and the presence of many other peptides and high-abundance proteins in biological samples increase the complexity of the method development process.

A review of almost 200 articles in the literature [11] shows that many different combinations of LC, MS, and sample preparation conditions have been used for bioanalysis of peptide therapeutics, making it challenging to identify a common starting point for method development. In addition, many of the published references utilize non-selective sample preparation methods such as protein precipitation (PPT) and reversed-phase solid-phase extraction (RP SPE). These techniques are often used either because they are inexpensive or because they require little to no method development. Though common, the use of less selective sample cleanup may necessitate the use of longer chromatographic runs to eliminate the co-elution of endogenous materials with the analyte and may result in methods which fail matrix effects or ISR criteria. Perhaps, the greatest difficulties arise during peptide handling. Adsorption, stability, and solubility are critical parameters to understand and control during the method development process.

This chapter describes the development of bioanalytical methods for therapeutic and endogenous (such as biomarkers) peptides in three segments: liquid chromatography, mass spectrometry, and sample preparation. Each segment will identify potential pitfalls, key parameters for consideration, and differences relative to small molecule analysis. Solutions and recommendations for successful method development will then be presented. Following the successful development of extraction and LC/MS/MS analytical methods, a logical, stepwise, and routine strategy to bioanalytical method development for peptides will be proposed. Special attention will be paid to developing simple, logical strategies which are practical for implementation in typical bioanalytical laboratories. A critical aspect of this proposal will be the development of more selective, sensitive, and reproducible methodologies and their validity with respect to current and evolving (future) regulatory requirements. Advanced techniques, requiring instrumentation not commonly found in bioanalytical laboratories, will be summarized only briefly and are not the focus of this chapter.

2.2 Liquid Chromatography

The challenges associated with chromatography of peptides stem in part from the diversity of this class of compounds. As a class, they span a broad range of sizes, molecular weights, isoelectric point (pI), three-dimensional structures, and polarity. Although a high degree of diversity is present, peptide composition is actually highly conserved—there are only a finite number of amino acids which comprise their sequences, ensuring the presence of multiple closely related species. Peptides are present in samples across an extensive linear dynamic range, 4–5 orders of magnitude or more. Whether derived from a protein digest or present as a naturally existing or synthetic peptide, peptides in biological matrices will need to be separated from numerous closely related interferences. For example, missed cleavages and secondary cleavages in protein digests result in peptides that are nearly identical to the target peptides. Chromatography of peptides is further complicated by other factors as well. These large molecules have multiple points of interaction with chromatographic surfaces, meaning that different parts of the molecule can interact in different ways, possibly yielding poor peak shape or peaks which elute in two places. In addition, larger molecules such as peptides exhibit slower diffusion properties which can also result in poor peak shape and/or carryover under conditions traditionally used in bioanalysis.

Although LC separation of peptides has been well documented in the context of proteomic analyses or peptide mapping for qualitative work, LC conditions for that type of work differ significantly from what is required for a bioanalytical workflow. Peptide mapping studies commonly utilize long, shallow gradients (60–120 min), low flow rates (relative to small molecule analyses), or nanoflow systems, and ion-pairing agents such as TFA. These conditions are not particularly attractive for a bioanalytical laboratory where throughput and MS sensitivity are key aspects of method development. LC systems for peptide quantitation in these types of laboratories must use MS-compatible buffers and additives, run times should be in the several minute range (approximately 2–10 min maximum, if possible), selectivity from endogenous interferences must be obtained, peak shapes for small and large peptides alike should be as Gaussian and narrow as possible (to improve signal-to-noise ratio), and the LC system should have characteristics, such as <2 μm particles, which have been shown to minimize the potential for matrix effects caused by co-elutions [12].

Several reviews have been published over the past 3 years on the topic of LC conditions for peptide bioanalysis. Surveying well over 250 journal articles, we find that the most common set of conditions consists of C_{18} chromatographic columns and water and acetonitrile mobile phases, most frequently modified with formic acid. Acidic conditions are typically used as the carboxyl groups on peptides are neutral at low pH, improving chromatographic retention and minimizing secondary interactions. Those peptides having strong basic quality (containing several arginine or lysine residues) may not exhibit ideal chromatographic behavior under these conditions; therefore, ion-pairing reagents, typically TFA, may be employed to improve chromatographic peak shape for strongly basic peptides. TFA, however, is known to cause significant ion suppression under electrospray conditions, and one desires to avoid its use where possible. The suppression observed is due to both the formation of strong ion pairs (which cannot be ionized) and the reduction in signal as a result of high droplet surface tension and conductivity [13].

In 2005, Garcia [14] published a thorough review of the impact of various modifiers on the sensitivity and resolution of peptides and proteins. Though many buffers and alternate volatile ion-pairing reagents were assessed, none were suitable replacements for TFA when peptide chemistry necessitated its use. The MS suppression experienced in the presence of TFA was outweighed by the resolution improvement it afforded. A recent review by Ewles and Goodwin [15] describes their findings with respect to balancing the drawbacks and benefits of TFA and optimal conditions. In their work, it appears that low concentrations of TFA (0.01–0.05 %) in both organic and aqueous mobile phases might provide the desired ion pairing without the degree of suppression associated with higher levels. In addition, mixtures of low concentrations of TFA with more standard MS modifiers and buffers (i.e., formic acid or ammonium formate) might provide both the peak shape and resolution benefits without the concomitant decrease in sensitivity that usually accompanies the use of TFA. The exact nature and composition of the mobile phase will be highly dependent on the peptide and its sequence as well as the desired retention, resolution, or sensitivity.

On rare occasion, the use of high-pH mobile phases has been reported either to neutralize basic groups or to provide improved solubility of the peptide. For the majority of peptide separations, acetonitrile is the organic solvent of choice, though Giorgianni et al. [16] reported improved detection limits for several peptides using methanol. On most modern LC systems, it is straightforward to screen ACN and MeOH and should perhaps be considered as part of routine method development for peptides. Acetone was also examined as an alternative to ACN [17]. Retention order for a set of test peptides remained the same; however, acetone resulted in wider peaks, increased tailing, and decreased retention relative to ACN. Peptide response by MS, however, was similar.

With respect to chromatography, the final topic worthy of mention is the use of hydrophilic interaction chromatography (HILIC). In HILIC separations, acetonitrile is typically the weak solvent and water the strong solvent. It is important to note that HILIC may only be used successfully for those peptides soluble in higher percentages of acetonitrile. Although its use for peptides has been reviewed in the past [18, 19], very few quantitative applications have been reported. One very recent example combines both HILIC SPE and HILIC chromatography to successfully quantitate several arginine-containing hexapeptides [20]. HILIC was also employed by Zhan et al. [21] to quantitate a tetrapeptide in plasma. In general, peptide separation by HILIC is employed successfully only for smaller, more polar peptides.

The use of end-capped silica-based or hybrid C_{18} stationary phases is most common for quantitative peptide applications. Naturally, materials which minimize interactions with surface silanols are normally used, primarily to improve peak shape. The relatively recent (1999) introduction of hybrid particles has enabled separations to be carried out over a broader pH range and with significantly reduced surface silanol interactions than traditional silica particles. Though C_{18} seems to be the preferred ligand, the use of shorter ligands may occasionally prove advantageous for particularly large or hydrophobic peptides. Figure 2.1 illustrates the improvement in peak shape and signal intensity that was obtained on a C_4 column for the HIV fusion inhibitor enfuvirtide when it was analyzed on columns having identical base chemistry, but with either C_4 or C_{18} ligands. From the figure, it is clear that peak width is narrower (0.4 min versus 0.6 min wide at base) and peak area and intensity are greater (25,560 area counts versus 15,900) on the C_4 column.

It is clear that resolution from closely related endogenous constituents, speed, and sensitivity are all absolutely critical for effective and efficient LC of peptides in bioanalysis. Sub-2 μm porous particle LC, or UHPLC, has been used extensively in both small molecule and peptide bioanalytical applications specifically for the benefits it provides in terms of these exact parameters [22–32]. As valuable as these characteristics are to small molecule analysis, they can be

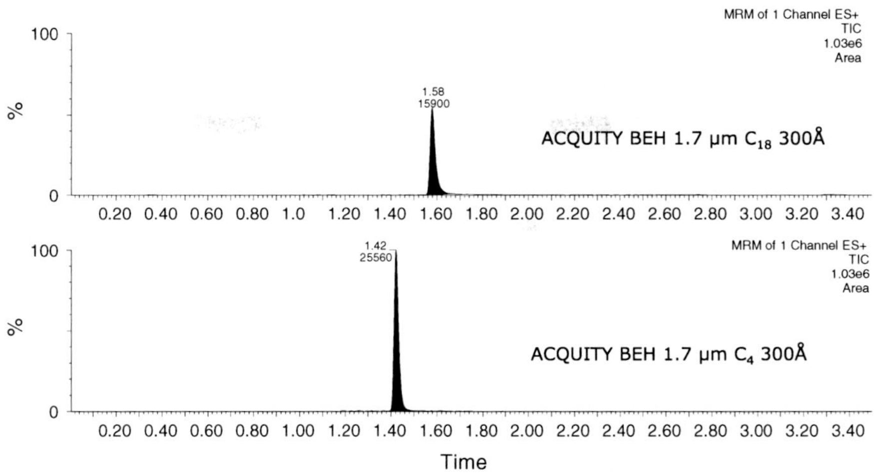

Fig. 2.1 Comparison of *peak shape* and intensity for the peptide enfuvirtide (MW 4492) chromatographed using a C4 and a C18 column

even more critical to successful peptide analysis. Peptide drugs are often modified versions of substances already in the body, meaning that there will almost always be a closely related interference present in the sample. The drug desmopressin and the endogenous hormone vasopressin upon which it is based are good examples. Desmopressin differs from human vasopressin by the loss of an amino group. Thus, the resolving power of the chromatography system used for this separation is critical, and sub-2 μm LC has demonstrated improvements in resolution and detection limits that are significantly better than conventional HPLC [22–24, 26–32].

It is important to understand the differences between small and large molecules as it pertains to the use of UHPLC. A van Deemter plot was constructed using flow rate instead of linear velocity to demonstrate how plate height for large molecules degrades much more rapidly than for small molecules at the higher flow rates that typically dominate most bioanalytical laboratories (Fig. 2.2). The figure was developed using the van Deemter equation below, where u = flow rate rather than linear velocity. In the below equation, HETP is equal to the height equivalent to a theoretical plate.

$$\mathrm{HETP} = 1.5\,(3.5) + 0.5/\mathrm{u} + 0.1666\,(3.5)^2\,\mathrm{u}$$

The plot clearly highlights how small molecules can be analyzed with much higher flow rates without a significant loss in column efficiency versus larger molecules like peptides which must be analyzed using lower flow rates in order to achieve maximum performance. This is primarily due to the lower diffusion rates of peptides in and out of the pores of the stationary phase. However, for high-throughput applications such as bioanalysis, it is not practical to use the very low flow rates needed for peptide separations.

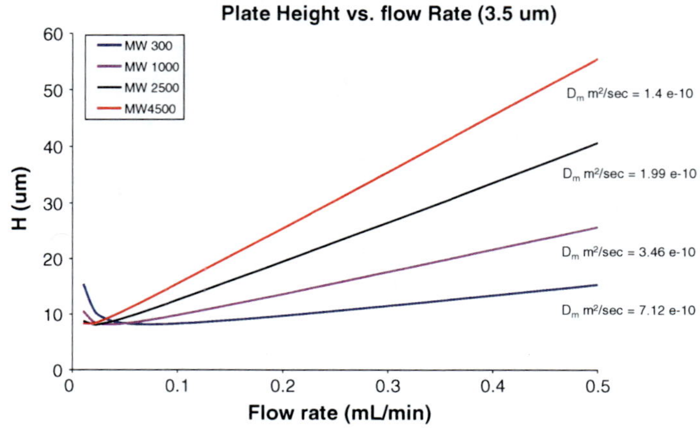

Fig. 2.2 Calculated van Deemter plot using flow rate for compounds of increasing molecular weight on a 2.1 mm ID, 3.5 μm particle size column. Assumptions are as follows: A, B, and C terms are equal to 1.5, 0.5, and 0.1666, respectively

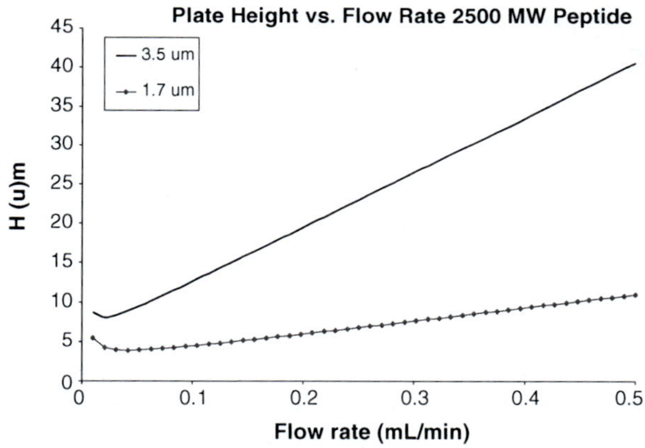

Fig. 2.3 van Deemter plot using flow rate for a model 2,500 MW peptide on 1.7 and 3.5 μm particle size 2.1 mm ID columns. Assumptions are as follows: A, B, and C terms are equal to 1.5, 0.5, and 0.1666, respectively, and the temperature used was 35 °C

In order to compensate for the slower diffusion of peptides, smaller particle sizes can be used. Figure 2.3 shows the calculated van Deemter plots for a peptide analyzed on 2.1-mm-diameter columns packed with both 1.7 and 3.5 μm particles.

While the optimum flow rate for this peptide is similar in both particles (~25–50 μL/min), the separation performance of the 3.5 μm particle column degrades more rapidly than the 1.7 μm particle column as the flow rate increases. From an implementation standpoint, this means that better resolution and peak shape can be obtained on sub-2 μm particles at higher flow rates. At 0.4 mL/min, a typical bioanalytical flow rate, the column packed with 3.5 μm particles has a 5X increase in plate height compared to its optimum flow rate, whereas the column packed with 1.7 μm particles exhibits only a 2X increase in plate height. This clearly indicates

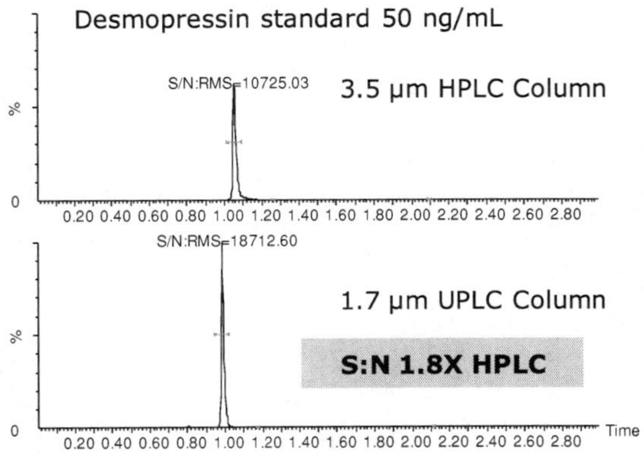

Peptide	S:N Ratio UPLC vs HPLC
Enfuvirtide	1.5X
Desmopressin	1.8X
Bivalirudin	2.5X
Vasopressin	5.8X
Angiotensin II	6.0X

Note: Both columns tested on a low dispersion system

Fig. 2.4 Influence of particle size on signal-to-noise ratio: 1.7 versus 3.5 μm columns

that the use of smaller particles is preferred for high-throughput bioanalysis of peptides.

From a practical standpoint, however, peptides are typically analyzed using a gradient rather than isocratic methods in order to reduce analysis times and to facilitate the separation of complex, diverse mixtures with a wide range of hydrophobicities. To illustrate the benefit of using small particles for peptide separations, 3.5 and 1.7 μm columns packed with the same stationary phase were compared (Fig. 2.4). Both columns were 2.1 × 50 mm, BEH C_{18}, run at 0.4 mL/min using formic acid and acetonitrile mobile phases. These columns have the same base particle and differ only in particle size. Both were run on a low dispersion system capable of operating up to 15,000 psi (ACQUITY UPLC system) using the same flow rate and gradient. The peptides run on the 1.7 μm particle column consistently elute as sharper, more efficient peaks, which translates into higher signal-to-noise ratio and the ability to achieve lower limits of detection. These data correlate well with previous findings by Gilar et al. [25] who demonstrated a marked increase in peak capacity for peptides using 1.8 μm particles compared to either 3.5 or 5 μm particles.

Another parameter of the chromatographic column to consider in peptide separations is the pore size. No concrete rule currently exists as to which pore size to use for peptides of a particular size, which means columns packed with particles of different pore sizes need to be tested to determine the effect on peak shape. Figure 2.5 shows a comparison of two columns packed with 1.7 μm particles but having two different pore sizes. If one is interested in a screening approach based on a single column, the larger pore-size column gives superior peak shape and results in better signal-to-noise ratio and lower detection limits, particularly for larger peptides. Smaller peptides generally perform equally well on columns with 130 or 300 Å pores. These data might also suggest that larger pores may mitigate some of the loss in efficiency observed for peptides at higher flow rates.

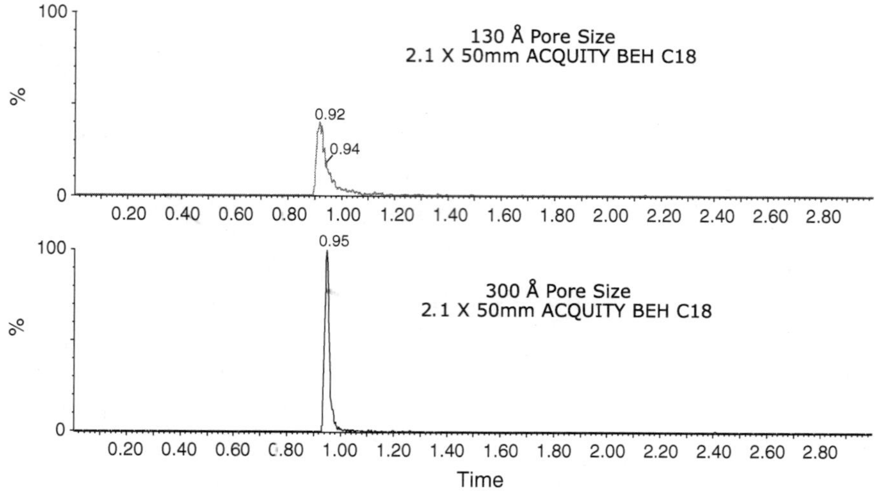

Fig. 2.5 The effect of chromatographic pore size on teriparatide, a 4118 MW peptide

Investigation of the particle diameter, pore size, and stationary-phase chemistry suggests that a simple screening method can be recommended for peptide bioanalysis. Initial method development for peptide chromatography should consist of a C_{18} column with 300 Å pores (2.1 × 50 mm) packed with <2 μm particles. 50 mm columns provide both the throughput required and often adequate separation. If improved separation is needed, longer columns may be used. A gradient from 15 to 75 % acetonitrile could be employed in a generalized screening regime as it represents the typical elution window for peptides. It is important to note that during gradient elution, gradient conditions should be set to ensure the elution of the peptide within the above-defined window. The use of ballistic gradients which could cause elution at very high percentages of organic may result in precipitation of the peptide on the column. Once a peptide has precipitated on the chromatographic column, it may be very difficult to remove and may result in poor chromatography in subsequent runs and ghost peaks.

A flow rate of 0.4 mL/min on a 2.1 mm diameter column correlates with previously published findings [33] and represents a starting point which balances speed with sensitivity arising from resolution and peak shape. Although the parameters defined here represent an appropriate starting point, they may be optimized during method development to achieve the desired resolution, run time, and limits of detection (LOD) for a particular assay.

In order to evaluate the feasibility and utility of the proposed chromatographic starting point for peptide bioanalysis, twelve therapeutic or endogenous peptides were selected for analysis. The peptides were chosen based on their diversity in molecular weights, acidity/basicity, and hydrophobicity. Details of each peptide along with their chemical properties are listed in Table 2.1. HPLC index is used

Table 2.1 Chemical properties for therapeutic and endogenous test peptides

Peptide	MW	pI	# of Residues	HPLC Index
Octreotide	1,019	9.3	8	40.8
Angiotensin II	1,046	7.35	8	38.3
Desmopressin	1,069	8.6	9	16.8
Vasopressin	1,084	9.1	9	7.6
Goserelin	1,270	7.3	10	31.7
Angiotensin I	1,296	7.51	10	56.2
Somatostatin	1,638	10.4	14	52.6
Neurotensin	1,673	8.93	13	44.4
Bivalirudin	2,180	3.87	20	46.2
BNP	3,464	12	32	15.9
Teriparatide	4,118	9.1	34	90.4
Enfuvirtide	4,492	4.06	36	155.9

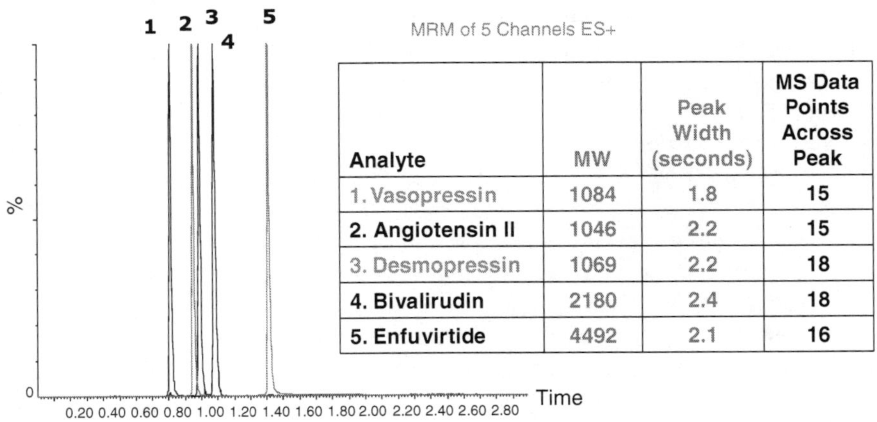

Fig. 2.6 Separation of five representative therapeutic peptides using a proposed LC screening method optimized for peptide bioanalysis. Data acquired on a Waters Xevo TQ MS triple quadrupole mass spectrometer using formic acid and acetonitrile at 35 °C

here as a measure of relative hydrophobicity. A low value indicates a more polar peptide, and a high value indicates a more hydrophobic peptide.

A separation of five representative therapeutic/endogenous peptides using the UHPLC screening conditions recommended above is shown in Fig. 2.6. Vasopressin, human antidiuretic hormone, (peak 1) and desmopressin (peak 3) are baseline-resolved using the proposed screening method. These compounds are analogous to each other, differing only in the loss of an amino group.

In addition to the challenges already addressed (selectivity, resolution, throughput, and peak shape), other common pitfalls encountered when analyzing peptides include carryover, adsorption, and issues related to solubility in the mobile phase and the injection solvent. In particular, solubility and adsorption problems can manifest themselves as any of the following: carryover, poor peak shape, poor linearity, poor reproducibility, and loss of sensitivity at low concentrations. In the case of suspected carryover, one needs to first determine whether the carryover is occurring in the chromatographic column or in the LC instrument itself (i.e., tubing, injector port, sample needle).

A simple test to determine the source of carryover involves performing an internal gradient. In other words, the gradient is repeated within the same run without performing a separate injection. If a peak appears in the second gradient at the expected retention time of the peptide of interest, then the carryover is suspected to be due to incomplete elution of the peptide from the stationary phase in the first gradient. This can be resolved in several ways. Column carryover (also called memory effect) is due to the inability of the chromatographic conditions to fully elute the peptide during the run, either due to slow and incomplete diffusion in and out of the chromatographic pores or due to poor solubility in the mobile phase with the modifier and flow rate being used. To improve the efficiency of diffusion and solubility in the mobile phase, the separation temperature can be increased and/or flow rate decreased. In addition, a higher strength or higher percentage of the mobile-phase additive can be used (i.e., increasing % of formic acid).

If the internal gradient test does not show carryover due to incomplete elution from the stationary phase, then the carryover is occurring in the injection fluidics, and adjustment of the needle washes and/or injection solvents may be required. A recent paper by Mitulovic et al. [29] identified an efficient wash solvent using trifluoroethanol (TFE) to clean not only the autosampler, but also a trap column if used. TFE can also be added to mobile phases for additional column cleaning for particularly troublesome peptides. An injection solvent that does not contain enough organic solvent and modifier can also cause carryover, since peptides precipitate out of solution during the injection process and "plate-out" on LC tubing and other components, resulting in adsorption and non-specific binding (NSB). Adsorption and NSB represent perhaps the greatest difficulty encountered when handling peptides and must be assessed as early in method development as possible, as they can affect not only the LC method but also the sample preparation. Care must be taken when choosing LC vials or plates. In general, plastic is better than glass, especially for basic peptides which interact with surface silanols in glass vials and plate inserts. Recently, so-called low-binding tubes and plates have been introduced by such manufacturers as Eppendorf. These tubes and plates are now widely used in peptide and protein analysis.

In general, peptide solutions should be made in a solvent containing a minimum of 5 % organic and 0.1 % formic acid or 0.05 % TFA, both of which help to keep peptides in solution. Injection solvents may contain even higher percentages of organic solvent and modifier in order to maintain solubility throughout the duration of overnight analytical runs and autosampler stability tests. A recent publication on the quantitation of amyloid-β peptides [34] demonstrates this quite nicely. In this work, the final injection solvent consisted of ~40 % organic and ~5 % NH_4OH, both of which were essential for maintaining solubility and minimizing adsorptive losses for this incredibly sticky class of peptides. Additional handling recommendations will be described in the sample preparation section of this chapter.

2.3 Mass Spectrometry

The vast majority of quantitative analysis in bioanalytical laboratories is performed on triple quadrupole mass spectrometers. This section will focus on the use of triple quadrupoles (TQs) as they represent the instrument configuration in most widespread use, though ion traps and quadrupole time-of-flight instruments are also employed. Recent advances in MS instrument design have resulted in the increased use of hybrid TOF instruments, but reduced sensitivity versus TQs in multiple reaction monitoring (MRM) mode. The existence of TQ MRM mode has limited the use of TOFs as a platform for low-level peptide quantitation. Triple quadrupole instruments, operated in MRM mode, still offer the highest sensitivity for targeted analyses and are the platform of choice for both small molecule and peptide quantitative applications.

MS of peptides differs significantly from that of small molecules, and in general, the overall MS signal for peptides is often lower than for small molecules for several reasons. First, peptides are multiply charged species, whereas small molecules are typically singly charged, and second, there may be several different multiply charged precursors present, both of which dilute the overall ion intensity across several species. Furthermore, peptides tend to form many low-abundance fragments rather than one or two intense ones, reducing overall signal for MRM experiments. An even greater loss of signal can be observed for large peptides that are not as efficiently transferred into the gas phase during ionization as small molecules. It may be advantageous to sum transitions to either improve signal intensity or reduce variability if the relative abundance of a specific precursor changes during the analysis. Clearly, one not only needs to consider the sensitivity of the MS specifically for large molecules, but also any additional aspects of the method (LC, sample concentration during extraction, etc.) that can be used to improve assay sensitivity.

During MS tuning, it is common to see several different precursor ions due to multiple charging. Whereas small molecules gain or lose a single proton, large biomolecules have multiple protonation sites and therefore can gain or lose several or many protons, generating what are called "multiply charged" species. For example, the N-terminus and the various amino acid side chains of peptides are common sites of protonation. If one considers that mass spectral detection is performed on the basis of mass-to-charge (m/z) ratio, the following equations can be used to calculate the expected m/z for the various possible multiply charged precursors a peptide may produce.

Singly charged $= M + H/1$, doubly charged $M + 2/2$, triply charged $= M + 3/3$, and so on correspond to doubly, triply, quadruply, and even higher charge states.

In electrospray positive ionization mode, the most common mode of MS analysis for peptides, peptides fragment in a very predictable manner. Primary fragmentation yields a series of ions corresponding to cleavage at the amide (or peptide) bonds between the amino acids that comprise the peptide sequence. If the charge is retained on the N-terminal fragment, the ions are classified as *b* ions;

if the charge is retained on the C-terminus, the ions are classified as *y*. There are other internal cleavage ions and immonium ions, but *b* and *y* ions are the most frequently observed [35]. Coincidently, these are also good choices for quantitation as they tend to be inherently specific for the peptide of interest. The final point related to fragmentation is that the most complete and the most useful fragmentation is typically obtained when the highest possible charge state is fragmented [35]. This reinforces the importance of tuning on multiple precursors.

Source conditions should be tuned for maximum transmission of the precursor without in-source fragmentation. As mentioned, it is good practice to tune and optimize several different MRM transitions. These may be multiple fragments from a single precursor or the same or different fragments from different precursors. These transitions can be used for both confirmation and to provide options for obtaining the best specificity and sensitivity, particularly for biological samples that contain many endogenous interferences. It is important to note that a transition that appears to be most intense during tuning of solvent standards may not be the transition with the highest signal for an extracted sample. When performing MS/MS tuning and optimization of the of the precursor ion signal intensity using collision-induced dissociation (CID), one must consider that fragments from multiply charged precursors may be multiply or singly charged. This requires MS/MS to be performed across a broad mass-to-charge range, often up to the maximum range of the quadrupoles. This can vary from 1,000 to 3,000 m/z, depending on the manufacturer and model. Careful choice of both precursor and fragment ions can be the critical factor in developing a robust and reliable MS method for peptide therapeutics. There are several guidelines that may be helpful in choosing the optimal fragment for example. Choice of fragment ion is fairly straightforward in the case of small molecule therapeutics, where the most intense fragments are often chosen for quantitation. Conversely, the most intense peptide fragments may not always be the most specific. In addition to non-specific water losses, it is quite common to see intense peptide fragments at low m/z values such as m/z 136, 110, 129. These specific fragments correspond to immonium ions, a result of secondary fragmentation, arising from individual amino acids in the sequence. The m/z ratios of 136, 120, 129, 110, and 86 indicate the presence of tyrosine, phenylalanine, arginine, histidine, and leucine/isoleucine, respectively. Transitions based on this type of fragment are often non-specific, resulting in increased baseline noise and multiple peaks from other isobaric peptides present in biological extracts, and should be avoided if possible. Occasionally, when a peptide does not fragment well (either no fragments are generated or many low-abundance fragments result, the "all or nothing" phenomenon), the use of a "pseudo-MRM" transition, such as a precursor-to-precursor transition (where both quadrupoles are set to the same m/z value), may be required. This approach may require extensive sample preparation and/or multidimensional LC to separate isobaric interferences that are not distinguished by unit mass resolving mass spectrometers such as quadrupoles. The ideal fragments for reliable, reproducible quantitation are *b* or *y*-sequence ions.

An additional consideration for both precursor and fragment ion choices is the use of higher m/z ions, for example, using a 4^+ instead of a 5^+ precursor and/or

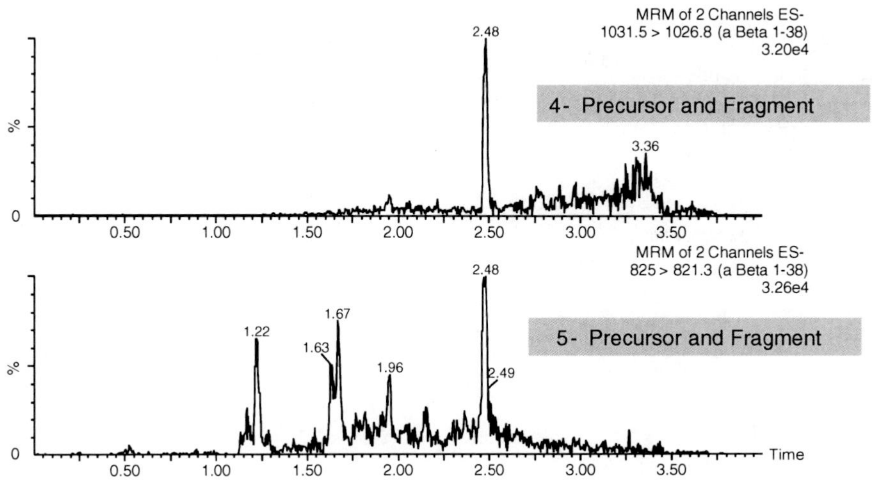

Fig. 2.7 Comparison of 4^- versus 5^- transitions used to monitor amyloid-β peptide 1–38 from a human plasma extract

the use of a fragment present at higher m/z than alternative choices. Transitions based on higher m/z ion pairs often benefit from reduced chemical noise relative to the equivalent pair (i.e., the same compound, same fragment, but different charge states) from a higher charge state present at lower m/z values. This is illustrated in Fig. 2.7 where two separate transitions for an amyloid-β peptide with a molecular weight of 4,132 are monitored during analysis of a human plasma extract. The transitions represent quadruply (4^-) or quintuply (5^-) charged versions of the same precursor to fragment pair.

An important aspect of MRM analysis of peptides is the mass range of the first and second quadrupoles. For example, the MS infusion of enfuvirtide (MW 4492) produced a dominant 3^+ precursor at approximately m/z 1,498, requiring an instrument with a mass range of at least 1,500 (Fig. 2.8). Similarly, MS/MS of the 2^+ precursor of bivalirudin (MW 2,180) at m/z 1,091 produced two major singly charged fragments at m/z 650 and m/z 1,531 (Fig. 2.9), again demonstrating the need for adequate mass range. In this case, a mass range of ≥2,000 dalton on both quadrupoles is desirable.

There are various factors that influence the nature and relative abundance of peptide precursors formed in the MS source. Chief among these are flow rate, pH, and concentration of the mobile-phase modifier. It is not uncommon, for example, to observe different charge-state precursors dominating at different flow rates. A recent publication [36] on the quantitation of angiotensin II describes the predominance of a triply charged precursor at 700 μL/min and that of the doubly charged under nanoflow conditions (250 nL/min). Figure 2.10 demonstrates this phenomenon when insulin is infused at either 10 μL/min or teed into the mobile-phase stream with the same composition at a flow rate of 200 μL/min. At the lower flow

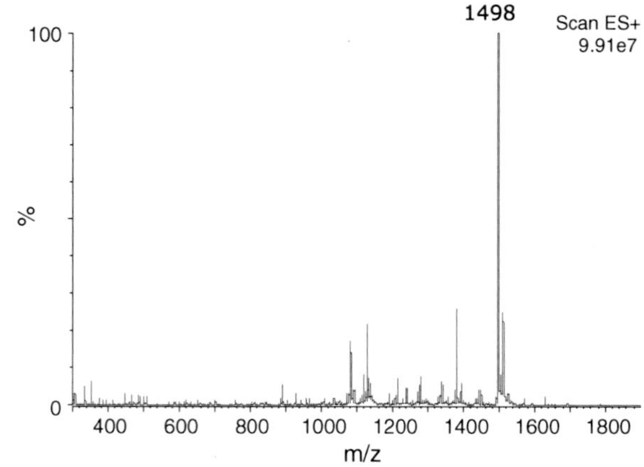

Fig. 2.8 MS infusion of a solution of enfuvirtide clearly demonstrates the formation of a dominant 3+ precursor at m/z 1,498, requiring adequate MS mass range on the first quadrupole. The 4+ precursor at m/z 1,124 is significantly less sensitive

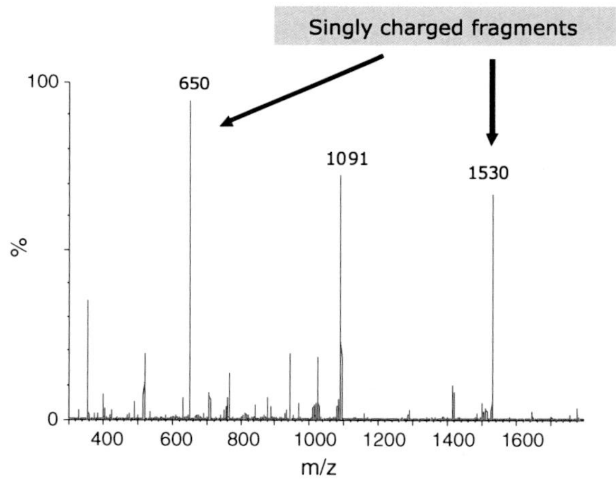

Fig. 2.9 CID of the 2+ precursor of bivalirudin (m/z 1,091), yielding intense singly charged fragments at m/z 650 and 1,530

rate, the 5^+ precursor is predominant (Fig. 2.10b), whereas the 4^+ precursor dominates the spectra when higher flow rate is used (Fig. 2.10a).

Another well-studied [37–39] phenomenon is the relationship between charge-state distribution and analyte concentration. Wang and Cole [40] demonstrated that the charge-state envelope shifts toward lower values as peptide concentration increases. For example, one might observe more doubly charged species than triply charged at higher analyte concentrations. This suggests that one should monitor transitions from several charge states during method development and assess any potential impact on quantitation.

From one instrument vendor to the next, it is common to see differences not only in sensitivity for large molecules, but also in the relative abundance of the various precursors formed. It is therefore good practice to evaluate multiple

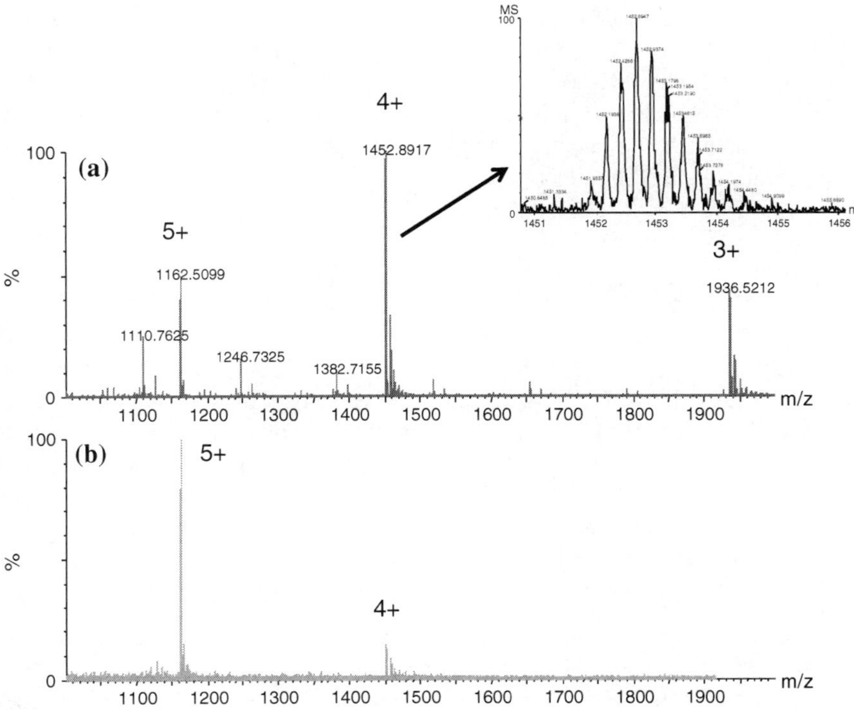

Fig. 2.10 MS infusion of insulin teed into LC flow at 200 μL/min (**a**) and at 10 μL/min (**b**) clearly demonstrates the influence of flow rate on relative abundance of precursors formed

vendors' MS systems as well as equip the laboratory with instruments from several vendors to ensure maximum flexibility when developing and validating methods for peptides.

2.4 Sample Preparation and Peptide Handling

2.4.1 Peptide Handling

Prior to discussing the utility of specific sample preparation techniques, one must first understand how the behavior of peptides, and the challenges associated with it, impacts the successful development of an extraction method. From the moment the peptide standard is dissolved (if it is in lyophilized powder form), one must address the issues of solubility and adsorption. In general, peptides are very poorly soluble, and many adsorb to the walls of vials, collection plates, LC surfaces, etc. This is particularly true for larger, more hydrophobic peptides.

Heavy emphasis is placed here on identifying the condition under which the peptide exhibits optimal solubility as this impacts every aspect of method development and forms the foundation for successful development of a reproducible, sensitive bio-analytical assay. In addition, failure to address adsorption/NSB can be likewise detrimental to the method as a whole and therefore must be assessed prior to extraction and preparation of test solutions.

2.4.2 Peptide Handling: Solubility

Peptide solubility is heavily dependent on its specific amino acid sequence, in particular, the hydrophobicity and acidity or basicity. There exist many sets of guidelines for solubilizing peptides, and all seem to agree on a few key points and rely on the use of well-understood characteristics of amino acids to help predict peptide solubility. Several good sets of guidelines, summarized below, can be found on web pages from Sigma-Aldrich, Thermo Scientific, GenScript, and Pierce. Table 2.2 summarizes the relevant amino acid properties.

A general set of solubility guidelines for consideration is as follows:

1. If a peptide is very small, <5 residues, it will likely dissolve in aqueous solutions unless the sequence is entirely comprised of hydrophobic residues.
2. Peptides containing >25 % charged residues and <25 % hydrophobic residues generally dissolve in aqueous solutions.
3. If the peptide is basic, acidic solutions (formic acid or TFA) with a low % organic (5 %) often work well. The converse is true for acidic peptides, try solubilizing in basic solutions (1–5 % NH_4OH, for example) with a low % organic.
4. Peptides containing >50 % hydrophobic residues may be only slightly soluble or insoluble in aqueous solutions. Hydrophobic peptides are best solubilized in DMSO, DMF, strong acid solutions (TFA, formic, acetic), or isopropanol. For cysteine-containing peptides, use DMF instead of DMSO.

Table 2.2 Useful amino acid properties

Property	Amino acid
Hydrophobic	A, F, I, L, M, P, V, W, Y
Moderate	C, G
Hydrophilic	D, E, H, K, N, Q, R, S, T, pyroglutamic acid
Positive charge	K, R, H, N-terminus
Negative charge	D, E, Y, C-terminus
Degradation likely	M, W
Prone to de-amidation	N,Q, C-terminal amides, N-terminal Q
Dehydration, cyclization to pGlu	
Prone to oxidation under mild conditions	C, M

5. Guanidine HCl or urea may be necessary for those peptides that tend to aggregate and can later be removed during sample preparation.
6. Peptides which contain >75 % of S, T, E, D, K, R, H, N, Q, or Y may form intramolecular hydrogen bonds and form gels in aqueous solutions. These peptides should be treated in the same manner as hydrophobic peptides (#4).

2.4.3 Peptide Handling: Adsorption

In addition to solubility, every effort must be made to minimize or eliminate adsorption. Peptides bind to vials, collection plates, pipette tips, and other surfaces. Many researchers have reported significant peptide losses, particularly to pipette tips, during solution preparation or sample handling [41–43]. This may be mitigated by "pre-treating" the tips through aspiration of the peptide solution up and down in the pipette tip prior to dispensing into the final vessel. Other factors that may influence adsorption include solvent composition, container material, temperature, and pH. Adsorption occurs primarily in solvent standards rather than in extracts of biological matrices. Most biological matrices contain residual proteins or other peptides at higher levels which can act as carriers, binding preferentially to surfaces rendering them "inert" to the low-level peptides of interest. Relative to vial/tube/plate composition, side chains of basic peptides can readily interact with surface silanols on glass and hydrophobic peptides may bind through hydrophobic interactions with polypropylene or other plastic surfaces. This effect is more pronounced at low peptide concentrations. Complete loss of peptide during serial dilution often results in the loss at the low end of the calibration curve or even absence of the peptide peak during analysis of low-level solvent standards. For this reason, when preparing standard curves, one should spike directly into plasma from the peptide stock solution (in which adsorption losses are negligible) and then prepare subsequent dilutions for the lower concentration points by diluting the plasma spiked with the peptide stock solution with additional plasma.

Issues related to adsorption are further exacerbated by inappropriate solvent choice. Peptide losses occur readily in aqueous solutions. Adsorption can be, in part, ameliorated using the proper solvent composition. Any information gathered during solubility testing should be applied to all subsequent solution preparation. For example, if the addition of acid, base, or organic improves solubility, the appropriate action should be taken to include these modifiers in dilutions of the concentrated standard. Peptides are naturally more soluble in aqueous solutions if they are charged and, conversely, more soluble in organic solutions when uncharged.

For particularly hydrophobic peptides, or low concentration aqueous solutions of peptides, one might also consider the addition of a commercially available protein such as serum albumin to help eliminate NSB to surfaces by blocking adsorption sites. This is not typically necessary for plasma or serum extracts. However, urine or cerebrospinal fluid (CSF) extracts may require the same treatment as solvent-based solutions.

Finally, concentration by evaporation and reconstitution should be avoided if at all possible as this frequently results in significant losses. Several options exist for addressing this problem. One may add a small volume of a viscous solvent such as dimethyl sulfoxide (DMSO) or glycerol prior to evaporation, which prevents complete dry down. Alternatively, certain SPE formats exist which enable up to 15-fold concentration of the sample without evaporation. This option has the additional benefits of not only increasing throughput by eliminating time-consuming evaporation, but also of reducing the number of handling steps while ensuring that the sample is in a solution which not only provides good solubility for the peptide (if SPE method development has been done properly) but also an injection solvent which is chromatographically compatible.

2.4.4 Peptide Handling: Stability

Another common concern is possible peptide instability or degradation occurring in vivo or ex vivo. Reubsaet et al. [44, 45] describe instability as falling into two categories: physical instability and chemical instability. Physical instability is primarily associated with unfolding (caused by temperature, pH extremes or guanidine HCl or urea denaturation) and aggregation (primarily due to hydrophobic interactions between partially unfolded species). Chemical instability is related to modifications in amino acids which can occur through oxidation, reduction, de-amidation, hydrolysis, arginine conversion, β-elimination, and racemization [44]. It is important to control conditions which may result in modification during all phases of method development. For example, the use of protease inhibitors to improve matrix stability has been widely accepted. Ewles and Goodwin [15] report testing numerous protease inhibitors. They concluded that 20 mM diisopropylfluorophosphate (DFP) or Pefabloc® (Roche Diagnostics, West Sussex, UK) was among the best options. In addition, it was noted that simple addition of acid (formic or hydrochloric) was often adequate to inhibit protease activity.

Chemical instability is primarily caused by hydrolysis, oxidation, pyroglutamic acid formation, and de-amidation. Peptides containing D (Asp) are most likely to undergo acid-catalyzed hydrolysis. De-amidation occurs under base-catalyzed conditions in the presence of N-G (Asn-Gly) or Q-G (Gln-Gly). Cysteine and methionine easily undergo oxidation, which is accelerated at high pH. The presence of an N-terminal Gln will most certainly result in pyroglutamic acid formation. It is important to note the sequence of the peptide you are working with and assess the potential for any chemical modifications that may occur either as a result of the storage or as a result extraction conditions. It may be necessary to eliminate extreme high-pH or low-pH conditions during an experiment, depending on the specific amino acid composition. A recent development is the increased use of dried blood spots (DBSs) in bioanalytical assays. In addition to the obvious benefits with respect to low sample volumes and less expensive storage and shipping, DBS has shown promise in stabilizing unstable compounds. This technique was successfully employed by Kehler et al. [46] for the analysis of the large peptide exendin-4.

2.4.5 Peptide Extraction

Once the issues of solubility, adsorption, and instability have been addressed, an extraction technique must be developed in order to isolate a peptide from a complex biological sample containing many closely related interferences. There are three main extraction techniques used in peptide bioanalysis: PPT, SPE, and liquid–liquid extraction (LLE), with SPE figuring most prominently in the literature [11]. The following section will review the pros and cons of each technique and then propose a broadly applicable screening method. In addition, due to the recent introduction of additional regulatory criteria that must be met during bioanalytical method validation (e.g., matrix effects, incurred sample reanalysis), a strong emphasis will be on selectivity of the various methods and their role in facilitating meeting evolving regulatory guidelines.

Depending on the specific matrix, components that must be separated include but are not limited to phospholipids, salts, proteins, other peptides, formulation agents, and dosing media, among others. This discussion will focus primarily on plasma and serum as these are the most common matrices in bioanalytical laboratories.

Prior to the extraction itself, one must disrupt binding between the therapeutic or endogenous peptides and proteins present in the biological matrix. This binding may be stronger than the binding between small molecules and proteins, necessitating additional pre-treatment alternatives. Common means of disrupting protein binding include pre-treatment with acid (4 % phosphoric, formic, 1–10 % TFA, or TCA) or base (5 % NH_4OH), or for particularly hydrophobic peptides, denaturation with guanidine HCl (dilute 1:1 with 6 M and shake for 30–45 min) or urea may be necessary. These reagents can later be removed during SPE without concern for peptide losses. Protein-binding problems typically manifest themselves as apparent "low recovery" during the extraction process. Peptides that are bound to proteins in the matrix either co-precipitate along with endogenous proteins during PPT or pass through an SPE device during sample loading. An easy test to confirm the presence of protein binding is to prepare a set of spiked samples in both the study matrix and phosphate-buffered saline (PBS). Recovery should be calculated according to Eq. 2.1 below. If recovery is higher in the PBS samples, then protein binding has likely occurred.

$$\%\ \text{SPE recovery} = \left(\frac{\text{average peak area in pre-spiked extracted samples}}{\text{average peak area in post-spiked extracted samples}} * 100\right) \tag{2.1}$$

In plasma, 22 proteins make up 99 % of the total protein content. Of these, albumins (MW ~65 kDa) make up about 45 % of the total protein content and are present at tens of mg/mL. Immunoglobulins (MW ~150 kDa) make up another 15–30 % [47]. These proteins are present at many orders of magnitude higher concentration than peptide therapeutics and biomarkers and are often even less soluble.

In addition to high levels of proteins, plasma phospholipids (PLs) are a major source of concern for both small and large molecule bioanalytical assays. The presence of high levels of residual PLs in sample extracts is particularly concerning, considering their role in matrix effects. Bennet and Van Horne identified PLs as the major source of matrix effects in plasma in 2003 (AAPS posters 2003), and discussions relating their removal continue to dominate industry-related conferences. A thorough investigation of various sample preparation techniques and their influence specifically on phospholipid removal clearly demonstrated several important differences between the techniques [12].

Overall, PPT is universally regarded as a quick, inexpensive technique. However, with respect to the removal of the various aforementioned interferences, the resultant extract is exceedingly "dirty" due to the entirely non-specific nature of the separation. As long as protein binding is disrupted, most high-abundance plasma proteins (typically proteins >~40 kDa) can be precipitated or separated using PPT. However, PPT has several serious drawbacks. For example, PPT does not remove PLs, salts, other peptides, metabolites, or dosing media, and formulation agents. In fact, PPT extracts PLs quite well, resulting in extracts with high concentrations of this undesirable class of compounds. During PPT, methanol solubilizes more lipids than ACN does, making it less desirable as an extraction solvent. However, the window to precipitate peptides coincidently is quite narrow when ACN is used; therefore, the concentration of ACN in a precipitation solvent should be carefully considered. Optimization of the final organic % and the nature of the organic used in the precipitation solvent and sample may balance precipitation of unwanted proteins with peptide solubility and hence peptide recovery. The addition of TFA or trichloroacetic acid (TCA) may help in cases where peptide solubility in the precipitation solvent is limited. The non-specificity which characterizes PPT often leads to severe matrix effects which cause variability, poor robustness, and poor reproducibility in the final assay.

The use of LLE to extract peptides from plasma has been reported only a handful of times and with lower than desired recovery. In general, the ionic nature of peptides and their poor solubility in very apolar solvents severely restrict the utility of this approach. Furthermore, typical LLE solvents such as methyl tertiary butyl ether (MTBE), hexane, and ethyl acetate also extract lipids efficiently, yielding final extracts saturated with PLs. LLE also does little to separate peptides of interest from other peptides present in the sample or many of the other common interferences.

Separation of a target peptide from other peptides in the sample is most readily accomplished by SPE, where manipulation of both organic content and nature and concentration of modifier can result in very selective final eluates. Neither PPT nor LLE possesses the degree of resolving power necessary to accomplish this as both rely on simple separation mechanisms: either physical precipitation only or distribution between aqueous and organic layers. SPE, historically silica-based C18 and more recently polymer-based reversed-phase (RP) or mixed-mode (MM) (having both RP and ion-exchange retention mechanisms) sorbents, seems to be the method of choice for peptide extraction. There are many reasons for the popularity of SPE for selective peptide isolation. In general, one can load in aqueous solutions rather than working with organics that may cause precipitation. In contrast to other techniques, any reagents used to disrupt protein binding, such as denaturants, acids, bases, will be eliminated during the process through a series of wash steps. In addition, the majority of unwanted proteins are eliminated during the sample loading step of an SPE method due primarily to their exclusion from the chromatographic pores of the sorbent. Dosing media and formulation agents may not be efficiently removed by PPT or LLE, but may, once again, be removed using SPE and judicious choice of wash and elution steps.

Most SPE methods can be automated or converted to online protocols such as that recently described by Calderon-Santiago et al. [48]. While online extraction methods eliminate many of the manual components of performing an extraction, extraction times of 7–12 min *per sample* (prior to LC) are commonplace. Method development may be more challenging when using online systems as risk of carryover increases with multiuse cartridges and recovery and matrix effects are more difficult to determine.

Extraction by SPE can be used with polar and non-polar, acidic and basic, and large and small peptides, making it an attractive platform for a universal method development approach. Solid-phase extraction, specifically MM SPE, is identified as the technique that provides the most selective final eluates [12]. In addition, a recent review of techniques specifically for peptide bioanalysis [15] also concluded that MM sorbents in conjunction with RP chromatography are the ideal platform for this application. MM SPE sorbents typically contain a RP backbone functionalized with a strong or weak cationic or anionic moiety. This allows one to bind analyte and/or interference molecules by either RP or ion exchange, providing dual orthogonal mechanisms with which to perform a separation. Furthermore, one can manipulate the organic content within each wash or elution step to further improve the selectivity and cleanliness of the final elution(s). Finally, extraction of peptides using the ion-exchange elution step and subsequent LC separation in the RP dimension ensures orthogonality of the bioanalytical method as a whole. Selective sample preparation in conjunction with a high-resolution chromatographic separation is critical as TQ MS systems operated at unit mass resolution could not differentiate between two isobaric peptides sharing an isobaric fragment. Unfortunately, this instance occurs more frequently than we would like due to the highly conserved chemical nature of peptide composition. There are only 20 naturally occurring amino acids that make up all of the peptides and proteins in the body, making separation challenging, thus requiring a multidimensional approach such as the one described herein which uses MM SPE to complement RP LC.

Properties such as pI may be useful in choosing an SPE sorbent and/or in optimizing wash and elution solvents. Additionally, the pH of wash and elution solutions as well as the specific nature of the SPE sorbent can be manipulated or changed to facilitate separation from other peptides based on knowledge of pI. For example, basic peptides such as desmopressin (pI 8.6) or octreotide (pI 8.3) are expected to bind to a (MM) cation-exchange sorbent, whereas an acidic peptide such as bivalirudin (pI 3.9) should bind to a (MM) anion-exchange sorbent. Depending on the exact % of organic required to elute the target peptide, wash solvents should contain as much organic as possible without eluting the peptide. The elution solvent should also contain the minimum % organic required to elute the peptide in order to minimize the more hydrophobic interferences in the extract. Not only can peptides of opposite ionic nature be separated from each other, but peptides of similar pIs can also be separated from each other through judicious choice of organic content in the wash and elution solutions. Using different wash and elution solvents allows for refinement of the SPE method to sequentially and systematically optimize the organic content so that the elution window of the peptide is tightly controlled. Finally, utilizing a small-volume elution SPE device can eliminate the evaporation and reconstitution steps, which frequently result in peptide losses due to adsorption to the plates or tubes used for evaporation or insolubility in the reconstitution solvent. This low elution volume format can effectively concentrate a sample up to 15-fold through well-researched plate designs which allow one to load several hundred microliters of sample and elute in as little as 25 μL. One such plate is the Waters μElution 96-well plate.

To further reinforce the benefits of MM SPE for peptide extraction, studies were performed in our laboratories to compare traditional sample preparation methods (PPT, LLE, and RP SPE) to MM SPE. Initially, this was performed using two of the 12 peptides from Table 2.1 (desmopressin and bivalirudin). Analyte recovery was calculated to compare and contrast extraction efficiency using generic methods for each technique, and matrix effect calculations were used to reflect sample cleanliness and as a representative measure of selectivity of the extraction types. Once the limitations of these techniques were characterized and understood, we endeavored to develop a simple screening method for peptide extraction.

The exercise of developing a universal sample preparation screening method for peptides serves here to clearly reinforce the importance of understanding and addressing adsorption, solubility, and stability through concrete examples and explanations.

In these studies, Eq. 2.1 (previously described) was used to calculate the recovery of each peptide from human plasma.

Matrix effects were calculated according to Eq. 2.2

$$\%\ \text{Matrix Effects (ME)} = \left(\left(\frac{\text{Response}_{\text{Post - Extracted Spiked Sample}}}{\text{Response}_{\text{Solvent Standard}}}\right) - 1\right) * 100 \qquad (2.2)$$

Results from the initial characterization experiments are shown in Fig. 2.11.

The data show that none of the techniques provided both the high recovery required to meet challenging detection limits and low matrix effects (which would indicate improved removal of interferences). In addition to the benefits previously

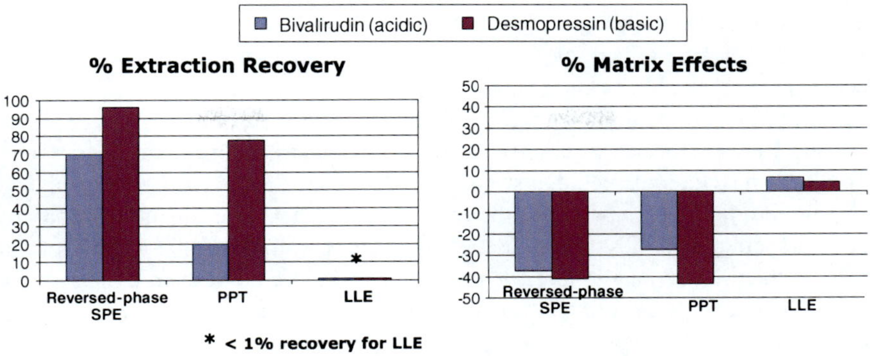

Fig. 2.11 *Extraction recovery* and *matrix effects* for an acidic and basic peptide using traditional sample preparation techniques

described, MM SPE has been shown to reduce matrix effects to a greater extent than other sample preparation techniques while still providing high analyte recovery [12]. Due to their zwitterionic nature, the behavior of peptide therapeutics under various SPE conditions can be difficult to predict. Therefore, initial proof-of-concept studies were performed on four different MM sorbents, using a generic set of conditions originally developed for small molecule screening [12]. Recall that each sorbent consists of a moiety that imparts RP behavior as well as an ion-exchange group (strong or weak cation or anion exchange) for additional selectivity and therefore is capable of producing two elutions: one contains compounds bound by RP (elute 1) and a second containing compounds bound by ion exchange (elute 2). Recovery was calculated for both elute 1 and elute 2 fractions on all four of the MM sorbents. As expected, using protocols designed for small molecules, recovery for test peptides was split between the two elutions and was on average <60 %, whereas these same protocols typically yield recoveries for small molecules that are >80 % on average. It was clear that knowledge and application of general peptide physiochemical properties were needed if the advantages of MM SPE were going to be successfully applied to develop a generic approach to peptide extraction. Conventional "small molecule" thinking and protocols incorporate steps and solvents used in a manner that yields poor results for peptides. For example, when one thinks of maximizing recovery for small molecules, the elution window is quite broad, ranging from 0 to 5 % organic required to elute very polar compounds to 100 % organic for hydrophobic small molecules. Generic SPE protocols for small molecules often use 100 % organic in the elution to ensure the highest recovery for many small molecules of potentially diverse properties. In contrast, if one considers peptides as a class, the elution window is much tighter, with most peptides eluting between approximately 30–65 % acetonitrile. This basic information should be applied during extraction method development as application of solutions containing higher than optimal organic content often results in peptide precipitation onto the SPE cartridges. It is also common to include modifiers such as formic acid in wash and elution steps for small molecules. However, stronger acids may be required if maximum solubility, and thus recovery, of peptides is to be obtained.

Further experimentation with additional peptides and examination of the resultant data were performed in order to formulate a construct for a peptide screening protocol. The data indicated that strong anion-exchange and weak cation-exchange sorbents produced higher recoveries on average in the ion-exchange elutions for the therapeutic peptides tested (data not shown). Subsequently, changes were made to the original small molecule protocols, including optimization of wash and elution solutions, to generate a protocol developed specifically for peptides which incorporates basic knowledge of peptide hydrophobicity, solubility, and their zwitterionic nature. Among these changes were to use a 75 % organic elution as opposed to a 100 % organic elution. In addition, TFA is added to improve solubility of hydrophobic peptides. TFA showed a significant improvement in recovery for certain peptides over formic acid without deleterious effects. The concentration of TFA was equally as important. Several concentrations in the final elution were tested, and 0.1 or 0.5 % TFA was not adequate to provide high recoveries for some of the larger or more hydrophobic peptides. A final concentration of 1 % TFA provided the optimal recovery for a diverse set of peptides. It is believed that a combination of improved solubility and ion pairing is responsible for the recovery increases. This elution solvent has the advantages of providing optimal solubility for a wide range of peptides, eliminating the majority of phospholipids (which typically require higher organic to elute) and producing an eluate ready for injection onto an LC/MS/MS system without further manipulation. The next section describes the method details for a universal screening protocol for MM SPE extraction of peptides from biological matrices. The method screens two SPE sorbents simultaneously to rapidly identify the best starting conditions. Although one may be able to predict the appropriate sorbent for a peptide based on pI, it is not always a definitive indication. The location of charged residues in the sequence and their accessibility to the sorbent influence retention may result in unpredictable behavior vis-a-vis pI and sorbent type. For example, two peptides with the similar pIs may interact differently with the same SPE sorbents, reinforcing the need for a screening protocol.

2.4.6 Proposed Peptide Extraction Screening Protocol

Based on discussions above, we have proposed a general screening protocol: sorbents: weak cation exchange and strong anion exchange

Format: 96-well reduced sorbent bed size

1. Condition with MeOH;
2. Equilibrate with H_2O;
3. Load diluted, pre-treated sample;
4. Wash with 5 % NH_4OH in H_2O;

5. Wash with 20 % ACN in H_2O;
6. Elute with 1 or 2 × 25 μL 1 % TFA in 75/25 ACN/H_2O;
7. Dilute with 25 or 50 μL H_2O if necessary.

Recovery for 12 peptides tested using this protocol is summarized in Fig. 2.12. Recovery for 9 out of the 12 peptides was acceptable (>80 % recovery) on a first pass using the screening method, clearly indicating that a single SPE platform can be successfully used for method development. Minor modifications to the method, including adjustment of pre-treatment to reduce protein binding and elimination of the high-pH wash in a single case, resulted in improved recovery for the remaining three peptides. Final SPE recovery and matrix effect values from this approach are summarized in Table 2.3. The minor modifications to the screening protocol necessary for three of the peptides serve here to highlight areas for troubleshooting should peptide recovery be low using the screening protocol. In general, there are a few primary reasons for actual or apparent low peptide recovery from a biological matrix: protein binding, inadequate solubility in the final elution solvent,

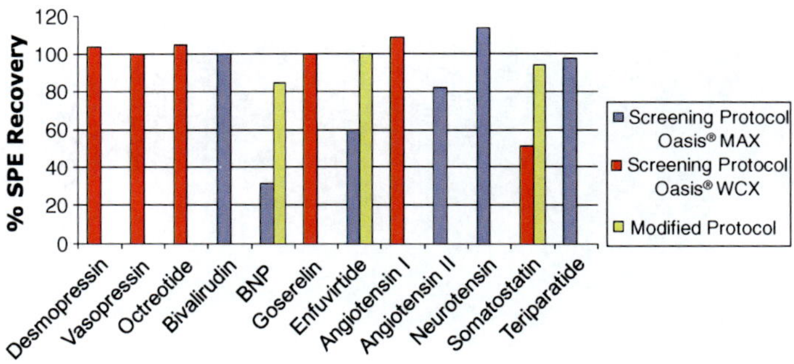

Fig. 2.12 SPE recovery for 12 therapeutic and endogenous peptides using a mixed-mode screening method optimized for peptides and with minor modifications

Table 2.3 Final SPE recovery and matrix effect values using SPE and UHPLC screening methods optimized for peptides

Peptide	% SPE recovery	% Matrix effects
Octreotide	88	<10 %
Angiotensin II	82	8 %
Desmopressin	104	<11 %
Vasopressin	100	−3 %
Goserelin	100	−2 %
Angiotensin I	109	a
Somatostatin	94	a
Neurotensin	114	6 %
Bivalirudin	100	10 %
BNP	84	a
Teriparatide	97	9 %
Enfuvirtide	102	a

[a]Not yet measured at a time of publication

chemical modification/instability (which changes the mass, rendering the original MRM incapable of quantifying the modified peptide), incomplete ionization of the peptide and sorbent during loading, non-specific binding, and insufficient solvent strength. In general, the larger and more hydrophobic a peptide is, the greater the likelihood of encountering one of these issues. In addition, one should examine the amino acid content of the target peptide and refer to Table 2.2 to identify any possible chemical modifications that could occur during processing or conditions which could cause instability.

As a first step toward improving recovery and/or determining the cause of low recovery, the following optional experiments may be performed:

1. Extract the sample in PBS + 10 μg/mL bovine serum albumin (BSA) and compare recovery to the target matrix. If recovery in the PBS solution is higher, this may indicate poor disruption of protein binding in the target matrix, indicating that a change in pre-treatment is required. If the generic acid pre-treatment is inefficient, higher concentrations of acid or base, or denaturation with guanidine HCl or urea, may be necessary.

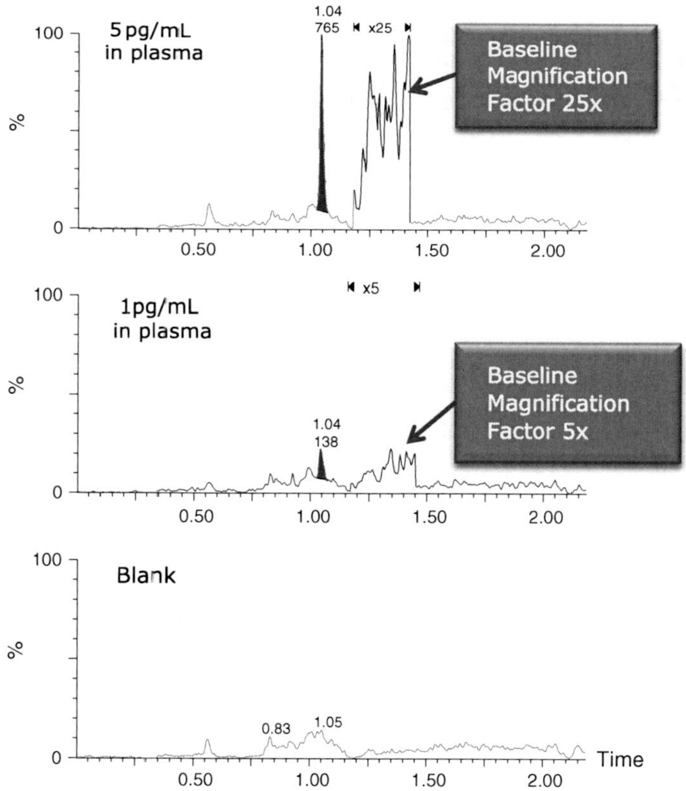

Fig. 2.13 LLOQ (between 1 and 5 pg/mL) for desmopressin extracted from human plasma using slightly modified versions of the proposed generalized UHPLC and SPE screening methods

2. Increase the concentration of TFA in the final elution to 5 or 10 %. This may improve solubility for larger or more hydrophobic peptides. Do not increase the organic % in the final elution, and this often results in precipitation. 75% acetonitrile is sufficient. Alternatively, different modifiers such as acetic acid may be assessed.
3. Exchange 10 mM ammonium acetate (pH ~6) for the NH_4OH wash in the generic protocol if using weak cation exchange. This may improve the ionization of the sorbent and peptide, facilitating complete binding upon sample loading. This modification also eliminates high-pH steps from the protocol, allowing one to accommodate base-labile peptides without loss.
4. For acid-labile basic peptides, a strong cation-exchange sorbent may be used with the ammonium acetate wash described in option 3 and the standard high-pH elution.

Korthals et al. (http://www.tno.nl/downloads/Poster_Peptides_BB1.pdf) reported a similar screening approach to that described here, based on four MM sorbents rather than the simpler two-sorbent method here.

The generic LC (as described in the chromatography section) and SPE conditions described in this chapter were used to quantify several therapeutic peptides in human plasma. Detection limits in extracted human plasma were determined for a subset of the test peptides. Blank human plasma and samples prepared at 0.001, 0.005, 0.01, 0.02, 0.05, and 0.1 ng/mL (approximately 1–100 fmol/mL) were extracted according to the generic screening method. Chromatographic results were evaluated to determine LOD and lower limit of quantification (LLOQ). In

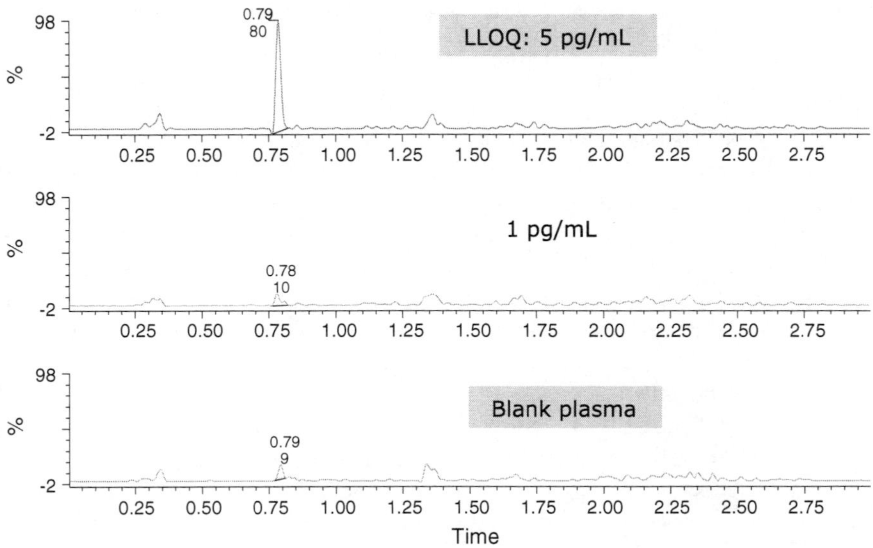

Fig. 2.14 LLOQ (5 pg/mL) for angiotensin II extracted from 350 μL of human plasma using the proposed generalized UHPLC and SPE screening methods

bioanalytical assays, the LOD is defined as the level which is three times that of an extracted blank matrix sample; similarly, LLOQ is defined as five times the level of the blank. Resulting representative chromatograms are shown in Figs. 2.13 and 2.14, demonstrating the successful application of this combination of techniques to attain detection limits in the single pg/mL range. Naturally, the exact detection limits achievable are dependent on many factors including size and hydrophobicity of the peptide, ionization and extraction efficiency, specificity of MS transition, chromatographic behavior, volume of sample used, and sensitivity of MS instrumentation employed, among others.

This platform and approach were also successfully applied in the development of a flexible, sensitive, and selective method for β-amyloid peptides, putative biomarkers for Alzheimer's disease [34]. β-amyloid peptides are considered one of the most difficult peptide classes to analyze due to their hydrophobicity, poor solubility, propensity to aggregate, high degree of non-specific binding, low circulating levels, poor MS sensitivity, and protein binding. Each aspect of peptide handling and each step of the extraction process were evaluated and optimized as per the recommendations in this chapter, ultimately yielding a method which overcame the inherent challenges faced. Lessons learned from this problematic group of peptides may be applied to other challenging peptides as well.

2.5 Alternative Techniques and Topics

The use of more advanced techniques such as 2D LC and nanoflow LC has been documented in cases where detection limits could not be reached with conventional approaches. This is typically due to ultralow levels in study samples or due to the presence of closely related endogenous and/or isobaric interferences that could not be resolved using more traditional instrumentation. Common configurations of 2D LC systems include trap and back elute, trap and forward elute, two-column approaches (RP–RP, RP–HILIC, etc.), parallel column regeneration, at column dilution and heart-cutting [49]. Trap and elute configurations enable one to load more sample at higher flow rates, focus the sample using a trapping column, and flush salts and other interferences to waste. Heart-cutting configurations maximize resolution by allowing one to take a narrow chromatographic band containing the peak of interest and "cut" it from the first column followed by loading of this greatly simplified sample onto the second column for further separation. An example of the benefit of 2D LC for a peptide separation is shown in Fig. 2.15.

A recent publication by Zhang et al. [50] details the development of a sensitive method for endogenous oxytocin, which reaches an LLOQ of 1 pg/mL in human plasma using SPE and 2D LC. RP was used in both dimensions. Oxytocin was eluted with a gradient on the first column; the peak was then heart-cut to the second column and eluted under isocratic conditions. The low flow rates used in nano-LC can provide significant improvements in MS ionization efficiency, resulting in dramatic sensitivity gains. However, nanoflow systems are often perceived as having poor robustness, requiring a very skilled operator and as being

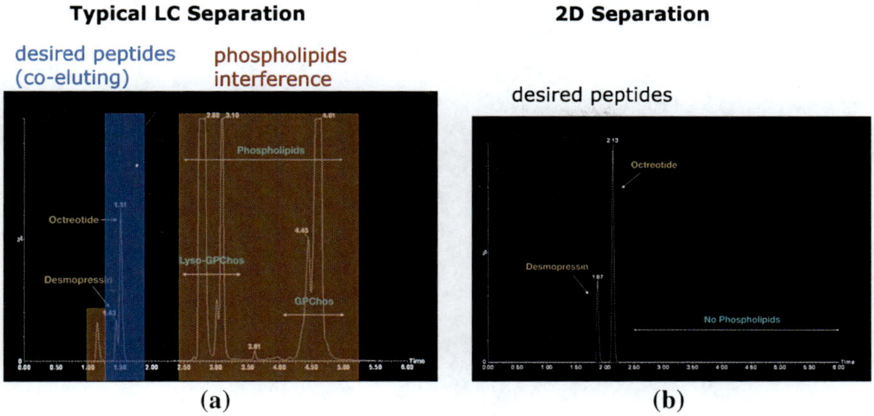

Fig. 2.15 Typical separation of 2 peptides on a 1D system **a**, the same separation following heart of the peptides onto a second column using a 2D system **b**. Data reproduced with permission from PPD Pharma

somewhat of an "art." The narrow diameter columns (typically 75–300 μm), sensitivity to integrity of connections and tubing cuts, and "finicky" nature of the spray from various tip types contribute to this perception and to the limited use of nanoflow in routine bioanalytical laboratories. In addition, the low flow rates required result in long chromatographic run times, severely restricting throughput.

Methods which include highly selective isolation and enrichment techniques based on affinity purification of peptides, such as immunoprecipitation (IP), prior to analysis can achieve even greater specificity. Li et al. [51] described such an approach for the quantitation of amyloid peptides during the 2009 AAPS meeting. Columns packed with anti-peptide antibodies have also been used to selectively enrich target peptides. This was successfully applied by Neubert et al. [52] to enrich signature tryptic peptides of pepsin/pepsinogen for protein quantitation. Though in more widespread use currently, these approaches are limited by the availability of commercial reagent kits or the internal resources needed to develop the highly specialized reagents or columns required.

Although this chapter focuses on therapeutic peptides, the extraction techniques described, and to a certain extent, the chromatography, can be applied to the more elaborate application area of protein quantitation using signature peptides once the digestion of the protein has been accomplished. Surfactants, denaturation, reduction, and alkylation reagents as well as digestion enzymes and other peptides can be removed during SPE of the digest mixture. Figure 2.16 demonstrates the potential benefit of SPE cleanup for a signature peptide from trastuzumab. Not only are many background peaks removed and the spectra simplified, but also signal intensity for the target peptide increased significantly as a result of cleanup using MM SPE. Recovery for the signature peptide was ~83 %. In this case, the extraction protocol also provided a 5X concentration of the sample, which was diluted during addition of the various reagents without evaporation.

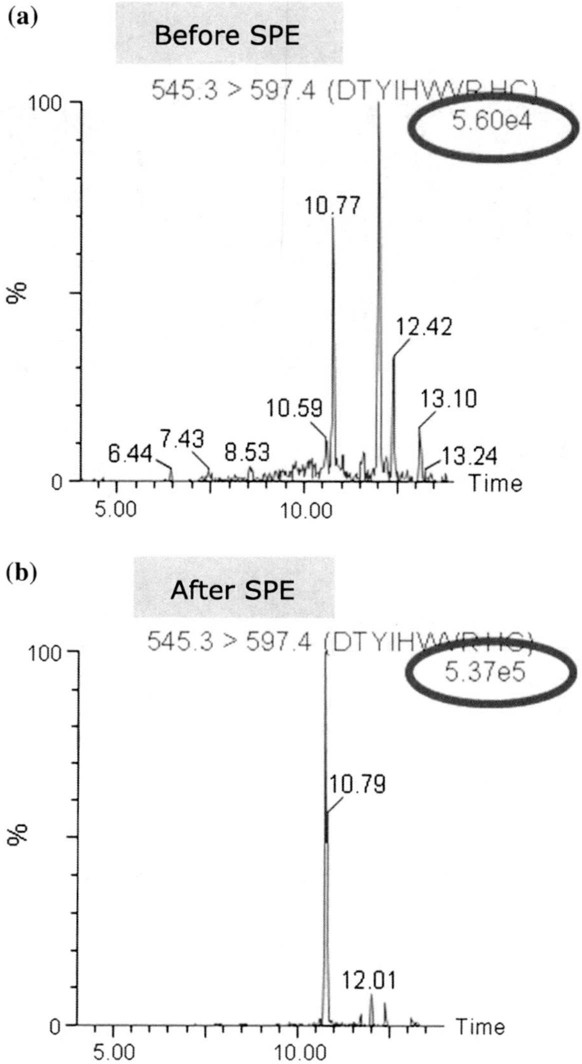

Fig. 2.16 LC/MS/MS analysis of a signature peptide from trastuzumab *before* **a** and *after* **b** cleanup of the digest mixture using strong cation-exchange SPE in reduced bed size format

2.6 Conclusions

The growing market for biotherapeutic peptides and the development of quantitative methods for those analytes have brought to light the challenges facing the analysis of this broad range of compounds. Regulatory requirements are encouraging development of methodologies that are time- and cost-effective while still producing assays that are sensitive enough to cope with biological matrices. This chapter identifies and discusses the challenges in detail, provides potential solutions, and then proposes a generally applicable platform to peptide bioanalysis method development. The generalized strategy incorporates an understanding of peptide challenges to produce methods which more readily achieve the development of robust, fast, and generally applicable assays. This was subsequently demonstrated using a relevant panel of therapeutic peptides extracted from a biologically relevant matrix. Highly targeted, specific assays can be developed individually for each of these peptides, but an approach that addresses at a diverse set of peptides serves to examine the multiple factors that need to be considered in detail for assay development. Overall, the data in this chapter combined with that of other researchers suggest that bioanalysis studies for peptide therapeutics are amenable to a platform-based approach to methods development when knowledge of peptide chemistry is carefully applied. Such standardized approaches for determining optimal SPE enrichment and MRM-based LC/MS analysis should permit companies to reduce development timelines and shorten time-to-market for peptide drugs. Where needed, advanced analytical techniques can provide the additional selectivity and/or sensitivity needed for exceedingly difficult or unique assays such as quantitation of certain endogenous biomarkers or low-level protein analysis through targeted enrichment and isolation of signature peptides.

Acknowledgments The author would like to acknowledge Dr. Guodong Chen, Dr. Thomas E. Wheat, and Kenneth J. Fountain for their invaluable help and contributions to this work.

References

1. Reichart J, Pechon P, Tartar A, Dunn M (2010) Development trends for peptide therapeutics: a comprehensive quantitative analysis of peptide therapeutics in clinical development. Summary Report by the Peptide Therapeutics Foundation
2. Vlieghe P, Lisowski V, Martinez J, Khrestcharisky M (2010) Synthetic therapeutic peptides: science and market. Drug Discov Today 15:40–56
3. Ho R, Gibaldi M (2003) Biotechnology and biopharmaceuticals. Wiley, New Jersey
4. Edwards C, Cohen M, Bloom S (1999) Peptides as drugs. QJM 92:1–4
5. Thayer A (2011) Improving peptides. Chem Eng News 89:13–20
6. Werle M, Bernkop-Schnürch A (2006) Strategies to improve plasma half life time of peptide and protein drugs. Amino Acids 30:351–367. doi:10.1007/s00726-005-0289-3
7. Nowatzke W, Rogers K, Wells E, Bowsher R, Ray C, Unger S (2011) Unique challenges of providing bioanalytical support for biological therapeutic pharmacokinetic programs. Bioanalysis 3:509–521
8. FDA (2001) Guidance for industry: bioanalytical method validation. In: CVM Ca (ed) http://www.fda.gov/downloads/Drugs/.../Guidances/ucm070107.pdf
9. Viswanathan CT, Bansal S, Booth B, DeStefano A, Rose M, Sailstad J, Shah V, Skelly JP, Swann P, Weiner R (2007) Workshop/conference report-quantitative bioanalytical methods validation and implementation: best practices for chromatographic and ligand binding assays. AAPS J 9(4):E30–E42
10. Fast DM, Kelley M, Viswanathan CT, O'Shaughnessy J, King P, Chaudhary A, Weiner R, DeStefano A, Tang D (2009) Workshop report and follow-up- AAPS workshop on current topics in GLP bioanalysis: assay reproducibility for incurred samples- implications of crystal city recommendations. AAPS J 11:238. doi:10.1208/s12248-009-9100-9
11. Van den Broek I, Sparidans R, Schellens J, Beijnen J (2008) Quantitative bioanalysis of peptides by liquid chromatography coupled to (tandem) mass spectrometry. J Chromatogr B 872:1–22

12. Chambers E, Wagrowski-Diehl D, Lu Z, Mazzeo J (2007) Systematic and comprehensive strategy for reducing matrix effects in LC/MS/MS analyses. J Chromatogr B 852:22–34
13. Apffel A, Fischer S, Goldberg G, Goodley PC, Kuhlmann FE (1995) Enhanced sensitivity for peptide mapping with electrospray liquid chromatography-mass spectrometry in the presence of signal suppression due to trifluoroacetic acid-containing mobile phases. J Chromatogr A 712:177–190. doi:10.1016/0021-9673(95)00175-m
14. Garcia MC (2005) The effect of the mobile phase additives on sensitivity in the analysis or peptides and proteins by high-performance liquid chromatography-electrospray mass spectrometry. J Chromatogr B 825:111–123
15. Ewles M, Goodwin L (2011) Bioanalytical approaches to analyzing peptides and proteins by LC-MS/MS. Bioanalysis 3:1379–1397
16. Giorgianni F, Capiello A, Beranova-Giorgianni S, Palma P, Trufelli H, Desiderio D (2004) LC-MS/MS Analysis of peptides with methanol as organic modifier: improved limits of detection. Anal Chem 76:7028–7038
17. Keppel T, Jacques M, Weis D (2010) The use of Acetone as a substitute for acetonitrile in analysis of peptides by liquid chromatography/electrospray ionization mass spectrometry. Rapid Commun Mass Spectrom 24:6–10
18. Zhu B, Mant C, Hodges R (1992) Mixed-mode hydrophilic interaction chromatography rivals reversed-phase liquid chromatography for the separation of peptides. J Chromatogr 594:75–86
19. Yoshida T (2004) Peptide Separation by hydrophilic interaction chromatography: a review. J Biochem Biophys Methods 60:265–280
20. Zhou W, Wang PG, Krynitsky AJ, Rader JI (2011) Rapid and simultaneous determination of hexapeptides (Ac-EEMQRR-amide and H2 N-EEMQRR-amide) in anti-wrinkle cosmetics by hydrophilic interaction liquid chromatography–solid phase extraction preparation and hydrophilic interaction liquid chromatography with tandem mass spectrometry. J Chromatogr A 1218:7956–7963. doi:10.1016/j.chroma.2011.08.091
21. Zhan Y, Chen X, Zhong D (2011) Quantifying Tetrapeptide SS-20 in rat plasma using hydrophilic interaction liquid chromatography coupled with electrospray ionization mass spectrometry. J Chromatogr B 879:3353–3359
22. de Bruijn P, Sleifer S, Lam L, Weimar E, Loos W (2010) Bioanalytical method for the quantification of sunitinib and its n-desethyl metabolite SU12662 in human plasma by ultra performance liquid chromatography/tandem triple-quadrupole mass spectrometry. J Pharm Biomed Anal 51:934–941
23. Denooz R, Mercerolle M, Lachatre G, Charlier C (2010) Ultra-performance liquid chromatography-tandem mass spectrometry method for the determination of bupropion and its main metabolites in human whole blood. J Anal Toxicol 35:280–286
24. Discenza L, D'Arienzo C, Olah T, Jemal M (2010) LC-MS/MS method using unbonded silica column and aqueous/methanol mobile phase for the simultaneous quantification of a drug candidate and co-administered metformin in rat plasma. J Chromatogr B 878:1583–1589
25. Gilar M et al (2005) Two-dimensional separation of peptides using RP-RP-HPLC system with different pH in first and second separation dimensions. J Sep Sci 28:1694–1703
26. Jemal M, Ouyang Z, Xia Y (2010) Systematic LC-MS/MS bioanalytical method development that incorporates plasma phospholipids risk avoidance, usage of incurred sample and well thought-out chromatography. Biomed Chromatogr 24:2–19
27. Kay R, Gregory B, Grace P, Pleasance S (2007) The application of ultra-performance liquid chromatography/tandem mass spectrometry to the detection and quantitation of apolipoproteins in human serum. Rapid Commun Mass Spectrom 21:2585–2593
28. Kay RG, Barton C, Velloso C, Brown P, Bartlett C, Blazevich A, Godfrey R, Goldspink G, Rees R, Ball G, Cowan D, Harridge S, Roberts J, Teale P, Creaser C (2009) High-throughput ultra-high-performance liquid chromatography/tandem mass spectrometry quantitation of insulin-like growth factor-I and leucine-rich alpha-2-glycoprotein in serum as biomarkers of recombinant human growth hormone administration. Rapid Commun Mass Spectrom 23:3173–3182

29. Mitulovic G et al (2009) Preventing carryover of peptides and proteins in nano LC-MS separations. Anal Chem 81:5955–5960
30. Pedraglio S, Rozio M, Misiano P, Reali V, Dondio G, Bigogno C (2007) New perspectives in bio-analytical techniques for preclinical characterization of a drug candidate: UPLC-MS/MS in in vitro metabolism and pharmacokinetic studies. J Pharm Biomed Anal 44:665–673
31. Plumb R, Rainville P, Potts W, Johnson K, Gika E, Wilson I (2009) Application of ultra performance liquid chromatography-mass spectrometry to profiling rat and dog bile. J Proteome Res 8:2495–2500
32. Shen J, Wang H, Tadros S, Hayes R (2006) Orthogonal extraction/chromatography and UPLC, two powerful new techniques for bioanalytical quantitation of desloratadine and 3-hydroxydesloratadine at 25 pg/mL. J Pharm Biomed Anal 40:689–706
33. Gilar M, Daly A, Kele M, Neue U, Gebler J (2004) Implications of column peak capacity on the separation of complex peptide mixtures in single- and two-dimensional high-performance liquid chromatography. J Chromatogr A 1061:183–192
34. Lame M, Chambers E, Blatnik M (2011) Quantitation of Amyloid Beta Peptides Aβ1-38, Aβ1-40, and Aβ1-42 in human CSF by ultra-performance liquid chromatography tandem mass spectrometry. Anal Biochem 419:133–139
35. Kinter M, Sherman NE (2000) Protein sequencing and identification using tandem mass spectrometry. Wiley, USA
36. Hunter C, Basa L Detecting a Peptide Biomarker for Hypertension in Plasma. Applied Biosystems Proteomics Technology Note. http://www3.appliedbiosystems.com/cms/groups/psm_marketing/documents/generaldocuments/cms_042573.pdf
37. Cole RB, Harrata AK (1993) Solvent effect on analyte charge state, signal intensity, and stability in negative ion electrospray mass spectrometry; implications for the mechanism of negative ion formation. J Am Soc Mass Spectrom 4:546–556. doi:10.1016/1044-0305(93)85016-q
38. John BF (1993) Ion formation from charged droplets: roles of geometry, energy, and time. J Am Soc Mass Spectrom 4:524–535. doi:10.1016/1044-0305(93)85014-o
39. Kebarle P, Tang L (1993) From ions in solution to ions in the gas phase—the mechanism of electrospray mass spectrometry. Anal Chem 65:972A–986A. doi:10.1021/ac00070a001
40. Wang G, Cole RB (1995) Mechanistic interpretation of the dependence of charge state distributions on analyte concentrations in electrospray ionization mass spectrometry. Anal Chem 67:2892–2900. doi:10.1021/ac00113a025
41. Granvogl B, Plöscher M, Eichacker L (2007) Sample preparation by in-gel digestion for mass spectrometry-based proteomics. Anal Bioanal Chem 389:991–1002. doi:10.1007/s00216-007-1451-4
42. Rappsilber J, Mann M, Ishihama Y (2007) Protocol for micro-purification, enrichment, pre-fractionation and storage of peptides for proteomics using StageTips. Nat Protoc 2:1896–1906
43. Speicher K, Harper S, Speicher D (2000) Systematic analysis of peptide recoveries from in-gel digestions for protein identifications in proteome studies. J Biomol Tech 11:74–86
44. Reubsaet JLE, Beijnen JH, Bult A, van Maanen RJ, Marchal JAD, Underberg WJM (1998) Analytical techniques used to study the degradation of proteins and peptides: chemical instability. J Pharm Biomed Anal 17:955–978. doi:10.1016/s0731-7085(98)00063-6
45. Reubsaet JLE, Beijnen JH, Bult A, van Maanen RJ, Marchal JAD, Underberg WJM (1998) Analytical techniques used to study the degradation of proteins and peptides: physical instability. J Pharm Biomed Anal 17:979–984. doi:10.1016/s0731-7085(98)00064-8
46. Kehler J, Bowen C, Boram S, Evans C (2010) Application of DBS for quantitative assessment of the peptide Exendin-4; comparison of plasma and DBS method by UHPLC-MS/MS. Bioanalysis 2:1461–1468
47. Issaq HJ, Xiao Z, Veenstra TD (2007) Chem Rev 107(8):3601–3620. doi:10.1021/cr068287r
48. Calderón-Santiago M, Mata-Granados JM, Priego-Capote F, Quesada-Gómez JM, de Castro MDL (2011) Analytical platform for verification and quantitation of target peptides in human serum: application to cathelicidin. Anal Biochem 415:39–45. doi:10.1016/j.ab.2011.04.005
49. Vooght-Johnson D (2010) The waters bioanalysis world tour. Bioanalysis 2:1931–1942

50. Zhang G, Zhang Y, Fast DM, Lin Z, Steenwyk R (2011) Ultra sensitive quantitation of endogenous oxytocin in rat and human plasma using a two-dimensional liquid chromatography–tandem mass spectrometry assay. Anal Biochem 416:45–52. doi:10.1016/j.ab.2011.04.041
51. Li W, Subbarao S, Neubert H, Johnson T, Lin J, Fountain S, Kuang B (2009) Development of a sensitive, high-throughput immunoaffinity-LC-Tandem mass spectrometry method for the quantification of Amyloid ß Peptides (Aβ) in Biological Matrices. American Association of Pharmaceutical Scientists, US
52. Neubert H, Gale J, Muirhead D (2010) Online high-flow peptide immunoaffinity enrichment and nanoflow LC-MS/MS: assay development for total salivary pepsin/pepsinogen. Clin Chem 56:1413–1423

Chapter 3
The Development and Implementation of LC/MS-Based Bioanalytical Methods for the Quantification of Protein Therapeutics in Drug Discovery

Timothy V. Olah, Asoka Ranasinghe, Hongwei Zhang, Richard L. Wong, John Mehl, Dieter M. Drexler, James Smalley, Steven Wu, Bogdan Sleczka, Yongxin Zhu, Yulia Benitex, Eric Shields and Baomin Xin

3.1 Introduction

The search to discover and develop viable therapies for the treatment of diseases continues to branch out in new directions and to improve and incorporate more efficient strategies to identify drug molecules in a cost-effective manner. Although proven treatments such as optimized small molecule drugs continue to provide an effective means for the management of certain medical conditions, alternatives such as engineered protein constructs have also been successful as therapeutic agents for treatment of a variety of diseases. Regardless of the type of drug molecule under consideration, optimized strategies and high-quality quantitative bioanalytical methods must be developed and applied throughout the drug discovery and development process in order to inform critical decisions during the selection and characterization of drug candidates.

For over 20 years, liquid chromatography coupled to mass spectrometry detection (LC/MS) has been the cornerstone technology in the pharmaceutical industry for definitive, quantitative analysis of small molecule drugs, drug metabolites, and corresponding endogenous entities in biological samples. The success of LC/MS as the predominant technology in the quantitative bioanalytical process for small molecule drugs stems from its ability to detect multiple analytes with high specificity and sensitivity within a single sample. This platform has been readily and widely applied throughout pharmaceutical research. As technology and instrumentation improves and novel and innovative research is conducted to

T. V. Olah (✉) · A. Ranasinghe · H. Zhang · R. L. Wong · J. Mehl · J. Smalley · S. Wu · B. Sleczka · Y. Zhu · Y. Benitex · E. Shields · B. Xin
Bristol-Myers Squibb, Drop code F12-04, Route 206 and Province Line Road, Princeton, NJ 08543, USA
e-mail: timothy.olah@bms.com

D. M. Drexler
Bristol-Myers Squibb, Drop code 4BC-308, 5 Research Parkway, Wallingford, CT 06492, USA

G. Chen (ed.), *Characterization of Protein Therapeutics using Mass Spectrometry*,
DOI: 10.1007/978-1-4419-7862-2_3,

better understand the mechanisms of diseases, LC/MS will continue to play an important role in this endeavor. The growing interest and expanded use of LC/MS-based methods for the quantitative analysis of proteins in early discovery stems from the experiences gained in small molecule bioanalysis. Some of the same advantages garnered in the analysis of small molecules: rapid development of methods on existing platforms, demonstration of wider linear dynamic ranges for assays, and discrimination of multiple target analytes on a molecular level have firmly positioned LC/MS as a widely applicable technology platform for protein bioanalysis and a viable complimentary technique to established ligand binding assays (LBA). Nevertheless, there are also significant challenges in isolating, concentrating, detecting, and quantifying specific proteins by LC/MS in the presence of an overwhelming abundance of endogenous proteins found in biological matrices.

In this chapter, we will describe the current development and refinement of LC/MS-based multiple-component bioanalytical methods to support the discovery and development of protein therapeutics in the pharmaceutical industry today, including a number of the integrated technical procedures used to develop a robust bioanalytical method capable of accurately and precisely measuring levels of different proteins. These steps in the bioanalytical process include the selection of targeted or surrogate peptides from in silico analysis, LC/MS parameter optimization for the detection of multiple peptides, sample preparation to isolate and/or enrich analyte proteins, optimization of proteolytic digestion conditions, and data processing and reporting on multiple analytes/peptides. LC/MS methods based upon the simultaneous detection of multiple peptides offer several advantages. These include the ability to obtain specific molecular information on protein modifications resulting from metabolism, assessment of the protein stability in biological matrices and the capability to simultaneously measure multiple proteins (target and therapeutic) in the same sample. Although still early in the application development phase, LC/MS has already begun to demonstrate utility as a viable bioanalytical technique for quantification of proteins in discovery laboratories.

3.2 Intact Protein Quantitation

The development of LC/MS-based methods for intact protein quantification, whether using triple-quadrupole mass spectrometers or high-resolution mass spectrometers, remains challenging, and there are only a few published examples of this application for small proteins [1–3]. Ji et al. [3] have developed an LC/MS/MS method for quantitative determination of rK5, a protein drug candidate with molecular weight of 10,464 Dalton and achieved a lower limit of quantitation (LLOQ) at 99.2 ng/mL in monkey plasma samples. In this work, a solid-phase extraction (SPE) method was developed to isolate and purify rK5 from endogenous proteins. For intact protein quantification, it is often critical to use trifluoroacetic acid as the mobile-phase modifier and to maintain a high column temperature between 40 °C and 60 °C to ensure good chromatographic peak shape.

There are several challenges for intact protein quantification using LC/MS. One major obstacle is the reduced sensitivity caused by significant signal splitting as the typical protein ionizes to form multiple "molecular" ions with different charge states. Given the relatively low abundance, sample extraction to isolate the protein of interest from endogenous proteins in the biological matrix can be challenging. In addition, linear dynamic ranges for ionized proteins are much narrower due to ionization competition at higher concentrations of the analyte of interest. Trifluoroacetic acid is often used as a mobile-phase modifier to improve peak shape but it can also result in significant suppression of mass spectrometric response. Therefore, quantification at the peptide level by LC/MS/MS is a more commonly used approach for protein bioanalysis.

3.3 Protein Quantitation Through Peptides

3.3.1 Selection of Surrogate Peptides

Currently, the most common approach to quantify a therapeutic protein using mass spectrometry involves digesting the protein into constituent peptide components which are then detected by either high-resolution or triple-quadrupole mass spectrometers. The formation of peptides after enzymatic digestion can be predicted in silico using a variety of commercially available software. The peptide sequence uniqueness compared to other potential interferences from endogenous proteins should be evaluated by a bioinformatics tool, such as BLAST (Basic Local Alignment Search Tool which compares primary biological sequence information, such as the amino acid sequences of different proteins) and further verified experimentally by the analysis of digested samples of neat protein and blank matrices. The digestion mixture of a neat protein, which contains multiple peptides, is used for multiple-component LC/MS-based method development for the protein.

In the case of monoclonal antibodies (mAbs), a major class of protein therapeutics, the peptides selected for bioanalysis preferably contain portions of the complementarily determining region (CDR) which is responsible for binding affinity and capacity of the mAb. Since this region of the mAb has been engineered to be unique from native proteins, there is a greater chance that peptides from the CDR region will be non-native and, therefore, less prone to interference from endogenous peptides. Peptides located in the heavy chain and the light chain of the mAb can also be a source for targeted analysis, as long as the selected peptides are unique and meet assay performance criteria. In addition, reproducible digestion efficiency is one of the critical elements of a rugged analytical method. To assess the progress of digestion, an equal molar amount of stable isotope–labeled (SIL) surrogate peptide can be added to the digestion mixture. Equivalent responses between the labeled and unlabeled analogs indicate complete digestion. To evaluate digestion efficiency, one can compare the post-spiked peptide standard curve at the corresponding protein concentration with the peptide standard curve generated after trypsin digestion of pre-spiked protein. If the two curves match well, then the digestion efficiency and reproducibility are high [4, 5].

The surrogate peptide is selected according to the following criteria [4, 6–8]: (a) It is a unique sequence in the intended species and its LC/MS response is not observed in endogenous components of the biological fluid; (b) It does not contain chemically reactive residues such as Met, Trp, or Cys; (c) It preferably does not contain unstable sequence such as Asn-Gly (to avoid N-deamidation) or Asp-Gly (to avoid isomerization to form isoaspartate-Gly); (d) It has a sequence length of 6–20 amino acids, preferably between 8 and 15, to facilitate MS ionization and to achieve sufficient chromatographic retention and resolution from other peptides; (e) When trypsin is used for digestion, the surrogate peptide does not contain a continuous sequence of arginine (RR), lysine (KK), alternating (RK) or (KR) sequences, or when the arginine or lysine is immediately followed by a proline (RP or KP) on the carboxyl side.

3.3.2 Assessment of Peptides Generated Through Enzymatic Digestion

As stated previously, LC/MS-based quantitative analysis of proteins relies on the detection and measurement of smaller peptides generated by enzymatic digestion of the target analyte. Among the proteases that are used for protein digestion for quantitative bioanalysis, trypsin is the enzyme of choice due to its relatively high cleavage specificity, its ability to digest insoluble substrates, and its relatively low cost that permits the use of higher amounts per experiment. Trypsin specifically cleaves the carboxyl side of lysine and arginine residues of peptides. The rate of digestion will be slowed if the arginine or lysine is followed by an acidic residue.

Depending on the nature of the protein, it may be possible to digest it directly in the biological matrix, for example, if the protein is already unfolded. However, the protein may need to be first chemically denatured using urea or guanidine HCl, reduced with dithiothreitol (DTT) or tris (2-carboxyethyl) phosphine (TCEP), and alkylated with iodoacetamide or iodoacetic acid prior to digestion. The optimal pH range for trypsin digestion is 7.0–9.0. Most published procedures indicate that trypsin digestion is usually conducted at 37 °C in a water bath overnight (~16 h), although there is a growing trend to develop faster conditions, such as using microwaves or elevated temperatures, to reduce incubation times. The digestion conditions for specific proteins are optimized by varying the enzyme–protein ratio, incubation time, and components of digestion buffer. To increase the trypsin digestion rate, the use of a mixed organic-aqueous solvent system, the addition of MS compatible surfactants to digestion buffer, and performing the digestion with excess amount of trypsin at elevated temperatures have been applied. These refinements have increased throughput significantly without loss of assay integrity [9–12]. The results are more consistent, and minimal digestion-induced deamidation or N-terminal glutamine cyclization products have been observed with more rapid digestion methods [13, 14]. The cleavage will not take place if a proline residue is on the carboxylic side of the cleavage site [15].

For current mass spectrometric analysis, peptides ranging from 6 to 20 amino acids are desired. Their physicochemical properties are amenable to mass spectrometric detection in terms of specific detection, linear response over a wide concentration range, and good reverse-phased chromatographic properties. Based on the typical frequency of occurrence of arginine and lysine in protein, the typical length of tryptic peptides are usually within the desired range to obtain sufficient signal response. When the size of peptides becomes larger, additional charges are needed to bring the mass-to-charge (m/z) ratio down to the range of current mass spectrometers. Some types of mass spectrometers, such as Time-of-flight (TOF), have a high upper mass range, which make it possible to detect intact proteins and larger peptides with higher m/z ratios. However, the higher the charge states of the analyte, the wider the spread of the charge distribution envelope. This makes it more difficult to achieve good quantitation limits, since the charged species selected for detection are only a fraction of the total charge population of the analyte of interest.

3.3.3 LC/Mass Spectrometric Characterization of Peptides: Charge States

Unlike small molecules, electrospray ionization (ESI) of peptides and proteins generates multiple charge states and displays various charge distribution envelopes. This phenomenon offers both advantages and disadvantages in the characterization and quantitation of proteins and peptides. Multiple charged states produce ions at a lower m/z range which are in the mass range of a variety of mass spectrometers to detect these entities. However, the effect of multiple charges increases the complexity of the mass spectra by overlapping an abundance of ions generated by a multitude of molecules into a narrow m/z window. Additionally, the charge state distribution envelope increases the population of ion species detected in the mass spectra, thus decreasing resolution and specificity. Moreover, the charge state distribution envelope spreads the ion signal across multiple ion species and decreases the detection limit. High-resolution accurate mass spectrometers can resolve these highly complex mass spectra and assign peptide identification based on their accurate masses.

Aside from solvent and instrument parameters, the sample matrix can also have a noticeable effect on charge state distribution. Matrix components compete for charges with analytes of interest causing the charge state distribution to downshift. This effect decreases the relative abundance of higher charge state species, while increasing the relative abundance of the lower charge state species. Consequently, different charge state species can yield different quantitation results when the matrices are different: for example, between the standard curve and the test samples; between samples collected from different animals; between samples collected at different time points, etc. In addition, analyte peptides can compete for charges from themselves and downshift when present in higher concentrations. This causes the standard calibration curve to bend downward for higher charge state species and bend upward for lower charge state species.

After peptides are generated through enzymatic digestion, LC/MS analysis should be performed to confirm the presence of peptides and to gauge their relative response by comparing signal strengths obtained from different MS scan types. A good choice of instrumentation for this step is a high-resolution accurate mass spectrometer. Full-scan high-resolution mass spectra (FS-HRMS) provide a clear pattern of charge states generated from multiple charged peptides and can be directly compared to the exact mass of the peptides predicted from in silico digestion. The exact mass of each charge state is calculated and data are mined for a match within a narrow mass tolerance range.

The mobile-phase pH, solvent makeup, and electrospray MS parameters will have an effect on charge state distribution. In the investigation of the effect of pH on the ionization of a motilin analog, a 22 amino acid polypeptide, revealed that higher charged ions are more abundantly produced (4+ , 5+ compared to 3+ ions) at lower pH [16]. In case of beta-endorphin (MW = 3466 Dalton with five lysine residues), the electrospray voltage was found to have no effect on the relative charge state distribution. However, when the declustering potential (DP) was increased, a shift from higher to lower charge states was observed through the loss of protons from the high charge state ions. In addition, the extent of deprotonation could also be manipulated by using solvents of varying proton affinity [17].

Triply and doubly charged species of peptides are the predominate ions that are usually observed following trypsin digestion of therapeutic proteins. Triply charged precursor ions might possess higher fragmentation efficiencies and, thus, may provide higher detection sensitivity. The evaluation of charge state distribution and enhanced formation of favorable charge state precursor ions should be a part of the optimization strategy for LC/MS/MS-based method development of peptides. Recently in our laboratory, a sensitive assay for a peptide-GLP-1 agonist was required. By optimizing the DP, a triply charged precursor ion with higher fragmentation efficiency was selected and a sensitive assay was developed based upon its detection [18].

3.3.4 Fragmentation Patterns of Surrogate Peptides; HRMS

Collision Induced Dissociation (CID) fragmentation patterns of surrogate peptides can also be readily obtained using HRMS. Additional injections using varying instrument parameters, different voltage settings in the source, and in the collision energy in a HCD (Higher-energy collisional dissociation) trap in an Orbitrap mass spectrometer will yield different responses. Programming experiments with low and high collision energy potential will provide MS/MS data that can be used to determine which product ions are increasing in intensity and can be used to improve assay sensitivity.

The CID of a peptide usually occurs by cleavage at amide bonds to generate "y" or "b" ions, if the COOH- or NH_2- terminal fragment retains the charge, respectively. When basic residues, especially Arg, are protonated, cleavage occurs selectively at the C-terminus of Asp or Glu. If the number of protons exceeds the number of Arg residues, cleavage occurs at the N-terminal side of Pro. The doubly charged precursor ion usually produces an abundance of "y" ions and weak "b" ions. The intensity of doubly charged "y" ions generated from a triply charged precursor may be higher than generated from the doubly charged precursor [19]. These product ions can be used for detection on a triple quadrupole under selected reaction monitoring (SRM) to transfer and optimize the LC/MS/MS method from the HRMS to a triple quadrupole. In some cases, LC/MS using exact mass can be utilized for quantification. However, when sensitivity is required, the triple-quadruple LC/MS/MS will most likely provide the more sensitive method for quantification.

3.3.5 HRMS in Protein/Peptide Quantitation: Emerging Trends

There is a growing interest in using HRMS for the quantitation of new chemical entities (NCE) in the industry because of its potential to provide additional qualitative information on untargeted endogenous analytes that are present in biological matrices [20–23]. High mass-resolving power of modern mass spectrometers (viz. TOF, Orbitrap) combined with superior chromatographic separation in ultra-high-pressure liquid chromatography (UHPLC) systems can provide the resolution and specificity required for more complex bioanalysis [24–27]. Although conventional SRM approaches provide high specificity and sensitivity, they also require some initial method development. This slows the process especially early in the discovery process when thousands of compounds are screened to select candidates for further development. The SRM work flow is less efficient due to the optimization of multiple ion source parameters and CID conditions required to select SRM transitions for each analyte of interest. SRM strategies will not detect untargeted compounds such as metabolites, matrix interferences, and potential endogenous biomarkers. HRMS full-scan data sets provide comprehensive details of each sample analyzed and can be probed further using post-acquisition software methods such as mass defect filter (MDF) to identify new analytes for more detailed evaluations [28–30]. These simultaneous quantitative and qualitative approaches are proving to be attractive options in drug discovery research, when screening for potential liabilities at early stages of the programs is cost-effective. Compared to small molecules, proteins present additional bioanalytical challenges due to the complexity of the matrix resulting from the post-translational modifications (PTM) and proteolytic cleavage of the therapeutic protein by peptidases. As described in the following section, HRMS offers valuable features for resolving some of these bioanalytical challenges [1, 20, 31–33].

Protein quantification using HRMS falls into two categories: direct quantification of therapeutic peptide/proteins (molecular weight typically < 20,000 Dalton) and the analysis of surrogate peptides derived from enzymatically digested proteins. Sample preparation and purification steps prior to LC/MS analysis can be quite different for peptides compared to that of proteins. However, the detection of both types of molecules using mass spectrometry presents similar bioanalytical challenges. Both types of molecules generally form multiply charged ions under ESI conditions which result in a "dilution effect" on the overall response of the molecules in the ion source, directly affecting the assay sensitivity. Narrow spacing between the multiply charged molecular ion clusters also may cause isotopic interferences particularly when conventional SRM-based methods are employed using low-resolution mass spectrometers. Furthermore, peptides typically produce multiple fragment ions during CID that result in further signal dilution and decreases in assay sensitivity because typically only one SRM transition, most often the one with the highest signal, is used for the quantification. The use of extracted ion chromatograms (XIC) from HRMS full-scan mass spectra for peptide quantification can overcome some of these limitations.

In certain cases, combining the response from several XICs from multiply charged isotopic clusters can be used to improve the detection limits of HRMS assays. This is illustrated in the processing of the TOF full-scan mass spectrum of an Endothelin-3 (MW = 2643 Dalton) standard (4,000 nM) in rat plasma, acquired using quadrupole TOF (QTOF) under positive ESI mode, 5-Hz acquisition rate, and a resolving power of 20 K (Fig. 3.1). The spectrum identifies ions with 3+ and 4+ charges, respectively, at *m/z* 881.714 and 661.539. The assay

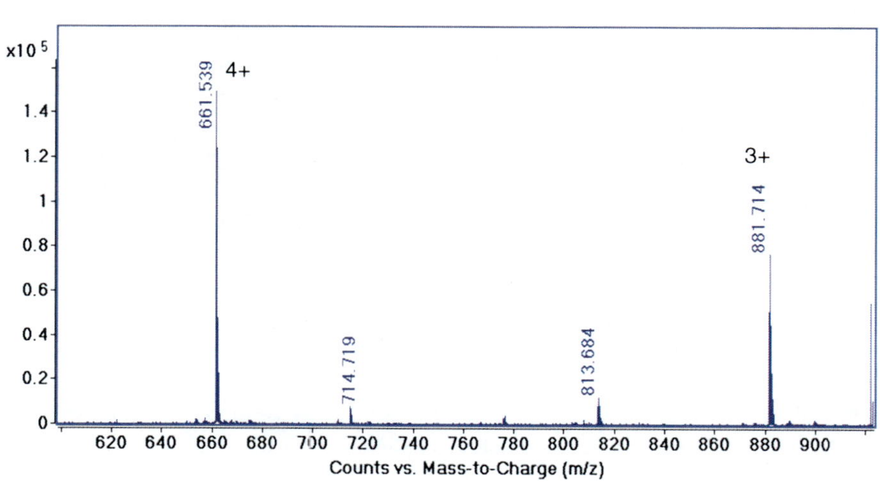

Fig. 3.1 UHPLC-Q-TOF full-scan mass spectrum of Endothelin-3 standard (4000 nM) in rat plasma obtained 5-Hz data acquisition rate and a resolving power of 20 K (Reproduced from Ref. [20] with permission of Future Science Ltd)

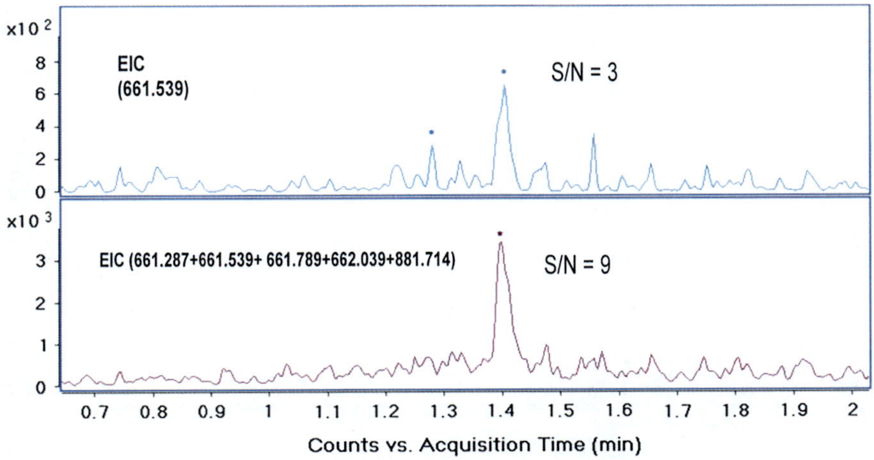

Fig. 3.2 Comparison of extracted ion chromatograms for endothelin-3 (20 nM) obtained with a single charge state and summing several charge states. (Reproduced from Ref. [20] with permission of Future Science Ltd)

LLOQ was 20 nM using a single XIC from the most abundant isotope (4+ charge, *m/z* 661.539). Summing XICs, in a defined Mass Extraction Window (MEW 10 millidaltons), from several ions improved the S/N three times of that obtained using single XIC (upper trace), providing the assay LLOQ of 7 nM (Fig. 3.2). However, this approach may not be universally applicable, especially when isotopic peaks are masked by matrix and chemical noise.

Common modifications such as deamidation may cause isotopic interferences particularly under SRM conditions at unit mass resolution. Each deamidation step changes the molecular weight of the parent peptide by only one Dalton, hence, the narrow spacing between the multiply charged molecular ion clusters results in multiple peaks in the specific SRM channel used for the quantitation. This is demonstrated in a recent rat PK study involving a proprietary compound, Peptide-1 (MW = 2615 Dalton). A UHPLC-QTOF–based quantitative method based upon XIC's from the full-scan mass spectra of corresponding MS peaks at 30 K resolving power was subsequently developed to support this PK study. In addition, the full-scan HRMS data unequivocally confirmed the presence of several deamidated metabolites as shown in the Fig. 3.3. The response of Metabolite-1 and Metabolite-2 increased with time at the expense of the parent (no metabolites were detected in the pre-dose samples) and resulted in metabolic interferences in the triple-quadrupole MS/MS-based SRM channels. This resulted in difficulties estimating dose-related responses since these metabolites may also contribute to the pharmacology. Because of the small changes in the molecular structure of the metabolites, LBA would have not likely distinguished metabolites from drug and measured the total combined concentration of the components.

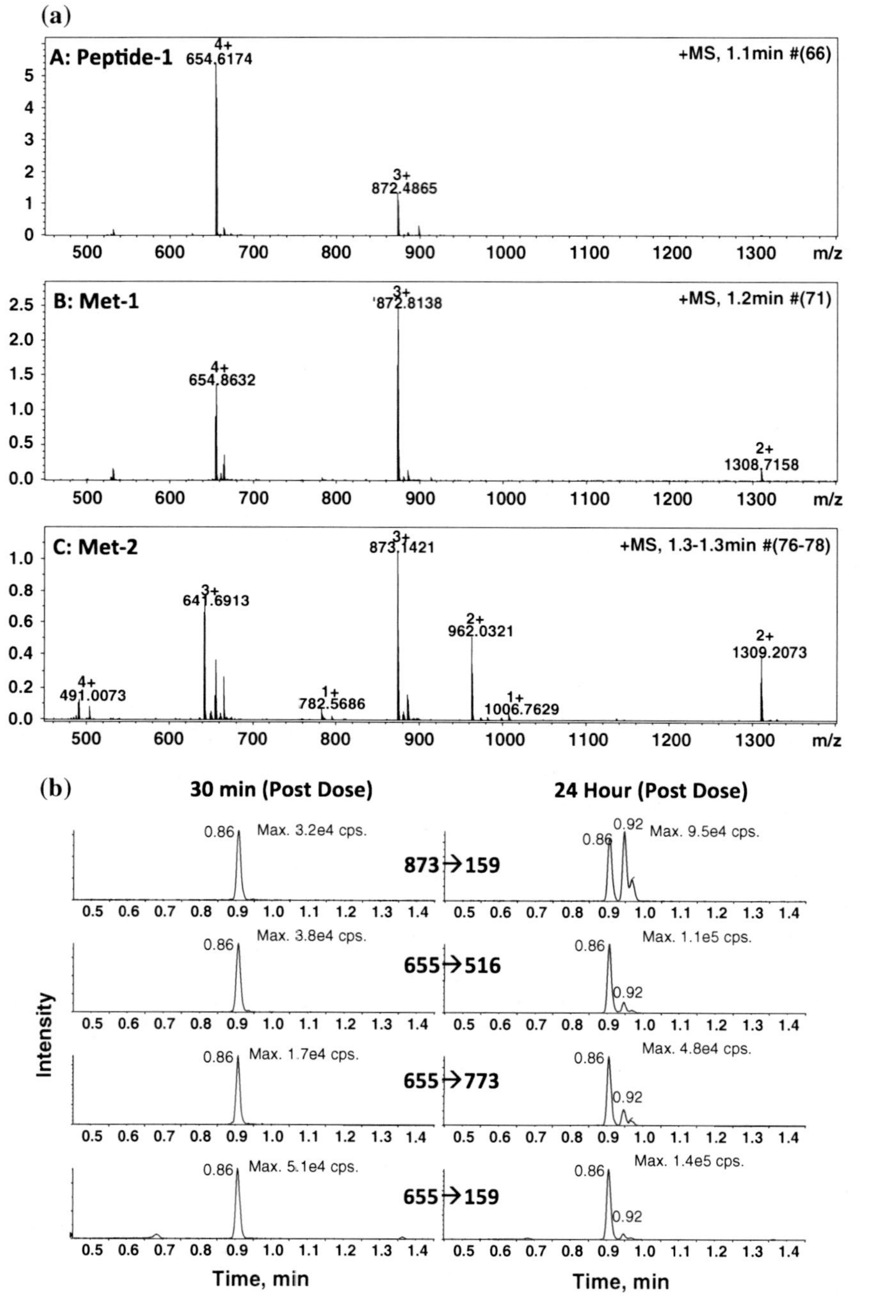

Fig. 3.3 **a** Mass spectra of Peptide-1, Met-1 and Met-2; Met-1 and Met-2 are formed by deamidation of one and two side-chain amide residues and caused the addition of one (Met-1) and two (Met-2) Da to Peptide-1's *m/z* value. **b** Comparison of UPLC/MS/MS (MRM) chromatograms of multiply charged ions (3+ and 4+) at 30 min and 24 h time points, obtained from rats dosed with Peptide-1(Reproduced from Ref. [20] with permission of Future Science Ltd)

3.3.6 High-Resolution Quantitation Using MRM

High-resolution accurate mass spectrometers such as QTOF can be operated in high-resolution MRM mode to enhance selectivity and reduce MRM background noise to improve overall S/N (signal-to-noise) ratios. MRM-HR mode offers another added level of selectivity compared with unit-resolution MRM by triple-quadrupole mass spectrometers, because product ions are detected with high-resolution and high-mass accuracy. MRM-HR data are obtained following post-acquisition data processing from MS/MS analysis. Full-scan MS/MS data are acquired at high speed and high resolution (>30, 000). High-resolution XIC of specific product ions at a mass window of 0.02 Dalton is generated post-acquisition and used for quantification. Fast acquisition ensures that sufficient data points are obtained across the chromatographic peak. The other advantages of using MRM-HR are that full-scan MS/MS data are collected throughout the analysis and target peptides can be confirmed by MS/MS sequencing. Alternative product ions can also be selected for quantitation when interferences are encountered in real biological samples or for the purpose of extending the linear dynamic range of the assay. Multiple MRM XICs can also be summed to improve S/N and data accuracy and precision.

We have observed that MRM-HR offers significant advantages for biologics quantitation. Sample preparation involving direct trypsin digestion of plasma/serum samples or on the protein pellet following precipitation of the plasma/serum sample is a commonly used approach for the initial analysis of proteins in early discovery. However, high-abundance endogenous proteins such as albumin, transferrin, immunoglobulin IgG, and IgA also generate high levels of peptides during trypsin digestion. It is not straightforward to remove high levels of endogenous tryptic peptides from the selected surrogate peptides by commonly used sample preparation methods, such as SPE, and they will also co-elute on HPLC or UHPLC columns. This could significantly compromise LLOQs due to high SRM background noise, co-eluting interference peaks, and ion suppression matrix effects. Operation at MRM-HR can reduce the background noise and eliminate interference when compared with SRM in unit resolution by triple-quadrupole mass spectrometers. This type of detection is becoming a viable alternative mass spectrometric platform for biologics quantitation.

3.4 Sample Preparation Techniques

Plasma (or serum) is a complex sample matrix containing a wide variety of soluble proteins that span a dynamic range of several orders in magnitude [34]. The complexity of this matrix presents a challenge for quantification of low-level protein and peptide biomarkers and therapeutics. The large matrix background not only can interfere with selective detection of analyte but can also perturb ionization in the form of variable signal suppression or enhancement. The physicochemical properties of the protein of interest will also determine how biological samples are to be prepared for LC/MS-based analysis. Since the physicochemical properties vary significantly among different proteins, there is no "one-size fits-all" method that can be employed effectively across the wide range of proteins encountered in discovery laboratories. Therefore, several different approaches, which have been explored and developed in authors' laboratories for a variety of protein therapeutics, are briefly described below.

3.4.1 Digestion of the Evaporation Residue of Supernatant After Precipitation with Water-Miscible Organic Solvents [4]

This approach has been developed and implemented for the analysis of therapeutic PEGylated proteins. Routinely, plasma samples were pipetted into a 96-well plate and to each sample, an analog protein (Internal Standard: IS) solution and a volume of formic acid in isopropyl alcohol were added for protein precipitation (PPT). The sample was mixed and centrifuged. The supernatant was transferred to a clean 96-well plate and evaporated to dryness. The residue was reconstituted with digestion buffer, and the trypsin enzyme reagent was added. The mixture was incubated at 37 °C overnight and then quenched with an acid. The sample was vortex-mixed and centrifuged. An aliquot was injected into column for LC/MS/MS analysis.

3.4.2 Direct Digestion of Plasma Samples

Plasma samples were thermally denatured, reduced, and alkylated prior to trypsin digestion, when necessary. To plasma samples in a 96-well plate, digestion buffer was added. The mixture was heated at 90 °C for denaturation, followed by cooling in an ice bath. Samples prone to gelling, like cyno plasma, were allowed to cool in room temperature while agitated on an Eppendorf Thermomixer. Reduction and alkylation procedures were then performed. The protein IS solution and trypsin dissolved in digestion buffer were added. The sample was heated at 60 °C for 1 h. Digestion was stopped by the addition of acid. After the addition of acetonitrile, the sample was mixed and centrifuged and an aliquot of supernatant was injected for LC/MS-based analysis.

3.4.3 Digestion of Protein Pellet Following Precipitation

To plasma samples pipetted into a 96-well plate, methanol was added, the plate vortex-mixed and centrifuged. The supernatant was removed and the pellet was resuspended with the addition of a digestion buffer solution (200 mM ammonium bicarbonate in 10 % methanol and 90 % water), and the IS was added. After vortex mixing, trypsin was added, and the sample was heated at 60 °C for 1 h. The digested plate was centrifuged to remove particulate matter, and an aliquot of the supernatant was injected for analysis [35].

3.4.4 Digestion of the Supernatant Following Precipitation

To plasma samples in a 96-well plate, methanol was added. The mixture was vortex-mixed, IS was added, and the samples were centrifuged. The supernatant was transferred to a microtube, and ammonium bicarbonate and trypsin were added. The mixture was heated at 60 °C with vortex mixing for 1 h. The mixture was cooled to room temperature, and quenched with 0.1 % formic acid. After centrifugation, an aliquot of supernatant was injected into column for analysis.

3.4.5 Direct Digestion of Proteins from Dried Blood Spot Samples

The sample preparation described by Sleczka et al. [36] may be suitable for analysis of proteins in small volumes of blood that have been collected and dried on

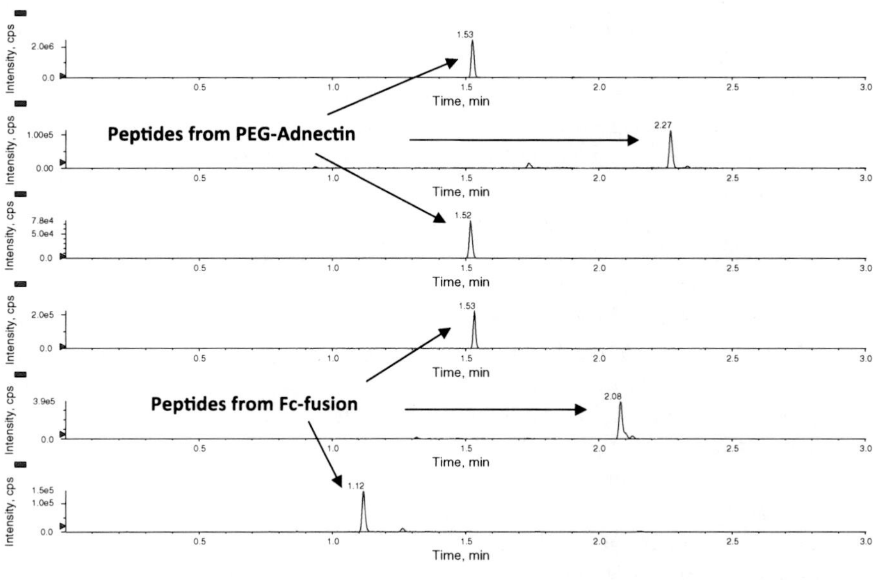

Fig. 3.4 LC-MS/MS chromatograms of surrogate peptides obtained from a digest of a dried blood spot standard containing both PEGylated Adnectin and FC-fusion protein

filter paper. 5 mm "punches" representing approximately 8 μl of whole blood were directly subjected to a basic trypsin digestion procedure and aliquots were analyzed by LC/MS-based methods that were programed to detect multiple specific tryptic surrogate peptides that were generated from two therapeutic proteins (Fig. 3.4). Although the sensitivity of these initial DBS-LC/MS methods was less than those obtained by standard LC/MS methods for the quantification of each protein separately in plasma and serum, it is analogous to what has been observed for small molecules. This is due, in part, to the smaller sampling volume eventually injected and analyzed by LC–MS and compromises made in procedures to accommodate the analysis of multiple analytes. Nonetheless, this strategy has the potential to be readily and more widely applied to a variety of studies quantifying specific proteins in blood or other biological matrices spotted on filter paper.

3.4.6 Immunoprecipitation (Immunocapture)

To address the removal of the significant background level of proteins that constitute serum and plasma, immunoprecipitation (IP) sample preparation combined with mass spectrometry has seen a growing number of reported applications. IP, or immunocapture (IC), utilizes a reagent antibody to selectively capture the protein or peptide analyte of interest from the biological sample. Following the immunoprecipitation step, unbound matrix components are washed away, and then the analyte is released or digested directly on-bead. In the case of protein analytes, digestion can be utilized prior to analysis by LC/MS/MS. Ackerman and Berna [37] published an excellent review article describing IP-MS techniques for low-abundance protein biomarkers. The authors describe two IP strategies that utilize capture antibodies directed against intact protein (anti-protein strategy), or directed against tryptic peptides (anti-peptide strategy). The later strategy has been described by Anderson [38–40] and is referred to as SISCAPA (stable isotope standards and capture by anti-peptide antibodies). The SISCAPA strategy requires antibodies to be raised against tryptic peptide epitopes, whereas, the intact protein strategy utilizes antibody reagents with immunoaffinity for the intact protein, many of which are commercially available.

IP-MS can dramatically reduce the plasma matrix background and has lead to reported LLOQs ranging from 4 to 130 pM utilizing the intact anti-protein strategy [41, 42] and LLOQs ranging from 0.1 to 59 pM utilizing the anti-peptide strategy with nanoflow HPLC [43]. Achievable LLOQs are dependent upon the analyte and starting plasma volume. LOQs in the low pM range require 1 mL of starting sample volume, whereas mid-pM LOQs can be achieved using 10–100 uL of starting plasma volume. The choice of strategy (anti-protein or anti-peptide) employed is primarily driven by the availability of IP reagents. Antibody reagents against protein and peptide biomarkers, as well as, therapeutics of interest are becoming increasingly available and can be employed for IP-MS method development. Anti-tryptic peptide antibodies need to first be generated for the tryptic peptide of interest.

Analyte recovery can be an important consideration in deciding what strategy to pursue. Ciccimaro et al. [44] developed an IP-MS method for determination of the absolute level of phosphorylation within a peptide region of a protein kinase. The authors determined that IP recovery was dependent upon phosphorylation state whereby the phosphorylated form had lower recovery than the unphosphorylated form. By using full-length isotopically labeled protein as the internal standard, the authors were able to compensate for differential recoveries during the isolation procedure. Berna et al. [45] demonstrated that when using SIL peptides added post-digestion as standards that a correction for the IP and digestion recovery of the protein should be used for accurate absolute quantification.

A variety of IP-MS work flows have been described. Magnetic beads coupled to capture antibody have been utilized by several groups and can be fully automated using magnetic sample manipulation equipment [43, 46]. Non-magnetic beads made from agarose or polystyrene have also been employed using centrifugation and filtration devices for sample processing. Online IP-MS can be employed when the analyte of interest is a peptide and digestion is not required [47]. 96-well ELISA format capture plates have also been utilized, and take advantage of widely available plate-washing equipment commonly found in ligand binding assay laboratories [48]. Microwave-assisted digestion has also been used to reduce the digestion time and enable sample processing to be completed within one day.

Gain in sensitivity is one advantage of using IP for sample preparation prior to LC/MS/MS. Improvement in selectivity is another advantage that IP can offer over less selective sample preparation methods, for example, PPT, prior to LC/MS/MS. Xu et al. [49] developed a selective IP-MS whole molecule assay for a PEGylated drug candidate. Despite being less sensitive than a PPT method, IP-MS utilizing anti-PEG antibody for capture provided greater selectivity for the intact drug compared with the PPT method. In other examples [40, 41], sensitive ELISA methods were in place for biomarkers; however, the ELISA methods could not discriminate between multiple forms of the analyte (fragmented and intact forms). IP-MS provides increased selectivity and can enable the detection of specific forms of an analyte of interest with high sensitivity.

3.5 UHPLC Optimization to Improve Assay Performance

Chromatography conditions are also an important step in the bioanalytical process and need to be optimized in order to obtain good peak shape, retention, and resolution from endogenous tryptic peptides [18]. The resolution from endogenous tryptic peptides is critical to reduce matrix effect and SRM background noise and to avoid interferences to improve overall S/N ratios. UHPLC with high column efficiency has been leveraged to improve sensitivity, selectivity, and the speed of analysis for surrogate peptides. With UHPLC, sharper chromatography peaks with increased peak heights are obtained due to smaller column particle sizes, higher column efficiency, and higher optimal flow rates can be applied. With small particle columns (sub-2 μm), the peak width at about 1 s full width at half maximum (FWHM) can be obtained, resulting in the increased peak heights. In addition, the analytes of interest can be better resolved from endogenous tryptic peptides and are, therefore, less prone to be subjected to matrix effects. As a result, the overall S/N ratio is increased. For surrogate peptides, between 6 and 21 amino acids, mobile phases composed of water and acetonitrile containing 0.1 % formic acid often generate good peak shape. This mobile-phase composition is mass spectrometry compatible and also helps to achieve high ionization efficiencies. Intact proteins on the other hand often require trifluoroacetic acid as mobile-phase modifier to achieve better peak shapes, but are subjected to ion suppression due to its strong acidic nature. As a result, mass spectrometry response is significantly suppressed. UHPLC column with a dimension of 2.1 × 50 mm are often used and operated under reasonable column pressures and run times. The column temperature is maintained between 40 °C and 60 °C to relieve column pressure, improve peak shape, separation, and speed of analysis. Slower gradients may be necessary, especially if samples are prepared by non-selective preparation methods, such as the direct trypsin digestion of plasma or serum samples, to allow better separation from endogenous tryptic peptides, minimize matrix effects, and improve the LLOQs. Two-step gradients are often run to allow slow gradient and reasonable total run time.

3.6 Selection of Internal Standards

The role of an Internal Standard (IS) is to compensate for potential analyte loss during sample preparation and variance during sample analysis as a result of matrix effects and ionization efficiencies (suppression/enhancement) [50]. Preferably the IS has similar physicochemical properties as the analyte, thus SIL analogs of the analyte are typically employed. In protein analysis, the efficiency of analyte isolation and enzymatic digestion during sample preparation is crucial for the analytical figures of merit of the surrogate peptide such as sensitivity, accuracy, precision, and reproducibility. Consequently, it would be advantageous to add an IS prior to any sample handling at the protein or peptide level either as a SIL analog or a non-labeled homolog. The SIL-IS is synthesized using amino acids containing stable isotopes (^{18}O, ^{13}C, ^{2}H or ^{15}N) which are then incorporated into the peptide or protein affording a defined mass difference which is easily distinguished by LC–MS. Most commonly, ^{13}C or ^{15}N are used as stable isotopes as ^{2}H causes a slight shift in retention time in reversed-phase HPLC, with the heavy form eluting slightly earlier than the light form [51].

3.7 Measurements of Free Versus Total Drug Concentrations

In immunoassay methodologies, it is generally assumed that the antigen captures the active or free target drug, while the LC/MS-based method, derived using the current surrogate peptide strategy, cannot distinguish peptides measured from the cleavage of intact protein or modified/different entities (i.e., the peptide can be generated from intact target protein, from bound target protein, or the partially degraded protein). The general notion is that the ligand binding assay measures the free form of target protein while the LC/MS method measures the total amount of protein in the sample. However, these assumptions may not always reflect reality. In a capture assay, therapeutic antibodies present in a bound form in the sample may dissociate during dilution and incubation with the capture reagent. Therefore, it is likely that the ligand binding assay may measure a mixture of bound and free form (in addition to molecular variants like post-translation modifications (PTM) and degradants that retain the relevant epitopes for binding) and it is highly dependent on assay parameters [52].

Additionally, the presence of Anti-Drug Antibodies (ADA), generated in vivo following the administration of a protein therapeutic, may also have an adverse effect on ligand binding assays. Recently in the authors' laboratories, a single subcutaneous administration of a PEGylated scaffold protein therapeutic to cynomolgus monkeys was conducted for a pharmacokinetic evaluation. The plasma samples were analyzed in parallel by established ELISA and LC/MS/MS-based methods. The drug concentration data obtained by the ELISA and LC–MS/MS methods are in excellent agreement at the initial time points through 96 h. However, the LC/MS/MS method generated significantly higher values than the ELISA method at the later time points. The result was investigated and the origin of the discrepancy was verified as due to the formation of ADA at the later time points in vivo. The last three time-point samples from monkey #1 were further quantified by western blot analysis. The comparison of drug concentration results determined by quantitative western blot analysis and LC/MS/MS revealed that after dissociation of protein-ADA complex, the released protein measured by western blot matches the MS data. The presence of ADA interfered with the drug's ability to bind to the target antigen that was used in the ELISA method [53]. Therefore, a clear understanding of the nature of a bioanalytical technique is vital in developing appropriate methods for measuring protein drugs. Free and total drug measurements by LBA have also been reviewed [54].

3.8 Summary

The expanding interest and escalating use of LC–MS-based methods for the quantitative analysis of proteins in early drug discovery research is a consequence of applying the collected experiences gained in small molecule bioanalysis over the past 20 years. The same advantages are noted for the analysis of small molecules and proteins: rapid development of methods on proven instrument platforms, demonstration of linear dynamic ranges of assays, and the provision of specific molecular details for multiple target analytes have established LC/MS as a viable orthogonal technique to LBA in protein laboratories. It is well acknowledged that additional development in each of the steps in the bioanalytical process must continue if analysts are to fully realize the potential of this application. However, as we continue to gain knowledge and experience in the use of proteins as viable therapeutic agents and targets, LC/MS will continue to demonstrate its versatility in the pharmaceutical industry.

References

1. Ruan Q, Ji QC, Arnold ME, Humphreys WG, Zhu M (2011) Strategy and its implications of protein bioanalysis utilizing high-resolution mass spectrometric detection of intact protein. Anal Chem 83:8937–8944
2. Ji QC, Gage EM, Rodila R, Chang MS, El-Shourbagy TA (2003) Method development for the concentration determination of a protein in human plasma utilizing 96-well solid-phase extraction and liquid chromatography/tandem mass spectrometric detection. Rapid Commun Mass Spectrom 17:794–799
3. Ji QC, Rodila R, Gage EM, El-Shourbagy TA (2003) A strategy of plasma protein quantitation by selective reaction monitoring of an intact protein. Anal Chem 75:7008–7014
4. Wu ST, Ouyang Z, Olah TV, Jemal M (2011) A strategy for liquid chromatography/tandem mass spectrometry based quantitation of pegylated protein drugs in plasma using plasma protein precipitation with water-miscible organic solvents and subsequent trypsin digestion to generate surrogate peptides for detection. Rapid Commun Mass Spectrom 25:281–290
5. Keshishian H, Addona T, Burgess M, Kuhn E, Carr SA (2007) Quantitative, multiplexed assays for low abundance proteins in plasma by targeted mass spectrometry and stable isotope dilution. Mol Cell Proteomics 6:2212–2229
6. Hagman C, Ricke D, Ewert S, Bek S, Falchetto R, Bitsch F (2008) Absolute quantification of monoclonal antibodies in biofluids by liquid chromatography-tandem mass spectrometry. Anal Chem 80:1290–1296
7. Kamiie J, Ohtsuki S, Iwase R, Ohmine K, Katsukura Y, Yanai K, Sekine Y, Uchida Y, Ito S, Terasaki T (2008) Quantitative atlas of membrane transporter proteins: development and application of a highly sensitive simultaneous LC/MS/MS method combined with novel in silico peptide selection criteria. Pharm Res 25:1469–1483
8. Han B, Copeland M, Geiser AG, Hale LV, Harvey A, Ma YL, Powers CS, Sato M, You J, Hale JE (2007) Development of a highly sensitive, high-throughput, mass spectrometry-based assay for rat procollagen type-I N-terminal propeptide (PINP) to measure bone formation activity. J Proteome Res 6:4218–4229
9. Strader MB, Tabb DL, Hervey WJ, Pan C, Hurst GB (2006) Efficient and specific trypsin digestion of microgram to nanogram quantities of proteins in organic-aqueous solvent systems. Anal Chem 78:125–134
10. Hervey WJt, Strader MB, Hurst GB (2007) Comparison of digestion protocols for microgram quantities of enriched protein samples. J Proteome Res 6:3054–3061
11. Russell WK, Park ZY, Russell DH (2001) Proteolysis in mixed organic-aqueous solvent systems: applications for peptide mass mapping using mass spectrometry. Anal Chem 73:2682–2685
12. Chen EI, Cociorva D, Norris JL, Yates JR 3rd (2007) Optimization of mass spectrometry-compatible surfactants for shotgun proteomics. J Proteome Res 6:2529–2538
13. Arsene CG, Ohlendorf R, Burkitt W, Pritchard C, Henrion A, O'Connor G, Bunk DM, Guttler B (2008) Protein quantification by isotope dilution mass spectrometry of proteolytic fragments: cleavage rate and accuracy. Anal Chem 80:4154–4160
14. Ren D, Pipes GD, Liu D, Shih LY, Nichols AC, Treuheit MJ, Brems DN, Bondarenko PV (2009) An improved trypsin digestion method minimizes digestion-induced modifications on proteins. Anal Biochem 392:12–21

15. Schuchert-Shi A, Hauser PC (2009) Peptic and tryptic digestion of peptides and proteins monitored by capillary electrophoresis with contactless conductivity detection. Anal Biochem 387:202–207
16. Yamaguchi K, Takashima M, Uchimura T, Kobayashi S (2000) Development of a sensitive liquid chromatography-electrospray ionization mass spectrometry method for the measurement of KW-5139 in rat plasma. Biomed Chromatogr 14:77–81
17. Hewavitharana AK, Herath HM, Shaw PN, Cabot PJ, Kebarle P (2010) Effect of solvent and electrospray mass spectrometer parameters on the charge state distribution of peptides—a case study using liquid chromatography/mass spectrometry method development for beta-endorphin assay. Rapid Commun Mass Spectrom 24:3510–3514
18. Zhang H, Xin B, Caporuscio C, Olah TV (2011) Bioanalytical strategies for developing highly sensitive liquid chromatography/tandem mass spectrometry based methods for the peptide GLP-1 agonists in support of discovery PK/PD studies. Rapid Commun Mass Spectrom 25:3427–3435
19. Wysocki VH, Resing KA, Zhang Q, Cheng G (2005) Mass spectrometry of peptides and proteins. Methods 35:211–222
20. Ranasinghe A, Ramanathan R, Jemal M, D'Arienzo CJ, Humphreys WG, Olah TV (2012) Integrated quantitative and qualitative workflow for in vivo bioanalytical support in drug discovery using hybrid Q-TOF-MS. Bioanalysis 4:511–528
21. Bateman KP, Kellmann M, Muenster H, Papp R, Taylor L (2009) Quantitative-qualitative data acquisition using a benchtop Orbitrap mass spectrometer. J Am Soc Mass Spectrom 20:1441–1450
22. O'Connor D, Mortishire-Smith R, Morrison D, Davies A, Dominguez M (2006) Ultra-performance liquid chromatography coupled to time-of-flight mass spectrometry for robust, high-throughput quantitative analysis of an automated metabolic stability assay, with simultaneous determination of metabolic data. Rapid Commun Mass Spectrom 20:851–857
23. Rousu T, Herttuainen J, Tolonen A (2010) Comparison of triple quadrupole, hybrid linear ion trap triple quadrupole, time-of-flight and LTQ-Orbitrap mass spectrometers in drug discovery phase metabolite screening and identification in vitro–amitriptyline and verapamil as model compounds. Rapid Commun Mass Spectrom 24:939–957
24. Ramanathan R, Jemal M, Ramagiri S, Xia YQ, Humpreys WG, Olah T, Korfmacher WA (2011) It is time for a paradigm shift in drug discovery bioanalysis: from SRM to HRMS. J Mass Spectrom 46:595–601
25. Goodenough AK (2011) High resolution mass spectrometry approaches for the quantification of proteins using stable isotopically labeled peptide. Am Pharm Rev 14:21–32
26. Wong RL, Xin B, Olah T (2011) Optimization of Exactive Orbitrap acquisition parameters for quantitative bioanalysis. Bioanalysis 3:863–871
27. Dillen L, Cools W, Vereyken L, Lorreyne W, Huybrechts T, de Vries R, Ghobarah H, Cuyckens F (2012) Comparison of triple quadrupole and high-resolution TOF-MS for quantification of peptides. Bioanalysis 4:565–579
28. Zhu M, Ma L, Zhang D, Ray K, Zhao W, Humphreys WG, Skiles G, Sanders M, Zhang H (2006) Detection and characterization of metabolites in biological matrices using mass defect filtering of liquid chromatography/high resolution mass spectrometry data. Drug Metab Dispos 34:1722–1733
29. Plumb RS, Johnson KA, Rainville P, Smith BW, Wilson ID, Castro-Perez JM, Nicholson JK (2006) UPLC/MS(E); a new approach for generating molecular fragment information for biomarker structure elucidation. Rapid Commun Mass Spectrom 20:1989–1994
30. Zhang H, Zhu M, Ray KL, Ma L, Zhang D (2008) Mass defect profiles of biological matrices and the general applicability of mass defect filtering for metabolite detection. Rapid Commun Mass Spectrom 22:2082–2088
31. Ramagiri S, Garofolo F (2012) Large molecule bioanalysis using Q-TOF without predigestion and its data processing challenges. Bioanalysis 4:529–540
32. Plumb RS, Fujimoto G, Mather J, Potts WB, Rainville PD, Ellor NJ, Evans C, Kehler JR, Szapacs ME (2012) Comparison of the quantification of a therapeutic protein using nominal and accurate mass MS/MS. Bioanalysis 4:605–615

33. Cuyckens F, Dillen L, Cools W, Bockx M, Vereyken L, de Vries R, Mortishire-Smith RJ (2012) Identifying metabolite ions of peptide drugs in the presence of an in vivo matrix background. Bioanalysis 4:595–604
34. Anderson NL, Anderson NG (2002) The human plasma proteome: history, character, and diagnostic prospects. Mol Cell Proteomics 1:845–867
35. Ouyang Z, Furlong MT, Wu S, Sleczka B, Tamura J, Wang H, Suchard S, Suri A, Olah T, Tymiak A, Jemal M (2012) Pellet digestion: a simple and efficient sample preparation technique for LC-MS/MS quantification of large therapeutic proteins in plasma. Bioanalysis 4:17–28
36. Sleczka BG, D'Arienzo CJ, Tymiak AA, Olah TV (2012) Quantitation of therapeutic proteins following direct trypsin digestion of dried blood spot samples and detection by LC-MS-based bioanalytical methods in drug discovery. Bioanalysis 4:29–40
37. Ackermann BL, Berna MJ (2007) Coupling immunoaffinity techniques with MS for quantitative analysis of low-abundance protein biomarkers. Expert Rev Proteomics 4:175–186
38. Anderson NL, Anderson NG, Haines LR, Hardie DB, Olafson RW, Pearson TW (2004) Mass spectrometric quantitation of peptides and proteins using Stable Isotope Standards and Capture by Anti-Peptide Antibodies (SISCAPA). J Proteome Res 3:235–244
39. Whiteaker JR, Zhao L, Zhang HY, Feng LC, Piening BD, Anderson L, Paulovich AG (2007) Antibody-based enrichment of peptides on magnetic beads for mass-spectrometry-based quantification of serum biomarkers. Anal Biochem 362:44–54
40. Young SA, Julka S, Bartley G, Gilbert JR, Wendelburg BM, Hung SC, Anderson WH, Yokoyama WH (2009) Quantification of the sulfated cholecystokinin CCK-8 in hamster plasma using immunoprecipitation liquid chromatography-mass spectrometry/mass spectrometry. Anal Chem 81:9120–9128
41. Kumar V, Barnidge DR, Chen LS, Twentyman JM, Cradic KW, Grebe SK, Singh RJ (2010) Quantification of serum 1–84 parathyroid hormone in patients with hyperparathyroidism by immunocapture in situ digestion liquid chromatography-tandem mass spectrometry. Clin Chem 56:306–313
42. Berna M, Ott L, Engle S, Watson D, Solter P, Ackermann B (2008) Quantification of NTproBNP in rat serum using immunoprecipitation and LC/MS/MS: a biomarker of drug-induced cardiac hypertrophy. Anal Chem 80:561–566
43. Whiteaker JR, Zhao L, Anderson L, Paulovich AG (2010) An automated and multiplexed method for high throughput peptide immunoaffinity enrichment and multiple reaction monitoring mass spectrometry-based quantification of protein biomarkers. Mol Cell Proteomics 9:184–196
44. Ciccimaro E, Hanks SK, Yu KH, Blair IA (2009) Absolute quantification of phosphorylation on the kinase activation loop of cellular focal adhesion kinase by stable isotope dilution liquid chromatography/mass spectrometry. Anal Chem 81:3304–3313
45. Berna MJ, Zhen Y, Watson DE, Hale JE, Ackermann BL (2007) Strategic use of immunoprecipitation and LC/MS/MS for trace-level protein quantification: myosin light chain 1, a biomarker of cardiac necrosis. Anal Chem 79:4199–4205
46. Xu K, Liu L, Saad OM, Baudys J, Williams L, Leipold D, Shen B, Raab H, Junutula JR, Kim A, Kaur S (2011) Characterization of intact antibody-drug conjugates from plasma/serum in vivo by affinity capture capillary liquid chromatography-mass spectrometry. Anal Biochem 412:56–66
47. Berna M, Schmalz C, Duffin K, Mitchell P, Chambers M, Ackermann B (2006) Online immunoaffinity liquid chromatography/tandem mass spectrometry determination of a type II collagen peptide biomarker in rat urine: Investigation of the impact of collision-induced dissociation fluctuation on peptide quantitation. Anal Biochem 356:235–243
48. Berna M, Ackermann B (2009) Increased throughput for low-abundance protein biomarker verification by liquid chromatography/tandem mass spectrometry. Anal Chem 81:3950–3956
49. Xu Y, Mehl JT, Bakhtiar R, Woolf EJ (2010) Immunoaffinity purification using anti-PEG antibody followed by two-dimensional liquid chromatography/tandem mass spectrometry for the quantification of a PEGylated therapeutic peptide in human plasma. Anal Chem 82:6877–6886

50. Bronsema KJ, Bischoff R, van de Merbel NC (2012) Internal standards in the quantitative determination of protein biopharmaceuticals using liquid chromatography coupled to mass spectrometry. J Chromatogr B Analyt Technol Biomed Life Sci 893–894:1–14
51. Zhang R, Sioma CS, Thompson RA, Xiong L, Regnier FE (2002) Controlling deuterium isotope effects in comparative proteomics. Anal Chem 74:3662–3669
52. Ezan E, Bitsch F (2009) Critical comparison of MS and immunoassays for the bioanalysis of therapeutic antibodies. Bioanalysis 1:1375–1388
53. Wang SJ, Wu ST, Gokemeijer J, Fura A, Krishna M, Morin P, Chen G, Price K, Wang-Iverson D, Olah T, Weiner R, Tymiak A, Jemal M (2012) Attribution of the discrepancy between ELISA and LC-MS/MS assay results of a PEGylated scaffold protein in post-dose monkey plasma samples due to the presence of anti-drug antibodies. Anal Bioanal Chem 402:1229–1239
54. Yang J, Quarmby V (2011) Free versus total ligand-binding assays: points to consider in biotherapeutic drug development. Bioanalysis 3:1163–1165

Chapter 4
Post-Translationally Modified Proteins: Glycosylation and Disulfide Bond Formation

Fotini N. Bazoti and Anthony Tsarbopoulos

Abbreviations

AEC	Anion-Exchange Chromatography
BEMAD	β-Elimination followed by Michael Addition with Dithiothreitol
CE	Capillary Electrophoresis
CHO	Chinese Hamster Ovary
CHO IL-4	CHO-derived Interleukin-4
CDG	Congenital Disorders of Glycosylation
CDT	Carbohydrate-Deficient Transferrin
CID	Collision-Induced Dissociation
ConA	Concanavalin A
CFG	Consortium for Functional Glycomics
CZE	Capillary Zone Electrophoresis
CREB	cAMP-responsive Element-Binding Factor
DHB	2,5-Dihydroxybenzoic acid
ECD	Electron-Capture Dissociation
ESI	Electrospray Ionization
ETD	Electron-Transfer Dissociation
FAB	Fast Atom Bombardment
FT ICR	Fourier Transform Ion Cyclotron Resonance
GCC	Graphitized Carbon Columns
HILIC	Hydrophilic Interaction Chromatography
HPA	3-hydroxypicolinic acid
HPLC	High-Performance Liquid Chromatography
HPV	Human Papillomavirus

A. Tsarbopoulos (✉)
Department of Pharmacology, University of Athens Medical School, 75 M. Asias street, Athens 115 27, Greece
e-mail: atsarbop@med.uoa.gr; atsarbop@gnhm.gr

F. N. Bazoti
Bioanalytical Department, The Goulandris Natural History Museum, 13 Levidou street, Kifissia 145 62, Greece
e-mail: fbazoti@gmail.com

G. Chen (ed.), *Characterization of Protein Therapeutics using Mass Spectrometry*,
DOI: 10.1007/978-1-4419-7862-2_4, © Springer Science+Business Media New York 2013

HRP	Horseradish Peroxidase
IL-4	Interleukin-4
IL-4R	Interleukin-4 receptor
IL-5Rα	Interleukin-5 receptor α-subunit
IA	Immunoaffinity
IM	Ion Mobility
IT	Ion Trap
KEGG	Kyoto Encyclopedia of Genes and Genomes
LAC	Lectin Affinity Chromatography
LC	Liquid Chromatography
LC–MS	Liquid Chromatography–Mass Spectrometry
LSI	Liquid Secondary Ion
mAb	Monoclonal Antibodies
MALDI	Matrix-Assisted Laser Desorption/Ionization
M-LAC	Multi-Lectin Affinity Chromatography
M_r	Molecular Weight
MS	Mass Spectrometry
MS/MS	Tandem Mass Spectrometry
m/z	Mass-to-charge ratio
nESI	Nano-Electrospray Ionization
NMR	Nuclear Magnetic Resonance
oTOF	Orthogonal Time-of-Flight
PAS	Periodate-acid-Schiff
PD	Plasma Desorption
PGC	Porous Graphitized Carbon Chromatography
PSA	Pisum Sativum Agglutinin
PSD	Post-Source Decay
PTMs	Post-Translational Modifications
rHuEPO	Recombinant Human Erythropoietin
r-RhCG	Recombinant human Chorionic Gonadotrophin
rtPA	Recombinant Tissue Plasminogen Activator
QTOF	Quadrupole Time-of-Flight
sAPP	Secreted Amyloid Precursor Protein
sDHB	2,5-Dihydroxybenzoic acid (DHB) with a 10 % admixture of 2-hydroxy-5-methoxybenzoic acid (super DHB)
SEC	Size-Exclusion Chromatography
SIM	Selected Ion Monitoring
SLAC	Serial Affinity Chromatography
SPE	Solid-Phase Extraction
TAS	Tagging-via-Substrate
Tf	Human Transferrin Glycoprotein
TOF	Time-of-Flight
uPAR	Urokinase-type Plasminogen Activator Receptor
UPLC	Ultra-Performance Liquid Chromatography
WGA	Wheat Germ Agglutinin

4.1 Introduction

Even though most medicines have historically been small molecules, many newly approved drugs over the last two decades have been derived from proteins. For the past few years, protein therapeutics have been enjoying the fastest growth within the global pharmaceutical industry. Protein-based therapeutics, such as insulin, interferons, monoclonal antibodies (mAb), growth hormones, erythropoietins, blood-clotting factors, colony-stimulating factors (CSFs), plasminogen activators, and reproductive hormones, play a significant role in the treatment of many major diseases, and protein therapies have revolutionized the methodology followed by drugs. These therapies exhibit high efficiency due to their targeted approach, which avoids side effects on healthy organs to a great extent. In recent years, the number of protein-based pharmaceuticals reaching the marketplace has increased exponentially, and they provide innovative as well as effective therapies for several chronic diseases which were previously not responsive to treatment. The global market for biologics or biotechnology therapeutics is one of the most prolific and fastest growing markets in the world, representing at least 24 and 22 % of all new chemical entities approved by the US and EU regulatory authorities, respectively [1]. Sales of biotech products in US showed an annual growth rate of 20 % between 2001 and 2006 compared with 6–8 % in the pharmaceutical market [2], and it is expected to grow at annual growth rate of around 13 % during the next three years (2012–2015), with the introduction of new protein therapeutics and enhanced investments contributing to this booming growth of this industry.

For protein therapeutics to be effective, they must be produced in biologically active forms, which require proper folding, and post-translational modifications (PTMs) with the extent of PTMs depending on the nature of the "host" cell and the conditions of the fermentation and recovery processes. Even though only a few biopharmaceutical proteins such as albumin (Recombumin) and insulin (Humulin N and Lispro) undergo simple modifications such that they can be manufactured using yeast or bacteria [3], most of the production platforms used to produce biopharmaceuticals comprise mammalian cells that have the ability to perform complex PTMs. The most prevalent modifications include variable glycosylation, formation of disulfide bonds, cysteine (C) and methionine (M) oxidation, phosphorylation, misfolding and aggregation, deamidation of asparagine (N) and glutamine (Q), and proteolysis at the C- and N-termini. Even though the presence of PTMs is often required for normal biological function or tissue disposition of the protein, in many cases, the role of the modification is as of yet unknown. Therefore, detailed characterization of these modifications is extremely important, because they may alter physical and chemical properties, folding, conformation distribution, stability, activity, which in turn may affect cellular processes, in which the protein is involved [4–6]. Examples of the latter can be regulation of signal transduction and a wide variety of cellular events such as growth, metabolism, proliferation and differentiation in case of protein phosphorylation [7–9], targeting, cell–matrix interaction, as well as pharmacokinetic and pharmacodynamic behavior in case of glycosylation [10, 11]. Therefore, a thorough verification of the protein's (amino acid) sequence, assessment of the purity and impurities in a recombinant protein drug product along with a detailed characterization of the existing PTMs, is a regulatory requirement prior to its approval for clinical use [12].

Full structural characterization of the existing PTMs in a recombinant protein often poses a considerable analytical challenge owing to their inherent complexity. The presence of PTMs often complicates or even prevents the use of classical tools for protein sequence analysis (e.g., automated Edman degradation). Moreover, the presence of lipid or carbohydrate covalent attachments on proteins can dramatically decrease the accuracy of the molecular weight (M_r) measurement when using sedimentation velocity, gel permeation, or SDS-PAGE analysis. Separation techniques such as high-performance liquid chromatography (HPLC) or capillary electrophoresis (CE) combined with a variety of mass spectrometry (MS) techniques are commonly employed for the profiling and quantitation of PTMs present in recombinant therapeutic proteins. The development of electrospray ionization (ESI) [13, 14] MS coupled with online liquid chromatographic (LC–MS) or electrophoretic separation (CE-MS) [15, 16] and matrix-assisted laser desorption/ionization (MALDI) [17, 18] has established MS as the technology of choice for protein mapping, localization, structure identification, and quantification of existing PTMs [19, 20]. Several MS-based approaches have been developed employing tailored tandem MS scanning methods diagnostic for specific PTMs, such as monitoring precursor/product-ion transitions and neutral loss scan [21–23].

Recently, online LC–MS combined with collision-induced dissociation (CID) and electron-capture dissociation (ECD) [24] or electron-transfer dissociation (ETD) [25] fragmentation has been used to elucidate disulfide linkages and site-specific glycosylation in recombinant therapeutic proteins and glycoproteins [26, 27]. Similarly, MS-based approaches can be employed in the production of a recombinant therapeutic protein in order to ensure the purity, the production yield, and the absence of chemical degradation and/or aggregation products in the protein formulations for clinical and eventually commercial use.

In this chapter, we discuss MS-based methodologies that are employed to detect, identify, and characterize two of the most prevalent PTMs in the production of therapeutic recombinant proteins, glycosylation and disulfide bond formation. These MS-based approaches discussed here are representative of those used for the comprehensive characterization and quantitation of other PTMs encountered in recombinant proteins intended for therapeutic use in humans.

4.2 Glycosylation

Glycosylation process, that is, the covalent attachment of oligosaccharide chains on the protein backbone, is considered as the most important and common PTM of proteins. It is estimated that over 70 % of all human proteins are glycosylated [28] and 90 % of protein therapeutics are glycosylated [10]. The carbohydrate moieties of glycoproteins (glycans) can modulate the biological functions of a glycoprotein such as circulation, cell-to-cell interactions, receptor binding, molecular and immune recognition, which in turn affect intracellular signaling, fertilization, embryonic development, immune defense, recognition of hormones, cell adhesion, and pathogenicity [4]. In addition, glycan-chain modification can significantly impact their physicochemical properties such as protein folding, solubility, stability, aggregation, and susceptibility to proteolysis [29]. Finally, carbohydrate modifications can also considerably alter protein conformation, which may consequently modulate the functional activity of the protein, especially in its interactions with other proteins or ligands. It has been established that altered glycosylation or variation of a protein's glycosylation pattern is associated with numerous diseases and disorders [30–32]. Therefore, detailed structural studies of the glycosylation and its inherent heterogeneity are also potentially vital toward understanding their function in complex physiopathological processes and establishing glycan profile changes between healthy and disease states [33, 34]. The latter has increased the potential of using glycan biomarkers for the diagnosis of several diseases [35], as well as for the design of new therapeutics [10, 36, 37]. Moreover, carbohydrate modification can be used toward the production of "custom-made" glycoproteins tailored, such as glycoproteins with defined homogeneous glycosylation structure, for specific therapeutic use [38].

Therefore, complete structural analysis of a glycoprotein end product will involve not only the determination of the primary peptide sequence, but also detailed analysis of the glycan structures including information on the individual glycosylation sites, the glycosylation patterns, and the structure elucidation of the attached carbohydrates (glycoproteome) [39–43]. As it has become obvious that many of the changes associated with disease and differentiation are due to the glycans attached to proteins (glycome), a thorough understanding of these glycan structures will be invaluable for gaining insight into their involvement in disease mechanisms and the potential for novel therapeutic interventions [44]. Characterizing the glycoproteome, however, is a challenging and daunting task because the structural heterogeneity of these glycans is vast, necessitating the development of highly sensitive and efficient analytical methods for detection, separation, and structural investigation of glycoproteins.

4.2.1 Intact Glycoprotein Analysis by Mass Spectrometry

An important preliminary step in the quality control and structure characterization of a therapeutic recombinant protein is the M_r determination of the protein product. On the intact glycoprotein level, non-spectrometric techniques such as SDS-PAGE, lectin affinity chromatography (LAC), isoelectric focusing (also in a capillary), or capillary zone electrophoresis (CZE) are generally used. In case of the two-dimensional (2D) gel electrophoresis separation of glycoproteins, characteristic spots reflecting their different isoelectric points and M_r of different glycoforms can be seen. The subsequent detection of the glycosylation pattern of the electroblotted glycoproteins may be performed by LAC [45, 46], where carbohydrate-specific lectins can be used to probe distinct oligosaccharide structures (motifs). In addition, this affinity purification can be employed as an enrichment method for the glycosylated peptides and proteins (see Sect. 4.2.2.2). Nevertheless, the low solubility of the membrane glycoproteins, resulting in their poor detection, is a significant drawback of the 2D gel electrophoresis approach. An alternative method of higher resolving potential is CZE or CE, where the various glycoforms are detected even though no information on the nature of the attached glycans is revealed [47]. These electrophoretic methods have been successfully used in the separation of sialic acid isoforms of endogenous and recombinant glycoproteins, and they have proved their usefulness in clinical diagnosis and product quality assessment [48].

In the late 1980s, the incorporation of ESI and MALDI MS, along with advances in electrophoretic separations and high-resolution MS, has provided a powerful analytical tool for the analysis and even quantitation of the intact individual glycoforms in glycoproteins [15]. ESI and MALDI MS are the premier methods of choice for M_r measurement and the ensuing protein mapping. In case of ESI MS analysis of therapeutic proteins, spraying of an aqueous protein solution at μL/min or nL/min flow rates generates multiply protonated signals with reduced mass-to-charge (m/z) ratios, thus making them readily detected by typical mass analyzers with a mass range up to 2,500 Da. This is demonstrated in the ESI mass spectrum of human recombinant interferon α-2b (INTRON A) (Fig. 4.1), which is used in the treatment of certain viral infections, including chronic hepatitis B, C, and D, malignant melanoma, follicular lymphoma, Kaposi's sarcoma caused by AIDS, and infections caused by human papillomavirus (HPV). The ESI mass spectrum

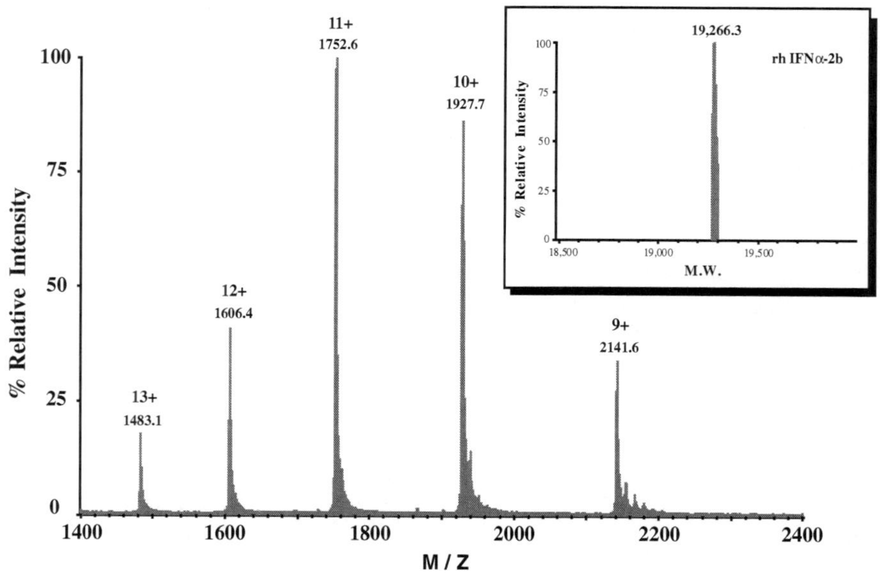

Fig. 4.1 Positive-ion ESI mass spectrum of human recombinant interferon α-2b (INTRON A)

exhibited a bell-shaped distribution of multiply charged ions ranging from the 9+ to the 13+ charge state, and the average M_r value derived from the five multiply charged ions present in the ESI mass spectrum was 19,266.3 (Fig. 4.1, inset). The excellent mass measurement accuracy, which is usually better than 0.01 % for masses up to 100 kDa [49], makes ESI MS an ideal preliminary method for monitoring the integrity of the therapeutic recombinant protein batches. In case of larger proteins, we observe greater charge states, often in the presence of a dilute acid, due to the presence of more available sites to carry the positive charge (i.e., K-, R-, H-, N-terminus). The simultaneous shift of the observed ion envelope distribution to lower m/z values is also accompanied by a decrease in the spacing between adjacent charge states, thus making the identification of the envelope's charge-state components difficult. This becomes more complicated in the analysis of a recombinant glycoprotein sample where the inherent complexity of the carbohydrate structure heterogeneity enhances the aforementioned analytical challenge. This complexity is shown in the ESI mass spectrum of the Chinese hamster ovary (CHO)-derived interleukin-4 (IL-4), a glycoprotein containing two N-linked glycosylation sites (Fig. 4.2) [50].

The ESI mass spectrum of CHO IL-4 (Fig. 4.2a) contained three envelopes of multiply charged ions ranging from 8+ to 10+ charge state, with each envelope containing several peaks corresponding to individual glycoforms of the glycoprotein and adducts thereof. This is better depicted in the deconvoluted mass spectrum (Fig. 4.2b), with the mono- and disialylated components (separated by 291 Da) representing the most abundant signals. Other higher M_r components indicated the presence of tri- and tetraantennary glycans containing up to three additional lactosamine units (in-chain mass of 365 Da), whereas satellite signals 98 Da higher were also observed (Fig. 4.2b). These signals probably arise from the attachment of sulfate groups, since sulfate salts were used in the protein isolation process and operating at a higher desolvation potential or using low pH solvents can minimize their formation [51]. Overall, the success of glycoprotein analysis by ESI MS depends on their relative carbohydrate content, with the success decreasing significantly with a relatively high percentage weight of the carbohydrate component. ESI MS analysis of complex glycoproteins by direct infusion often results in broad unresolved signals arising from the large number of different glycoforms and potential salt adducts, along with the ESI multiple charging phenomenon that spreads the signals over a large m/z region. In agreement with that, ESI MS analysis of the 44 kDa ovalbumin containing 4 % carbohydrate was successful [52], whereas glycoproteins with higher carbohydrate content such as the CHO IL-5 (M_r ~ 31 kDa; 15 % carbohydrate) and CHO IL-4 receptor (IL-4R; M_r ~ 38 kDa with 35 % carbohydrate) did not give any ESI signals [53]. Another contribution to the unsuccessful ESI MS analysis is the poor ionization efficiency in the positive ion mode due to the presence of negatively charged glycans, as this is demonstrated in the comparative analysis of recombinant human erythropoietin (rHuEPO) and its asialo counterpart [54]. The use of nano-electrospray ionization (nESI) overcomes this drawback and improves the sensitivity of analysis due to the generation of smaller-sized droplets [55]. Moreover, the interfacing of the nESI source with orthogonal time-of-flight

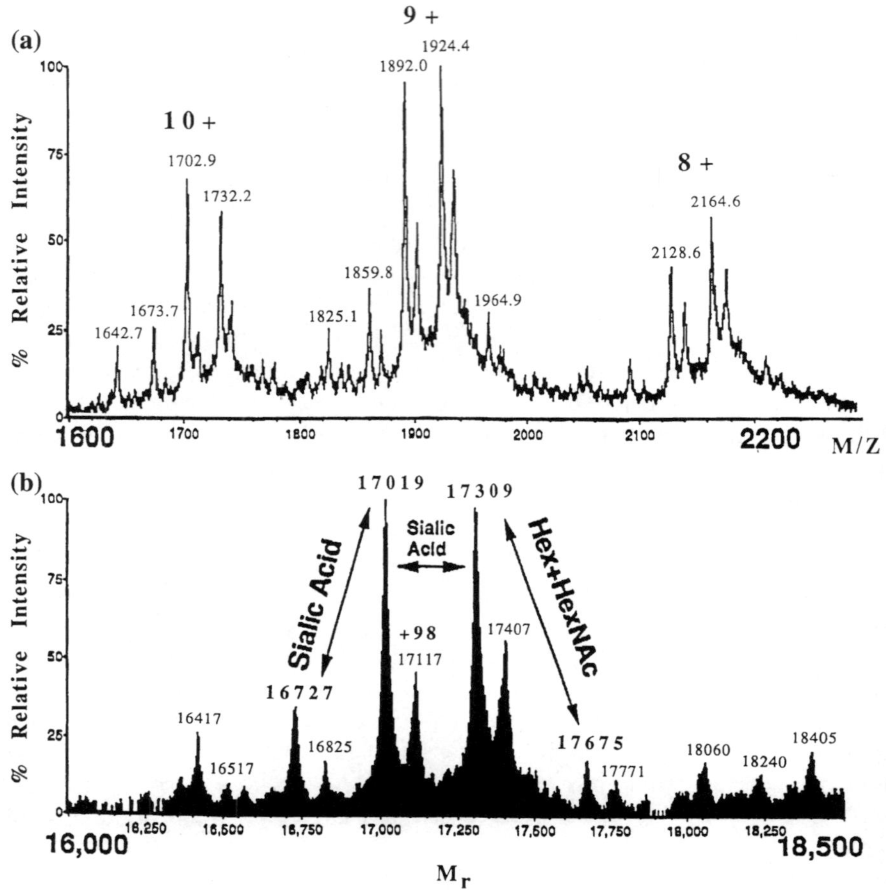

Fig. 4.2 Positive-ion ESI mass spectrum of CHO-derived interleukin-4 (IL-4): Raw spectrum (**a**) and deconvoluted spectrum (**b**) showing the individual glycoforms of the glycoprotein. (Reprinted with permission from Wiley [50])

(oTOF) instrumentation [56] has led to better mass measurement accuracy and increased analytical mass range, thus offering new momentum to the ESI MS analysis of glycoproteins. It should be emphasized that the commonly used quadrupole, quadrupole ion trap, and even Orbitrap [57] analyzers have mass range of analysis limited to m/z 2,000 and 4,000 (Orbitrap), which is a significant drawback when larger glycoproteins or non-covalent complexes thereof must be detected; thus, an upper mass limit greater than even m/z 10,000 may be required [58, 59]. This is shown in the nESI mass spectrum of Sf9-derived IL-4R (M_r ~ 30.2 kDa) obtained on an oTOF mass spectrometer, where an extensive series of multiply charged ions up to m/z 3,000 corresponding to two sets of high-mannose glycoforms separated by a fucosylated $Man_3(GlcNAc)_2$ structure (in-chain mass of 1,039 Da) were observed [51]. Therefore, the improved mass resolving power, sensitivity and extended mass range, has made the oTOF, hybrid quadrupole TOF (QTOF), and recently the ion mobility (IM) [60] TOF as the analyzers of choice for nESI MS analysis of glycoproteins. The use of the IM TOF analyzer is nicely shown in the nESI mass spectrum of the intact therapeutic mAb trastuzumab (Herceptin), which is a humanized monoclonal immunoglobulin γ-1 (IgG1) antibody directed against the HER2/neu receptor, which is over-expressed in about 25 % of all breast cancer patients [61]. In the ESI mass spectrum of trastuzumab obtained on an IM TOF mass spectrometer [62] (Fig. 4.3), an extensive series of multiply charged ions ranging from the 35+ up to the 75+ charge state were observed, and the separation between successive charge states was sufficient to reveal the presence of six glycoform variants. The illustration of these glycoforms is portrayed in the zoomed spectrum for the 53+ charge state (Fig. 4.3b), while their respective assignment is shown in the deconvoluted mass spectrum (Fig. 4.3c). The spectrum clearly reveals the glycoprofile difference between trastuzumab antibodies from different batches (shown in different colors) where the intensity of each glycoform varies.

It should be mentioned that the mass measurement accuracy of the main glycoform is within 1.5–2 Da (~10 ppm) from its theoretical mass value (148,057 Da), an unthinkable achievement prior to the advent of ESI and MALDI MS. The latter is an essential attribute of this method and renders it suitable to distinguish the lot-to-lot heterogeneity in glycosylation profile of the commercially available glycoprotein biopharmaceutical. Glycoprotein heterogeneity can result in an enhancement or loss of the protein's biological activity, as this has been demonstrated in the case of rHuEPO, where desialylation causes complete loss of its hormonal activity in vivo [63]. In particular, intravenously administered rHuEPO consisting of highly branched sialylated oligosaccharide structures has been shown to result in a plasma half-life of 5–6 h as compared to desialylated rHuEPO, which is cleared within minutes [64].

On the other hand, glycoprotein analysis by MALDI MS yields signals corresponding to protonated molecules (MH^+) of the individual glycoforms and allows the determination of the heterogeneity for glycoproteins with M_r less than 30 kDa and a relatively low percentage of carbohydrate content. This is clearly shown in the screening of the glycosylation profile for the human soluble urokinase-type plasminogen activator receptor (uPAR) expressed in CHO cells, where the extent and type of glycosylation in its three domains was assessed by MALDI MS [65]. On the contrary, MALDI MS analysis of the Sf9-derived interleukin-5 receptor α-subunit (IL-5Rα) [66] and CHO IL-4R [53] with 17 and 35 % carbohydrate content, respectively, did not provide any information on the type of the glycosylation. In addition, the choice of an appropriate MALDI matrix is very important toward achieving the optimum mass resolving power and separation of the individual glycoform signals [67, 68]. This is shown in the MALDI mass spectra of the Sf9-derived IL-5Rα (in a reflectron and a linear TOF instrument) using different matrices, where the use of the sDHB matrix (2,5-dihydroxybenzoic acid with a 10 % admixture of 2-hydroxy-5-methoxybenzoic acid) in a linear TOF instrument resulted in a more reliable mass measurement due to minimized metastable fragmentation [69] (Fig. 4.4).

Fig. 4.3 Positive-ion ESI ion mobility (IM) TOF mass spectrum of the intact therapeutic monoclonal antibody trastuzumab (Herceptin) (**a**); The 53+ charged ion with the signals corresponding to various glycoforms is annotated (**b**); The glycoform assignments and the glycosylation heterogeneity of the monoclonal antibody arising from variations in the hexose and fucose content are shown in the deconvoluted mass spectrum (**c**). (Reprinted with permission from Springer [62])

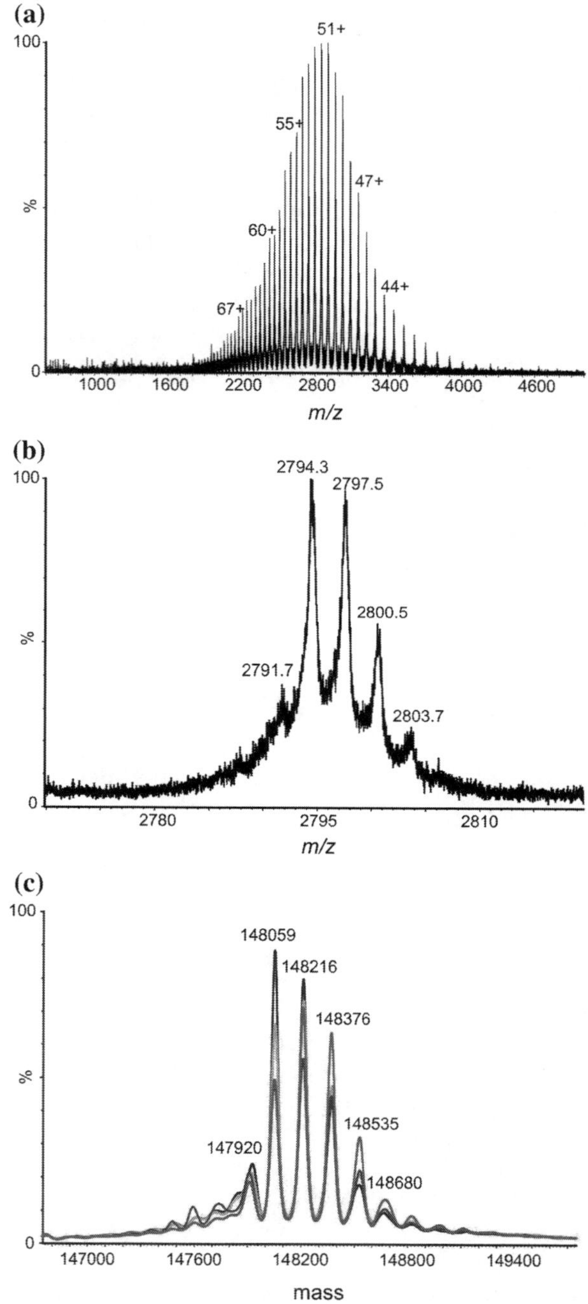

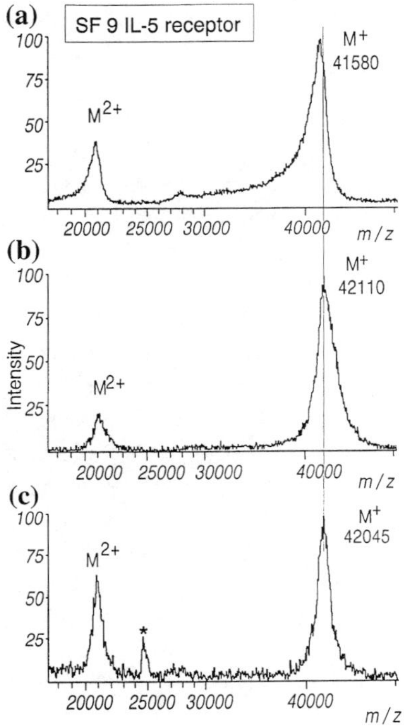

Fig. 4.4 Positive-ion MALDI-TOF mass spectrum of Sf9-derived interleukin-5 receptor α (IL-5Rα) obtained with reflectron TOF with sDHB (**a**) and HPA (**b**) matrix, and linear TOF instrument using sDHB matrix (**c**). The *asterisk* denotes an internal calibrant. (Reprinted with permission from Wiley [69])

Overall, ESI MS analysis of intact glycoproteins has better success over MALDI MS for surveying the individual glycoforms in a glycoprotein biotherapeutic sample and ensuring the homogeneity of the manufacturing batches. Nevertheless, the biggest challenge for the analysis of glycoproteins is their low abundance compared to that of unmodified proteins and the resulting low intensities of the mass spectral signals. This is mainly due to the low ionization efficiency of glycoproteins and the distribution of their signal among the various glycoforms sharing a common peptide sequence, thus rendering their detection an overwhelming task. This can be overcome by performing an enrichment step for the glycoproteins, which eliminates the most abundant unmodified proteins from competing for charge during the ionization process and results in higher ionization efficiencies and increased probability for detecting glycoproteins. The commonly used analytical methods for glycoprotein/glycopeptide enrichment are discussed in Sect. 4.2.2.2. Another promising solution to this problem is coupling of ESI MS with a separation device such as nano-LC [70], CE [71] or CZE [72] that can definitely improve the chances for a successful analysis. This is shown in the analysis of intact rHuEPO and bovine α1-acid glycoproteins by a developed CZE-ESI MS method without any complicated sample treatment, where characterization of the intact glycoforms was provided along with their relative intensities [73, 74]. In addition to the efficient separation of the intact glycoforms, small glycan modifications such as acetylation, oxidation, and sulfation could be successfully characterized. Similarly, high-resolution CE-Fourier transform ion cyclotron resonance (FT ICR) MS analysis was used for the profiling of the intact glycoforms of recombinant human chorionic gonadotrophin (r-RhCG) produced in a murine cell line, which resulted in the identification of over 60 different glycoforms with up to nine sialic acids [75]. These studies suggest that CE-MS can be an important tool for rapid assessment of the recombinant product quality either for product release or for in-process control, and even for demonstrating comparability of a glycoprotein therapeutic biosimilar to the innovator product being replicated.

Moreover, the rapid assessment of glycosylation at the molecular level is invaluable in glycoform screening of glycoproteins involved in certain diseases, such as the human transferrin (Tf) model glycoprotein for congenital disorders of glycosylation (CDG) diagnosis. CE-ESI MS was used successfully for carbohydrate-deficient transferrin (CDT) detection and CDG-type characterization [76]. Comparative analysis of serum samples from healthy and CDG patients by CE-ESI MS (Fig. 4.5) provided partial separation of Tf glycoforms and identification of the carbohydrate-deficient Tf glycoforms in the CDG patients' serum. It is clearly shown that the Tf glycoforms in the CDG serum correspond to a disialoform containing one free N-glycosylation site (Fig. 4.5e) and another one occupied by a biantennary instead of a triantennary N-linked sialylated glycan (Fig. 4.5f), thus confirming that the sample belongs to a patient who has CDG of type I [76].

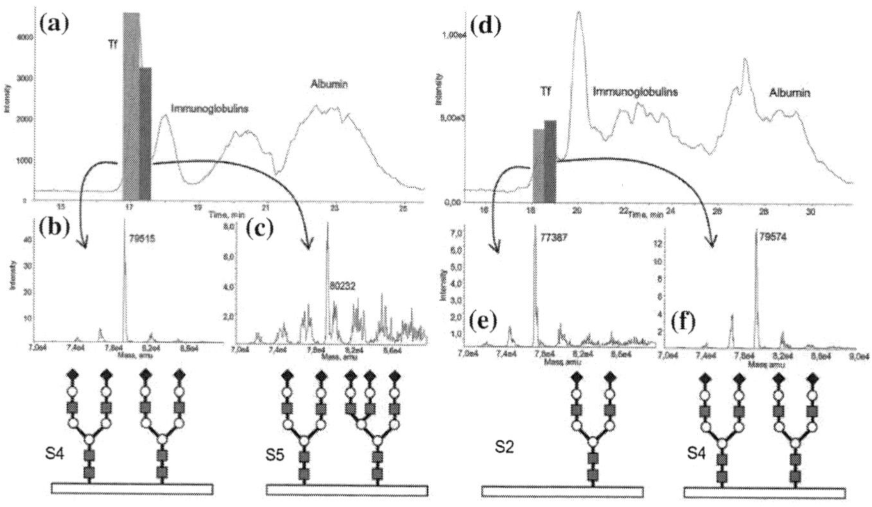

Fig. 4.5 Total ion electropherogram obtained for a serum from a healthy individual (**a**) and a CDG patient (**d**) under optimized CE-ESI MS conditions. The deconvoluted spectra obtained from the beginning and the end of the Tf peak are shown in (**b**) and (**c**), respectively. In case of the CDG patient, the deconvoluted spectra obtained from the two partial resolved glycoforms of Tf provided identification of the carbohydrate-deficient Tf glycoforms in the CDG patients' serum (**e**) and (**f**). The most probable glycan composition is displayed below the deconvoluted mass spectra. (Reprinted with permission from Wiley [76])

4.2.2 Mass Spectrometry and Glycoproteomics

Glycoproteomics involves the study of the glycosylation of proteins, including the structures of the attached oligosaccharide moieties and the identification of the glycosylation sites. There are two distinct classes of protein glycosylation in nature depending on the linkage site. First, the "O-linked" are the ones that are linked to serine (S), threonine (T), or hydroxyproline residues in the protein backbone, and then the "N-linked" which are linked to N residues through an N-acetylglucosamine residue (GlcNAc). Regarding O-glycosylation, a number of monosaccharides attached to S and T have been found, most commonly N-acetylgalactosamine (GalNAc), GlcNAc, xylose, mannose, and fucose [29]. O-glycans are synthesized in a stepwise process that involves single monosaccharide transfer steps, and their biosynthesis takes place after protein N-glycosylation, folding, and oligomerization. O-glycosylation may occur at any S or T residue with no single common core structure or consensus protein sequence. Extended structures from a core GalNAc that are called mucin-type O-glycans are the most frequently occurring [77, 78]. In contrast to O-glycans, N-glycosylation sites can be predicted by the tripeptide sequon Asn-Xaa-Ser/Thr (N-X-S/T, where X is any amino acid except P) [79, 80] (Fig. 4.6). All three types of N-glycans found in mature glycoproteins share a pentasaccharide core (i.e., the trimannosyl core with two N-acetylglucosamine residues ($Man_3GlcNAc_2$)) because of a common biosynthetic pathway in the endoplasmic reticulum compartment of the cell. This N-glycan $Man_3GlcNAc_2$ core is common to complex, high-mannose, and hybrid structures as shown in Fig. 4.6.

The high-mannose-type glycoproteins (e.g., ovalbumin) contain two to eight mannose residues added to the pentasaccharide core. Glycoproteins containing complex-type N-structures (e.g., fetuin) exhibit the highest structural variation by having a number of GlcNAc, Gal, Fuc and NeuAc (sialic acid) residues attached to the N-glycan $Man_3GlcNAc_2$ core, as well as possible extension and/or branching of the outer chains through lactosamine repeats and sialylation. Finally, the hybrid-type glycoproteins combine features from both high-mannose- and complex-type glycans [79]. At this point, it should be emphasized that for both N- and O-glycosylation, there is an inherent *microheterogeneity* resulting from the array of glycan structures associated with each glycosylation site. Moreover, there is *macroheterogeneity* due to the fact that not all N-glycan sequons or S/T residues present in the glycoproteins are glycosylated. The end result is a diverse degree of occupancy at different O- or N-linked glycosylation sites with a wide array of structurally different oligosaccharides that generate a complex mixture of glycosylated variants (*glycoforms*). The variety of these glycoforms depends not only on the polypeptide backbone and the number of putative glycosylation sites but also on the cell type, in which the glycoprotein is expressed, and its development stage. Therefore, characterizing the glycoproteome is a demanding task because of the inherent macro- and microheterogeneity of glycans along with the complex nature of this modification.

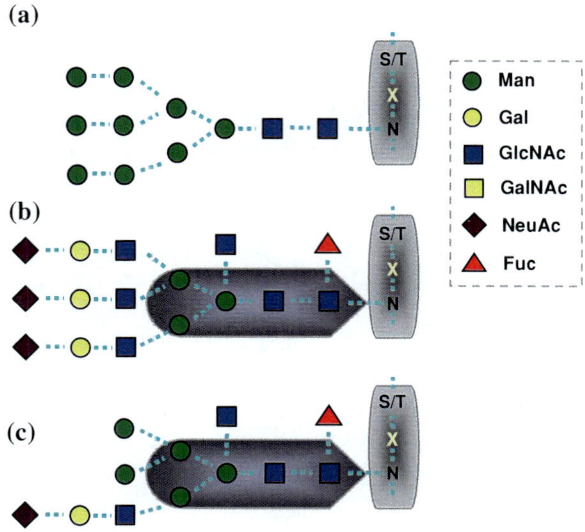

Fig. 4.6 Classes of N-linked carbohydrate structures sharing a common pentasaccharide core structure, that is, the trimannosyl core with two N-acetylglucosamine residues ($Man_3GlcNAc_2$). **a** High-mannose-type; **b** Complex-type (triantennary); **c** Hybrid-type. The sugar symbols used throughout this chapter are those adopted by the consortium for functional glycomics (CFG). *Circles* represent hexoses (Hex) [*yellow*: Galactose (Gal), *green*: Mannose], *squares* represent N-acetylhexosamines (HexNAc) [*blue*: N-acetylglucosamine (GlcNAc), *yellow*: N-acetylgalactosamine (GalNAc)], *red triangle*: fucose, *purple diamond*: N-acetylneuraminic acid (NeuAc). (Gal: , Man: , GlcNAc: , GalNAc: , Fuc: , NeuAc:)

4.2.2.1 Top-Down and Bottom-Up Analytical Approaches

Complete structural characterization of a glycoprotein includes the following tasks: (1) characterization of glycans in intact glycoproteins (2) determination of the protein primary sequence and the glycosylation attachment sites (3) characterization of glycopeptides, and (4) structural analysis of chemically or enzymatically released glycans. The M_r determination of intact glycoproteins by either ESI or MALDI MS analysis is successful only for glycoproteins up to 20–30 kDa with a relatively low percentage of carbohydrate content as demonstrated above (Sect. 4.2.1). Even though this accurate M_r measurement is valuable for profiling of intact glycoproteins and providing very useful information on the type and extent of glycosylation, there is no information on the nature and the attachment sites of the glycan chains. Therefore, one needs to cleave the protein into smaller fragments before attempting MS analysis. In the *top-down* approach, the intact molecule is introduced into the mass spectrometer where limited fragmentation of the ionized protein is induced (i.e., in the gas phase) and the resulting product-ion mass spectra can provide information on the location of the glycosylation sites (or other PTMs) [81, 82]. Even though there are several mass analyzers capable of measuring intact proteins and large ionic fragments (such as TOF, QTOF, FT ICR), the unusually high resolving power (>10^5) of FT ICR has made possible accurate assignments of ESI charge state and mass, even for MS/MS of intact proteins [83, 84]. Such *top-down* methods have proven especially powerful in stability and formulation studies of intact antibodies with M_r ~ 150 kDa used as therapeutics in the biopharmaceutical industry. Nevertheless, FT ICR instruments have not become standard analytical tools for the characterization of recombinant biopharmaceuticals mainly due to the high cost of acquisition and maintenance. For that reason, the most commonly followed MS-based approach for characterization of a recombinant biopharmaceutical involves the enzymatic digestion of the glycoprotein (usually with trypsin or another endoprotease) followed by the separation/analysis of the resulting peptide digests by LC–MS/MS [41, 85, 86] or CE-MS/MS [87] and MALDI MS [88] (*bottom-up* approach). In case of purified proteins or simple mixtures thereof, LC–MS or MALDI MS analysis of the proteolytic mixture provides M_r information on the peptide components. The advantage of MALDI-TOF MS is the simplicity of the spectra, which contain usually intense protonated (MH^+) or sodiated signals corresponding to the enzyme-generated peptides. Further, protein structural information can be deduced by carrying out LC–MS/MS analysis of the enzyme-generated peptides. Peptide identification is performed through a direct search of the M_r measured values and the tandem MS-derived fragment ions (sequence tags) [89] against a protein sequence database (*peptide fingerprinting*). The general experimental workflow comprising the commonly employed approaches in glycoproteomic analysis is shown in Fig. 4.7. Of course, the nature of the glycoprotein sample determines the number of the necessary steps needed in order to determine site-specific glycosylation and heterogeneity.

In general, MS mapping of the enzyme-generated peptide mixtures provides not only confirmation of the expected protein sequence but also identification of any existing modifications, including the glycosylation attachment sites. In addition, unexpected mass spectral signals can provide insights into the glycosylation profile of the protein, taking into consideration the known N-glycan structures (Fig. 4.6). Nevertheless, there are several problems associated with the *bottom-up* approach. The major problem arises from the fact that many glycoproteins are resistant to enzymatic proteolysis (e.g., trypsin or *S. aureus* V8 protease) due to the presence of the attached glycans near the proteolytic site, thus requiring an additional specific enzymatic proteolysis. In addition, the resulting mixture of peptides and glycopeptides could complicate the analysis because glycosylation strongly diminishes the ionization efficiency of the peptide [90, 91], especially when the glycans are terminated with the negatively charged sialic acid moiety [47]. This problem becomes more significant considering that the glycopeptides are in much lower abundance than the peptides from the same glycoprotein, and the glycopeptide signals are distributed over several peaks due to the glycan heterogeneity and multiple adduct ion formation. However, several enrichment methods (either in parallel or sequentially) prior to glycoprotein analysis can be used to compensate for the low abundance of glycopeptides (and glycoproteins) and the presence of multiple glycan structures (heterogeneity) [92]. The use of

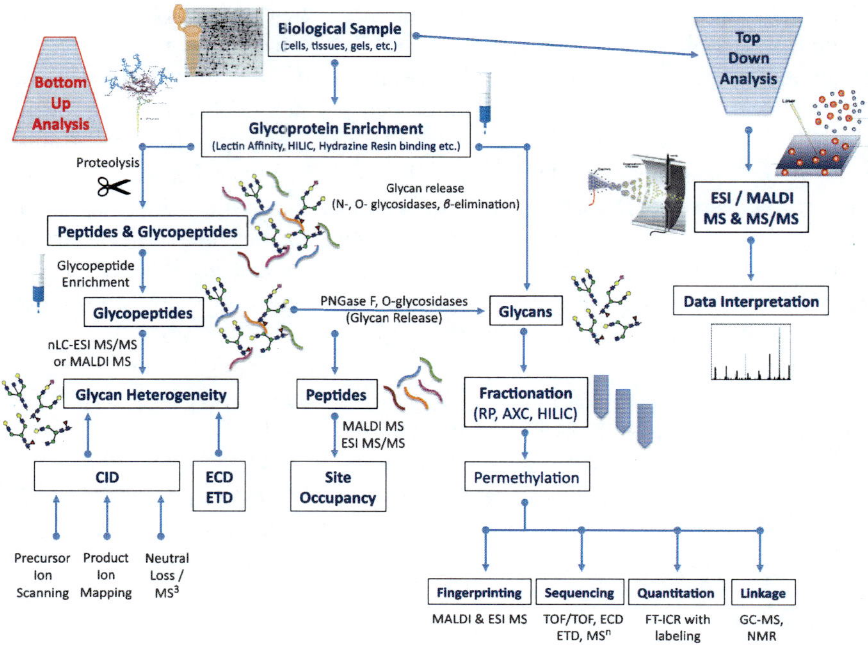

Fig. 4.7 The general experimental workflow comprising the commonly employed approaches in glycoproteomic analysis. The *top-down* approach starts with the analysis of intact glycoproteins, while the *bottom-up* analytical approach begins with proteolytic or chemical cleavage of the glycoprotein, followed by mapping of the generated glycopeptides by an assortment of LC–MS and tandem MS techniques

glycoprotein enrichment methods can bypass the aforementioned obstacles in glycoprotein analysis by achieving exclusion or reduction of the most abundant unmodified peptides from the analysis, thus improving the ionization efficiency of the low-abundance glycopeptides, which do not have to compete for charge during the ionization process with unmodified peptides.

4.2.2.2 Glycopeptides Enrichment Methods

Enrichment of glycoproteins and glycopeptides can be achieved by using the natural affinity of lectins for their glycan "handles" [93], whereas other analytical methodologies based on general physical and chemical properties of glycopeptides have been employed, such as size-exclusion chromatography (SEC) [94], hydrophilic interaction chromatography (HILIC) [95–97] or graphitized carbon columns (GCC) [98, 99]. A rough classification of the commonly used enrichment techniques in glycopeptides analysis can be made into chemical [100, 101] and chromatographic methods (such as affinity chromatography [102–104], LAC [93], immunoaffinity chromatography [105], SEC [94], hydrophilic phases [96, 97] and GCC [99]).

Lectin Affinity Chromatography

Lectins are proteins originating from plants, fungi, bacteria, or animals that express a special affinity toward glycans [106] and thus are used for glycopeptide/glycoprotein isolation from complex mixtures after being immobilized onto various solid supports such as silica [107], agarose [46], resins, magnetic beads, and affinity membranes. These are used in different arrangements, such as columns [108–110], tubes [46], and microfluidic chips [111]. Lectins generally interact with specific motifs in a glycan and demonstrate selectivity for different oligosaccharides and broad range of specificity [112], thus enabling glycoprotein/glycopeptide isolation from a complex protein mixture along with glycoform pre-fractionation. Widely used lectins include concanavalin A (ConA) [113, 114], which binds glycan residues containing mannose and glucose and affords broad selectivity (i.e., high-mannose, hybrid, complex biantennary [115]), wheat germ agglutinin (WGA), which presents selectivity for GlcNAc and NeuAc, and Jacalin (JCA), which expresses affinity against galactose (b1-3) GalNAc and O-linked glycoproteins.

Various analytical strategies have been proposed for the isolation and preconcentration of glycoproteins/glycopeptides prior to MS analysis [93]. In summary, the sample enrichment using lectin columns can be performed before or after the protein mixture digestion by loading the sample onto the columns under high-ionic-strength buffers to prevent non-specific retention. The same loading buffer containing a displacer (haptene saccharide) is used to elute the captured glycopeptides/glycoproteins, which can then be subjected to MS analysis.

There are two principal enrichment methodologies based on LAC: Serial Affinity Chromatography (SLAC) [116] and Multi-Lectin approach (M-LAC) [117]. The first one uses a serial set of lectin columns with different specificity, thus enabling the sequential selective binding of various glycan moieties of a peptide or protein mixture. SLAC has proven to be a powerful tool for rapid and primary elucidation of glycans' structural features, especially when columns with broad (ConA, WGA, or JCA) and narrow selectivities (also known as "structure-specific affinity selectors," i.e., Sambucus nigra agglutinin, SNA) are combined [118]. The SLAC approach was used in the characterization of a prostate-specific antigen in human prostate cancer [119]. Furthermore, coupling LAC with advances in stable isotopic labeling has been successfully applied for the comparative analysis of sialylated proteins [120], thus providing a valuable tool for exploring the glycosylation sites of the whole proteome as well as an excellent tool for biomarker discovery. On the other hand, the M-LAC approach uses a single column (multi-lectin column) containing various lectins with broad specificity (i.e., ConA, WGA, JCA), thus enabling the comprehensive isolation of glycoproteins/glycopeptides from a complex mixture covering an extended dynamic range. This approach was used in the study of glycoproteins in human serum [117, 121] and plasma [122].

Integrated analytical platforms combining LAC with various separation techniques have been developed lately in order to overcome the low-throughput drawback of the off-line procedures. Such methodologies include a microfluidic chip [111] containing a polymeric monolithic column with immobilized pisum sativum agglutinin (PSA), an integrated glass microchip [123] for online tryptic digestion of glycoproteins in the first channel, followed by selective enrichment of resulting peptides through ConA in the second channel. The eluted fractions were subjected to CE and nano-LC–MS analysis employing capillary polymethacrylate monolithic columns with immobilized ConA and WGA [109] allowing large volume injection and adequate sensitivity. Similarly, a fully automated LAC system coupled online to ESI MS with silica-based lectin microcolumns [108] demonstrated high-binding capacity and excellent reproducibility, whereas a variation of this platform with SLAC [124] was proved to be superior over the M-LAC approach for the selective enrichment of small volumes of blood serum [115].

Immunoaffinity Chromatography

Immunoaffinity (IA) enrichment protocols for glycoproteins/glycopeptides rely on the unique specificity of the antibody–antigen interaction and enable the highly selective adsorption of a target analyte through the covalent attachment on a properly functionalized solid support containing an affinity ligand [125]. This can be performed either by the covalent attachment of antibody fragments via proper chemistries that provide correct orientation of the fragment, or by immobilization of a secondary binder molecule. The elution of the bound ligands is achieved by lowering the pH of the eluting buffer to pH 1–3, by using chaotropic salts, or by using polarity-reducing agents in order to weaken the antibody–antigen hydrophobic interactions. Although the IA enrichment approach has been mainly used for off-line targeted glycoproteomics [105], an online integration of this technique with CE was employed for the pre-concentration of rHuEPO in diluted solutions [126]. This integrated platform has demonstrated high loading sample capacity and good separation efficiency of the glycoforms.

Porous Graphitized Carbon Chromatography

Porous graphitized carbon chromatography (PGC) has been employed for the separation of oligosaccharides, in their native form as well as after derivatization, based on a retention mechanism driven mainly by hydrophobic and electrostatic stacking interactions. The oligosaccharide analytes are eluted in order of increased size, and structural isomer resolution is often provided [47]. In addition to separation, PGC has been used for the selective enrichment of glycans and glycopeptides. An off-line approach combining solid-phase extraction (SPE) with PGC cartridges was used to concentrate and pre-fractionate pronase glycopeptides and glycans prior to MALDI-TOF MS analysis [127]. An automated variation of the aforementioned approach for glycoprotein analysis has been reported combining digestion, extraction, and separation processes in one analysis [128]. This integrated platform employs a pronase-based chromatographic bioreactor for the *in situ* rapid digestion of glycoproteins, an online SPE of the produced glycopeptides with a PGC trap column, and separation by LC–MS/MS. This system allowed the direct sequencing of the glycans and peptides along with simultaneous characterization of the glycan composition and localization of the glycosylation site.

Chemical Derivatization Methods

In addition to the affinity techniques described above, that do not change the structure of the modification and the peptides/proteins, several chemical methods specific to the glycan moieties have been used for the detection and the purification of glycosylated proteins. Most of the chemical derivatization strategies use two basic reactions: (1) the Schiff base reaction of aldehydes with a hydrazine [129–131], and (2) a Staudinger ligation between a phosphine and an azide [132, 133]. However, most of these derivatization methods provide peptide/protein identification without much information about the site or the structure of glycosylation, mainly due to inadequate search algorithms and the occasional modification of the glycan structure [112].

One of the strategies using the Schiff base reaction is the O-GlcNAc ketone enrichment method [134], where a chemo-enzymatic approach using an engineered β-1,4-galactosyl transferase is employed to transfer a ketone containing substrate onto O-GlcNAc-modified proteins. A Schiff base reaction was used to biotinylate the ketones with biotin-hydrazine and subsequently the O-GlcNAcylated peptides/proteins were captured on a streptavidin affinity column. This methodology was successfully used for the identification of the cAMP-responsive Element-Binding Factor (CREB), a low-abundance protein with two known O-GlcNAc sites, in a whole cell lysate [135]. Another derivatization enrichment approach for the glycoproteome and especially for N-glycosylation, is the Periodate-acid-Schiff (PAS) reaction using an iminobiotin hydrazide via the Schiff base reaction [129]. The derivatized peptides/proteins are affinity purified on a streptavidin column and analyzed by MS. This reaction exploits the unique vicinal diol functionality of glycans, thus oxidizing these diols to aldehydes without affecting any other amino acid apart from M, which is oxidized to its sulfoxide analog. This approach provides important information regarding N-glycosylation site modifications and has been used for high-throughput quantitative analyses [136]. In addition, it is an extremely versatile process for proteins and peptides, as different coupling agents such as biotin hydrazides and digoxigenin hydrazides can be incorporated. However, the major disadvantage of the PAS strategy is the heterogeneous modification of the glycan structures by an undefined number of hydrazide tags, thus necessitating PNGase cleavage of the glycans in order to sequence the peptide backbone. In this way, all information pertaining to the glycan structure is lost and only N-glycosylation sites can be determined.

In a modification [133] to the standard Staudinger reaction (a reaction of an azide with a phosphine), the intermediate aza-ylide formed in the standard Staudinger reaction reacts with an electrophilic trap to form an amide bond with a compound that is biotin tagged. This reaction is biologically unique as neither phosphines nor azides occur in biomolecules and also offers the possibility to design phosphines in order to incorporate a wide variety of tags, such as fluorescent probes and affinity tags [137, 138]. A tagging-via-substrate (TAS) strategy based on a tag attached to the modification substrate was used for the identification of O-GlcNAc glycosylated proteins [117, 124], as well as for the detection and isolation of other PTMs in proteins, such as farnesylation [139]. Another derivatization method that has been used for the enrichment of O-linked β-GlcNAc is β-elimination followed by Michael addition with dithiothreitol (BEMAD) [140]. BEMAD has also been used to quantitate both O-glycosylated and O-phosphorylated peptides [141].

4.2.3 Determination of Site-Specific Glycosylation and Heterogeneity

The complete characterization of a glycoprotein biopharmaceutical involves the analysis of the glycan structures that are expressed on the glycoprotein of a given organism or cell line, the identification of the proteins that express these glycans, as well as the individual glycosylation sites on each protein [39, 41]. MS and tandem MS analysis of glycopeptides usually after chromatographic or electrophoretic separation, either online or off-line, holds a central role in all the strategies for glycoproteomic analysis [142] (Fig. 4.7). The most commonly followed experimental approach for providing a detailed glycoprotein mapping involves the analysis of enzymatically derived glycopeptides by fast atom bombardment (FAB) [22, 143], MALDI [144] or LC-ESI MS and MS/MS [50, 85]. This MS mapping identifies most of the expected peptide signals (*peptide fingerprinting*), whereas any new, unexpected mass spectral signals may correspond to glycopeptides. In a similar off-line strategy, the isolated fractions are mapped by ESI or MALDI MS and MS/MS approaches. Nevertheless, there are several issues related to these MS approaches, such as the potential deglycosylation of glycopeptides in the gas phase combined with the low ionization efficiency and low abundance of the glycopeptides compared with the peptides derived from the proteolytic digestion. One of the remedies to ensure the appearance of glycoproteomic information within the copious proteomic data is enrichment of glycoproteins and/or glycopeptides prior to analysis (as discussed above). Another way to overcome this difficulty is the carbohydrate removal from the glycoprotein by base-catalyzed β-elimination for O-linked glycans or digestion with PNGase F (N-Glycanase) for N-linked glycans. The former leads to the conversion of S and T residues to A and α-aminobutyric acid sequences, respectively (i.e., loss of 16 Da), whereas the latter converts the glycosylated N residues to D (i.e., increase of 1 Da). In the MS mapping of the enzyme-generated peptide mixture of the deglycosylated protein, the former O- and N-glycosylated peptides can be readily identified by the appearance of new mass spectral signals at lower m/z (for O-linked sugars) or higher m/z (for N-linked sugars) than those of the unglycosylated peptides [145]. This mass difference can be magnified by carrying out the N-Glycanase reaction in fully or partially (50 %) ^{18}O-labeled glycosylated N residues, which results in characteristic doublets separated by 2 Da [146]. These doublets can be used to locate the modification site and to determine the degree of occupancy at each N-linked glycosylation site. Another approach for N-linked glycans involves the release of the high-mannose- and hybrid-type oligosaccharides by digestion of the glycoprotein with endoglycosidase H, leaving a GlcNAc residue attached to the peptide's N residue. That results in the detection of peptides having M_r values 203 Da higher than that of the respective unglycosylated peptides. Glycosylation sites containing complex-type glycans are unaffected by the endoglycosidase H treatment. This approach was employed in the FAB carbohydrate mapping of the major envelope glycoprotein gp120 of HIV type 1 [147] and recombinant tissue plasminogen activator (rtPA) [148]. On a similar approach, glycoprotein mapping of CHO rHuEPO was facilitated by removal of terminal NeuAc residues with neuraminidase followed by LC-ESI MS analysis of the enzyme-generated peptide fragments of asialo CHO rHuEPO [54]. rHuEPO contains three N-glycosylation sites at N-24, N-38, and N-83 and a single O-glycosylation site at S-126; the glycans account for up to 40 % of the total molecular mass. This LC–MS mapping provided information on the microheterogeneity of the carbohydrate structures, which is associated with the presence or absence of lactosamine extensions and varying levels of O-acetylated NeuAc residues. Similarly, comparative LC-ESI MS tryptic mapping of untreated and neuraminidase-treated rtPA allowed the identification of the attachment site of two hybrid-type carbohydrates on one of the tryptic peptides [149]. The same analytical protocol was applied in the characterization of a rtPA mutant with an additional glycosylation site (T103N), where two new complex-type carbohydrate chains have been observed [149]. An analogous LC-ESI MS/MS approach combined with a multi-enzymatic digestion strategy was employed for the characterization of the glycosylation occupancy in the generic variant of rtPA (TNK-tPA), which was approved for treatments of acute myocardial infarction and ischemic stroke [150]. TNK-tPA has the same amino acid sequence as natural human tPA except for the three substitutions: T103N, N117Q, and AAAA for KHRR (296–299) which lead to longer half-life and higher fibrin activity than those of tPA. Nevertheless, differences in the glycosylation occupancy at N184 along with different extents of deamidation at N184 and oxidation at M207 have been observed between the therapeutic biosimilar and the innovator product, thus raising concerns as to its bioequivalence.

In the case of CHO IL-4, comparative LC–MS tryptic and V8 protease mapping of CHO IL-4 and its N-Glycanase-treated protein revealed that the N residue in the sequon $N^{38}TT$ was glycosylated rather than the other potential site at $N^{105}QS$. We should point out that the presence of carbohydrate often provides shielding of a neighboring proteolytic site, thus leading to the incorporation of the adjacent peptide fragment, as demonstrated by the incorporation of the T5 tryptic glycopeptide into the adjacent disulfide-linked peptide T4–T10 of CHO IL-4 [50]. When ESI MS/MS approaches are incorporated in the analysis of the LC- or CE-separated enzymatic fragments of a glycoprotein, the identification of glycopeptide-containing chromatographic fractions is facilitated by the appearance of several diagnostic fragment ions. CID product-ion spectra of ESI generated glycopeptides in a variety of instruments, such as triple quadrupole, ion trap (IT), and QTOF, are dominated by fragmentation of glycosidic linkages thereby revealing predominantly information on the composition and sequence of the glycan moiety. Glycopeptide marker ions under CID conditions are low-mass sugar-specific oxonium ions (B-type fragmentation in the Domon and Costello nomenclature [151]) of m/z 162 for Hex^+, m/z 204 for $HexNAc^+$, m/z 274 and 292 for $NeuAc^+$, m/z 366 for $Hex\text{-}HexNAc^+$, and m/z 657 for $NeuAc\text{-}Hex\text{-}HexNAc^+$. Scanning for these diagnostic fragment ions in the "precursor ion" mode on triple-quadrupole mass spectrometers can selectively identify the glycopeptides within the enzymatic digest mixture, whereas screening of constant neutral losses of terminal monosaccharides could also pinpoint the glycopeptides. Selected ion monitoring (SIM) experiments can also be performed for glycopeptide identification with IT and QTOF mass analyzers. In cases, where MS/MS is not available, these low-mass glycopeptides marker ions can be generated by either "in-source" fragmentation of ESI-produced ions [50, 149, 152] or post-source decay (PSD) of MALDI-produced ions [153]. In the former, increasing the source entrance potential into the mass spectrometer, which controls the collision excitation and the extent of fragmentation, induces the fragmentation. This online LC–MS "in-source" CID mapping of glycopeptides utilizes both low and high source potentials and monitoring of the resulting sugar-specific oxonium ions. In case of complex/hybrid or high-mannose structures, monitoring of the oxonium ions at m/z 204, 274/292, 366 and 657 has allowed the fast glycan profiling in the LC-ESI MS analysis of the trypsin-treated CHO rTPA [154] and CHO IL-4 [50] without having to search each individual mass spectrum for glycopeptide-characteristic patterns. In the case of rTPA, this method allowed the identification of a low-level novel N-glycosylation at N142, which is part of an atypical N-Y-C consensus motif. Although this site is only 1 % occupied by predominantly biantennary hybrid structures, it was readily detected by this sensitive LC-ESI MS tryptic mapping approach. In the case of CHO IL-4, the observation of the glycopeptides marker ions at m/z 274, 366 and 657 revealed the presence of sialylated complex-type N-glycans in the specific chromatographic fraction. In addition, the mass separation of the signals within the triply and quadruply multiply charged ion envelopes revealed the presence of mono- and di-sialylated glycoforms (291 Da apart) along with higher M_r components containing additional lactosamine units (365 Da apart) owing to the presence of extended arms or branching. Similarly, this rapid glycopeptide screening approach was applied to other mammalian-cell-derived proteins, such as the Sf9-derived IL-5Rα, where this low/high "in-source" fragmentation allowed the identification of all glycopeptide-containing fractions in the LC-ESI MS tryptic peptide map of Sf9 IL-5Rα (Fig. 4.8). This method allowed the identification of four glycosylation sites in Sf9 IL-5Rα out of the six potential sites fulfilling the N-glycosylation consensus sequence [66].

The ESI mass spectrum of one glycopeptide-containing fraction (Fig. 4.8, peak 10) showed signals corresponding to doubly and triply charged tryptic glycopeptides containing a $Man_9(GlcNAc)_2$ high-mannose carbohydrate (Fig. 4.9). All these

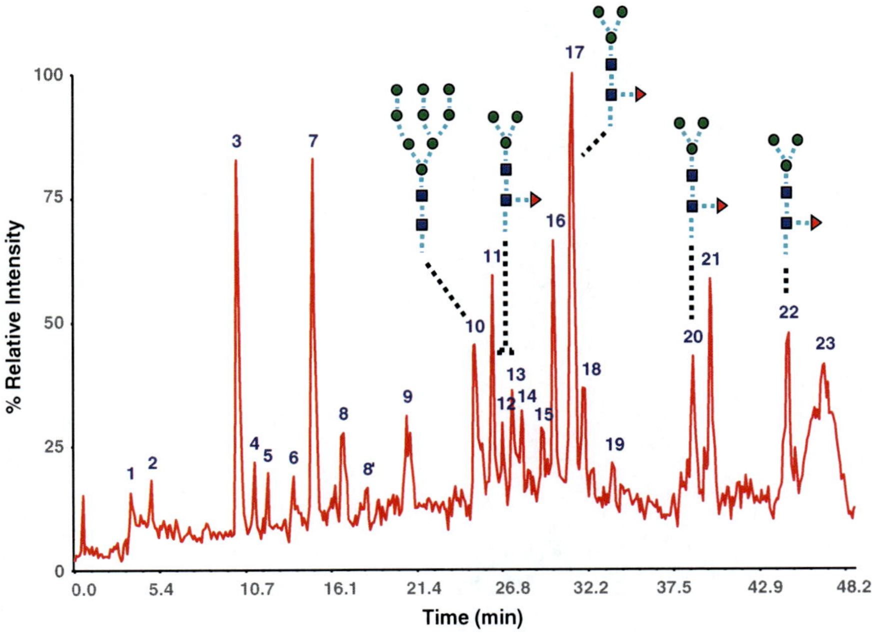

Fig. 4.8 Total ion current (TIC) chromatogram obtained from the reversed-phase LC-ESI MS analysis of the Sf9 IL-5Rα tryptic digest using "in-source" fragmentation. The tryptic fractions containing glycopeptides are labeled, as inferred by the detection of the low-mass diagnostic ions at m/z 162, 204, 274, 366, and 657

glycopeptides contain the N196 glycosylation site and the M_r values of the respective glycoforms differ by 162 Da due to an extensive heterogeneity in the Man (●) content, as shown in the deconvoluted mass spectrum (Fig. 4.9, inset).

The assignment of the putative glycan structures to the experimental masses with a high degree of confidence is made possible by the excellent mass measurement accuracy provided by ESI MS analysis. Corroborative information on the composition and sequence of the attached glycans can be attained from MS/MS analysis of the glycopeptides, because CID tandem mass spectra of glycopeptides contain mainly fragments arising from glycosidic bond cleavage [155]. In the analysis of the therapeutic glycoprotein BRP 3 EPO by a combined anion-exchange chromatography (AEC)—ultra-performance liquid chromatography (UPLC) MS/MS approach, tetra-antennary glycans with up to four NeuAc and up to five poly-N-acetyl lactosamine extensions were observed at the glycosylation sites N24 and N83, whereas biantennary glycans were the major structures at N38 [156]. The presence of these large repeating glycan motifs although at low levels may infer additional functional interactions for EPO and may be beneficial in terms of immunogenicity. A more detailed characterization of N-glycopeptides, especially in terms of the peptide sequence, can be obtained by an alternative approach combining MS/MS and MS^3 experiments in an IT MS [142]. The glycopeptide ion is

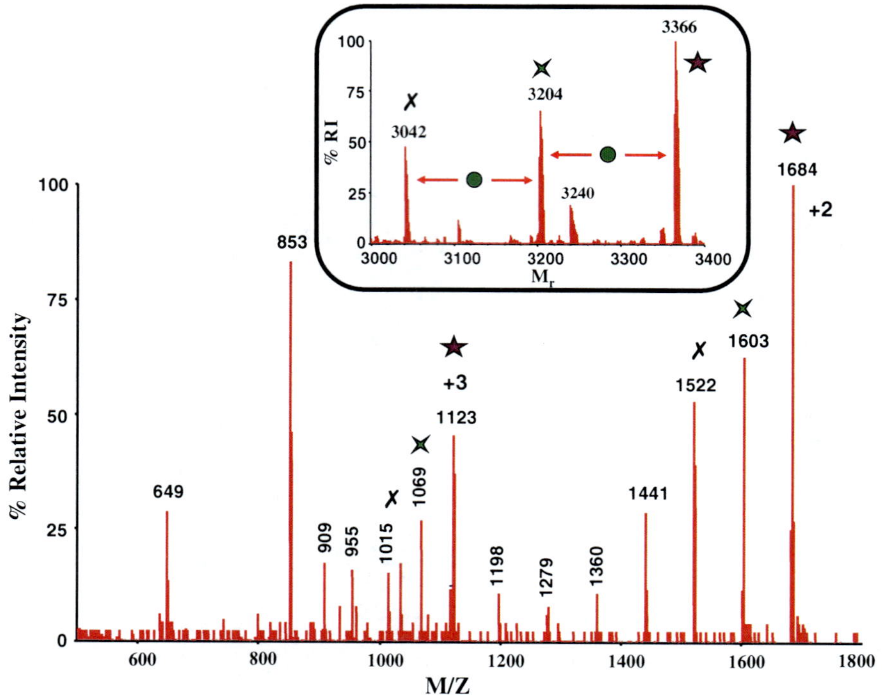

Fig. 4.9 Positive-ion ESI mass spectrum of Sf9 IL-5Rα tryptic glycopeptide component (Fig. 4.8, TIC *peak* 10). The deconvoluted mass spectrum (shown in the *inset*) indicates the presence of a high-mannose carbohydrate component with M_r value of 3366. The *asterisk* and *cross*-denoted peaks in the doubly and triply charged ESI ion envelopes correspond to two glycoforms with M_r values 3204 and 3042, respectively, arising from glycoform heterogeneity due to variations in the mannose (●) content

selected and fragmented, and the peptide ion carrying a single GlcNAc (which is often the most abundant ion) is subjected to a second fragmentation cycle resulting in extended fragmentation of the peptide moiety into *b*- and *y*-series ions, thus allowing the deduction of the glycan attachment site. MS/MS analysis of N-glycopeptides with QTOF mass analyzers at low collision energy exhibited mostly cleavages of glycosidic linkages providing information on the glycan moiety [157]. Nevertheless, CID mass spectra at elevated collision energies resulted in a significant level of *b*-type and *y*-type peptide fragmentation, thus allowing identification of the glycosylation site. The potential of the nESI QTOF MS/MS in the characterization of O-glycopeptides has also been demonstrated in the analysis of mucin-type glycopeptides with S- or T-linked O-glycans [88, 158] where information on the structure and the attachment site of the O-glycan has been provided based on the *b*-type and *y*-type peptide ions comprising the glycan attachment site.

Alternatively, the development of the complementary mass spectrometric fragmentation techniques of electron-capture dissociation (ECD) [24, 159] and electron-transfer dissociation (ETD) [25] has expanded the analytical options for mapping the modification sites of both N-glycosylation and O-glycosylation. In the ECD technique, which is mainly restricted to FT ICR analyzers, multiply protonated peptide ions are irradiated with low-energy electrons (<0.2 eV) and undergo fragmentation. On the other hand, ETD can be combined with IT, QIT, and Orbitrap analyzers and peptide fragmentation is generated through gas-phase electron-transfer reactions from singly charged anions (e.g., anions of fluoranthene, sulfur dioxide) to a multiply charged peptide/glycopeptide. Unlike the traditional MS/MS techniques, both ECD and ETD appear to retain labile PTMs and induce fragmentation of the peptide backbone with minimal loss of the glycan moiety. ECD and ETD of glycopeptides result in the cleavage of the amine backbone (N–C_α) to generate preferentially c′ and z• fragments ions (nomenclature of Zubarev and co-workers [160]). The intact oligosaccharide moieties are retained in the fragment ions containing the site of glycosylation. Consequently, ECD and ETD represent excellent tools for the localization of modification sites in post-translationally modified proteins [161–163], and there have been few reports of using theses techniques in the characterization of N-linked [142, 164] and O-linked glycopeptides [162, 165]. This is nicely shown in the ESI tandem MS analysis of a tryptic glycopeptide (S295-R313) from horseradish peroxidase (HRP) containing a core-fucosylated and core-xylosylated trimannosyl N-glycan attached to the N298 residue (Fig. 4.10) [142]. The $[M+3H]^{3+}$ ion at m/z 1119 was subjected to CID fragmentation which led to preferential cleavage of glycosidic linkages rather than polypeptide bonds (Fig. 4.10a), thus providing information primarily on the composition and sequence of the glycan moiety. On the contrary, ETD ion activation of the $[M+3H]^{3+}$ ion yielded the cleavage of the peptide backbone with no loss of the glycan moiety, thus leaving the N-glycan modification on the N298 residue intact and providing complete peptide backbone sequence through the observed c′- and z• -ion series (Fig. 4.10b).

Therefore, the use of both CID and ETD ion activation in the LC–MS analysis of glycopeptides has allowed the characterization of both glycan structure (CID-MS/MS) and peptide sequence/site attachment (ETD-MS/MS) within the same LC–MS run. Similarly, use of LC–MS and the ETD and CID fragmentation techniques allowed the identification of two distinct O-glycopeptide structures and three glycosylation sites from the secreted amyloid precursor protein (sAPP695) expressed in CHO cells [166]. This de novo characterization of unknown O-glycosylation sites was extremely challenging due to the large number of S and T residues (27 S and 39 T residues) contained in the protein sequence of the APP fragment. In a modified strategy, LC–MS combined with CID, ETD, and CID of an isolated charge-reduced species derived from ETD was employed to determine the peptide backbone sequence and the site of modification for an O-linked glycosylated peptide fragment of rtPA at the low femtomol level [167].

In case of glycoprotein mapping by MALDI MS, the intense protonated (MH^+) glycopeptides signals are much more stable in CID than the multiply protonated glycopeptide species obtained by ESI. Although PSD, as well as CID, is used for MS/MS of glycopeptides, precise analysis of fragment ion peaks often seems to

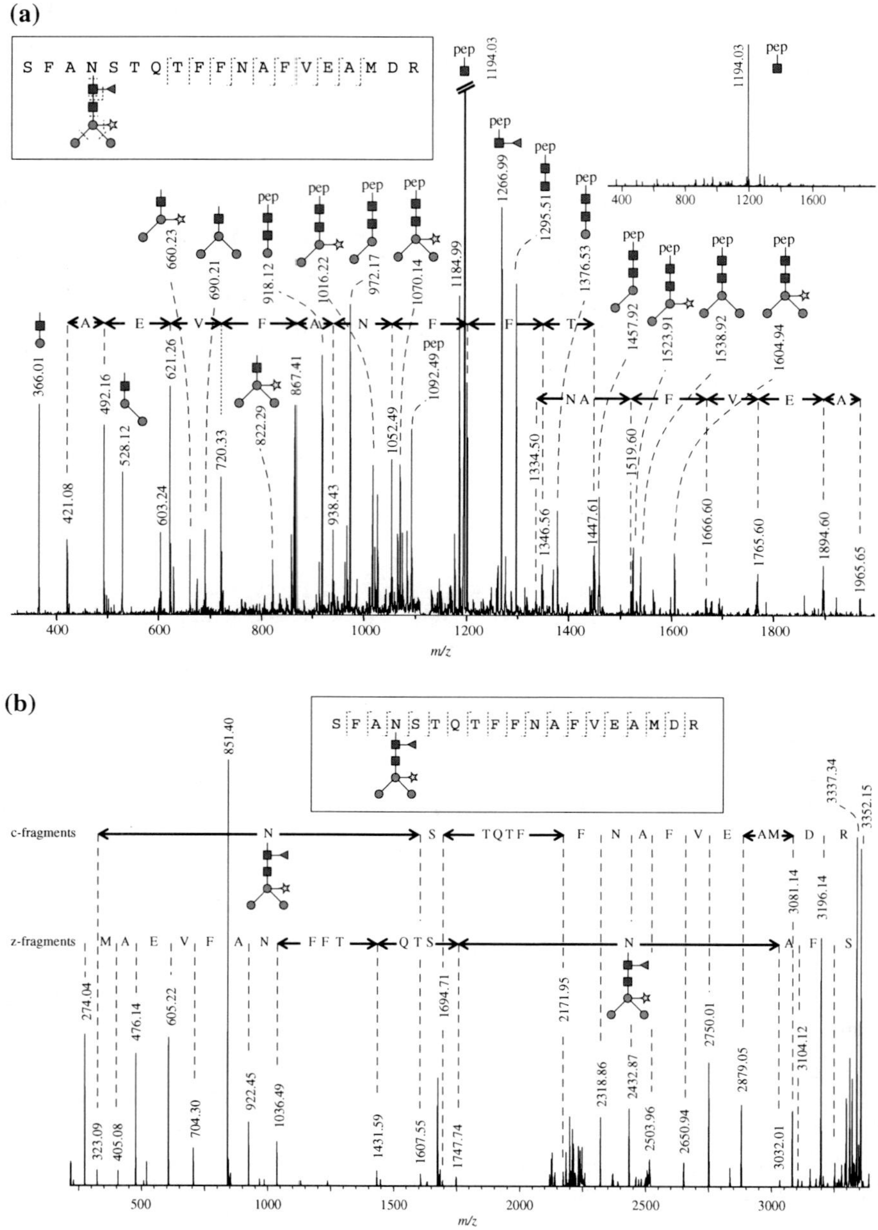

Fig. 4.10 Positive-ion nESI tandem mass spectrum of the RP-HPLC-purified tryptic glycopeptide Ser295-Arg313 from horseradish peroxidase (HRP) containing a core-fucosylated and core-xylosylated trimannosyl N-glycan attached to N298 residue. The $[M+3H]^{3+}$ ion at m/z 1119 was subjected to CID (**a**) and ETD (**b**) fragmentation which led to cleavage of the glycosidic bonds and the peptide backbone, respectively. (Reprinted with permission from Elsevier [142])

be difficult because of preferential and fast deglycosylation, and the limited peptide sequence information [168]. Therefore, fragmentation of these glycopeptide ions by metastable dissociation in a MALDI-TOF/TOF MS or by CID in a MALDI QTOF instrument is performed at higher energies. MALDI-TOF/TOF MS of N-glycopeptides results in a set of cleavages at or near the innermost GlcNAc residue, with the peptide moiety retained in all the fragment ions. In addition, peptide bond cleavages next to the fragmentation of glycosidic bonds are observed (predominantly *b*-type and *y*-type ions), which provide useful peptide sequence tags [169, 170]. All these fragments comprising the N-glycosylation site retain the attached glycan, thus confirming the glycan attachment site. Similarly, MALDI-TOF/TOF MS of O-glycopeptides generate fragmentation patterns from the glycopeptides precursor ions (*b*- and *y*-series ions), which can be used for identification of O-glycosylation sites as it was demonstrated in the case of mucin-type glycopeptide derivatives [171].

At this point, we should point out that parallel *glycomic* analyses for providing information on the linkage, branching points, and configuration of the constituent monosaccharides (microheterogeneity) are also essential in the whole glycoproteomic strategy. In general, the glycans are released by enzymatic or chemical digestion of the glycoprotein or the glycopeptide mixture, undergo permethylation and then subjected to a range of techniques, selected upon the level of analysis to be carried out, that is, fingerprinting, linear sequencing, linkage, branching, or quantitation of monosaccharides [172, 173]. In one of the followed approaches, the permethylated glycans are subjected to LC–MS analysis and the supplied mass spectral information on the specific glycans and their relative amounts can be compared and matched with data at the glycopeptide and overall glycoprotein levels (Fig. 4.7). Incorporation of MALDI-TOF and ESI tandem MS can definitely enhance the analytical potential for tackling complex glycobiology structural issues [43]. Further information on the carbohydrate secondary structures can be provided by well-established methods in structural glycobiology such as X-ray crystallography and especially 2D nuclear magnetic resonance (NMR) analysis [174, 175], albeit the requirement for highly purified glycans and large amounts of sample.

4.2.4 Bioinformatics Tools for Glycoprotein Analysis

Because of the extreme glycan heterogeneity, interpretation of the data produced from the aforementioned glycoproteomic approaches and glycopeptide identification through a comprehensive large-scale data analysis is a challenging task. The development and use of informatics tools and databases for glycobiology research has increased considerably in recent years [176], even though the progress of these tools for glycobiology and glycomics is still in its infancy compared to those already used in genomics and proteomics. Even though, the automated identification of proteins from MS and MS/MS spectra is now almost routine by using informatics tools such as Mascot (http://www.matrixscience.com/), there is lack of rapid and accurate automated tools for retrieving structural information from MS data in case of glycoproteomics. The MS and MS/MS-derived information should be searched for putative glycopeptides predicted by comparison with other glycoconjugate structures derived through the same biosynthetic machinery in other closely related organism, cell line, or tissue. Nevertheless, the inherent complexity of the glycan structures combined with the wide range of techniques employed in their study renders the development of similar automated computational tools a formidable task [177]. In addition, the lack of libraries of glycan sequences similar to the SWISS-PROT protein databank makes matters more challenging. It should be emphasized that more than half of all proteins are glycosylated, based on the analysis of well-characterized proteins deposited in the SWISS-PROT databank [28].

In case of proteomics, bioinformatics tools essentially utilize sequences of the building blocks of proteins (20 amino acids), which are always linked in a predicted linear way in order to provide automated protein identification from MS and tandem MS data. On the contrary, carbohydrates are structurally diverse as their building blocks, the monosaccharides, may be connected in various ways to form branched structures, thus complicating their digital encoding. Moreover, in contrast to protein expression, glycosylation is a non-template-driven synthetic process where multiple enzymes are involved and the final glycoprotein product depends on the type of enzymes expressed in the cell that synthesizes the glycoprotein. The development of bioinformatics methods has mainly found applications in glycosylation analysis, glycomics, glycan structure analysis, glycan biomarker prediction, and glycan structure mining (e.g., using lectins that recognize a certain glycan [178]). In the glycosylation analysis and the prediction of glycosylation binding sites on proteins, the first step is the selective search of protein databases for proteins containing only the consensus sequence for N-linked glycosylation. Several software platforms have been developed for the identification of intact N-linked glycopeptides, such as *GlycoMod* [179], *GlycoPep DB* [180], *Cartoonist* [181], *Peptoonist* [182], and *Glyco-Miner* [183]. These methods can be used mainly for glycopeptides generated from specific enzymes, for example, trypsin or endoproteinase Glu-C, whereas *GlycoX* [184] can be used for interpretation of mass spectra obtained from non-specific proteases, such as protease K. *Cartoonist* is one of the earlier developed glycomic MS interpretation approaches containing a library of several hundred archetype glycans derived from information about biosynthetic pathways and employing a set of rules to modify these structures. *Cartoonist* incorporates the same assumptions used by human expert in the annotation of MS data, and it is used to automatically annotate N-glycans in MALDI mass spectra with diagrams or cartoons of the most possible glycans consistent with the observed mass values. *Peptoonist* [182] uses MS/MS data to identify glycosylated peptides in LC-ESI MS runs of enzymatically digested glycoproteins and MS data to identify the N-glycans present on each of those peptides. On the other hand, the *GlyDB* [185] approach has been developed to address the need for structure annotation of N-linked glycopeptides in the LC-ESI MS analysis of glycoprotein proteolytic digests. The annotation of low-resolution tandem MS spectra of N-linked glycopeptides arising from low-energy CID, where cleavage along the glycosidic bonds occurs preferentially, is based on matching experimental spectra to theoretical spectra generated by a linearized database of glycan structures using the established search engine SEQUEST. Similarly, *GlycoPep ID* [186] is a web-based tool used to identify the peptide moiety of either sialylated, sulfated, or both sialylated and sulfated glycopeptides, by correlating the product ions of suspected glycopeptides to a peptide composition. Following the identification of the peptide portion, the mass of the remaining segment can be attributed to the carbohydrate component.

Even though the development and use of informatics tools and databases for glycobiology and glycomics research has increased significantly in recent years, it has lagged behind the development of similar tools for genomics and proteomics. This drawback arises from the lack of comprehensive and well-organized compilations of glycan sequences and efficient automatic assignment procedures for high-throughput analysis of glycans. Most of the aforementioned library-based sequencing and N-glycopeptide identification tools for MS data interpretation are not publicly available; they have their own standards, databases and/or run on a special hardware platform. Moreover, the independently developed database with their own format and language along with the absence of publicly available databases with carefully assigned MS spectra of glycans hinders the development of efficient scoring algorithms. Therefore, rules should be established for the standardization of the structural description of glycans and the deposit of glycan structures and the associated glyco-related data in databases of complex glycan structures. In addition, the deposit of complex glycan structures and glyco-related data in generally accepted databases should be maintained by well-recognized international institutions such as NCBI (www.ncbi.nlm.nih.gov) and European Bioinformatics Institute (EMBL-EBI, www.ebi.ac.uk), which house genome sequencing data (GenBank) and protein related databases, respectively. It is also essential to ensure the intercompatibility of the related data formats, in order to facilitate data exchange between different databases and efficient cross-linking and referencing thereof between various projects.

Toward this direction, the EU FP6-funded EUROCarbDB project (http://www.ebi.ac.uk/eurocarb/home.action) was an initiative to create the technical framework where interested research groups could feed in their complex glycan structural data, which would be archived and maintained at the EMBL-EBI. Other most prominent publically available glycan-related databases are the Consortium for Functional Glycomics (CFG) relational database (http://www.functionalglycomics.org/glycomics/common/jsp/firstpage.jsp), the Kyoto Encyclopedia of Genes and Genomes glycome informatics resource (KEGG GLYCAN) (http://www.genome.jp/kegg/glycan/), and Glycosciences.de (http://www.dkfz.de/spec/glycosciences.de/sweetdb/index.php). Finally, genomic/proteomic findings need to be integrated with biomedical studies where glycan structures can serve as biomarkers for specific diseases or malfunctions [187], like the ones provided by the KEGG resources [188–190].

4.3 Disulfide Bond Formation

4.3.1 MS Determination of Disulfide Bonds

Even though glycosylation enjoys more popularity in the PTM literature, disulfide bond formation is one of the most common PTMs playing a critical role in establishing and stabilizing the three-dimensional structure of proteins [6, 191]. The physiological and pathological relevance of disulfide bonds to diseases has been recognized in several cases, such as tumor immunity [192], neurodegenerative diseases [193], and G-protein receptors [194]. These cross-linkages between the sulfhydryl groups of two C residues can be either *intramolecular or intermolecular.* The former stabilize the tertiary structures of proteins, while the latter are involved in stabilizing quaternary structures of proteins [195, 196]. For protein therapeutics, the generation of correctly folded recombinant proteins is of paramount importance. Difficulties in folding recombinant protein products are common from *E.coli* cell line, thus resulting in loss of specific activity compared to the native material. Similarly, over-expression of proteins in CHO cell line leads to disulfide scrambling. Therefore, there are significant efforts to develop reliable methods for mapping disulfide bonds in therapeutic proteins, thus ensuring drug quality. The determination of disulfide bond arrangements of proteins not only provides insights into protein activity relationships but also guides further structural determination by NMR and X-ray crystallography. The first step in disulfide mapping is the determination of the number of disulfides in a given protein, which can be readily deduced by a simple MS analysis before and after protein reduction. This is nicely illustrated in the ESI MS analysis of recombinant interferon α-2b and GM-CSF, where reduction resulted in a 4 Da shift in the measured M_r, thus indicating the presence of two disulfide bonds [197]. In case of GM-CSF, the ESI mass spectrum prior to and after treatment with β-mercaptoethanol clearly showed a 4 Da shift in the measured M_r (Fig. 4.11 insets), hence confirming the presence of two disulfide bonds in the recombinant protein product.

4.3.2 Disulfide Mapping

Following the determination of the number of disulfide linkages, mapping of the protein's primary sequence by proteolytic cleavage of the protein between half-cystine residues to produce disulfide-linked peptides and MS analysis of the resulting peptide fragments allows the identification of the existing disulfide arrangement [198]. The potential of MS in this disulfide mapping approach was first realized with the implementation of soft ionization techniques, such as FAB/liquid secondary ion (LSI) [199–201], plasma desorption (PD) [202, 203], and later by the more sensitive method of MALDI [18, 204]. That was nicely illustrated in the disulfide mapping of several therapeutic proteins, such as recombinant human interferon

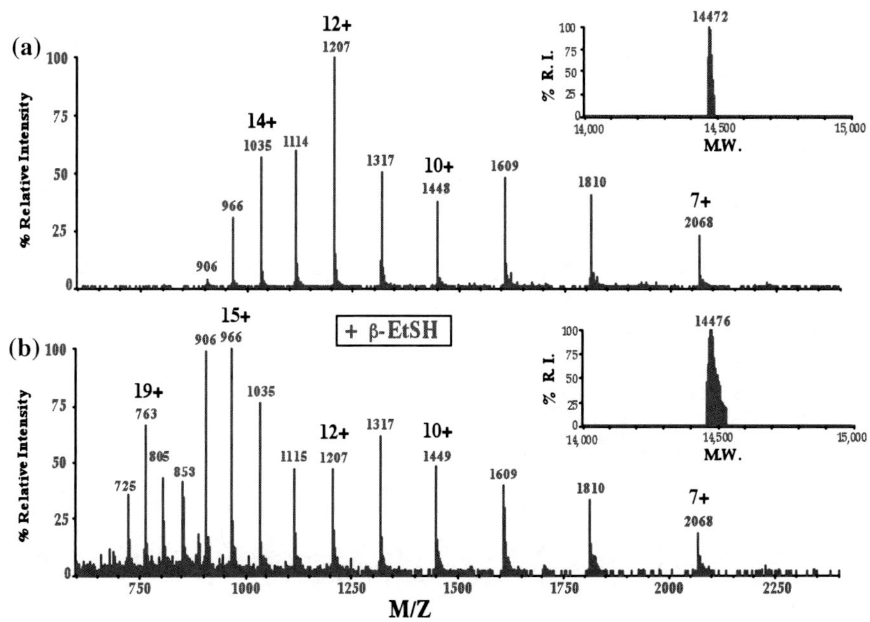

Fig. 4.11 Positive-ion ESI mass spectrum of recombinant human granulocyte–macrophage colony-stimulating factor (rhGMCSF) in 1 % HCOOH (**a**) and after treatment with β-mercaptoethanol (**b**). The deconvoluted spectra are shown in the *insets*. (Reprinted with permission from Wiley [197])

α-2b (INTRON A) [205, 206], human growth hormone [203] and IL-4 [207] by FAB, PD, and MALDI mapping. It should be noted that weak ion signals corresponding to the MH^+ of the constituent C-containing peptides were also present in the FAB, LSI, PD, and MALDI mass spectra arising from fragmentation of disulfide-linked peptides during the ionization process [208]. This is shown in the LSI mass spectrum of the disulfide-linked tryptic core peptide of rhGM-CSF (expected M_r 7,613) (Fig. 4.12), where additional signals at 5665.2 and 4412.4 Da were also observed due to the presence of the partially reduced peptides T_5-S–S-T_{11} and T_{11}-S–S-T_{13}, respectively (Fig. 4.12, inset) [197].

Even though the disulfide-linked peptides yield unique mass spectral signals, the protein fragmentation should be carefully controlled to avoid rearrangement of disulfide bonds (disulfide scrambling), which can take place at neutral and alkaline pH [209]. Therefore, protein cleavage methods performed in aqueous solvents at acidic pH are preferred, such as cyanogen bromide [210] and pepsin [200]. This acidic pH is also optimum for disrupting the protein conformation and making the cleavage sites between half-cystine residues more accessible. That was nicely illustrated in the first report on the disulfide mapping of insulin where FAB MS of peptic digest peptides was combined with *Edman* analysis for disulfide bond analysis [200]. The intramolecularly linked peptides are identified by the 2 Da increase

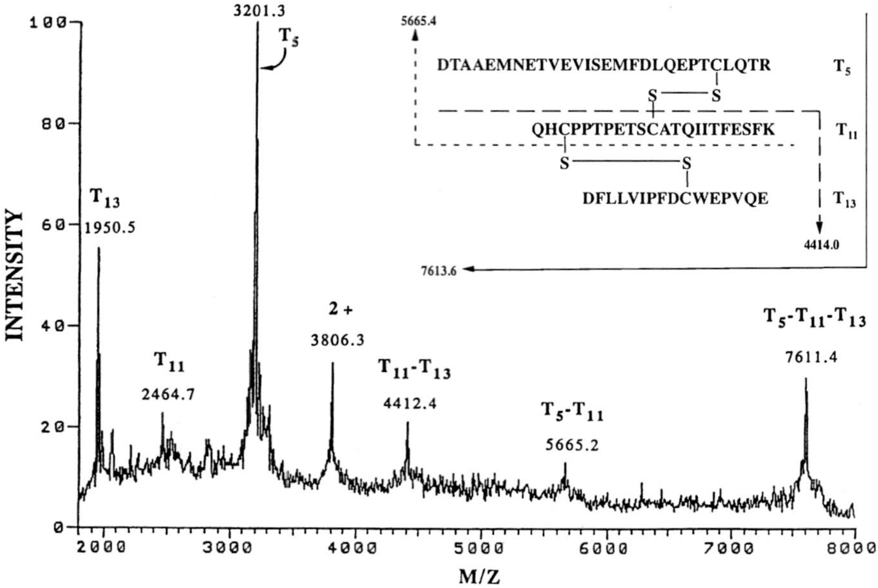

Fig. 4.12 Positive-ion Cs^+ LSI mass spectrum of the HPLC-isolated fraction containing the disulfide-linked tryptic core peptide T_5-T_{11}-T_{13} of rhGM-CSF. (Reprinted with permission from Wiley [197])

upon reduction in their constituent half-cystines with β-mercaptoethanol or dithiothreitol, whereas intermolecularly bridged peptides yield protonated MH^+ signals of the constituent half-cystine-containing peptide fragments.

The advent of ESI [13] has made LC-ESI MS the favorite approach for analyzing the enzyme-generated protein fragments and mapping disulfide linkages in recombinant proteins [198, 206]. Analysis of the peptide mixtures before and after reduction generally allows the identification of the C residues involved in disulfide bonding, taking all aforementioned precautions to minimize disulfide scrambling. It should be noted that ESI MS analysis of disulfide-linked peptides is not conducive to peptide signals arising from partial disulfide bond reduction, as shown in the ESI mass spectrum of the disulfide-linked tryptic peptide T_{20}-$T_{25,26}$ of IL-5Rα (Fig. 4.13).

When protein chains are disulfide-linked and proteolysis between half-cystine residues is not possible, identification of the exact location of the disulfide linkage often requires (1) successive proteolytic digestions, such as the ones demonstrated for interleukin-13 (chymotrypsin plus *S. aureus* V8 protease) [211] and rtPA (Lys-C plus trypsin) [26] or (2) chromatographic separation of the enzyme-derived protein fragments coupled with online MS/MS analysis (e.g., LC-ESI MS/MS), and/or off-line MS/MS analysis and *Edman* sequencing [212, 213]. This is essential for proteins where three proteolytic fragments are linked by intermolecular disulfides or where two peptide chains contained an intramolecular disulfide

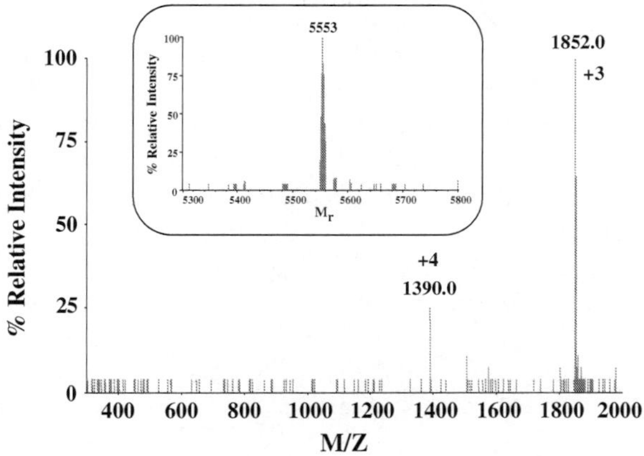

Fig. 4.13 Positive-ion ESI mass spectrum of Sf9 IL-5Rα tryptic fraction (Fig. 4.8, TIC *peak* 19) containing the disulfide-bonded peptides T_{20} and $T_{25,26}$ with M_r value of 5553

and no further proteolysis is possible. The existence of disulfide bonds is usually confirmed by fragmentation of putatively disulfide-linked peptides by MS/MS analysis following ionization by FAB [214], ESI [215], or MALDI PSD [216]. In the MALDI PSD approach, the characteristic ion triplet separated by 33 Da, arising from cleavage at the C–S bond with a concomitant proton transfer [168], can be used as a diagnostic tool for the location and identification of disulfide-paired peptides, even from complex digest mixtures of proteins.

The LC-ESI MS and tandem MS approach is especially valuable in the disulfide mapping of protein receptors and therapeutic proteins having high M_r, such as rtPA and mAb. In mAb, the inter- and intrachain disulfides are responsible for maintaining the characteristic three-dimensional antibody structure, which allows the highly specific antigen binding. Therefore, complete disulfide mapping in mAb is critical for ensuring its therapeutic activity, because incomplete disulfide linkages and/or free sulfhydryl groups can lead to antibody fragments with no antigen-binding activity [217]. In case of the anti-HER2 mAb (Herceptin) that interferes with the HER2/neu receptor and used for the treatment of early-stage breast cancer, the disulfides were completely mapped by LC-ESI MS with the combination of ETD and CID fragmentation [218]. Using ETD cleaves preferentially the disulfides into two polypeptides while CID generates mainly peptide backbone cleavage (with the disulfides intact). This approach was successful in mapping a total of 16 disulfides, 12 intra- and 4 intermolecular, in anti-HER2 mAb and a similar therapeutic mAb. This ETD fragmentation strategy can be further enhanced by CID-MS^3 on the dissociated peptides (after ETD) in order to provide corroborating information on the linkage assignment. The same multi-fragmentation approach in combination with multi-enzyme digestion scheme (Lys-C followed by trypsin and Glu-C) was employed in the mapping of the 17 disulfide linkages in human growth hormone [26] and rtPA, as well as for the identification of the unpaired C residue in rtPA [219]. The ETD-MS^2 spectrum of the disulfide-linked tryptic peptide T_7-T_8-T_9 clearly showed that the unassigned C residue (C83) was found to be paired with either a glutathione or C molecule, which could shed light into the activation or signaling pathway of rtPA. A novel approach based on IM MS was also employed for the rapid characterization of disulfide variants in intact IgG2 mAb [220]. IM MS revealed two to three gas-phase conformer populations for IgG2, compared to only one conformer for IgG1 mAb and a C232S mutant of IgG2, thus indicating that the observed conformers are apparently related to disulfide variants. Therefore, IM MS is a new powerful tool for the characterization of intact mAb and may be useful for fingerprinting higher-order structures of these protein therapeutics.

Finally, disulfide mapping combined with *stable isotope-labeling of peptides with* ^{18}O greatly facilitated the identification and characterization of disulfide-linked peptides [221]. Isotope profiles of enzymatically generated peptides produced in 50 % $H_2{}^{18}O$ (v/v) in $H_2{}^{16}O$ would produce unique doublets separated by 2 Da, whereas the disulfide-linked peptides should be distinctly different than single-chain peptides [222]. Therefore, the disulfide-linked peptides could be identified in complex peptic digests or chromatographic fractions thereof by MS analysis, and especially MALDI-TOF MS. This procedure is ideally performed in acidic solutions (e.g., peptic digestion) in order to preclude disulfide rearrangement and it may also be used to aid the interpretation of product-ion spectra of disulfide-linked peptides.

4.4 Future Prospects and Challenges

In the past two decades, recombinant protein therapeutics have changed the face of modern medicine as they provide innovative and effective therapies for numerous previously incurable diseases. Protein therapeutics have already a significant role in almost every field of medicine, even though this role is still only in its infancy. The number of recombinant proteins in clinical trials for new and existing therapeutic targets continues to increase annually, as does the total number of protein-based pharmaceuticals reaching the marketplace. The acceptance of the various protein therapies can be attributed to the increasing prevalence of chronic diseases, such as cancer, diabetes, cardiovascular diseases, and neurological/neurodegenerative disorders. In addition, the rising penetration of medical insurance industry has made protein therapeutics available to a wider population. The global protein therapeutics market is expected to grow at an annual rate of 13 % during 2012–2015, arising from the introduction of new protein therapeutics in the major sectors of protein therapeutics market, which include mAb, insulin, interferons, G-CSF, tPA, EPO, coagulation factors, etc.

Recombinant therapeutic proteins for human use must be characterized thoroughly prior to clinical development in order to satisfy the rigorous regulatory requirements (ICH Q6B guidance) [12]. In addition, the manufactured final product should be comparable to that used in preclinical and clinical studies, and its purity, potency, safety, stability, and batch-to-batch consistency should be established. Advances in MS techniques, especially MALDI and ESI, have made MS-based mapping approaches powerful and essential analytical tools for structure characterization of therapeutic proteins and evaluation of recombinant protein heterogeneity including identification of PTMs, sequence variants, and degradation products in recombinant proteins. Structure characterization of all PTMs in a protein is of a great concern for regulatory agencies, such as glycosylation and disulfide linkages. Glycosylation, the most common form of PTM, plays a crucial role in the stability and therapeutic potency of the glycoprotein, as it was demonstrated for rHuEPO. Moreover, changes in levels and types of glycosylation can be associated with certain diseases, such as aggressive breast cancer [223], thus making glycoprotein screening invaluable, not only for diagnostic purposes, but also for design of novel therapeutic drugs. In addition, glycan profiling of normal and diseased forms of a glycoprotein has provided new insights into future research in rheumatoid arthritis, prostate cancer, and congenital disorders of glycosylation [224–226].

In general, LC–MS and tandem MS peptide mapping is the standard and well-accepted approach by the regulatory agencies (FDA, EMA) for identifying PTMs and establishing the recombinant product purity. Nevertheless, a variety of tandem MS experiments should be performed in order to provide insights into the glycan structure (low-energy CID) and peptide backbone sequence/site attachment (ETD and/or high-energy CID) within the same LC–MS run. These MS fragmentation approaches are ideally suited with higher-resolution mass spectrometers, for example, QTOF, IM TOF, and Orbitrap analyzers. The interpretation of the complex and abundant data generated from these experiments undoubtedly requires the support of the growing resources of bioinformatics tools for automated search and identification of glycopeptides and the attached glycans. The advantages of this multi-fragmentation approach (ETD, CID) combined with these high-resolution mass analyzers are also essential in the mapping of disulfide linkages in recombinant protein therapeutics. Even though disulfide linkages are assigned in the initial development stage of the protein, they often need to be reassigned in large-scale production or when the cell production conditions change. Therefore, confirmation of disulfide linkages and identification of any unpaired C location needs to be provided by the aforementioned mapping approach, thus ensuring the proper folding and biological activity of the protein therapeutic product. The latter is especially critical in case of developing innovative treatments using mAbs, which are expected to top the global market in protein therapeutics in the near future. Fast growth in protein therapeutics will also strengthen the emerging segment of biogenerics (biosimilars), which is a key future growth sector due to patent expirations of the branded innovator products. In that case, a thorough characterization of the biosimilar product in terms of glycosylation occupancy and identification of disulfide linkages will be essential for evaluating the comparability between the innovator and biosimilar products. In case of a generic variant of rtPA (TNK-tPA) [150], the analysis strategy was focused on regions that could impact the clot lysis activity such as the glycosylation occupancy at the N184 site and the different extent of oxidation at several M sites. Finally, the advent of more accurate and sensitive instrumentation will enable the development of novel methodologies for the structural characterization of recombinant protein therapeutics and shed some light into the role of specific carbohydrates in many complex biological interactions. That, in turn, will incite the development of novel glycosylated therapeutics for treating infectious, chronic, and other diseases, as well as the improvement of the immunogenicity and pharmacokinetic profiles of existing protein therapeutics.

Acknowledgments We acknowledge the kind permission of the Schering-Plough Research Institute to reproduce previously reported, but unpublished data regarding the IL-5Rα.

References

1. Walsh G (2006) Biopharmaceutical benchmarks 2006. Nat Biotechnol 24(7):769–776
2. Aggarwal S (2007) What's fueling the biotech engine? Nat Biotechnol 25(10):1097–1104
3. Roach P, Woodworth JR (2002) Clinical pharmacokinetics and pharmacodynamics of insulin lispro mixtures. Clin Pharmacokinet 41:1043–1057
4. Dwek RA (1996) Glycobiology: toward understanding the function of sugars. Chem Rev 96:683–720
5. Collins MO, Yu L, Choudhary JS (2007) Analysis of protein phosphorylation on a proteome-scale. Proteomics 7:2751–2768
6. Wedemeyer WJ, Welker E, Narayan M et al (2000) Disulfide bonds and protein folding. Biochemistry 39:4207–4216
7. Graves JD, Krebs EG (1999) Protein phosphorylation and signal transduction. Pharmacol Ther 82:111–121
8. Hunter T (2000) Signaling-2000 and beyond. Cell 100:113–127
9. Cohen P (2002) The origins of protein phosphorylation. Nat Cell Biol 4:E127–E130
10. Walsh G, Jefferis R (2006) Post-translational modifications in the context of therapeutic proteins. Nat Biotechnol 24:1241–1252
11. Li H, d'Anjou M (2009) Pharmacological significance of glycosylation in therapeutic proteins. Curr Opin Biotechnol 20:678–684
12. CPMP/ICH harmonised tripartite guideline Q6B (1999) Specifications: test procedures and acceptance criteria for biotechnological/biological products. March 1999 and EMA guideline (2010) requirements for quality documentation concerning biological investigational medicinal products in clinical trials. February 2010
13. Whitehouse CM, Dreyer RN, Yamashita M et al (1985) Electrospray interface for liquid chromatographs and mass spectrometers. Anal Chem 57:675–679
14. Fenn JB, Mann M, Meng CK et al (1989) Electrospray ionization for mass spectrometry of large biomolecules. Science 246:64–71
15. Smith RD, Udseth H (1988) Capillary zone electrophoresis-MS. Nature 331:639–640
16. Kelly JF, Locke SJ, Ramaley L et al (1996) Development of electrophoretic conditions for the characterization of protein glycoforms by capillary electrophoresis-electrospray mass spectrometry. J Chromatogr A 720:409–427
17. Karas M, Bachmann D, Bahr U et al (1987) Matrix-assisted ultraviolet-laser desorption of nonvolatile compounds. Int J Mass Spectrom Ion Process 78:53–68
18. Karas M, Hillenkamp F (1988) Laser desorption ionization of proteins with molecular masses exceeding 10,000 daltons. Anal Chem 60:2299–2301
19. Hancock WS, Wu SL, Shieh P (2002) The challenges of developing a sound proteomics strategy. Proteomics 2:352–359
20. Larsen MR, Trelle MB, Thingholm TE et al (2006) Analysis of posttranslational modifications of proteins by tandem mass spectrometry. Biotechniques 40:790–798
21. Covey T, Shushan B, Bonner R, Schröder W, Hucho F (1991) Methods in protein sequence analysis. In: Jörnvall H, Höög JO, Gustavsson AM (eds) LC/MS and LC/MS/MS screening of the sites of posttranslational modification in proteins. Birkhäuser Press, Basel

22. Dell A, Morris HR (2001) Glycoprotein structure determination by mass spectrometry. Science 291:2351–2356
23. Bateman RH, Carruthers R, Hoye JB et al (2002) A novel precursor ion discovery method on a hybrid quadrupole orthogonal acceleration time-of-flight (Q-TOF) mass spectrometer for studying protein phosphorylation. J Am Soc Mass Spectrom 13:792–803
24. Zubarev RA, Kelleher NL, McLafferty FW (1998) Electron capture dissociation of multiply charged protein cations. A nonergodic process. J Am Chem Soc 120:3265–3266
25. Syka JE, Coon JJ, Schroeder MJ et al (2004) Peptide and protein sequence analysis by electron transfer dissociation mass spectrometry. Proc Natl Acad Sci USA 101:9528–9533
26. Wu SL, Jiang H, Lu Q et al (2009) Mass spectrometric determination of disulfide linkages in recombinant therapeutic proteins using online LC-MS with electron-transfer dissociation. Anal Chem 81:112–122
27. Wang D, Hincapie M, Rejtar T et al (2011) Ultrasensitive characterization of site-specific glycosylation of affinity-purified haptoglobin from lung cancer patient plasma using 10 μm i.d. porous layer open tubular liquid chromatography-linear ion trap collision-induced dissociation/electron transfer dissociation mass spectrometry. Anal Chem 83(6):2029–2037
28. Apweiler R, Hermjakob H, Sharon N (1999) On the frequency of protein glycosylation, as deduced from analysis of the SWISS-PROT database. Biochim Biophys Acta 1473:4–8
29. Spiro RG (2002) Protein glycosylation: nature, distribution, enzymatic formation, and disease implications of glycopeptides bonds. Glycobiology 12:43R–56R
30. Schachter H (2001) The clinical relevance of glycobiology. J Clin Invest 108:1579–1582
31. Dwek MV, Brooks SA (2004) Harnessing changes in cellular glycosylation in new cancer treatment strategies. Curr Cancer Drug Targets 4:425–442
32. Wuhrer M (2007) Glycosylation profiling in clinical proteomics: heading for glycan biomarkers. Expert Rev Proteomics 4:135–136
33. Dube DH, Bertozzi CR (2005) Glycans in cancer and inflammation—potential for therapeutics and diagnostics. Nat Rev Drug Discov 4:477–488
34. Fuster MM, Esko JD (2005) The sweet and sour of cancer: glycans as novel therapeutic targets. Nat Rev Cancer 5:526–542
35. An HJ, Kronewitter SR, de Leoz ML et al (2009) Glycomics and disease markers. Curr Opin Chem Biol 13:601–607
36. Niwa T (2006) Mass spectrometry for the study of protein glycation in disease. Mass Spectrom Rev 25:713–723
37. Morelle W, Canis K, Chirat F et al (2006) The use of mass spectrometry for the proteomic analysis of glycosylation. Proteomics 6:3993–4015
38. Bennett CS, Dean SM, Payne RJ et al (2008) Sugar-assisted glycopeptide ligation with complex oligosaccharides: scope and limitations. J Am Chem Soc 130:11945–11952
39. Novotny MV, Mechref Y (2005) New hyphenated methodologies in high sensitivity glycoprotein analysis. J Sep Sci 28:1956–1968
40. Wuhrer M, Deedler AM, Hokke CH (2005) Protein glycosylation analysis by liquid chromatography-mass spectrometry. J Chromatogr B 825:124–133
41. Geyer H, Geyer R (2006) Strategies for analysis of glycoprotein glycosylation. Biochim Biophys Acta 1764:1853–1869
42. Mariño K, Bones J, Kattla JJ et al (2010) A systematic approach to protein glycosylation analysis: a path through the maze. Nat Chem Biol: 713–723
43. North SJ, Hitchen PG, Haslam SM et al (2009) Mass spectrometry in the analysis of N-linked and O-linked glycans. Curr Opin Struct Biol 19:498–506
44. Axford J (2001) The impact of glycobiology on medicine. Trends Immunol 22:237–239
45. Mortz E, Sareneva T, Haebel S et al (1996) Mass spectrometric characterization of glycosylated interferon-gamma variants separated by gel electrophoresis. Electrophoresis 17:925–931
46. Nawarak J, Phutrakul S, Chen ST (2004) Analysis of lectin-bound glycoproteins in snake venom from the elapidae and viperidae families. J Proteom Res 3:383–392
47. Mechref Y, Novotny MV (2002) Structural investigations of glycoconjugates at high sensitivity. Chem Rev 102:321–369

48. Ramdani B, Nuyens V, Codden T et al (2003) Analyte comigrating with trisialotransferrin during capillary zone electrophoresis of sera from patients with cancer. Clin Chem 49:1854–1864
49. Smith RD, Loo JA, Edmonds CG et al (1990) New developments in biochemical mass spectrometry: electrospray ionization. Anal Chem 62:882–899
50. Tsarbopoulos A, Pramanik BN, Nagabhushan TL et al (1995) Structural analysis of the CHO-derived interleukin-4 by liquid-chromatography/electrospray ionization mass spectrometry. J Mass Spectrom 30:1752–1763
51. Tsarbopoulos A, Bahr U, Karas M, Pramanik BN (2002) Structural analysis of glycoproteins by electrospray ionization mass spectrometry. In: Pramanik BN, Ganguly AK, Gross ML (eds) Applied electrospray mass spectrometry. Marcel Dekker, New York
52. Duffin KL, Welply JK, Huang E et al (1992) Characterization of N-linked oligosaccharides by electrospray and tandem mass spectrometry. Anal Chem 64:1440–1448
53. Rajan N, Tsarbopoulos A, Kumarasamy R et al (1995) Characterization of recombinant human interleukin-4 receptor from CHO cells: Role of N-linked oligosaccharides. Biochem Biophys Res Commun 206:694–702
54. Rush RS, Derby PL, Smith DM et al (1995) Microheterogeneity of erythropoietin carbohydrate structure. Anal Chem 67:1442–1452
55. Wilm M, Mann M (1996) Analytical properties of the nanoelectrospray ion source. Anal Chem 68:1–8
56. Verentchikov AN, Ens W, Standing KG (1994) Reflecting time-of-flight mass spectrometer with an electrospray ion source and orthogonal extraction. Anal Chem 66:99–107
57. Makarov A (2000) Electrostatic axially harmonic orbital trapping: a high-performance technique of mass analysis. Anal Chem 72:1156–1162
58. Olivova P, Chen W, Chakraborty AB et al (2008) Determination of N-glycosylation sites and site heterogeneity in a monoclonal antibody by electrospray quadrupole ion-mobility time-of-flight mass spectrometry. Rapid Commun Mass Spectrom 22:29–40
59. Benesch JLP, Robinson CV (2006) Mass spectrometry of macromolecular assemblies: preservation and dissociation. Current Opin Struct Biol 16:245–251
60. Clemmer DE, Jarrold MF (1997) Ion mobility measurements and their applications to clusters and biomolecules. J Mass Spectrom 32:577–592
61. Carter P, Presta L, Gorman CM et al (1992) Humanization of an Anti-p185HER2 antibody for human cancer therapy. Proc Natl Acad Sci USA 89:4285–4289
62. Damen CWN, Chen W, Chakraborty AB et al (2009) Electrospray ionization quadrupole ion-Mobility time-of-flight mass spectrometry as a tool to distinguish the lot-to-lot heterogeneity in N-Glycosylation profile of the therapeutic monoclonal antibody Trastuzumab. J Amer Soc Mass Spectrom 20:2021–2033
63. Dube S, Fisher JW, Powell JS (1988) Glycosylation at specific sites of erythropoietin is essential for biosynthesis, secretion and biological function. J Biol Chem 263:17516–17521
64. Delorme E, Lorenzini T, Giffin J et al (1992) Role of glycosylation on the secretion and biological activity of erythropoietin. Biochemistry 31:9871–9876
65. Ploug M, Rahbek-Nielsen H, Nielsen PF et al (1998) Glycosylation profile of a recombinant urokinase-type plasminogen activator receptor expressed in Chinese hamster ovary cells. J Biol Chem 273(22):13933–13943
66. Tsarbopoulos A, Prongay A, Baldwin S et al (1996) Mass spectrometric analysis of the Sf9 cell-derived interleukin-5 Receptor. In: Proceedings of the 44th ASMS conference on mass spectrometry and allied topics, Portland: 12–16 May
67. Karas M, Bahr U, Strupat K et al (1995) Matrix dependence of metastable fragmentation of glycoproteins in MALDI TOF mass spectrometry. Anal Chem 67:675–679
68. Giménez E, Benavente F, Barbosa J et al (2007) Towards a reliable molecular mass determination of intact glycoproteins by matrix-assisted laser desorption/ionization time-of-flight mass spectrometry. Rapid Commun Mass Spectrom 21:2555–2563
69. Tsarbopoulos A, Pramanik BN, Karas M et al (1995) Factors affecting the choice of matrix in matrix-assisted laser desorption/ionization time-of-flight mass spectrometry of glycoproteins. J Mass Spectrom: S207–S209

70. Liu CL, Bowers LD (1997) Mass spectrometric characterization of the β-subunit of human chorionic gonadotropin. J Mass Spectrom 32:33–42
71. Neusüb C, Demelbauer U, Pelzing M (2005) Glycoform characterization of intact erythropoietin by capillary electrophoresis-electrospray-time of flight-mass spectrometry. Electrophoresis 26:1442–1450
72. Demelbauer UM, Plematl A, Kremser L et al (2004) Characterization of glyco isoforms in plasma-derived human antithrombin by on-line capillary zone electrophoresis-electrospray ionization-quadrupole ion trap-mass spectrometry of the intact glycoproteins. Electrophoresis 25:2026–2032
73. Balaguer E, Demelbauer U, Pelzing M et al (2006) Glycoform characterization of erythropoietin combining glycan and intact protein analysis by capillary electrophoresis–electrospray–time-of-flight mass spectrometry. Electrophoresis 27:2638–2650
74. Balaguer E, Neususs C (2006) Glycoprotein characterization combining intact protein and glycan analysis by capillary electrophoresis-electrospray ionization-mass spectrometry. Anal Chem 78:5384–5393
75. Thakur D, Rejtar T, Karger BL et al (2009) Profiling the glycoforms of the intact α subunit of recombinant human chorionic gonadotropin by high-resolution capillary electrophoresis-mass spectrometry. Anal Chem 81:8900–8907
76. Sanz-Nebot V, Balaguer E, Benavente F et al (2007) Characterization of transferrin glycoforms in human serum by CE-UV and CE-ESI-MS. Electrophoresis 28:1949–1957
77. Hang HC, Bertozzi CR (2005) The chemistry and biology of mucin-type Olinked glycosylation. Bioorg Med Chem 13:5021–5034
78. Wopereis S, Lefeber DJ, Morava E et al (2006) Mechanisms in protein O-glycan biosynthesis and clinical and molecular aspects of protein O-glycan biosynthesis defects: a review. Clin Chem 52:574–600
79. Kornfeld R, Kornfeld S (1985) Assembly of asparagine-linked oligosaccharides. Annu Rev Biochem 54:631–664
80. Vance BA, Wu W, Ribaudo RK et al (1997) Multiple dimeric forms of human CD69 result from differential addition of Nglycans to typical (Asn–X–Ser/Thr) and atypical (Asn–X–Cys) glycosylation motifs. J Biol Chem 272:23117–23122
81. Kelleher NL, Lin H, Valaskovic G et al (1999) Top down versus bottom up protein characterization by tandem high-resolution mass spectrometry. J Am Chem Soc 121:806–812
82. Kelleher NL (2004) Top-down proteomics. Anal Chem 76:196A–203A
83. Reid GE, McLuckey SA (2002) 'Top down' protein characterization via tandem mass spectrometry. J Mass Spectrom 37:663–675
84. Siuti N, Kelleher NL (2007) Decoding protein modifications using top-down mass spectrometry. Nat Methods 4:817–821
85. Ling V, Guzzetta AW, Canova-Davis E et al (1991) Characterization of the tryptic map of recombinant DNA derived tissue plasminogen activator by high-performance liquid chromatography-electrospray ionization mass spectrometry. Anal Chem 63:2909–2915
86. Huddleston MJ, Bean MF, Carr SA (1993) Collisional fragmentation of glycopeptides by electrospray ionization LC/MS and LC/MS/MS: methods for selective detection of glycopeptides in protein digests. Anal Chem 65:877–884
87. Amon S, Alina D, Zamfir AD et al (2008) Glycosylation analysis of glycoproteins and proteoglycans using capillary electrophoresis-mass spectrometry strategies. Electrophoresis 29:2485–2507
88. Alving K, Körner R, Paulsen H et al (1998) Nanospray-ESI low-energy CID and MALDI post-source decay for determination of O-glycosylation sites in MUC4 peptides. J Mass Spectrom 33:1124–1133
89. Hunt DF, Shabanowitz J, Yates JR et al (1986) Tandem quadrupole Fourier-transform mass spectrometry of oligopeptides and small proteins. Proc Natl Acad Sci USA 83:6233–6237
90. Mechref Y, Madera M, Novotny MV (2009) Assigning glycosylation sites and microheterogeneities in glycoproteins by liquid chromatography/tandem mass spectrometry. Methods Mol Biol 492:161–180

91. Annesley TM (2003) Ion suppression in mass spectrometry. Clin Chem 49:1041–1044
92. Temporini C, Calleri E, Massolini G et al (2008) Integrated analytical strategies for the study of phosphorylation and glycosylation in proteins. Mass Spectrom Rev 27:207–236
93. Drake RR, Schwegler EE, Malik G et al (2006) Lectin capture strategies combined with mass spectrometry for the discovery of serum glycoprotein biomarkers. Mol Cell Proteomics 5:1957–1967
94. Alvarez-Manilla G, Atwood J III, Guo Y et al (2006) Tools for glycoproteomic analysis: size exclusion chromatography facilitates identification of tryptic glycopeptides with N-linked glycosylation sites. J Proteome Res 5:701–708
95. Tajiri M, Yoshida S, Wada Y (2005) Differential analysis of site-specific glycans on plasma and cellular fibronectins: Application of a hydrophilic affinity method for glycopeptides enrichment. Glycobiology 15(12):1332–1340
96. Wada Y, Tajiri M, Yoshida S (2004) Hydrophilic affinity isolation and MALDI multiple-stage tandem mass spectrometry of glycopeptides for glycoproteomics. Anal Chem 76:6560–6565
97. Hägglund P, Bunkenborg J, Elortza F et al (2004) A new strategy for identification of N-glycosylated proteins and unambiguous assignment of their glycosylation sites using HILIC enrichment and partial deglycosylation. J Proteome Res 3:556–566
98. Liu X, Li X, Chan K et al (2007) One- pot methylation in glycomics application: esterification of sialic acids and permanent charge construction. Anal Chem 79:3894–3900
99. Larsen MR, Højrup P, Roepstorff P (2005) Characterization of gel-separated glycoproteins using two-step proteolytic digestion combined with sequential microcolumns and mass spectrometry. Mol Cell Proteomics 4:107–119
100. Brittain SM, Ficarro SB, Brock A et al (2005) Enrichment analysis of peptide subsets using fluorous affinity tags and mass spectrometry. Nat Biotechnol 23:463–468
101. Mirzaei H, Regnier F (2005) Affinity chromatographic selection of carbonylated proteins followed by identification of oxidation sites using tandem mass spectrometry. Anal Chem 77:2386–2392
102. Zhang W, Zhou G, Zhao Y et al (2003) Affinity enrichment of plasma membrane for proteomics analysis. Electrophoresis 24:2855–2863
103. Zhang H, Yi EC, Li XJ (2005) High throughput quantitative analysis of serum proteins using glycopeptide capture and liquid chromatography mass spectrometry. Mol Cell Proteomics 4:144–155
104. Zhao Y, Zhang W, Kho Y et al (2004) Proteomic analysis of integral plasma membrane proteins. Anal Chem 76:1817–1823
105. Bailey MJ, Hooker AD, Adams CS et al (2005) A platform for high-throughput molecular characterization of recombinant monoclonal antibodies. J Chromatogr B 826:177–187
106. Bundy JL, Fenselau C (2001) Lectin and carbohydrate affinity surfaces for mass spectrometric analysis of microorganisms. Anal Chem 73:751–757
107. Xiong L, Andrews D, Regnier F (2003) Comparative proteomics of glycoproteins based on lectin selection and isotope coding. J Proteome Res 2:618–625
108. Madera M, Mechref Y, Novotny MV (2005) Combining lectin microcolumns with high-resolution separation techniques for enrichment of glycoproteins and glycopeptides. Anal Chem 77:4081–4090
109. Bedair M, El Rassi Z (2005) Affinity chromatography with monolithic capillary columns II. Polymethacrylate monoliths with immobilized lectins for the separation of glycoconjugates by nano-liquid affinity chromatography. J Chromatogr A 1079:236–245
110. Okanda FM, El Rassi Z (2006) Affinity chromatography with monolithic capillary columns for glycomics/proteomics: 1. polymethacrylate monoliths with immobilized lectins for glycoprotein separation by affinity capillary electrochromatography and affinity nano-liquid chromatography in either a single column or columns coupled in series. Electrophoresis 27:1020–1030
111. Mao X, Luo Y, Dai Z et al (2004) Integrated lectin affinity microfluidic chip for glycoform separation. Anal Chem 76:6941–6947

112. Budnik BA, Lee RS, Steen JA (2006) Review Global methods for protein glycosylation analysis by mass spectrometry. Biochim Biophys Acta 1764:1870–1880
113. Wang L, Li F, Sun W et al (2006) Concanavalin A-captured glycoproteins in healthy human urine. Mol Cell Proteomics 5:560–562
114. Kaji H, Saito H, Yamauchi Y et al (2003) Lectin affinity capture, isotope-coded tagging and mass spectrometry to identify N-linked glycoproteins. Nat Biotechnol 21:667–672
115. Madera M, Mechref Y, Klouckova I et al (2007) High-sensitivity profiling of glycoproteins from human blood serum through multiple- lectin affinity chromatography and liquid chromatography/tandem mass spectrometry. J Chromatogr B 845:121–137
116. Cummings RD, Kornfeld S (1984) The distribution of repeating [Gal beta 1, 4GlcNAc beta 1, 3] sequences in asparagine-linked oligosaccharides of the mouse lymphoma cell lines BW5147 and PHAR 2.1. J Biol Chem 259:6253–6260
117. Yang Z, Hancock WS (2004) Approach to the comprehensive analysis of glycoproteins isolated from human serum using a multi-lectin affinity column. J Chromatogr A 1053:79–88
118. Qiu R, Regnier FE (2005) Use of multidimensional lectin affinity chromatography in differential glycoroteomics. Anal Chem 77:2802–2809
119. Sumi S, Arai K, Kitahara S et al (1999) Serial lectin affinity chromatography demonstrates altered asparagine-linked sugar-chain structures of prostate-specific antigen in human prostate carcinoma. J Chromatogr B 727:9–14
120. Xiong L, Regnier FE (2002) Use of a lectin affinity selector in the search for unusual glycosylation in proteomics. J Chromatogr, B: Anal Technol Biomed Life Sci 782:405–418
121. Yang Z, Hancock WS (2005) Monitoring glycosylation pattern changes of glycoproteins using multi-lectin affinity chromatography. J Chromatogr A 1070:57–64
122. Wang Y, Wu S, Hancock WS (2006) Approaches to the study of N-linked glycoproteins in human plasma using lectin affinity chromatography and nano-HPLC coupled to electrospray linear ion trap Fourier transform mass spectrometry. Glycobiology 16:514–523
123. Yue GE, Roper MG, Balchunas C et al (2006) Protein digestion and phosphopeptides enrichment on glass microchip. Anal Chim Acta 564:116–122
124. Madera M, Mechref Y, Klouckova I et al (2006) Semiautomated high-sensitivity profiling of human blood serum glycoproteins through lectin preconcentration and multidimensional chromatography/tandem mass spectrometry. J Proteome Res 5:2348–2363
125. Guzman NA, Phillips TM (2005) Immunoaffinity CE for proteomics studies. Anal Chem 77:60A–67A
126. Benavente F, Hernández E, Guzman NA et al (2007) Determination of human erythropoietin by on-line immunoaffinity capillary electrophoresis: a preliminary report. Anal Bioanal Chem 387:2633–2639
127. An HJ, Peavy TR, Hedrick JL et al (2003) Determination of N- glycosylation sites and site heterogeneity in glycoproteins. Anal Chem 75:5628–5637
128. Temporini C, Perani E, Calleri E et al (2007) Pronase-immobilized enzyme reactor: an approach for automation in glycoprotein analysis by LC/LC-ESI/MSn. Anal Chem 79:355–363
129. Jebanathirajah J, Steen H, Roepstorff P (2003) Using optimized collision energies and high resolution, high accuracy fragment ion selection to improve glycopeptide detection by precursor ion scanning. J Am Soc Mass Spectrom 14:777–784
130. Zhang H, Li XJ, Martin DB et al (2003) Identification and quantification of N-linked glycoproteins using hydrazide chemistry, stable isotope labeling and mass spectrometry. Nat Biotechnol 21:660–666
131. Khidekel N, Arndt S, Lamarre-Vincent N et al (2003) A chemoenzymatic approach toward the rapid and sensitive detection of O-GlcNAc posttranslational modifications. J Am Chem Soc 125:16162–16163
132. Sprung R, Nandi A, Chen Y et al (2005) Tagging-via-substrate strategy for probing O-GlcNAc modified proteins. J Proteome Res 4:950–957
133. Saxon E, Bertozzi CR (2000) Cell surface engineering by a modified Staudinger reaction. Science 287:2007–2010

134. Khidekel N, Ficarro SB, Peters EC et al (2004) Exploring the O-GlcNAc proteome: direct identification of O-GlcNAc-modified proteins from the brain. Proc Natl Acad Sci USA 101:13132–13137
135. Lamarre-Vincent N, Hsieh-Wilson LC (2003) Dynamic glycosylation of the transcription factor CREB: a potential role in gene regulation. J Am Chem Soc 125:6612–6613
136. Zhang Y, Wolf-Yadlin A, Ross PL et al (2005) Time-resolved mass spectrometry of tyrosine phosphorylation sites in the epidermal growth factor receptor signaling network reveals dynamic modules. Mol Cell Proteomics 4:1240–1250
137. Vocadlo DJ, Hang HC, Kim EJ et al (2003) A chemical approach for identifying O-GlcNAc-modified proteins in cells. Proc Natl Acad Sci USA 100:9116–9121
138. Prescher JA, Dube DH, Bertozzi CR (2004) Chemical remodelling of cell surfaces in living animals. Nature 430:873–877
139. Kho Y, Kim SC, Jiang C et al (2004) A tagging-via- substrate technology for detection and proteomics of farnesylated proteins. Proc Natl Acad Sci USA 101:12479–12484
140. Wells L, Vosseller K, Cole RN et al (2002) Mapping sites of O-GlcNAc modification using affinity tags for serine and threonine post-translational modifications. Mol Cell Proteomics 1:791–804
141. Vosseller K, Hansen KC, Chalkley RJ et al (2005) Quantitative analysis of both protein expression and serine/threonine post-translational modifications through stable isotope labeling with dithiothreitol. Proteomics 5:388–398
142. Wuhrer M, Catalina MI, Deelder AM et al (2007) Glycoproteomics based on tandem mass spectrometry of glycopeptides. J Chromatogr B Analyt Technol Biomed Life Sci 849:115–128
143. Carr SA, Hemling ME, Bean MF et al (1991) Integration of mass spectrometry in analytical biotechnology. Anal Chem 63:2802–2824
144. Burlingame AL (1996) Characterization of protein glycosylation by mass spectrometry. Curr Opin Biotechnol 7:4–10
145. Carr SA, Roberts GD (1986) Carbohydrate mapping by mass spectrometry: a novel method for identifying attachment sites of Asn-linked sugars in glycoproteins. Anal Biochem 157:396–406
146. Küster B, Mann M (1999) 18O-labeling of N-glycosylation sites to improve the identification of gel-separated glycoproteins using peptide mass mapping and database searching. Anal Chem 71:1431–1440
147. Leonard CK, Spellman MW, Riddle L et al (1990) Assignment of intrachain disulfide bonds and characterization of potential glycosylation sites of the type 1 recombinant human immunodeficiency virus envelope glycoprotein (gp120) expressed in Chinese hamster ovary cells. J Biol Chem 265:10373–10382
148. Carr SA, Roberts GD, Jurewicz A et al (1998) Structural fingerprinting of Asn-linked carbohydrates from specific attachment sites in glycoproteins by mass spectrometry: application to tissue plasminogen activator. Biochimie 70:1445–1454
149. Guzzetta AW, Basa LJ, Hancock WS et al (1993) Identification of carbohydrate structures in glycoprotein peptide maps by the use of LC/MS with selected ion extraction with special reference to tissue plasminogen activator and a glycosylation variant produced by site directed mutagenesis. Anal Chem 65:2953–2962
150. Jiang H, Wu SL, Karger BL et al (2010) Characterization of the glycosylation occupancy and the active site in the follow-on protein therapeutic: TNK-tissue plasminogen activator. Anal Chem 82:6154–6162
151. Domon B, Costello CE (1988) A systematic nomenclature for carbohydrate fragmentations in FAB-MS/MS spectra of glycoconjugates. Glycoconj J 5:397–409
152. Carr SA, Huddleston MJ, Bean MF (1993) Selective identification and differentiation of N- and O-linked oligosaccharides in glycoproteins by liquid chromatography-mass spectrometry. Protein Sci 2:183–196
153. Harvey DJ, Bateman RH, Bordoli RS et al (2000) Ionization and fragmentation of complex glycans with a quadrupole time-of-flight mass spectrometer fitted with a matrix-assisted laser desorption/ionization ion source. Rapid Commun Mass Spectrom 14:2135–2142

154. Borisov OV, Field M, Ling VT et al (2009) Characterization of Oligosaccharides in recombinant tissue plasminogen activator produced in Chinese hamster ovary cells: Two decades of analytical technology development. Anal Chem 81:9744–9754
155. Demelbauer UM, Zehl M, Plematl A et al (2004) Determination of glycopeptide structures by multistage mass spectrometry with low-energy collision-induced dissociation: comparison of electrospray ionization quadrupole ion trap and matrix-assisted laser desorption/ionization quadrupole ion trap reflectron time-of-flight approaches. Rapid Commun Mass Spectrom 18(14):1575–1582
156. Bones J, McLoughlin N, Hilliard M et al (2011) 2D-LC Analysis of BRP 3 Erythropoietin N-Glycosylation using anion exchange fractionation and hydrophilic interaction UPLC reveals long Poly-N-Acetyl lactosamine extensions. Anal Chem 83:4154–4162
157. Harazono A, Kawasaki N, Itoh S et al (2006) Site-speciWc N-glycosylation analysis of human plasma ceruloplasmin using liquid chromatography with electrospray ionization tandem mass spectrometry. Anal Biochem 348:259–268
158. Schmitt S, Glebe D, Alving K et al (1999) Analysis of the Pre-S2 N- and O-Linked Glycans of the M surface protein from human hepatitis B virus. J Biol Chem 274:11945–11957
159. Zubarev RA, Horn DM, Fridriksson EK et al (2000) Electron capture dissociation for structural characterization of multiply charged protein cations. Anal Chem 72:563–573
160. Kjeldsen F, Haselmann KF, Budnik BA et al (2002) Dissociative capture of hot (3–13 eV) electrons by polypeptide polycations: an efficient process accompanied by secondary fragmentation. Chem Phys Lett 356:201–206
161. Kjeldsen F, Haselmann KF, Budnik BA et al (2003) Complete characterization of posttranslational modification sites in the bovine milk protein PP3 by tandem mass spectrometry with electron capture dissociation as the last stage. Anal Chem 75(10):2355–2361
162. Mikesh LM, Ueberheide B, Chi A et al (2006) The utility of ETD mass spectrometry in proteomic analysis. Biochim Biophys Acta 1764(12):1811–1822
163. Schroeder MJ, Webb DJ, Shabanowitz J et al (2005) Methods for the detection of paxillin post-translational modifications and interacting proteins by mass spectrometry. J Proteome Res 4(5):1832–1841
164. Hogan JM, Pitteri SJ, Chrisman PA et al (2005) Complementary structural information from a tryptic N-linked glycopeptide via electron transfer ion/ion reactions and collision-induced dissociation. J Proteome Res 4(2):628–632
165. Mirgorodskaya E, Roepstorff P, Zubarev RA (1999) Localization of O-glycosylation sites in peptides by electron capture dissociation in a Fourier Transform mass spectrometer. Anal Chem 71:4431–4436
166. Perdivara I, Petrovich R, Allinquant B et al (2009) Elucidation of O-Glycosylation structures of the β-Amyloid precursor protein by liquid chromatography-mass spectrometry using electron transfer dissociation and collision-induced dissociation. J Proteom Res 8:631–642
167. Wu SL, Huhmer AF, Hao Z et al (2007) On-line LC-MS approach combining collision-induced dissociation (CID), electron-transfer dissociation (ETD), and CID of an isolated charge-reduced species for the trace-level characterization of proteins with posttranslational modifications. J Proteome Res 6(11):4230–4244
168. Tsarbopoulos A, Bahr U, Pramanik BN et al (1997) Glycoprotein Analysis by Delayed extraction and post-source decay MALDI TOF MS. Int J Mass Spectrom Ion Process 169(170):251–261
169. Wuhrer M, Hokke CH, Deelder AM (2004) Glycopeptide analysis by matrix-assisted laser desorption/ionization tandem time-of-flight mass spectrometry reveals novel features of horseradish peroxidase glycosylation. Rapid Commun Mass Spectrom 18:1741–1748
170. Bykova NV, Rampitsch C, Krokhin O et al (2006) Determination and characterization of site-specific N-Glycosylation using MALDI-Qq-TOF tandem mass spectrometry: case study with a plant protease. Anal Chem 78:1093–1103
171. Kurogochi M, Matsushita T, Nishimura SI (2004) Post-translational modifications on proteins: facile and efficient procedure for the identification of O-Glycosylation sites by MALDI-LIFT-TOF/TOF mass spectrometry. Angew Chem Int Ed Engl 43:4071–4075

172. Harvey DJ (1999) Matrix-assisted laser desorption/ionization mass spectrometry of carbohydrates. Mass Spectrom Rev 18:349–451
173. Zaia J (2010) Mass spectrometry and glycomics. OMICS 14(4):401–418
174. Wormald MR, Petrescu AJ, Pao Y-L et al (2002) Conformational studies of oligosaccharides and glycopeptides: complementarity of NMR, X-ray crystallography, and molecular modelling. Chem Rev 102:371–386
175. Koerner TA, Yu RK, Scarsdale JN et al (1988) Analysis of complex carbohydrate primary and secondary structure via two-dimensional proton nuclear magnetic resonance spectroscopy. Adv Exp Med Biol 228:759–784
176. Perez S, Mulloy B (2005) Prospects for glycoinformatics. Curr Opin Struct Biol 15:517–524
177. Aoki-Kinoshita KF (2008) An introduction to bioinformatics for glycomics research. PLoS Comput Biol. doi:10.1371/journal.pcbi.1000075
178. von der Lieth CW, Lütteke T, Frank M (2006) The role of informatics in glycobiology research with special emphasis on automatic interpretation of MS spectra. Biochim Biophys Acta 1760:568–577
179. Cooper CA, Gasteiger E, Packer NH (2001) GlycoMod-a software tool for determining glycosylation compositions from mass spectrometric data. Proteomics 1:340–349
180. Go EP, Rebecchi KR, Dalpathado DS et al (2007) GlycoPep DB: a tool for glycopeptide analysis using a "smart search". Anal Chem 79:1708–1713
181. Goldberg D, Sutton-Smith M, Paulson J et al (2005) Automatic annotation of matrix-assisted laser desorption/ionization N-glycan spectra. Proteomics 5:865–875
182. Goldberg D, Bern M, Parry S et al (2007) Automated N-glycopeptide identification using a combination of single- and tandem-MS. J Proteome Res 6:3995–4005
183. Ozohanics O, Krenyacz J, Ludanyi K et al (2008) GlycoMiner: a new software tool to elucidate glycopeptide composition. Rapid Commun Mass Spectrom 22:3245–3254
184. An HJ, Tillinghast JS, Woodruff DL et al (2006) A new computer program (GlycoX) to determine simultaneously the glycosylation sites and oligosaccharide heterogeneity of glycoproteins. J Proteome Res 5:2800–2808
185. Ren JM, Rejtar T, Li L et al (2007) N-Glycan structure annotation of glycopeptides using a linearized glycan structure database (GlyDB). J Proteome Res 6:3162–3173
186. Irungu J, Go EP, Dalpathado DS et al (2007) Simplification of mass spectral analysis of acidic glycopeptides using GlycoPep ID. Anal Chem 79:3065–3074
187. Hizukuri Y, Yamanishi Y, Nakamura O et al (2005) Extraction of leukemia specific glycan motifs in humans by computational glycomics. Carbohydr Res 340:2270–2278
188. Aoki K, Yamaguchi A, Ueda N et al (2004) KCaM (KEGG Carbohydrate Matcher): a software tool for analyzing the structures of carbohydrate sugar chains. Nucleic Acids Res 32:W267–W272
189. Aoki K, Mamitsuka H, Akutsu T et al (2005) A score matrix to reveal the hidden links in glycans. Bioinformatics 21:1457–1463
190. Hashimoto K, Goto S, Kawano S et al (2006) KEGG as a glycome informatics resource. Glycobiology 6:63R–70R
191. Creighton TE (1984) Disulfide bond formation in proteins. In: Wold F, Moldave K (eds) Methods in enzymology, vol 107. Academic Press, San Diego, p 305
192. Dranoff G (2009) Targets of protective tumor immunity. Ann NY Acad Sci 1174:74–80
193. Nakamura T, Lipton SA (2009) Cell death: protein misfolding and neurodegenerative diseases. Apoptosis 14:455–468
194. Wess J, Han SJ, Kim SK et al (2008) Conformational changes involved in G-protein-coupled-receptor activation. Trends Pharmacol Sci 29:616–625
195. Thornton JM (1981) Disulphide bridges in globular proteins. J Mol Biol 151:261–287
196. Welker E, Raymond LD, Scheraga HA et al (2002) Intramolecular versus intermolecular disulfide bonds in prion proteins. J Biol Chem 277:33477–33481
197. Tsarbopoulos A, Pramanik B, Labdon J et al (1993) Isolation and characterization of a resistant core peptide of recombinant human Granulocyte-Macrophage colony-stimulating factor (GM-CSF); confirmation of the GM-CSF amino acid sequence by mass spectrometry. Protein Sci 2:1948–1958

198. Gorman JJ, Wallis TP, Pitt JJ (2002) Protein disulfide bond determination by mass spectrometry. Mass Spectrom Rev 21:183–216
199. Barber M, Bordoli RS, Sedgwick RD et al (1981) Fast atom bombardment of solids (FAB): A new ion source for mass spectrometry. J Chem Soc, Chem Commun 7:325–327
200. Morris HR, Pucci P (1985) A new method for rapid assignment of S-S bridges in proteins. Biochem Biophys Res Commun 126:1122–1128
201. Smith DL, Zhou Z (1990) Strategies for locating disulfide bonds in proteins. In: McCloskey JA (ed) Methods in enzymology, vol 193. Academic Press, New York, p 374
202. Sundqvist B, Roepstorff P, Fohlman J et al (1984) Molecular weight determination of proteins by californium plasma desorption mass spectrometry. Science 226:696–698
203. Tsarbopoulos A, Becker GW, Occolowitz JL et al (1988) Peptide and protein mapping by 252Cf-Plasma desorption mass spectrometry. Anal Biochem 171:113–123
204. Robertson JG, Adams GW, Medzihradszky KF et al (1994) Complete assignment of disulfide bonds in bovine dopamine beta-hydroxylase. Biochemistry 33:11563–11575
205. Pramanik BN, Tsarbopoulos A, Labdon JE et al (1991) Structural analysis of biologically active peptides and recombinant proteins and their modified counterparts by mass spectrometry. J Chromatogr 562:377–389
206. Chen G, Liu YH, Pramanik BN (2007) LC/MS analysis of proteins and peptides in drug discovery. In: Kazakevich Y, LoBrutto R (eds) HPLC for pharmaceutical scientists. Wiley, New York
207. Tsarbopoulos A, Karas M, Strupat K et al (1994) Comparative mapping of recombinant proteins and glycoproteins by plasma desorption and matrix-assisted laser desorption/ionization mass spectrometry. Anal Chem 66:2062–2070
208. Patterson SD, Katta V (1994) Prompt fragmentation of disulfide-linked peptides during matrix-assisted laser desorption ionization mass spectrometry. Anal Chem 66:3727–3732
209. Sanger F (1953) A disulphide interchange reaction. Nature 171:1025–1026
210. Yazdanparast R, Andrews PC, Smith DL et al (1987) Assignment of disulfide bonds in proteins by fast atom bombardment mass spectrometry. J Biol Chem 262:2507–2513
211. Tsarbopoulos A, Varnerin J, Cannon-Carlson S et al (2000) Mass spectrometric mapping of disulfide bonds in recombinant human Interleukin-13. J Mass Spectrom 35:446–453
212. Sun Y, Bauer MD, Keough TW et al (1996) Disulfide bond location in proteins. Methods Mol Biol 61:181–210
213. Bauer M, Sun Y, Degenhardt C et al (1993) Assignment of all four disulfide bridges in echistatin. J Prot Chem 12:759–764
214. Bean MF, Carr SA (1992) Characterization of disulfide positions in proteins and sequence analysis of cystine-bridged peptides by tandem mass spectrometry. Anal Biochem 201:216–226
215. Pitt JJ, Da Silva E, Gorman JJ (2000) Determination of the disulfide bond arrangement of new castle disease virus hemagglutinin neuraminidase, correlation with a beta-sheet propeller structural fold predicted for paramyxoviridae attachment proteins. J Biol Chem 275:6469–6478
216. Gorman JJ, Ferguson BL, Speelman D et al (1997) Determination of the disulfide bond arrangement of human respiratory syncytial virus attachment (G) protein by matrix assisted laser desorption/ionization time-of-flight mass spectrometry. Protein Sci 6:1308–1315
217. Angal S, King DJ, Bodmer MW et al (1993) A single amino acid substitution abolishes the heterogeneity of chimeric mouse/human (IgG4) antibody. Mol Immunol 30:105–108
218. Wang Y, Lu Q, Wu SL et al (2011) Characterization and comparison of disulfide linkages and scrambling patterns in therapeutic monoclonal antibodies: using LC-MS with electron transfer dissociation. Anal Chem 83:3133–3140
219. Wu SL, Jiang H, Hancock WS et al (2010) Identification of the unpaired cysteine status and complete mapping of the 17 disulfides of recombinant tissue plasminogen activator using LC-MS with Electron transfer dissociation/collision induced dissociation. Anal Chem 82:5296–5303
220. Bagal D, Valliere-Douglass JF, Balland A et al (2010) Resolving disulfide structural isoforms of IgG2 monoclonal antibodies by ion mobility mass spectrometry. Anal Chem 82:6751–6755

221. Wallis TP, Pitt JJ, Gorman JJ (2001) Identification of disulfide-linked peptides by isotope profiles produced by peptic digestion of proteins in 50 % (18) O water. Protein Sci 10:2251–2271
222. Rose K, Savoy LA, Simona MG et al (1988) C-terminal peptide identification by fast atom bombardment mass spectrometry. Biochem J 250:253–259
223. Dwek MV, Ross HA, Leathem AJ (2001) Proteome and glycosylation mapping identifies post-translational modifications associated with aggressive breast cancer. Proteomics 1:756–762
224. Rudd PM, Elliott T, Cresswell P et al (2001) Glycosylation and the immune system. Science 291:2370–2376
225. Peracaula R, Tabares G, Royle L et al (2003) Altered glycosylation pattern allows the distinction between prostate-specific antigen (PSA) from normal and tumor origins. Glycobiology 13:457–470
226. Butler M, Quelhas D, Critchley AJ et al (2003) Detailed glycan analysis of serum glycoproteins of patients with congenital disorders of glycosylation indicates the specific defective glycan processing step and provides an insight into pathogenesis. Glycobiology 13:601–622

Chapter 5
Mass Spectrometric Characterization in Protein Therapeutics Discovery

Jingjie Mo, Adrienne A. Tymiak and Guodong Chen

5.1 Introduction

5.1.1 Protein Therapeutics

With advances in recombinant DNA and hybridoma technologies, including chimerization and humanization, which allow customized proteins to be produced in large quantities, pharmaceutical applications of proteins as human therapeutics have expanded rapidly in the last 30 years [1]. Protein therapeutics now represent the second largest biopharmaceutical product category after vaccines [2]. Advantages of protein therapeutics include high specificity, efficacy, and fewer side effects. They have been applied in the treatment of many life-threatening diseases, such as cancer, infectious diseases, inflammation, and genetic disorders, and have a higher regulatory approval rate compared to small molecules in drug development [3]. Advances in formulation science that deliver proteins in vivo with improved pharmacokinetic (PK) and pharmacodynamic (PD) properties are also enabling rapid development of new protein therapeutics [4]. Pharmaceutical companies are now dedicating more and more of their pipeline efforts to protein therapeutics. According to global business intelligence (GBI) research, the therapeutic protein market is forecast to grow to $141.5 billion by 2017, which represents a growth rate of 6.2 % between 2010 and 2017 [5].

In protein drug discovery, there are typically two categories of protein molecules that need to be characterized: the target protein (found in the body) and the therapeutic protein (drug candidate). Target proteins normally present as a set of molecules that include the primary human target molecule(s) and other close analogs (i.e., chemically or biochemically modified forms and structurally homologous sequences). Understanding and characterizing the physiologically relevant protein target(s) can facilitate the identification of selectively acting lead drug

J. Mo (✉) · A. A. Tymiak · G. Chen
Bristol-Myers Squibb, Route 206 and Province Line Road,
Princeton, NJ 08543, USA
e-mail: Jingjie.Mo@bms.com

G. Chen (ed.), *Characterization of Protein Therapeutics using Mass Spectrometry*,
DOI: 10.1007/978-1-4419-7862-2_5, © Springer Science+Business Media New York 2013

candidates [6]. Once the target proteins are produced, the process of discovering a therapeutic protein involves screening for a lead sequence followed by sequence optimization and characterization of the final development candidate. Several methods can be used to identify lead sequences, such as traditional immunization or molecular display technologies [7, 8]. The goal during the optimization phase is to identify a candidate protein with superior biophysical, PK, and PD properties to maximize its chances of success during downstream development. It can take up to 15 years to develop one new therapeutic protein from the earliest stages of discovery to the time it is available on the market [5]. Production of therapeutic proteins begins with the development of a suitable cell line, determined by the selection of cell type. *Escherichia coli* (*E. coli*) is the most commonly used microbial protein expression system; and Chinese hamster ovary (CHO) is the most commonly used mammalian protein expression system, although murine myeloma (NS0), baby hamster kidney (BHK), human embryonic kidney (HEK-293), and human retina–derived PER-C6 cell lines are also widely used [9, 10].

Recombinant proteins, in particular monoclonal antibodies (mAbs), constitute the fastest growing sector within the therapeutic protein industry [11–13], with an estimated market value of around US$48.5 billion (over 47 % of protein therapeutics) at the end of 2011 [14]. Antibodies, also known as immunoglobulins (Igs), are represented by five distinct structural classes: IgA, IgD, IgE, IgG, and IgM. IgGs are the most abundant class, and all antibody drugs approved for clinical use today are based on IgG antibodies [15, 16]. Polyclonal antibodies harvested from immunized animals are mixtures of structurally diverse proteins with different binding properties, while mAbs are antibodies that are produced from immune cells that are all clones of one single parent cell and are therefore identical in primary sequence.

Typical IgGs are Y-shaped protein molecules composed of two identical light chains and two identical heavy chains. Each light chain contains one variable domain (V_L) and one constant domain (C_L), and each heavy chain contains one variable domain (V_H) and three constant domains (C_H1, C_H2, and C_H3). IgGs are linked together by different numbers of disulfide bonds (S–S), depending on the subtype. For example, in subtype IgG1 molecules, each light chain is linked to a heavy chain through one S–S, and the two heavy chains are connected in the hinge region through two S–S (Scheme 5.1). The enzyme papain can cleave IgGs at the hinge region to produce two Fab (antigen binding) fragments and one Fc (crystallizable) fragment. The Fab fragment contains

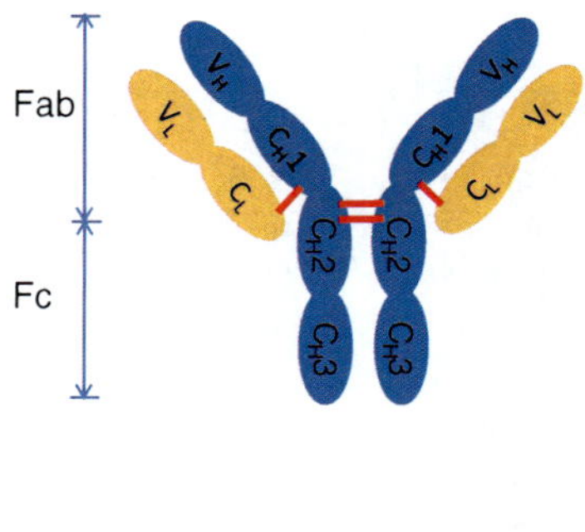

Scheme 5.1 Structure of IgG1 antibody

complementarity-determining regions (CDRs) in V_L and V_H that are hypervariable and responsible for the diversity and specificity of antigen binding [17]. The Fc fragment, consisting of the C_H2 and C_H3 domains, has roles in recruiting cytotoxic effector functions such as complement activation, antigen-dependent cellular cytotoxicity, binding to the phagocyte Fc receptors, and providing long serum half-lives (typically two to three weeks) through interaction with the neonatal Fc receptor (FcRn) [18, 19]. There is also an N-linked biantennary oligosaccharide chain located within each C_H2 domain that can influence the structure and function of the antibody.

Full-length mAbs have been used in a myriad of therapeutic applications. However, in applications where binding to the target alone is sufficient for therapeutic efficacy and Fc-induced effector functions are not required and/or may even cause unwanted side effects (e.g., cytokine inactivation, receptor blockade, viral neutralization, targets requiring fast clearance rates, and targets requiring protein cross-linking), use of mAbs is limited [20–22]. There are different approaches to eliminate the Fc-induced effector functions while still retaining the targeted function; examples include mutating the amino acids (AAs) in Fc that are critical for receptor binding [23], removing the N-linked glycosylation in the C_H2 domain [24] or mutating the hinge region sequence [25]. The use of modified antibodies that do not have Fc domains or that have been dissected to recombinant antibody fragments is increasing as an alternative to full-length mAbs, since they are smaller yet highly specific proteins possess properties advantageous for some medical applications, can be produced more economically, and are easily amenable to genetic manipulation [8, 20, 26]. Proteolytic treatment with papain and pepsin can remove the Fc region and generate Fab and Fab_2 fragments, respectively [27]. In addition, there are many other forms of antibody fragments that have been studied, including monovalent fragments (e.g., Fab, scFv, and single V_H and V_L domains), bivalent fragments (e.g., Fab_2, diabodies and minibodies), and multivalent fragments (e.g., triabodies and tetrabodies) [26].

The single-chain variable fragment (scFv) is the most commonly used antibody truncation, and it has been engineered into several types of scFv-based fragments for a range of therapeutic and diagnostic applications [8]. scFv fragments contain both the V_H and V_L domains (antigen-binding sites) that can be linked by a flexible polypeptide linker (e.g., $(GGGGS)_3$) or associated covalently through a S–S [28]. They can also be used to form multivalent and multispecific molecules with increased target-binding affinity and in vivo persistence [8]. There are several FDA-approved antibody fragment therapeutics, and many others are undergoing clinical evaluation [26]. However, due to the absence of the Fc domain and a low molecular weight (MW), antibody fragments generally have short in vivo half-lives, and chronic treatments require high doses and frequent administration.

Strategies to extend the circulating half-life and improve bioavailability of an antibody fragment include conjugating the antibody fragment with another protein or polymer such as albumin [29, 30], with a heavy chain fragment (e.g., Fc fusion protein) [31, 32], or with a water-soluble polymer polyethylene glycol (PEG) [33,

34]. PEG is a class of polymers that is non-toxic, non-immunogenic, non-antigenic, highly soluble in water and FDA approved. It is typically eliminated from the body in the urine (for PEGs < 20 kDa) or in feces (for PEGs > 20 kDa) [35]. PEGylation is the process of covalently attaching one or more PEG chains to the protein drug. PEGylated protein conjugates have several advantages: prolonged in vivo circulating half-lives due to increased molecular mass, decreased degradation by metabolic enzymes due to the steric shielding of cleavage sites by PEG [36], improved PK and PD properties due to better solubility [37, 38], and reduced protein immunogenicity [39]. Because of these favorable properties, PEGylation now plays an important role in drug delivery, increasing the potential of peptides and proteins to become useful therapeutic agents [40]. Several classes of protein drugs, such as enzymes, cytokines, and antibodies, are significantly improved by PEGylation [41]. In some cases, PEGylation may be accompanied by decreased biological activities; however, the prolonged body-residence times and increased stabilities can offset this effect [42–46].

5.1.2 Mass Spectrometry in Protein Therapeutics Discovery

Therapeutic proteins produced using recombinant DNA technologies are generally complex and heterogeneous molecules that are further subject to a variety of enzymatic or chemical modifications, such as glycosylation, deamidation, oxidation, and S–S formation [47]. Although they are highly specific molecules, their efficacy, clearance, and immunogenicity properties are highly dependent on the AA sequence, presence/absence of specific modifications, conformational changes upon modifications, non-covalent interactions with receptor proteins, and aggregation caused by misfolding and modifications. For these reasons, there is a growing need for more precise protein characterization methods, particularly during the discovery phase of drug development when a large number of candidates are being investigated. As one of the most commonly used analytical techniques in pharmaceutical research and development, mass spectrometry (MS) has become an essential analytical tool for the characterization of protein therapeutics because of its analytical sensitivity, resolution, selectivity, and molecular specificity [6, 16, 47–51]. When coupled to online liquid chromatography (LC) separation, LC/MS can provide detailed information about the primary structure of a protein, such as its MW, AA sequence, post-translational modifications (PTMs), and degradation products. In addition, MS can be used to investigate higher-order structures when combined with hydrogen/deuterium exchange (HDX) and ion mobility spectrometry (IMS) (Scheme 5.2). MS is routinely used to support selection of host expression systems, identification of clones with the most favorable quality attributes, and evaluation of molecular stability under different stress conditions. In the following sections, MS characterization of protein MW, AA sequence, common PTMs, and higher-order structure, as well as its applications in clone selection and stability evaluation, will be described.

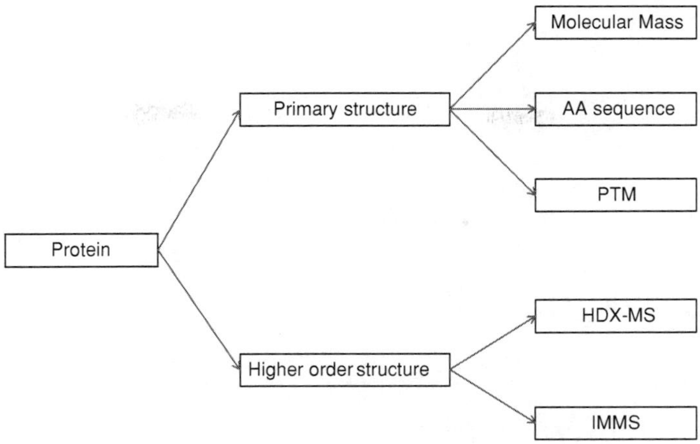

Scheme 5.2 Application of MS in structural characterization of recombinant protein therapeutics

5.2 Structural Characterization of Protein Therapeutics by Mass Spectrometry

5.2.1 General Approach

With advances in ionization methods, including electrospray ionization (ESI) [52] and matrix-assisted laser desorption/ionization (MALDI) [53], the development of novel MS instrumentation and ion activation methods, MS has expanded significantly the capability of protein characterization with enhanced accuracy, throughput and detection limits below femtomoles of material. The general approach of MS-based protein characterization typically involves analysis of either intact protein (top-down) or peptide fragments generated from enzymatic digestion of the protein (bottom-up), as shown in Scheme 5.3. Samples are generally separated by LC prior to MS analysis because non-MS compatible salts or other matrix components can interfere with the protein analyte. Ultra-high-performance liquid chromatography (UHPLC) utilizing small stationary phase particles in a packed column increases the speed of analysis with superior resolution, sensitivity, and peak capacity [54]. If the protein of interest comes from cell culture supernatants, an additional "capture" step, such as affinity purification, may be required to concentrate the protein from the cell culture medium prior to the LC separation step [55]. The "top-down" approach of analyzing intact proteins involves minimum sample treatment and avoids artificial modifications that may arise during sample handling. Direct tandem MS fragmentation analysis (MS/MS or MS^n) can provide information on overall protein heterogeneity, MW, and PTMs, such as

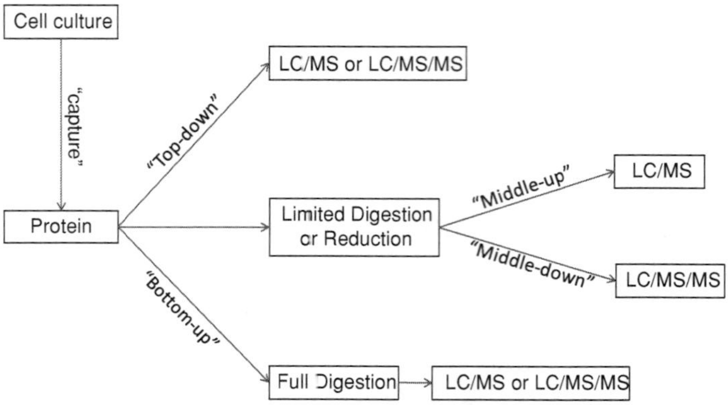

Scheme 5.3 General approach of MS characterization for protein therapeutics

glycosylation and N-/C-terminal sequences. However, the "top-down" approach is limited by protein size [56]. "Bottom-up" approaches, on the other hand, are not limited by protein size as the large molecules are digested to produce small peptides for analysis; however, this approach consumes more material, requires more sample treatment, and may introduce artificial modifications or degradations [57–59]. There are also methods that fall between the "top-down" and "bottom-up" approaches. If intact protein is cleaved into a few large fragments for MS analysis, it is called a "middle-up" approach; if these large fragments undergo further MS/MS fragmentation, that is referred to as a "middle-down" approach.

5.2.2 *Molecular Mass Analysis*

The first step in characterizing a protein is often determination of its MW. It is an important physical parameter that can be used to confirm primary structure and identity of the protein, characterize PTMs, and detect degradations resulting from clipping of terminal AAs. Analysis of intact large proteins requires the use of an appropriate ionization technique and a mass analyzer with suitable resolving power and mass accuracy [47]. ESI and MALDI are the most commonly used ionization methods for proteins, with ESI preferable, when the mass spectrometer is coupled to LC. ESI generates multiply charged ions requiring data deconvolution [52], while MALDI produces mostly singly charged ions [60]. Time-of-flight (TOF)-type mass analyzers have been widely accepted as the standard instrumentation for intact mass analysis of large molecules as they deliver higher resolving power and mass accuracy, and a wider m/z range (up to 5,000 in a hybrid ESI-qTOF instrument) compared to other mass analyzers [61]. Fourier transform ion cyclotron resonance (FT-ICR) MS provides the greatest resolving power, mass accuracy and peak capacity [62]; however, it is not as common in industrial laboratories as the TOF-type instruments due to its high cost and more stringent requirements for routine maintenance.

For an ESI-qTOF instrument, the mass accuracy of intact mAbs (~150 kDa) is usually within 100 ppm (~15 Da) with daily external calibration. When experimental conditions are optimized to minimize adduct formation and deconvolution parameters are carefully controlled, the mass accuracy can approach 25–50 ppm [63–65] or even 10 ppm [66]. Higher throughput for mass determination can be achieved by coupling MS to LC [16]. Typically, reversed-phase high-performance liquid chromatography (RP-HPLC) is preferred due to mobile phase compatibility with MS. However, due to the large size and relatively high hydrophobicity of antibodies, high column temperatures are often used and trifluoroacetic acid (TFA) is included in the mobile phases to improve elution and peak-shaped profiles [67, 68]. HPLC performance can also be improved by utilizing the appropriate type of column and using organic solvents with high eluotropic strength coefficients, such as isopropyl and n-propyl alcohols [68, 69]. Using optimized RP-HPLC methods, scientists have successfully separated an intact mAb from its C-terminal Lys variants [67], degradation products [68], cysteinylated forms [70], and disulfide variants [71, 72]. Size-exclusion chromatography (SEC) can also be coupled to MS for intact mass analysis of large molecules if the mobile phase is modified to exclude salt. In some cases, SEC/MS is advantageous over RP-HPLC/MS in terms of sensitivity and overall quality of the mass spectra [65].

5.2.2.1 Mass Analysis of mAbs

LC/MS analysis of a typical intact mAb on an UHPLC system coupled to an ESI-qTOF-type mass spectrometer with 10 k resolving power is shown in Fig. 5.1. In

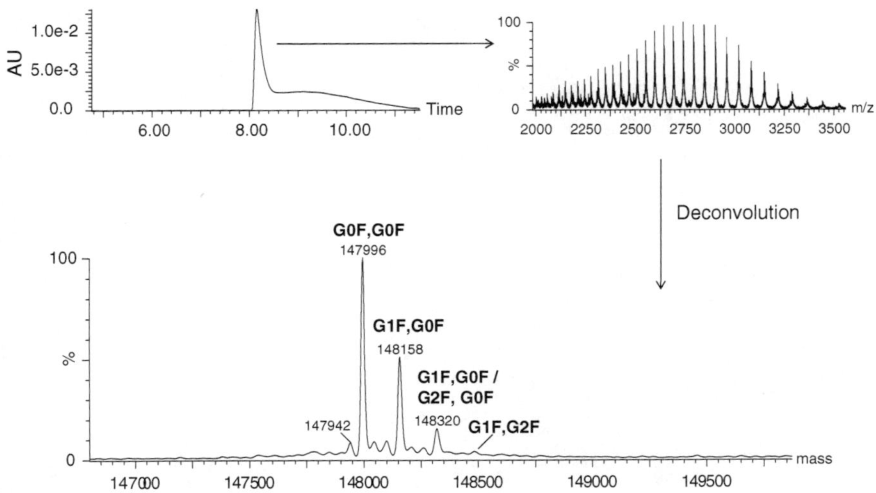

Fig. 5.1 LC/MS analysis of an intact mAb

this case, different charge states of the intact molecule occur over the m/z range of 2,000–3,500, and spectral deconvolution of these charge states can provide the molecular mass of the protein. For large molecules such as mAbs, it is extremely difficult to resolve the isotopic peaks and reliably determine the monoisotopic mass, and even if the monoisotopic peak is resolved, its relative abundance is virtually zero. Therefore, for large molecules, the average mass is normally used to confirm the MW. As shown in Fig. 5.1, the observed mass for the (G0F, G0F) glycoform is 147,996 Da (other glycoforms will be discussed later). The theoretical mass is 147,992 Da, assuming that the AA sequence has the C-terminal Lys cleaved and the N-terminal Gln modified to pyro-glutamic acid (pyro-E) in both heavy chains (both of these modifications are commonly observed in mAbs and will be discussed later). This mass difference results in a mass error of 4 Da (<30 ppm). If the LC/MS experiment is performed on a mass spectrometer with modest resolving power and mass accuracy, the mass analysis can be facilitated by limited digestion using papain or other enzymes to generate Fab, Fab_2, and Fc fragments [73–78], or partial or complete reduction of S–S to generate separate heavy chains and light chains [78, 79]. Compared to intact mAbs, these fragments are much smaller and easier to analyze and therefore have reduced requirements for the mass range and resolving power of the mass spectrometer. As mentioned earlier, this is referred to as a "middle-up" approach (Scheme 5.3).

Figure 5.2 shows the mass spectrum of the same mAb analyzed in Fig. 5.1 after partial reduction by dithiothreitol (DTT). Under controlled temperature and time, and in the absence of denaturant, the inter-chain S–S can be selectively reduced, so that the intra-chain S–S remains intact [79]. The constituent heavy chains and light

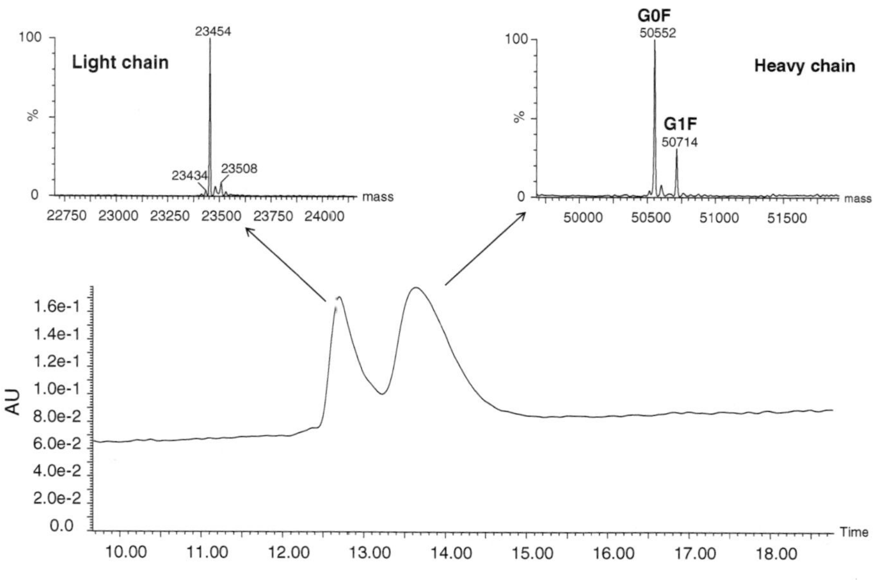

Fig. 5.2 LC/MS analysis of a reduced mAb: heavy chain and light chain

chains in the reduced sample are then separated by RP-UHPLC with a C4 column. Observed masses for both heavy chains and light chains are within 5 Da of the theoretical masses. Issues with this approach include tailing of the heavy chain component and the potential for carryover problems [16]. These issues can be solved by the inclusion of blank runs between samples to increase confidence in the results.

5.2.2.2 Mass Analysis of PEGylated Proteins

PEGylated proteins present significant challenges for structural characterization due to the heterogeneity caused by the number of PEG moieties attached and the sites of PEGylation. PEGylated proteins generally show a distribution of oligomeric masses with a difference of 44 Da between each oligomer. Most published studies use MALDI-TOF MS for PEG characterization, as MALDI is a soft ionization process [80] and generates singly charged ions [60] that can greatly reduce the complexity of the mass spectra for easier and more confident data interpretation [81–84]. ESI–MS with high resolution mass accuracy can better elucidate structural information for larger PEGylated proteins; however, the ESI–MS approach has limitations as the PEG charge state distribution is often convoluted over the size distribution, resulting in many isobaric ions and broad mass spectral features [85]. A conventional way to solve this problem is dePEGylation before mass analysis [86]. Other strategies include charge reduction to reduce the complexity associated with multiple charges [87–90], incorporation of high-resolution FT-ICR MS to resolve multiply charged species [91], using ion mobility mass spectrometry (IMMS) to resolve multiple charge states [85], and using a gas-phase proton-transfer reaction approach [92]. Applying these methodologies for characterization of PEGylated proteins can map PEGylation sites, structural variants, degradation products, modifications and potential PEG cleavage sites.

5.2.2.3 Mass Analysis of Glycoproteins

Glycosylation represents the most pronounced and complex type of protein PTM. It can significantly change protein conformation and consequently modulate the functional activity of proteins as well as protein/protein interactions [93–96]. There are mainly two types of protein glycosylation: N-linked glycosylation, in which the glycan is attached to the amide group of Asn in the consensus sequence of Asn-X-Ser/Thr (sometimes Asn-X-Cys), where X can be any AA except Pro [97, 98]; and O-linked glycosylation, in which the glycan is linked to the oxygen on a Ser or Thr residue [99]. Recombinant IgG antibodies produced in CHO cells possess a conserved N-linked glycosylation site in the Fc C_H2 domain on each heavy chain. These are primarily fucosylated biantennary complex structures that differ in the number of terminal Gal residues (G0F, G1F, and G2F), as shown in Scheme 5.4. Other more immature structures have also been reported, such as high mannose (Man5) or hybrid types [100–102]; however, these structures are typically present at very low levels.

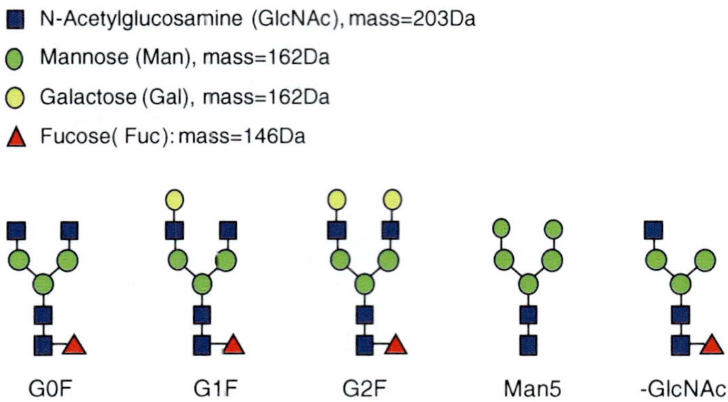

Scheme 5.4 Glycans with fucosylated biantennary core structures

Approximately 20 % of human IgG also contain N-linked glycosylation in the Fab region in addition to the conserved Fc site [103]. Structural changes of antibodies caused by glycan variants have been shown to impact antigen-binding and antibody effector functions [104–107]. In early discovery research, cataloging and controlling glycosylation patterns of recombinant proteins are important activities to support many decision points during expression and assay development [108]. The high resolution and extended mass range of intact mass analysis allows the distribution of the major glycoforms and their relative abundance to be monitored [109, 110]. In cases where glycosylation is present at sites beyond the Fc (e.g., O-glycosylation or N-glycosylation in the Fab region), site-specific glycosylations are typically analyzed at the glycopeptide level to allow the glycans attached to the different sites to be characterized separately [111, 112]. Figures 5.1 and 5.2 display intact and reduced mass spectra of a typical mAb, illustrating G0F, G1F, and low level of G2F glycoforms. The relative peak abundance can be used to estimate the relative abundance of each glycoform, not counting the differences in ionization efficiency. MS can also provide detailed structural analysis of the glycans; however, it is not necessary to carry out this type of analysis during early drug discovery as further process optimization and downstream purification will likely remove most low-level impurities. For proteins with more complicated glycosylation profiles, removing the carbohydrate portion before MS analysis can significantly reduce the molecular heterogeneity and facilitate mass spectral interpretation. N-linked glycosylations can be completely removed with peptide-N-glycosidase F (PNGase F) treatment or hydrolysis [47]. O-linked oligosaccharides are typically removed using β-elimination, hydrolysis, or O-glycosidase [47]. However, the complete removal of O-linked glycosylation is very challenging.

5.2.2.4 Applications: Assessment of Protein Fragmentation/Degradation

Fragmentation is one of the major degradation pathways of therapeutic proteins in liquid formulations [113]. To assist in clone selection, the susceptibility of proteins to fragmentation can be assessed under stressed conditions such as elevated temperatures, exposure to chemicals, light, or a combination of these. For example, incubating at high temperature can lead to a progressive loss of intact antibodies due to non-enzymatic hydrolysis [114–116], with the hinge region being most susceptible to hydrolysis, generating Fab, Fc, and antibodies missing one Fab arm [68, 116–119]. Protein fragmentation often occurs at the C-terminal side of an acidic residue, or near a Ser residue, and the predominant cleavage site occurs between Asp and Pro [120]. With high mass accuracy, the fragments (degradants) generated from hydrolysis can be assigned. These fragments usually can be resolved in UHPLC separation, and their relative abundance to intact molecules can be estimated from either the chromatographic peak abundance or MS peak abundance (especially for low abundance fragments). By measuring susceptibility to fragmentation, clones with similar sequences can be differentiated based on stability in simple buffer screening conditions.

5.2.3 Amino Acid Sequence Analysis

5.2.3.1 Tandem Mass Spectrometry

It is essential to perform sequence confirmation for therapeutic proteins to establish product identity and integrity and to ensure that the protein has the correct AA sequence as predicted by DNA sequencing. To analyze protein AA sequences, tandem mass spectrometry (MS/MS or MS^n) is needed to cleave the peptide backbone to generate fragments, ideally at each AA residue. There are many different types of MS/MS techniques, including collisionally activated dissociation (CAD) or collision-induced dissociation (CID) [121], infrared multiphoton dissociation (IRMPD) [122], blackbody infrared dissociation (BIRD) [123], electron capture dissociation (ECD) [124], and electron transfer dissociation (ETD) [125]. CID involving cleavage at amide bonds of the peptide backbone to generate characteristic *b* and *y* ions (see Scheme 5.5 for nomenclature of fragment ions) is the most widely used fragmentation method. It is more efficient to fragment the backbone for highly

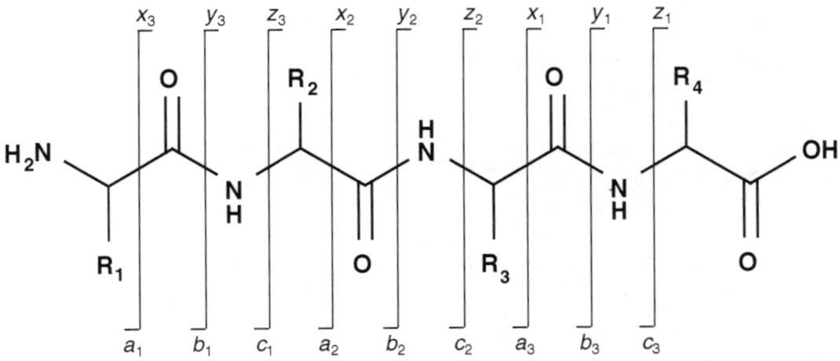

Scheme 5.5 Nomenclature of peptide fragment ions

charged ions (e.g., 2+, 3+) compared to singly charged ions [126], and complete sequence coverage can be achieved for peptides less than 5 kDa [127]. CID is also AA dependent as it preferentially cleaves at the N-termini of Pro residues [128]. IRMPD and BIRD are similar to CID as they are all "slow heating" methods [129], which activate ions to increase the internal energy to above the dissociation threshold for various bonds in the ions. Therefore, these methods cause weak bonds, such as non-covalent interactions, and labile side chain modifications, such as phosphorylation and glycosylation, to dissociate before the backbone amide bonds. ECD and ETD techniques fragment peptides in a very different manner from CID. They introduce a low energy electron (<1 eV) into the molecular ion from either an electron source (ECD) or another chemical reagent (ETD) to generate a radical ion, which has large charge recombination energy (>5 eV) that can quickly dissociate the backbone and form characteristic *c* and *z* ions (see Scheme 5.5) before this energy is randomized over the molecular ion to dissociate weak bonds. Therefore, labile PTMs and non-covalent interactions can be preserved in ECD and ETD [130–136]. Compared to CID, ECD, and ETD are less dependent on AA sequence and thus provide more extensive and complementary sequence coverage [137–140].

Normally, the MS/MS techniques described above are applied in data-dependent analysis (DDA), where MS detection switches between two modes: a full MS scan to obtain masses of precursor ions and several MS/MS scans to obtain masses of fragment ions from these precursors, one precursor per scan. One disadvantage of DDA is the loss of data in one scan mode while performing detection in the other mode, which results in poor duty cycles and slow analysis. A new form of MS and MS/MS data acquisition that maximizes the instrument duty cycle and collects information for both precursor and fragment ions in the same mode is called MS^E, which was first introduced in 2005 [141–143]. MS^E is a CID-based fragmentation technique that utilizes parallel alternating scans acquired at either low collision energy to obtain precursor ion information or high collision energy to obtain both precursor and fragment ions information. Multiple peptides can be detected and analyzed in the same scan. When combined with UHPLC, it provides both chromatographic and MS efficiencies that are ideal for fast analysis, narrow and rapid eluting peak analysis, and complex mixture analysis. Figure 5.3 illustrates an example of MS^E data collected on a Synapt G2 HDMS instrument. In the low collision energy scan, two peptides were observed (P1 and P2) with different charge states. In the high collision energy scan, fragment ions from both peptides were observed (red series ions from P1 and blue series ions from P2).

5.2.3.2 "Bottom-up" and "Top-down" Sequencing

"Bottom-up" is the most versatile and frequently used approach for AA sequence analysis of proteins [144]. In the "bottom-up" approach, the protein undergoes the following steps: denaturation using a denaturant such as urea, guanidine, or RapiGest SF [145]; reduction of S–S normally using DTT; alkylation of free Cys using reagents such as iodoacetamide (IAM), iodoacetic acid (IAA), or

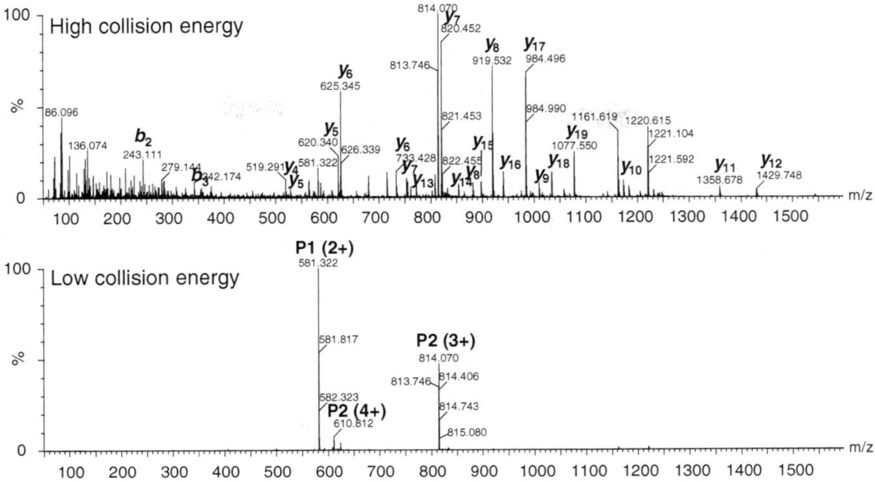

Fig. 5.3 MS^E fragmentation of two peptides (P1 and P2), *top* high collision energy scan, *bottom* low collision energy scan

4-vinylpyridine (4-VP); and digestion using a protease such as trypsin, Lys-C, Glu-C, or Asp-C. The digested peptides are then separated by RP-HPLC or UHPLC and analyzed by MS, MS/MS, or MS^E, which is often referred to as LC/MS/MS peptide mapping analysis. After generating the MS and MS/MS data, the sequence of each peptide can be determined, and by putting all the peptide sequences together, the sequence of the entire protein can be confirmed. This data analysis can be done either manually or preferably by utilizing a database that contains protein AA sequence information predicted from DNA sequences [146–150].

Figure 5.4 shows an example of peptide mapping analysis from a typical mAb. The top chromatogram contains peaks for digested tryptic peptides separated in RP-UHPLC. Each of these peaks represents one or several peptides, and the MS spectrum of each peak (e.g., the one circled by red line) gives initial confirmation of the peptide based on accurate mass (<2 ppm); the MS/MS spectrum further confirms the identity of these peptides. It is important to note that very small peptides (2–3 AA residues) are often undetected due to their short retention times in RP columns [16]. Nevertheless, with the use of UHPLC/MS at high resolving power and mass accuracy as well as multiple enzyme options when necessary, typical peptide mapping experiment can often achieve a sequence coverage above 90 %.

Although the "bottom-up" approach provides high sequence coverage and is applicable for very large proteins, it often requires relatively large sample amounts and is both time consuming and labor intensive [16, 47]. The "top-down" approach is a viable alternative as it directly sequences the protein in the gas phase and involves minimum sample handling [127]. For complex mixtures, the "top-down" approach can be used to isolate a single component that generates

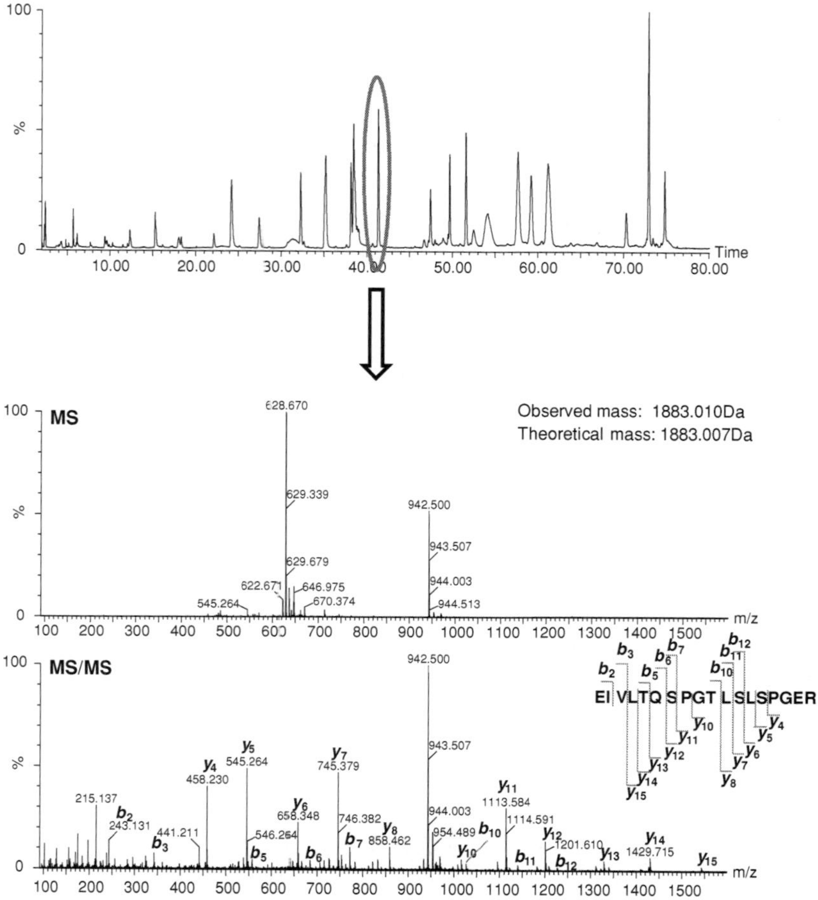

Fig. 5.4 "Bottom-up"/peptide mapping approach for protein sequencing. *Top* base peak chromatogram of RP-UHPLC/MS analysis of a protein tryptic digest (peptide map). *Bottom* an example of using MS and MS/MS spectra to assign one of the eluted peaks in the peptide map

molecular ions for MS/MS fragmentation, provides molecular identity with far higher reliability, and allows direct characterization of sequence errors and PTMs [151–154]. Mass spectrometers with high resolving power and mass accuracy, such as FT-ICR and Orbitrap or their hybrid instruments [62, 155–162], are preferred in the "top-down" approach. However, currently "top-down" MS is feasible only for rapid characterization of small- to medium-sized proteins (<50 kDa) [163]. For larger proteins, "top-down" structural analysis remains a challenge with only limited success due to the decreased MS/MS efficiency of larger ions [56]. For example, "top-down" sequencing of a mAb provides only information about AA sequence in the variable regions and terminal regions [164]. As a rule of thumb, the size of a protein that can be analyzed by "top-down" MS is on the same order as the resolution of the instrument [16]. In other words, an instrument with a resolution of 10 k should yield a significant amount of structural information for proteins of up to 10 kDa.

One way to increase the effective mass range for current MS/MS instrumentation is to use the "middle-down" approach, in which limited proteolysis or S–S reduction is performed on large proteins to generate a few large fragments (3–20 kDa) which can then be subjected to MS/MS analysis [16, 47]. An example is the MS/MS analysis of separated light and heavy chains on a Q-TOF instrument [165]. Considerable effort is being invested for improving the utility of "top-down" sequencing [166, 167]. The optimal technique for solving a structural question depends on the nature of the problem, throughput requirements, available sample amount, and other factors. Due to the complementary nature of these methods, complete molecular characterization often requires a combination of techniques [168].

5.2.3.3 Applications: Identification and Quantification of Terminal Variants

C-terminal Lys cleavage (sometimes the second C-terminal residue, usually Gly, is also partially cleaved) [169–171] and N-terminal pyro-E formation [79, 172] are commonly observed in recombinant mAbs produced from CHO cells. A "bottom-up" approach is useful to characterize and quantify these terminal variants. In a typical analysis, the antibody is first digested using trypsin and the peptide mixture is separated by RP-UHPLC and analyzed by MS and MS/MS on a QTOF instrument. From the protein AA sequence, the monoisotopic masses of C-terminal peptides [e.g., SLSLSLG(K)] generated from the trypsin digest can be calculated: 803.4753 Da with Lys and 675.3803 Da without Lys. By using extracted ion chromatograms (EICs), MS signals from these two peptides can be extracted from the peptide map, with observed monoisotopic masses of 803.4672 and 675.3751 Da and a mass error of 10.0 and 7.7 ppm, respectively (Fig. 5.5). This high mass accuracy provides an initial identity of these two peptides, and further confirmation can be achieved through MS/MS analysis (Fig. 5.6). By integrating the EIC signals of these two peptides, the relative abundance of the C-terminal variant (Lys cleavage) can be calculated. It is important to note that such semi-quantitative analysis does not take into consideration potential differences in ionization efficiencies for these two peptides (with and without Lys).

In a similar fashion, N-terminal variants can be monitored. For example, the unmodified N-terminal peptide, EVQLLESGGGLVQPGGSLR, has a theoretical mass of 1,895.0112 Da and an observed mass of 1,895.0218 (5.6 ppm, spectrum not shown here). When N-terminal Glu is modified to pyro-E, the calculated mass is 1,877.0006 Da and observed mass is 1,877.0080 Da (3.9 ppm, spectrum not shown here). MS/MS spectra of the modified and unmodified peptides further suggest that the modification is at the N-terminus (Fig. 5.7): all of the *y* ions observed

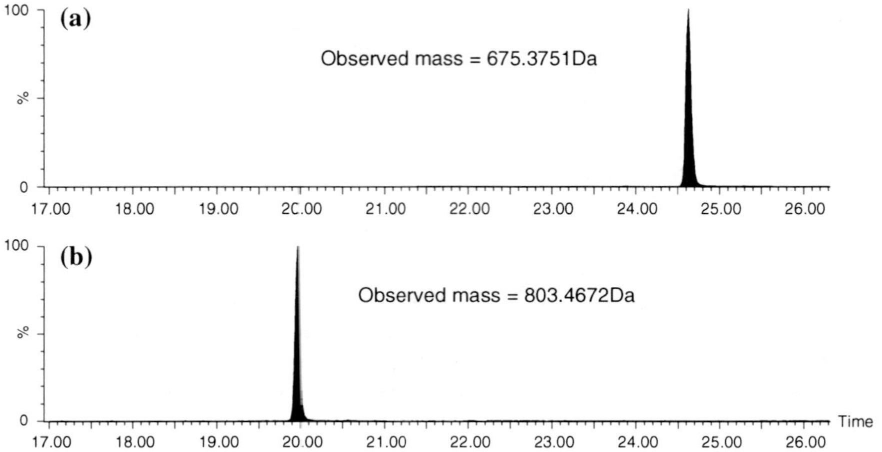

Fig. 5.5 Extracted ion chromatograms of *C*-terminal peptides without Lys (**a**) and with Lys (**b**)

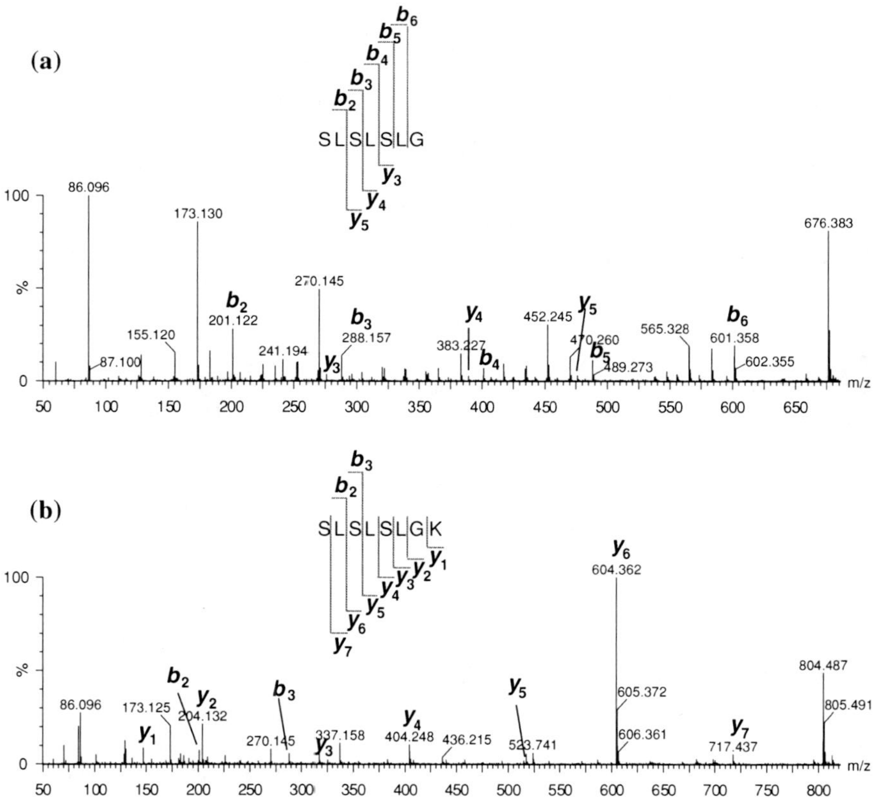

Fig. 5.6 MS/MS spectra of *C*-terminal peptides: without Lys (**a**) and with Lys (**b**)

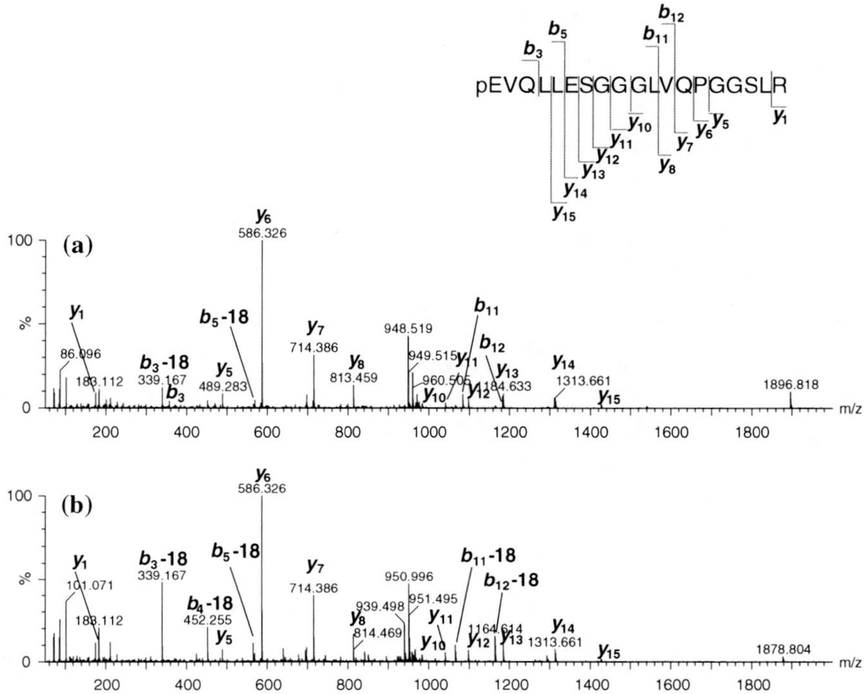

Fig. 5.7 MS/MS spectra of *N*-terminal peptides: unmodified (**a**) and modified Glu to pyro-E (**b**)

for the modified peptide (bottom) have the same mass as those observed for the unmodified peptide (top), and all of the *b* ions observed for the modified peptide have masses that are 18 Da lower compared to those observed for the unmodified peptide. Although some *b* ions from the unmodified peptide are also 18 Da lower in mass, their abundance is much lower and the difference in mass can probably be attributed to neutral water loss (−18 Da).

5.2.4 Disulfide Bond Linkage Analysis

Disulfide bond (S–S) formation is a common covalent PTM of proteins and is critical for stabilizing protein structure and function [173]. For example, the four subtypes of IgG antibodies have homologous intra-chain S–S and characteristically different inter-chain S–S linkages [72]. Mis-linkages or scrambling of these S–S can lead to structural isoforms of the antibodies, resulting in increased sample heterogeneity and potentially changing antigen-binding affinities [71, 72]. Therefore, localization and characterization of S–S and S–S scrambling are important aspects of protein structural analysis.

5.2.4.1 Non-Reduced and Reduced Peptide Mapping Approach

A common strategy for characterizing S–S linkages involves identification of S–S linked peptides in non-reduced digests of proteins and characterization of their half-cystinyl peptide constituents in reduced digests of proteins using LC/MS and LC/MS/MS analysis [174]. Instruments with high resolving power and mass accuracy are preferred for this purpose as the S–S linked peptides normally have high masses. The challenge associated with this approach is that S–S scrambling/shuffling can occur during sample preparation, which results in incorrect assignment of S–S linkages. Although the exact mechanism of S–S scrambling is unclear, it is likely that the presence of free sulfhydryl groups (SH) in basic pH conditions causes thiol–disulfide exchanges, which result in S–S scrambling [175, 176].

There are a number of approaches to minimizing S–S scrambling. The use of pepsin for protein digestion at low pH has been demonstrated to reduce S–S scrambling; however, ragged peptide products resulting from pepsin digests (pepsin is a non-specific enzyme for proteins) increase the complexity of peptide mapping [177]. Using a solvent that contains 50 % $H_2{}^{18}O$ during pepsin digestion can simplify the identification of S–S linked peptides because these peptides have distinct isotope profiles compared to the same peptides with ^{16}O [174]. To date, the most practical way to minimize S–S scrambling is to alkylate SH groups with suitable reagents, such as N-ethylmaleimide (NEM), prior to enzymatic digestion and to split the digest into two aliquots: one aliquot subjected to peptide mapping analysis directly (non-reduced peptide mapping) and the other aliquot reduced before peptide mapping analysis (reduced peptide mapping), as shown in

Scheme 5.6 Procedure for S–S linkage analysis using non-reduced and reduced peptide mapping methods

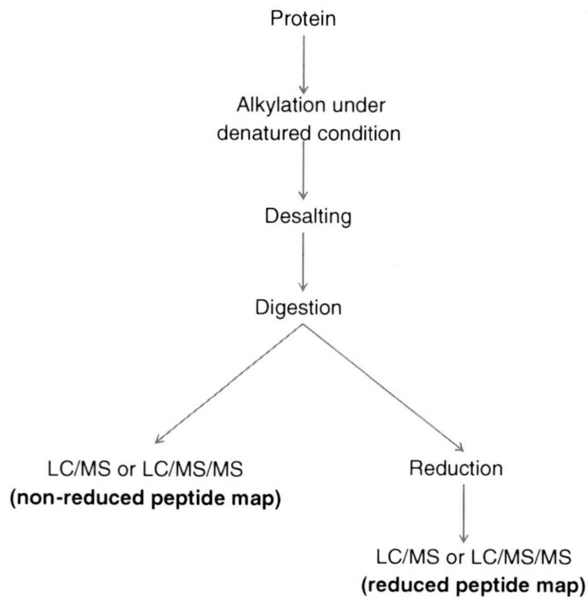

Scheme 5.6. This approach has been used to minimize S–S scrambling, to provide a quantitative measurement of free SH groups and to assign S–S linkages correctly [72, 175, 176, 178].

5.2.4.2 ETD/CID MS^3 Approach

Another way to perform S–S linkage analysis is to directly identify the S–S linked peptides produced from a non-reduced digestion by MS/MS or MS^n. This approach involves minimum sample preparation and sample consumption, but requires special instrumentation. The fragmentation efficiency of CID or other "slow heating" methods on S–S linked peptides is poor and generates insufficient backbone fragmentation, making it very difficult to accurately and confidently identify S–S linked peptides. ETD, with its unique fragmentation pathway, has been shown to preferentially break apart S–S linkages in peptides and generate dissociated half-cystinyl peptide constituents, which can be identified with further fragmentation analysis (MS^3). Many research groups have applied ETD/CID MS^3 to S–S analysis, where ETD is utilized first to break S–S linked peptides (MS^2) and then CID is applied to sequence the dissociated peptides [179–182].

5.2.4.3 N-Terminal Edman Sequencing Approach

For S–S linked peptides containing two closely spaced and symmetrical S–S linkages (e.g., in the hinge region of an IgG1), there is no enzymatic cleavage site between the Cys residues, and therefore, the linkage pattern cannot be identified using the above two approaches. For this type of analysis, N-terminal Edman sequencing has been successful in identifying the correct arrangement of S–S linkages [176]. Although two bonding patterns are possible for the disulfides, only one can produce the di-phenylthiohydantoin (PTH)-cystine and, subsequently, a signal in Edman degradation.

5.2.4.4 Applications: Examples of Disulfide Bond Linkage Analysis

Figure 5.8 shows an example of S–S linkage analysis of a typical IgG1 mAb using the non-reduced and reduced peptide mapping approach. The top and bottom traces show the non-reduced and reduced peptide mapping chromatograms, respectively. When these two traces are compared, peaks that appear only in the top trace indicate presence of S–S linked peptides, for example, P1–P2 and P3–P4. Accurate mass measurements of these peaks provide initial identification of these peptides (Table 5.1). Peaks that appear only in the bottom trace represent the half-cystinyl peptide constituents generated from the reduction of S–S linked peptides, for example, P1, P2, and P3 (P4 is not detected as it contains two

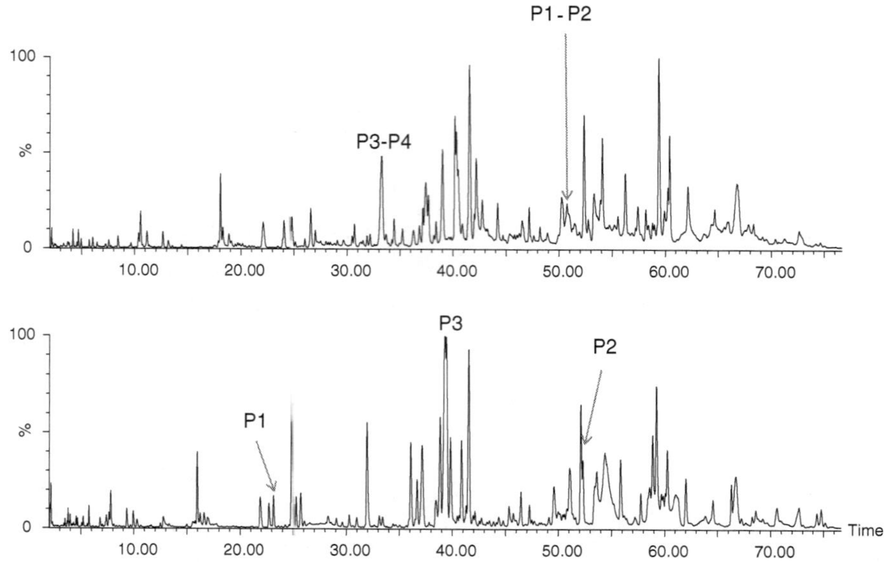

Fig. 5.8 Non-reduced (*top*) and reduced (*bottom*) peptide mapping of a mAb: two S–S linked peptides (P1–P2 and P3–P4) and their corresponding reduced peptides (P1, P2 and P3, P4 were not observed)

Table 5.1 Comparison of theoretical and observed masses of disulfide linked peptides and their corresponding reduced peptides[a]

	Theoretical mass (Da)	Observed mass (Da)
P1–P2	3422.53	3422.53
P3–P4	2328.11	2328.12
P1[a]	1331.59	1331.61
P2[a]	2138.03	2138.06
P3[a]	2207.00	2207.02

[a]Reduced peptides are alkylated with iodoacetamide

AA residues and elutes very early with salts). MS and MS/MS analysis of these reduced peptides can provide AA sequence confirmation. For example, P1 and P2 in the reduced peptide mapping was first identified by matching observed and theoretical masses (Table 5.1) and subsequently confirmed by MS/MS spectra as shown in Fig. 5.9.

Figure 5.10 illustrates how an S–S linked peptide (P3–P4) can be identified using ETD/CID MS^3. The observed mass of the S–S linked peptide (2328.12 Da) is consistent with the mass calculated from the sum of the mass of P3 and the mass of P4 (2328.12 Da), which provides the initial identification. After applying ETD, the S–S linkage breaks, yielding separate P3 and P4 peptides (in some

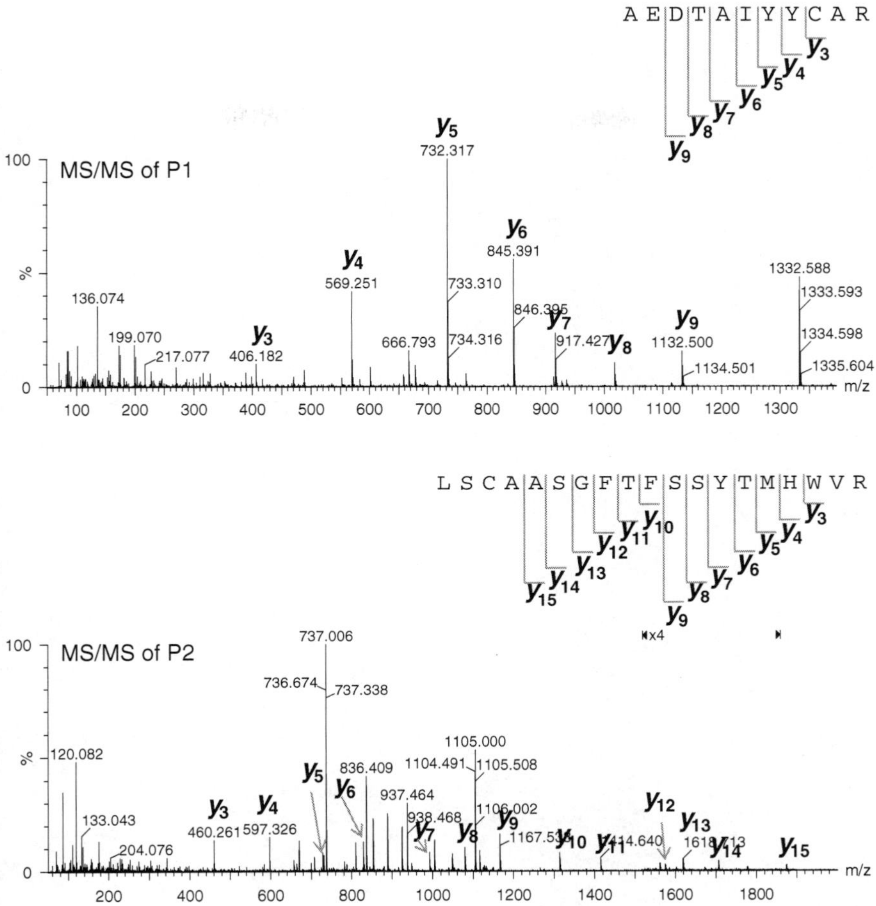

Fig. 5.9 MS/MS spectra of reduced peptides: P1 and P2

cases, P3 and P4 are held together by non-covalent interactions and additional "supplemental activation" is needed to break them apart [183]), and the individual masses of P3 and P4 are consistent with the masses calculated from their AA sequences. Further CID fragmentation of P3 confirms its AA sequence (P4 has only two AA residues and is not subject to CID for further identification).

When there are three peptides linked by two S–S linkages, the ETD/CID MS^3 approach can be used to determine the correct linkage pattern. As displayed in Scheme 5.7, the CID fragments of the dipeptide (P1–P2 or P2–P3) generated from ETD are different depending on the patterns of S–S linkages in the original peptide. For example, in pattern 1, the b_3 up to b_7 ions generated from CID of P2 peptide in P1–P2 all contain the linked P1 peptide; in pattern 2, only the b_6 and b_7 ions contain the linked P1 peptide.

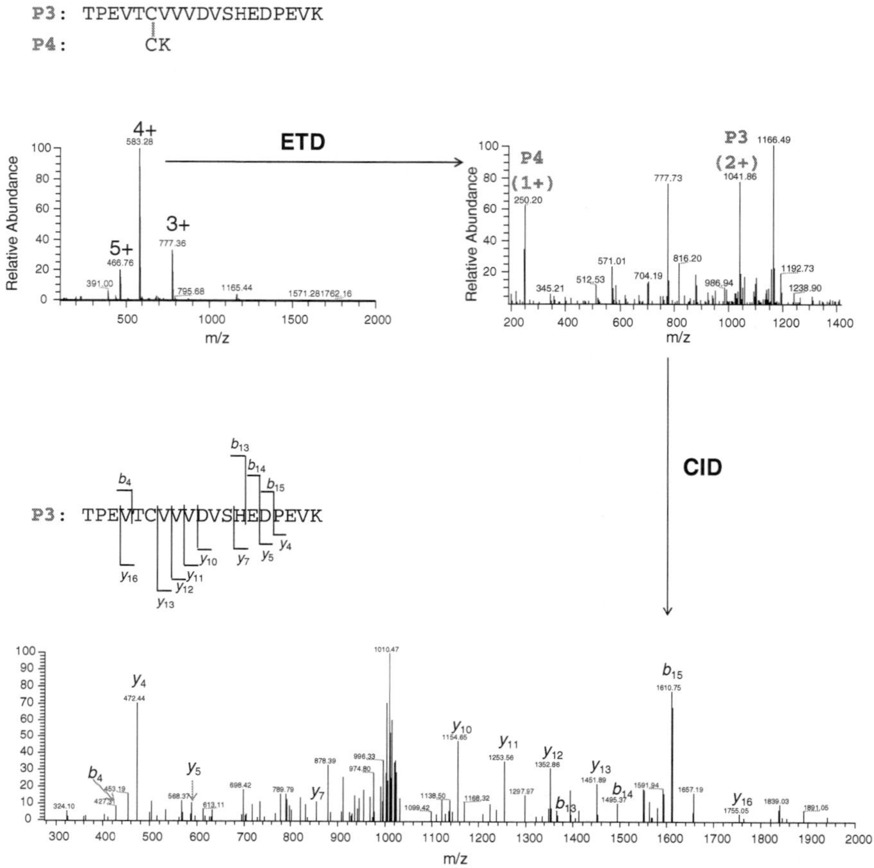

Fig. 5.10 Identification of S–S linked peptide (P3–P4) by ETD/CID MS^3

5.2.5 Common Chemical Modification Analysis

5.2.5.1 Oxidation

Protein oxidation can be induced by the presence of oxidants, transition metal ions, heat and light [184, 185]. All AAs are susceptible to oxidation, although their susceptibilities vary greatly [186]. Due to their high reactivity with various reactive oxygen species, free Cys residues that are not involved in S–S linkages, Met, Trp, His, and Tyr residues, in that order, are most prone to oxidation [185]. The thiol group (RSH) in free Cys is a very reactive functional group and is involved in many biological pathways; hence, very few pharmaceutical proteins contain free Cys [185]. As summarized in Table 5.2, Cys can incorporate up to three oxygen atoms to form sulfenic acid (RSOH), sulfinic acid (RSO2H), or sulfonic

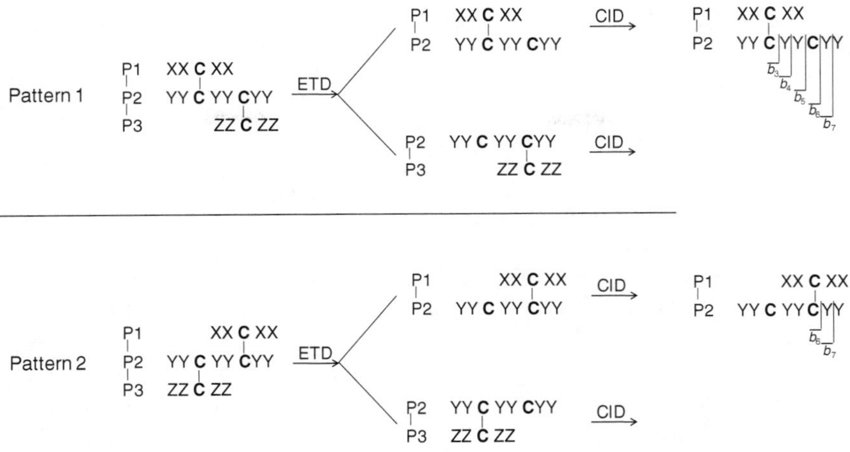

Scheme 5.7 Differentiation of S–S linkage patterns among three peptides using ETD/CID MS^3

Table 5.2 Common oxidation products of labile amino acids in protein

AA residue	Oxidation product	Mass change (Da)
Cys	Sulfenic acid (RSOH)	+16
	Sulfinic acid (RSO2H)	+32
	Sulfonic acid (RSO3H)	+48
Met	Met sulfoxide	+16
	Met sulfone	+32
Trp	Kynurenine	+4
	Hydroxy-Trp	+16
	3-hydroxykynurenine	+20
	N-formylkynurenine	+32
	Hydroxy-N-formylkynurenine	+48
His	2-oxo-His	+16
	Asp/Asn	−22/−23
Tyr	3,4-dihydroxyphenylalanine	+16

acid (RSO3H) [187], which correspond to mass increases of 16, 32 or 48 Da, respectively. Met oxidation is most common in protein therapeutics and will be discussed in more detail in the next paragraph. Trp oxidation is most likely to occur when proteins are exposed to UV light [188–190], and under such conditions, Trp can be oxidized to several forms, including kynurenine, hydroxy-Trp, 3-hydroxykynurenine, N-formylkynurenine, or hydroxy-N-formylkynurenine, with mass increases of 4, 16, 20, 32 or 48 Da, respectively [191]. Trp oxidation in the CDR region has been showed to result in a loss of protein-binding affinity and bioactivity [189, 190]. His and Tyr oxidation are less common than the others mentioned above, and their oxidation products are listed in Table 5.2 [184, 185, 192].

Met oxidation is very common in proteins as it contains a sulfur residue on the side chain that is susceptible to oxidation (Scheme 5.8). When oxidized, Met sulfoxide forms, resulting in a mass increase of 16 Da. Under harsh oxidative conditions, further oxidation of Met sulfoxide into Met sulfone can occur and result in a mass increase of 32 Da [184]. Met oxidation has been demonstrated to change protein structure, stability and function [193–200]. In humanized and fully human mAbs, the two Met residues in the Fc constant regions, one in the C_H2 and one in the C_H3 domain, are the most susceptible to oxidation [201–203]. Although Met oxidation in the Fc region can change antibody conformation [204–206], it is not expected to affect antigen-binding affinity due to the presence of the highly flexible hinge region [202]. However, in the 3D structure, these two Met residues are located close to the C_H2–C_H3 interface, and oxidation of these residues can decrease the binding affinity of antibody to Protein A, Protein G, FcRn and Fcγ and reduce the circulation half-life of the antibody [207–209]. Oxidation of these two Met residues can also influence the antibody-dependent complement cascade [210], and complement dependent cytotoxicity and phagocytosis [204].

Since protein oxidation can result in the loss of biological activity and/or other changes that are detrimental to therapeutic potency, monitoring the level of protein oxidation is critical for biopharmaceutical research and development. There are a wide array of tests such as AA analysis [211], capillary electrophoresis [212], immunoassays [213, 214], or HPLC/UV [188, 202, 215, 216] to detect and monitor protein oxidation. MS-based assays at either the intact protein or peptide level can be used; however, intact mass analysis of oxidized proteins requires high resolving power of the mass spectrometer since the expected mass increases (e.g., 16 or 32 Da) needs to be discriminated from molecular adducts such ammonia (+17 Da), water (+18 Da), sodium (+22 Da), and potassium (+38 Da). LC/MS/MS peptide mapping is a more practical approach to obtain both site-specific information and an estimation of the level of oxidation [185, 205, 206, 217, 218].

Figure 5.11 shows an example of identification of Met oxidation by MS and MS/MS analysis at the peptide level (peptides are generated from protein tryptic digestion). Oxidized peptides containing Met sulfoxide can be separated from non-oxidized peptides using RP-UHPLC, as oxidized peptides normally elute earlier than their corresponding non-oxidized peptides on a C18 column. The observed

Scheme 5.8 Methionine oxidation pathway

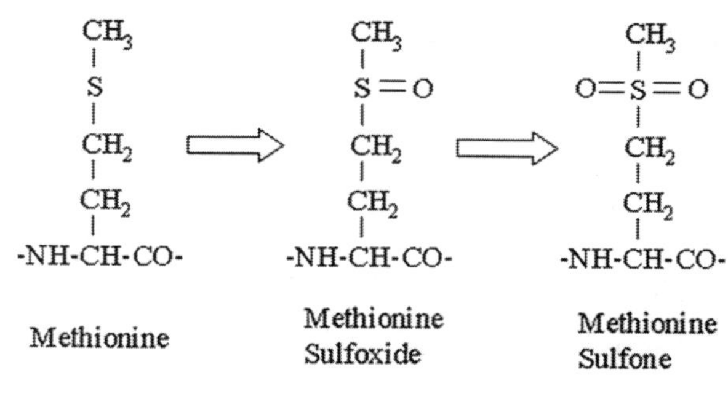

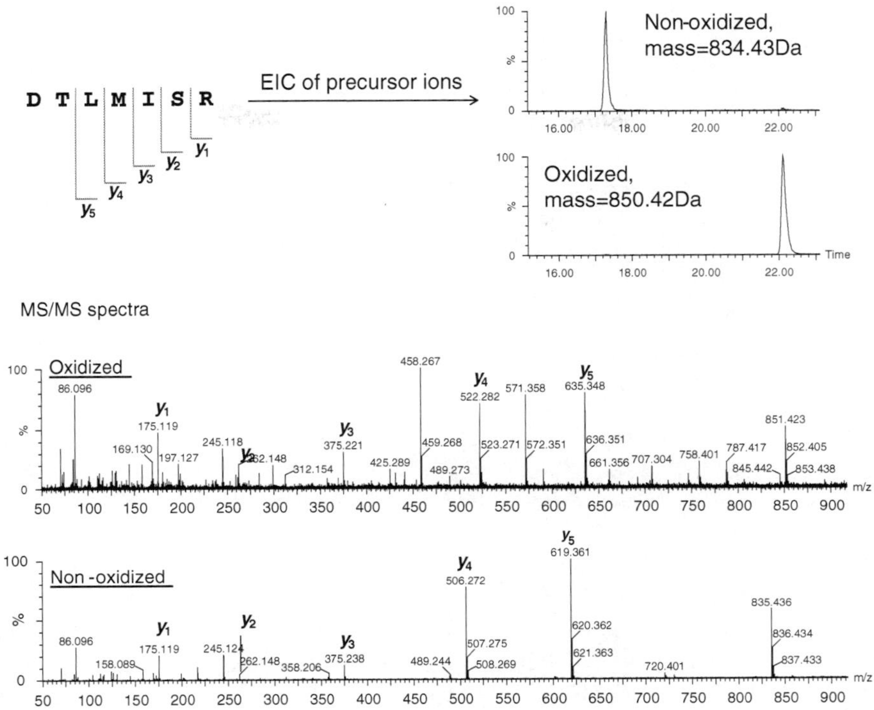

Fig. 5.11 Identification and quantification of Met oxidation products

precursor ion of the oxidized peptide (DTLMISR with Met oxidized to Met sulfoxide) is 16 Da heavier than the non-oxidized peptide (850.42 vs. 834.43 Da). A comparison of their MS/MS spectra shows that all the y ions that contain the Met site (y_4 and y_5) have a 16 Da shift in mass, while the y ions that do not contain Met (y_1, y_2, and y_3) have the same mass in both spectra. Quantitation of Met oxidation can be achieved by using either the UHPLC/UV signal or EIC MS signal of the precursor ions. MS quantitation is more sensitive than UV quantitation, but it is influenced by the difference in ionization efficiencies between the non-oxidized and oxidized peptides; thus, MS measurements without appropriate standards provide only a semi-quantitative estimate of oxidation levels. UV quantitation is reliable only when there is no co-eluted peptide for both oxidized and non-oxidized peptides.

5.2.5.2 Deamidation and Isomerization

Deamidation is a chemical degradation of protein and peptides that mainly occurs at Asn residues. As shown in Scheme 5.9, deamidation can be the product of two pathways [219, 220]: under acidic conditions, Asn deamidates by direct hydrolysis of the amide side chain and forms Asp (pathway B); under neutral and basic

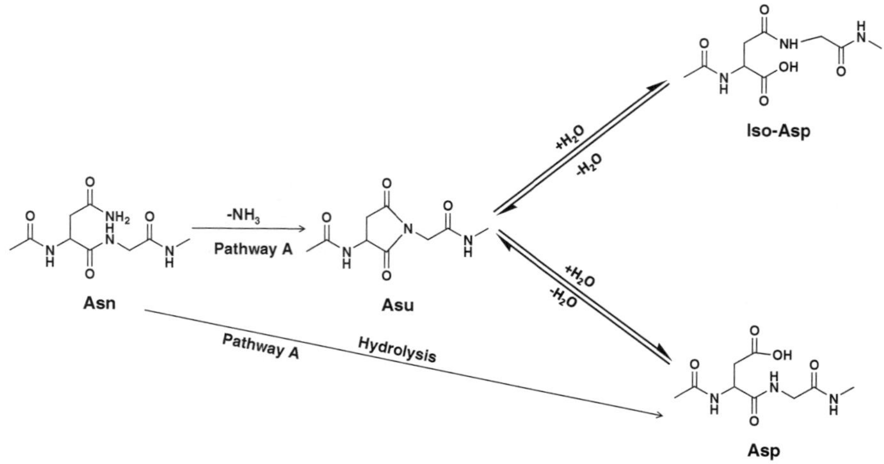

Scheme 5.9 Deamidation pathways

conditions, Asn deamidates by forming a cyclic imide intermediate (succinimide or Asu) and produces iso-aspartic acid (iso-Asp) and Asp (pathway A) with a ratio of approximately 3:1 [221]. Asp and iso-Asp are isoforms with the same mass, which is 1 Da heavier than the mass of Asn. Deamidation can also occur on Gln residues in a similar manner, but to a lesser degree and at a much slower rate [222, 223]. Isomerization is a process in which Asp converts to iso-Asp through an Asu intermediate (Scheme 5.10) [224]. Isomerization of Glu is rarely observed, except when it occurs at the N-terminus of a protein [172]. One of the major concerns regarding the deamidation and isomerization of therapeutic proteins is that the unnatural AA iso-Asp can potentially increase immunogenicity [15]. However, deamidation is known to occur in vivo [225, 226] and under physiological temperature (37 °C) and pH (7.4), so it seems that deamidation is a naturally occurring

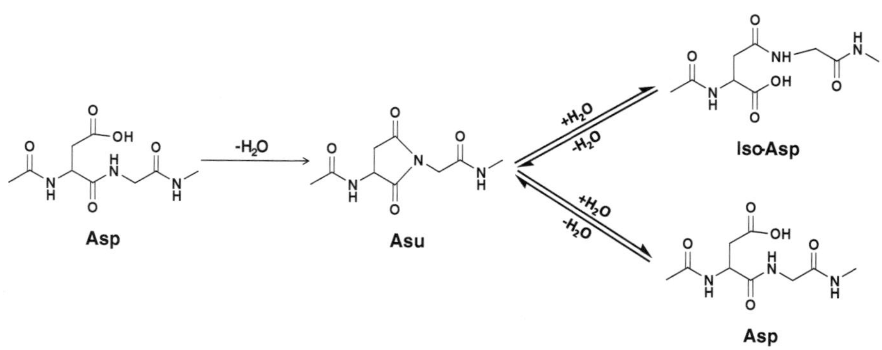

Scheme 5.10 Isomerization pathways

process for endogenous proteins. The potential immunogenicity of iso-Asp in a given protein can only be assessed in clinical studies. Regardless, deamidation and isomerization can change protein conformation and significantly affect the potency and binding efficiency of therapeutic proteins [227–235].

There are many studies investigating the factors that influence the rates of deamidation and isomerization in proteins. Both processes are dependent on pH, temperature, buffer composition, the C-terminal neighboring AA, local polypeptide conformation, and tertiary structure. When undergoing pathway A (above pH 5), deamidation rates increase with increasing pH; when undergoing pathway B (below pH 3), deamidation rates increase with decreasing pH [219, 236]. Therefore, deamidation is typically minimal at pH 4–5. Isomerization is favored under slightly acidic conditions (pH 5–6) [236]. Higher temperatures favor both deamidation and isomerization [219, 224]. The relationship between primary sequence and deamidation/isomerization has been extensively studied. Although the N-terminal AA of Asn has little or no effect on deamidation rates, the C-terminal AA of Asn has a significant effect under neutral and basic conditions (pathway A), with a small AA (e.g., Gly, Ser, or His) favoring deamidation as the low steric hindrance can facilitate the formation of Asu [237–240]. A similar effect was observed for isomerization, and therefore, Asp followed by a Gly is typically the most favorable site for isomerization [240, 241]. Secondary and tertiary structures also influence deamidation rates. For example, the rate of deamidation in peptides is inversely proportional to the extent of α-helicity [242], and a molecule in its native (folded) conformation deamidates more slowly than when it is in a denaturated conformation [243, 244]. Buffer composition has an effect on deamidation and isomerization rates through its impact on the pH, dielectric constant and viscosity of the buffer solution as well as on the protein conformation [245, 246].

Since deamidation shifts the pI of a protein toward the acidic end, any charge-based assays, such as ion exchange chromatography and isoelectric focusing, can be used to monitor overall deamidation. There is also a commercial IsoQuant kit that can be used for quantitative detection of iso-Asp residues in proteins and peptides [247]. However, this method only detects the presence of iso-Asp residues, and an adjustment, based on the approximate 3:1 ratio of iso-Asp to Asp, needs to be made to account for Asp residues in order to get an accurate assessment of deamidation levels. All of these methods, however, only provide overall deamidation levels. To obtain site-specific information, the most commonly used technique is LC/MS/MS peptide mapping [243, 248]. Briefly, after the protein is digested, the unmodified peptide can be separated from its deamidated forms, and MS or MS/MS detection can further differentiate the deamidated peptides from the unmodified peptides because deamidation results in a 1 Da increase in mass. As shown in Fig. 5.12, the precursor ion (3+) of a deamidated peptide is 1 Da higher than the precursor ion of the unmodified peptide (1,808.03 vs. 1,807.03 Da), and MS/MS fragments containing the deamidation site (N) also show a 1 Da increase in mass compared to those generated from unmodified peptides (e.g., y_3 is at m/z 319.2 in the MS/MS spectrum of deamidated peptide, while it is at m/z 318.2 in the MS/MS spectrum of unmodified peptide). The level of deamidation can be

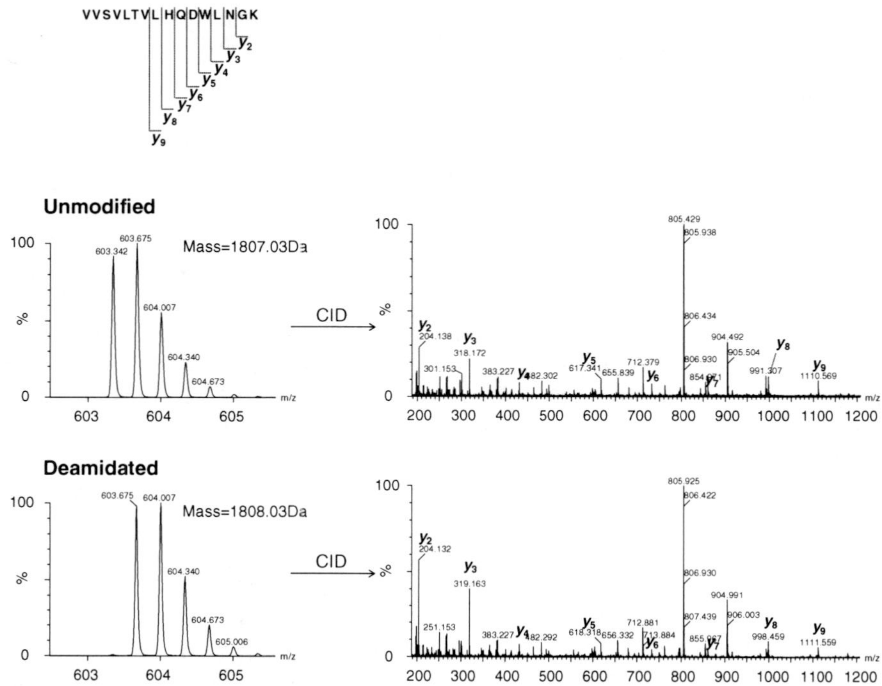

Fig. 5.12 MS and MS/MS spectra of unmodified and deamidated peptides (VVSVLTVLHQDWLNGK)

assessed using either LC–UV or EIC signals, similar to the quantitation of Met oxidation mentioned earlier. It should be noted that enzymatic digestion under alkaline conditions can induce deamidation; therefore, digestion is normally controlled at pH 7.5 for a short period of time with a high enzyme-to-protein ratio in order to minimize experimentally induced deamidation. There are studies using $H_2{}^{18}O$ labeling to differentiate between deamidation that is present in the original sample from deamidation that occurs during sample preparation [59].

For characterization of isomerization, peptide mapping is the most commonly used approach. HPLC methods, such as using RP-HPLC [234] and hydrophobic interaction chromatography (HIC) [229, 246], have been developed to separate peptides/proteins with iso-Asp from those with Asp residues. Further differentiation between Asp-containing peptides and iso-Asp-containing peptides by MS requires the use of specialized MS instrumentation and alternative MS/MS approaches because iso-Asp and Asp residues are isobaric. Some groups have successfully differentiated peptides containing either Asp or iso-Asp by using $H_2{}^{18}O$ labeling [249]. Some reports have suggested that the different intensity ratios of *b*:*y* ions from CID fragmentation of the Asp-containing peptides and the iso-Asp-containing peptides can be used to differentiate them [250]. However, other reports suggest that this is not a general phenomenon and may be dependent on both peptide sequence and the type of MS instrumentation used. ECD implemented on an FT-ICR mass spectrometer has demonstrated the ability to differentiate peptides containing Asp from those containing iso-Asp residues since it produces unique diagnostic ions for both peptides, that is, side chain loss (−60 Da) for Asp-containing peptides, and (c_r+58 Da) and (z_{l-r}–57 Da) for iso-Asp-containing peptides where l is the total number of AA residues in the peptide, and r is the position of iso-Asp residue [251–254]. Because of its similar fragmentation pathway to ECD, ETD can produce the same diagnostic ions and therefore is also capable of differentiating between peptides containing Asp and iso-Asp residues [255, 256]. With a low-resolution instrument such as an ion trap, identification of diagnostic peaks of Asp-containing peptides is difficult due to interference from side chain fragment ions of Arg and Glu residues [255, 256]. However, the diagnostic peaks for iso-Asp-containing peptides can still be clearly defined. For example, ETD fragmentation of deamidated peptides with an iso-Asp residue from VVSVLTVLHQDWLNGK using an ion trap instrument exhibits a diagnostic product ion (z_3–57 Da) at m/z 246 indicative of the presence of iso-Asp residue.

5.2.5.3 Applications: Evaluation of Protein Stability

During protein drug discovery, once a lead molecule is selected, its stability profile is evaluated to ensure that the molecule has acceptable biophysical properties for development into a final product. Typically, the molecule is tested under different stress conditions, such as elevated temperatures, increased humidity, different pH, exposure to chemicals, light, or a combination of these, to monitor its susceptibility to different types of modifications. Normally, the "hot spots" for modifications, such as Met oxidation, Asn deamidation in Asn-Gly sequence and Asp isomerization in Asp-Gly sequence, are monitored, and their levels are calculated from EIC MS signals of the modified and unmodified peptides using LC/MS/MS as described above. The results can be used as guidelines for candidate optimization, buffer optimization, etc. For example, if a high level of deamidation is observed at a particular site, mutation of that Asn or its neighboring residues may be needed to prevent or minimize deamidation. Another phenomenon commonly observed is that the pH of the formulation buffer plays an important role in controlling the level of deamidation.

5.2.6 Higher-Order Structure Analysis

Since the biological activity of all proteins is dictated by higher-order structure and dynamics in solution, the therapeutic properties of protein biopharmaceuticals are uniquely determined by their conformation [257]. Techniques that allow for molecular investigation of protein structure and conformational dynamics are invaluable to fully understand how proteins drive and contribute to basic biological and biochemical events. In this section, MS-based approaches for protein structural analysis, hydrogen/deuterium exchange mass spectrometry (HDX–MS) and ion mobility mass spectrometry (IMMS) will be briefly described. More details are also provided in Chaps. 8 and 10.

5.2.6.1 Hydrogen/Deuterium Exchange Mass Spectrometry

Over the last decade, HDX–MS has been used extensively to study protein conformational behavior in solution [258–262]. HDX–MS offers several advantages over other techniques such as X-ray crystallography and NMR spectroscopy: it is more sensitive; sample preparation is simpler; it can analyze protein mixtures that are more complex; and it offers high structural resolution with minimal sample preparation. In HDX–MS, the exchange rates of protein amide hydrogens with deuteriums from deuteriated buffer are monitored by MS, and a 1 Da increase in mass reflects the exchange of one hydrogen with a deuterium. Because the rate of HDX is dependent on protein exposure to the solvent and on inter-/intra-molecular hydrogen bonding, the information obtained from HDX–MS can be correlated to the protein structure in solution at the intact molecular level (global conformation) as well as at the peptide level (local conformation), and sometimes even at the single AA residue level. HDX–MS has been widely used to probe protein conformation and locate changes in protein structure [263, 264], to study the impact of PTMs, such as glycosylation and Met oxidation, on protein conformation and function [265, 266], and to investigate complex protein interactions at their binding sites [265, 267–270]. With automated instrumentation and data analysis software becoming commercially available, the application of HDX–MS in bioanalytical laboratories for exploring protein or protein complex higher-order structures is likely to become routine practice in the near future.

5.2.6.2 Ion Mobility Mass Spectrometry

With the recent introduction of commercially available instrumentation in several forms, IMMS has been extensively integrated into life sciences research programs [271]. When coupled to LC, IMMS has been used for the structural characterization of peptides, proteins, complex mixtures, and heterogeneous macromolecular assemblies [272–274]. Compared to LC/MS alone, IMMS provides an additional separation step based on protein size, shape, and conformation that offers significant simplification of the spectra of complex biological samples. For example, IMMS has been utilized to separate the heavy chains and light chains from a reduced antibody, which reduces molecular complexity, yields more accurate mass measurements, and facilitates quantitation of different glycoforms [275]. It has also been used to quickly resolve disulfide structural isoforms of IgG2 antibodies to assist in S–S heterogeneity analysis [276]. IMMS has been particularly useful for characterizing PEGylated proteins since it can separate PEGs based on their size and charge [85, 92], and it offers enhanced dynamic range, increased sensitivity, and increased specificity for characterizing large PEGylated molecules.

There are also efforts to use computational approaches to interpret and discover additional structural details from spectra generated by IMMS, for example, correlating drift time (ion mobility parameter) with mass to determine which class (lipidomics, proteomics, glycomics, and metabolomics) the analyte belongs to, or to reveal specific structural motifs in the analyte [277]. IMMS is a powerful new methodology for characterizing intact biomolecules and may be used to fingerprint the higher-order structure of therapeutic proteins in the near future.

5.3 Conclusions

MS is an essential tool for candidate selection and lead molecule stability evaluation in protein therapeutics discovery. Key structural features of proteins, including MW, AA sequence, S–S linkages, glycosylation structure and profile, and many different PTMs as well as higher-order structures can be characterized by utilizing MS. Continued technical improvements in instrumentation will provide new mass spectrometers with higher sensitivity, resolving power, and mass accuracy, which will further improve the performance of MS in protein structural analyzes. Although "bottom-up" methods still remain the most popular approaches for protein characterization, especially for large proteins, "top-down" approaches incorporating advanced separation techniques, such as UHPLC and efficient MS/MS techniques such as ETD for large biomolecules, are evolving as promising alternatives that involve less sample handing prior to structural analyzes. Increased demands for analyzing protein higher-order structures will lead to further advances in HDX–MS and IMMS for probing protein conformational dynamics, as well as protein/protein (drug/target) interactions for a better understanding of mechanisms of action at the molecular level. With its sensitivity and versatility, it is clear that MS will continue to play a major role in biopharmaceutical research and development of protein therapeutics.

References

1. Walsh G (2000) Biopharmaceutical benchmarks. Nat Biotechnol 18:831–833
2. Chapman AP (2002) PEGylated antibodies and antibody fragments for improved therapy: a review. Adv Drug Deliv Rev 54:531–545
3. Reichert JM, Rosensweig CJ, Faden LB et al (2005) Monoclonal antibody successes in the clinic. Nat Biotechnol 23:1073–1078
4. Roberts MJ, Bentley MD, Harris JM (2002) Chemistry for peptide and protein PEGylation. Adv Drug Deliv Rev 54:459–476
5. GBI Research (2011) Therapeutic proteins market to 2017—high demand for monoclonal antibodies will drive the market. Report code: GBIHC080MR. http://gbiresearch.com
6. Kim YJ, Doyle ML (2010) Structural mass spectrometry in protein therapeutics discovery. Anal Chem 82:7083–7089
7. Jackel C, Kast P, Hilvert D (2008) Protein design by directed evolution. Annu Rev Biophys 37:153–173
8. Weisser NE, Hall JC (2009) Applications of single-chain variable fragment antibodies in therapeutics and diagnostics. Biotechnol Adv 27:502–520

9. Meyer HP, Brass J, Jungo C et al (2008) An emerging star for therapeutic and catalytic protein production. BioProcess Int 6(Suppl. 6):10–21
10. Rita Costa A, Elisa Rodrigues M, Henriques M et al (2010) Guidelines to cell engineering for monoclonal antibody production. Eur J Pharm Biopharm 74:127–138
11. Reichert JM (2010) Antibodies to watch in 2010. MAbs 2:84–100
12. Pavlou AK, Belsey MJ (2005) The therapeutic antibodies market to 2008. Eur J Pharm Biopharm 59:389–396
13. Maggon K (2007) Monoclonal antibody "gold rush". Curr Med Chem 14:1978–1987
14. Maheshwari S (2011) Global protein therapeutics market: beefing up towards futuristic growth. http://www.pharmaphorum.com
15. Correia IR (2010) Stability of IgG isotypes in serum. MAbs 2:221–232
16. Zhang Z, Pan H, Chen X (2009) Mass spectrometry for structural characterization of therapeutic antibodies. Mass Spectrom Rev 28:147–176
17. Janeway C, Travers P, Walport M et al (2001) Immunobiology. Garland Publishing, New York
18. Roopenian DC, Akilesh S (2007) FcRn: the neonatal Fc receptor comes of age. Nat Rev Immunol 7:715–725
19. Carter PJ (2006) Potent antibody therapeutics by design. Nat Rev Immunol 6:343–357
20. Labrijn AF, Aalberse RC, Schuurman J (2008) When binding is enough: nonactivating antibody formats. Curr Opin Immunol 20:479–485
21. Newman R, Hariharan K, Reff M et al (2001) Modification of the Fc region of a primatized IgG antibody to human CD4 retains its ability to modulate CD4 receptors but does not deplete CD4(+) T cells in chimpanzees. Clin Immunol 98:164–174
22. Langer F, Ingersoll SB, Amirkhosravi A et al (2005) The role of CD40 in CD40L- and antibody-mediated platelet activation. Thromb Haemost 93:1137–1146
23. Shields RL, Namenuk AK, Hong K et al (2001) High resolution mapping of the binding site on human IgG1 for Fc gamma RI, Fc gamma RII, Fc gamma RIII, and FcRn and design of IgG1 variants with improved binding to the Fc gamma R. J Biol Chem 276:6591–6604
24. Bolt S, Routledge E, Lloyd I et al (1993) The generation of a humanized, non-mitogenic CD3 monoclonal antibody which retains in vitro immunosuppressive properties. Eur J Immunol 23:403–411
25. Dall'Acqua WF, Cook KE, Damschroder MM et al (2006) Modulation of the effector functions of a human IgG1 through engineering of its hinge region. J Immunol 177:1129–1138
26. Holliger P, Hudson PJ (2005) Engineered antibody fragments and the rise of single domains. Nat Biotechnol 23:1126–1136
27. Porter RR (1959) The hydrolysis of rabbit y-globulin and antibodies with crystalline papain. Biochem J 73:119–126
28. Worn A, Pluckthun A (2001) Stability engineering of antibody single-chain Fv fragments. J Mol Biol 305:989–1010
29. Smith BJ, Popplewell A, Athwal D et al (2001) Prolonged in vivo residence times of antibody fragments associated with albumin. Bioconjug Chem 12:750–756
30. Holt LJ, Basran A, Jones K et al (2008) Anti-serum albumin domain antibodies for extending the half-lives of short lived drugs. Protein Eng Des Sel 21:283–288
31. Hu S, Shively L, Raubitschek A et al (1996) Minibody: a novel engineered anti-carcinoembryonic antigen antibody fragment (single-chain Fv-CH3) which exhibits rapid, high-level targeting of xenografts. Cancer Res 56:3055–3061
32. Kenanova V, Olafsen T, Crow DM et al (2005) Tailoring the pharmacokinetics and positron emission tomography imaging properties of anti-carcinoembryonic antigen single-chain Fv-Fc antibody fragments. Cancer Res 65:622–631
33. Krinner EM, Hepp J, Hoffmann P et al (2006) A highly stable polyethylene glycol-conjugated human single-chain antibody neutralizing granulocyte-macrophage colony stimulating factor at low nanomolar concentration. Protein Eng Des Sel 19:461–470
34. Yang K, Basu A, Wang M et al (2003) Tailoring structure-function and pharmacokinetic properties of single-chain Fv proteins by site-specific PEGylation. Protein Eng 16:761–770
35. Yamaoka T, Tabata Y, Ikada Y (1994) Distribution and tissue uptake of poly(ethylene glycol) with different molecular weights after intravenous administration to mice. J Pharm Sci 83:601–606

36. Cunningham-Rundles C, Zhuo Z, Griffith B et al (1992) Biological activities of polyethylene-glycol immunoglobulin conjugates. Resistance to enzymatic degradation. J Immunol Methods 152:177–190
37. Karr LJ, Donnelly DL, Kozlowski A et al (1994) Use of poly(ethylene glycol)-modified antibody in cell extraction. Methods Enzymol 228:377–390
38. Delgado C, Francis GE, Fisher D (1994) Bioextraction of low abundance cells by affinity partitioning. Methods Enzymol 228:395–402
39. Abuchowski A, van Es T, Palczuk NC et al (1977) Alteration of immunological properties of bovine serum albumin by covalent attachment of polyethylene glycol. J Biol Chem 252:3578–3581
40. Veronese FM, Pasut G (2005) PEGylation, successful approach to drug delivery. Drug Discov Today 10:1451–1458
41. Pasut G, Veronese FM (2004) Protein, peptide and non-peptide drug PEGylation for therapeutic application. Exp Op Ther Patents 14:859–894
42. Bailon P, Palleroni A, Schaffer CA et al (2001) Rational design of a potent, long-lasting form of interferon: a 40 kDa branched polyethylene glycol-conjugated interferon alpha-2a for the treatment of hepatitis C. Bioconjug Chem 12:195–202
43. Yoshioka Y, Tsutsumi Y, Ikemizu S et al (2004) Optimal site-specific PEGylation of mutant TNF-alpha improves its antitumor potency. Biochem Biophys Res Commun 315:808–814
44. Shibata H, Yoshioka Y, Ikemizu S et al (2004) Functionalization of tumor necrosis factor-alpha using phage display technique and PEGylation improves its antitumor therapeutic window. Clin Cancer Res 10:8293–8300
45. Li YP, Pei YY, Ding J et al (2001) PEGylated recombinant human tumor necrosis factor alpha: preparation and anti-tumor potency. Acta Pharmacol Sin 22:549–555
46. Li YP, Pei YY, Zhou ZH et al (2001) PEGylated recombinant human tumor necrosis factor alpha: pharmacokinetics and anti-tumor effects. Biol Pharm Bull 24:666–670
47. Srebalus Barnes CA, Lim A (2007) Applications of mass spectrometry for the structural characterization of recombinant protein pharmaceuticals. Mass Spectrom Rev 26:370–388
48. Domon B, Aebersold R (2006) Mass spectrometry and protein analysis. Science 312:212–217
49. Chen G, Warrack BM, Goodenough AK et al (2011) Characterization of protein therapeutics by mass spectrometry: recent developments and future directions. Drug Discov Today 16:58–64
50. Chen G, Pramanik BN (2009) Application of LC/MS to proteomics studies: current status and future prospects. Drug Discov Today 14:465–471
51. Chen G, Pramanik BN (2008) LC-MS for protein characterization: current capabilities and future trends. Expert Rev Proteomics 5:435–444
52. Fenn JB, Mann M, Meng CK et al (1989) Electrospray ionization for mass spectrometry of large biomolecules. Science 246:64–71
53. Karas M, Bachmann D, Bahr U et al (1987) Matrix-assisted ultraviolet laser desorption of non-volatile compounds. Int J Mass Spectrom Ion Processes 78:53
54. Plumb R, Castro-Perez J, Granger J et al (2004) Ultra-performance liquid chromatography coupled to quadrupole-orthogonal time-of-flight mass spectrometry. Rapid Commun Mass Spectrom 18:2331–2337
55. Battersby JE, Snedecor B, Chen C et al (2001) Affinity-reversed-phase liquid chromatography assay to quantitate recombinant antibodies and antibody fragments in fermentation broth. J Chromatogr A 927:61–76
56. Han X, Jin M, Breuker K et al (2006) Extending top-down mass spectrometry to proteins with masses greater than 200 kilodaltons. Science 314:109–112
57. Xiao G, Bondarenko PV, Jacob J et al (2007) 18O labeling method for identification and quantification of succinimide in proteins. Anal Chem 79:2714–2721
58. Fodor S, Zhang Z (2006) Rearrangement of terminal amino acid residues in peptides by protease-catalyzed intramolecular transpeptidation. Anal Biochem 356:282–290
59. Gaza-Bulseco G, Li B, Bulseco A et al (2008) Method to differentiate asn deamidation that occurred prior to and during sample preparation of a monoclonal antibody. Anal Chem 80:9491–9498

60. Karas M, Gluckmann M, Schafer J (2000) Ionization in matrix-assisted laser desorption/ionization: singly charged molecular ions are the lucky survivors. J Mass Spectrom 35:1–12
61. Rouse JC, McClellan JE, Patel HK et al (2005) Top-down characterization of protein pharmaceuticals by liquid chromatography/mass spectrometry: application to recombinant factor IX comparability- a case study. Methods Mol Biol 308:435–460
62. Bogdanov B, Smith RD (2005) Proteomics by FTICR mass spectrometry: top down and bottom up. Mass Spectrom Rev 24:168–200
63. Gadgil HS, Pipes GD, Dillon TM et al (2006) Improving mass accuracy of high performance liquid chromatography/electrospray ionization time-of-flight mass spectrometry of intact antibodies. J Am Soc Mass Spectrom 17:867–872
64. Lyubarskaya Y, Houde D, Woodard J et al (2006) Analysis of recombinant monoclonal antibody isoforms by electrospray ionization mass spectrometry as a strategy for streamlining characterization of recombinant monoclonal antibody charge heterogeneity. Anal Biochem 348:24–39
65. Brady LJ, Valliere-Douglass J, Martinez T et al (2008) Molecular mass analysis of antibodies by on-line SEC-MS. J Am Soc Mass Spectrom 19:502–509
66. Nichols AC, Corley S, Bourell JH (2005) Accurate mass analysis of intact IgG1 and IgG2 antibodies using ESI-Q-TOF mass spectrometers. In: 53rd ASMS conference on mass spectrometry allied topics, San Antonio, TX
67. Dillon TM, Bondarenko PV, Speed Ricci M (2004) Development of an analytical reversed-phase high-performance liquid chromatography-electrospray ionization mass spectrometry method for characterization of recombinant antibodies. J Chromatogr A 1053:299–305
68. Dillon TM, Bondarenko PV, Rehder DS et al (2006) Optimization of a reversed-phase high-performance liquid chromatography/mass spectrometry method for characterizing recombinant antibody heterogeneity and stability. J Chromatogr A 1120:112–120
69. Ren D, Pipes GD, Hambly DM et al (2007) Reversed-phase liquid chromatography of immunoglobulin G molecules and their fragments with the diphenyl column. J Chromatogr A 1175:63–68
70. Ren D, Pipes G, Xiao G et al (2008) Reversed-phase liquid chromatography-mass spectrometry of site-specific chemical modifications in intact immunoglobulin molecules and their fragments. J Chromatogr A 1179:198–204
71. Dillon TM, Ricci MS, Vezina C et al (2008) Structural and functional characterization of disulfide isoforms of the human IgG2 subclass. J Biol Chem 283:16206–16215
72. Wypych J, Li M, Guo A et al (2008) Human IgG2 antibodies display disulfide-mediated structural isoforms. J Biol Chem 283:16194–16205
73. Bennett KL, Smith SV, Truscott RJ et al (1997) Monitoring papain digestion of a monoclonal antibody by electrospray ionization mass spectrometry. Anal Biochem 245:17–27
74. Adamczyk M, Gebler JC, Wu J (2000) Sequencing of anti-thyroxine monoclonal antibody Fab fragment by ion trap mass spectrometry. Rapid Commun Mass Spectrom 14:999–1007
75. Remmele RL Jr, Callahan WJ, Krishnan S et al (2006) Active dimer of Epratuzumab provides insight into the complex nature of an antibody aggregate. J Pharm Sci 95:126–145
76. Hagmann ML, Kionka C, Schreiner M et al (1998) Characterization of the F(ab')2 fragment of a murine monoclonal antibody using capillary isoelectric focusing and electrospray ionization mass spectrometry. J Chromatogr A 816:49–58
77. Gadgil HS, Bondarenko PV, Pipes G et al (2007) The LC/MS analysis of glycation of IgG molecules in sucrose containing formulations. J Pharm Sci 96:2607–2621
78. Yan B, Valliere-Douglass J, Brady L et al (2007) Analysis of post-translational modifications in recombinant monoclonal antibody IgG1 by reversed-phase liquid chromatography/mass spectrometry. J Chromatogr A 1164:153–161
79. Yu L, Remmele RL Jr, He B (2006) Identification of N-terminal modification for recombinant monoclonal antibody light chain using partial reduction and quadrupole time-of-flight mass spectrometry. Rapid Commun Mass Spectrom 20:3674–3680
80. Karas M, Hillenkamp F (1988) Laser desorption ionization of proteins with molecular masses exceeding 10,000 daltons. Anal Chem 60:2299–2301
81. Lee KC, Moon SC, Park MO et al (1999) Isolation, characterization, and stability of positional isomers of mono-PEGylated salmon calcitonins. Pharm Res 16:813–818

82. Na DH, Youn YS, Lee KC (2003) Optimization of the PEGylation process of a peptide by monitoring with matrix-assisted laser desorption/ionization time-of-flight mass spectrometry. Rapid Commun Mass Spectrom 17:2241–2244
83. Foser S, Schacher A, Weyer KA et al (2003) Isolation, structural characterization, and antiviral activity of positional isomers of monopegylated interferon alpha-2a (PEGASYS). Protein Expr Purif 30:78–87
84. Yoo C, Suckau D, Sauerland V et al (2009) Toward top-down determination of PEGylation site using MALDI in-source decay MS analysis. J Am Soc Mass Spectrom 20:326–333
85. Trimpin S, Plasencia M, Isailovic D et al (2007) Resolving oligomers from fully grown polymers with IMS-MS. Anal Chem 79:7965–7974
86. Vestling MM, Murphy CM, Keller DA et al (1993) A strategy for characterization of polyethylene glycol-derivatized proteins. A mass spectrometric analysis of the attachment sites in polyethylene glycol-derivatized superoxide dismutase. Drug Metab Dispos 21:911–917
87. Ebeling DD, Westphall MS, Scalf M et al (2000) Corona discharge in charge reduction electrospray mass spectrometry. Anal Chem 72:5158–5161
88. Lennon JD 3rd, Cole SP, Glish GL (2006) Ion/molecule reactions to chemically deconvolute the electrospray ionization mass spectra of synthetic polymers. Anal Chem 78:8472–8476
89. Liang X, McLuckey SA (2007) Transmission mode ion/ion proton transfer reactions in a linear ion trap. J Am Soc Mass Spectrom 18:882–890
90. Huang L, Gough PC, Defelippis MR (2009) Characterization of poly(ethylene glycol) and PEGylated products by LC/MS with postcolumn addition of amines. Anal Chem 81:567–577
91. O'Connor PB, McLafferty FW (1995) Oligomer characterization of 4–23 kDa polymers by electrospray Fourier transform mass spectrometry. J Am Chem Soc 117:12826–12831
92. Bagal D, Zhang H, Schnier PD (2008) Gas-phase proton-transfer chemistry coupled with TOF mass spectrometry and ion mobility-MS for the facile analysis of poly(ethylene glycols) and PEGylated polypeptide conjugates. Anal Chem 80:2408–2418
93. Slawson C, Housley MP, Hart GW (2006) O-GlcNAc cycling: how a single sugar post-translational modification is changing the way we think about signaling networks. J Cell Biochem 97:71–83
94. Zachara NE, Hart GW (2002) The emerging significance of O-GlcNAc in cellular regulation. Chem Rev 102:431–438
95. Helenius A, Aebi M (2004) Roles of N-linked glycans in the endoplasmic reticulum. Annu Rev Biochem 73:1019–1049
96. Haltiwanger RS, Lowe JB (2004) Role of glycosylation in development. Annu Rev Biochem 73:491–537
97. Satomi Y, Shimonishi Y, Takao T (2004) N-glycosylation at Asn(491) in the Asn-Xaa-Cys motif of human transferrin. FEBS Lett 576:51–56
98. Vance BA, Wu W, Ribaudo RK et al (1997) Multiple dimeric forms of human CD69 result from differential addition of N-glycans to typical (Asn-X-Ser/Thr) and atypical (Asn-X-cys) glycosylation motifs. J Biol Chem 272:23117–23122
99. Hang HC, Bertozzi CR (2005) The chemistry and biology of mucin-type O-linked glycosylation. Bioorg Med Chem 13:5021–5034
100. Bailey MJ, Hooker AD, Adams CS et al (2005) A platform for high-throughput molecular characterization of recombinant monoclonal antibodies. J Chromatogr B Analyt Technol Biomed Life Sci 826:177–187
101. Kamoda S, Ishikawa R, Kakehi K (2006) Capillary electrophoresis with laser-induced fluorescence detection for detailed studies on N-linked oligosaccharide profile of therapeutic recombinant monoclonal antibodies. J Chromatogr A 1133:332–339
102. Chen X, Flynn GC (2007) Analysis of N-glycans from recombinant immunoglobulin G by on-line reversed-phase high-performance liquid chromatography/mass spectrometry. Anal Biochem 370:147–161
103. Jefferis R (2005) Glycosylation of recombinant antibody therapeutics. Biotechnol Prog 21:11–16
104. Matsumiya S, Yamaguchi Y, Saito J et al (2007) Structural comparison of fucosylated and nonfucosylated Fc fragments of human immunoglobulin G1. J Mol Biol 368:767–779
105. Arnold JN, Wormald MR, Sim RB et al (2007) The impact of glycosylation on the biological function and structure of human immunoglobulins. Annu Rev Immunol 25:21–50

106. Burton DR, Dwek RA (2006) Immunology. Sugar determines antibody activity. Science 313:627–628
107. Jefferis R (2009) Recombinant antibody therapeutics: the impact of glycosylation on mechanisms of action. Trends Pharmacol Sci 30:356–362
108. Hossler P, Khattak SF, Li ZJ (2009) Optimal and consistent protein glycosylation in mammalian cell culture. Glycobiology 19:936–949
109. Wan HZ, Kaneshiro S, Frenz J et al (2001) Rapid method for monitoring galactosylation levels during recombinant antibody production by electrospray mass spectrometry with selective-ion monitoring. J Chromatogr A 913:437–446
110. Huang L, Mitchell CE (2005) High-throughput LC/MS methodology for a(1-->3)Gal determination of recombinant monoclonal antibodies: a case study. Methods Mol Biol 308:411–419
111. Mechref Y, Muzikar J, Novotny MV (2005) Comprehensive assessment of N-glycans derived from a murine monoclonal antibody: a case for multimethodological approach. Electrophoresis 26:2034–2046
112. Roberts GD, Johnson WP, Burman S et al (1995) An integrated strategy for structural characterization of the protein and carbohydrate components of monoclonal antibodies: application to anti-respiratory syncytial virus MAb. Anal Chem 67:3613–3625
113. Wang W, Singh S, Zeng DL et al (2007) Antibody structure, instability, and formulation. J Pharm Sci 96:1–26
114. Liu H, Gaza-Bulseco G, Sun J (2006) Characterization of the stability of a fully human monoclonal IgG after prolonged incubation at elevated temperature. J Chromatogr B Analyt Technol Biomed Life Sci 837:35–43
115. Liu H, Gaza-Bulseco G, Lundell E (2008) Assessment of antibody fragmentation by reversed-phase liquid chromatography and mass spectrometry. J Chromatogr B Analyt Technol Biomed Life Sci 876:13–23
116. Cohen SL, Price C, Vlasak J (2007) Beta-elimination and peptide bond hydrolysis: two distinct mechanisms of human IgG1 hinge fragmentation upon storage. J Am Chem Soc 129:6976–6977
117. Cordoba AJ, Shyong BJ, Breen D et al (2005) Non-enzymatic hinge region fragmentation of antibodies in solution. J Chromatogr B Analyt Technol Biomed Life Sci 818:115–121
118. Xiang T, Lundell E, Sun Z et al (2007) Structural effect of a recombinant monoclonal antibody on hinge region peptide bond hydrolysis. J Chromatogr B Analyt Technol Biomed Life Sci 858:254–262
119. Gaza-Bulseco G, Liu H (2008) Fragmentation of a recombinant monoclonal antibody at various pH. Pharm Res 25:1881–1890
120. Manning MC, Patel K, Borchardt RT (1989) Stability of protein pharmaceuticals. Pharm Res 6:903–918
121. Senko MW, Speir JP, McLafferty FW (1994) Collisional activation of large multiply charged ions using Fourier transform mass spectrometry. Anal Chem 66:2801–2808
122. Little DP, Speir JP, Senko MW et al (1994) Infrared multiphoton dissociation of large multiply charged ions for biomolecule sequencing. Anal Chem 66:2809–2815
123. Price WD, Schnier PD, Williams ER (1996) Tandem mass spectrometry of large biomolecule ions by blackbody infrared radiative dissociation. Anal Chem 68:859–866
124. Zubarev RA, Kelleher NL, McLafferty FW (1998) Electron capture dissociation of multiply charged protein cations. A nonergodic process. J Am Chem Soc 120:2365–3266
125. Syka JE, Coon JJ, Schroeder MJ et al (2004) Peptide and protein sequence analysis by electron transfer dissociation mass spectrometry. Proc Natl Acad Sci U S A 101:9528–9533
126. Reid GE, McLuckey SA (2002) 'Top down' protein characterization via tandem mass spectrometry. J Mass Spectrom 37:663–675
127. Kelleher NL (2004) Top-down proteomics. Anal Chem 76:197A–203A
128. Qin J, Chait BT (1999) Collision-induced dissociation of singly charged peptide ions in a matrix-assisted laser desorption ionization ion trap mass spectrometer. Int J Mass Spectrom 190(191):313–320

129. McLuckey SA, Goeringer DE (1997) Slow heating methods in tandem mass spectrometry. J Mass Spectrom 32:461–474
130. Kelleher NL, Zubarev RA, Bush K et al (1999) Localization of labile posttranslational modifications by electron capture dissociation: the case of gamma-carboxyglutamic acid. Anal Chem 71:4250–4253
131. Cooper HJ, Hakansson K, Marshall AG (2005) The role of electron capture dissociation in biomolecular analysis. Mass Spectrom Rev 24:201–222
132. Mirgorodskaya E, Roepstorff P, Zubarev RA (1999) Localization of O-glycosylation sites in peptides by electron capture dissociation in a Fourier transform mass spectrometer. Anal Chem 71:4431–4436
133. Haselmann KF, Jorgensen TJ, Budnik BA et al (2002) Electron capture dissociation of weakly bound polypeptide polycationic complexes. Rapid Commun Mass Spectrom 16:2260–2265
134. Wu SL, Huhmer AF, Hao Z et al (2007) On-line LC-MS approach combining collision-induced dissociation (CID), electron-transfer dissociation (ETD), and CID of an isolated charge-reduced species for the trace-level characterization of proteins with post-translational modifications. J Proteome Res 6:4230–4244
135. Wiesner J, Premsler T, Sickmann A (2008) Application of electron transfer dissociation (ETD) for the analysis of posttranslational modifications. Proteomics 8:4466–4483
136. Mikesh LM, Ueberheide B, Chi A et al (2006) The utility of ETD mass spectrometry in proteomic analysis. Biochim Biophys Acta 1764:1811–1822
137. Zubarev RA, Horn DM, Fridriksson EK et al (2000) Electron capture dissociation for structural characterization of multiply charged protein cations. Anal Chem 72:563–573
138. Olsen JV, Haselmann KF, Nielsen ML et al (2001) Comparison of electron capture dissociation and collisionally activated dissociation of polycations of peptide nucleic acids. Rapid Commun Mass Spectrom 15:969–974
139. Pitteri SJ, Chrisman PA, Hogan JM et al (2005) Electron transfer ion/ion reactions in a three-dimensional quadrupole ion trap: reactions of doubly and triply protonated peptides with SO2*. Anal Chem 77:1831–1839
140. Srikanth R, Wilson J, Bridgewater JD et al (2007) Improved sequencing of oxidized cysteine and methionine containing peptides using electron transfer dissociation. J Am Soc Mass Spectrom 18:1499–1506
141. Silva JC, Denny R, Dorschel CA et al (2005) Quantitative proteomic analysis by accurate mass retention time pairs. Anal Chem 77:2187–2200
142. Plumb RS, Johnson KA, Rainville P et al (2006) UPLC/MS(E); a new approach for generating molecular fragment information for biomarker structure elucidation. Rapid Commun Mass Spectrom 20:1989–1994
143. Silva JC, Denny R, Dorschel C et al (2006) Simultaneous qualitative and quantitative analysis of the Escherichia coli proteome: a sweet tale. Mol Cell Proteomics 5:589–607
144. Bongers J, Cummings JJ, Ebert MB et al (2000) Validation of a peptide mapping method for a therapeutic monoclonal antibody: what could we possibly learn about a method we have run 100 times? J Pharm Biomed Anal 21:1099–1128
145. Yu YQ, Gilar M, Lee PJ et al (2003) Enzyme-friendly, mass spectrometry-compatible surfactant for in-solution enzymatic digestion of proteins. Anal Chem 75:6023–6028
146. Hernandez P, Muller M, Appel RD (2006) Automated protein identification by tandem mass spectrometry: issues and strategies. Mass Spectrom Rev 25:235–254
147. Forner F, Foster LJ, Toppo S (2007) Mass spectrometry data analysis in the proteomics era. Curr Bioinf 2:63–93
148. Matthiesen R (2007) Methods, algorithms and tools in computational proteomics: a practical point of view. Proteomics 7:2815–2832
149. Henzel WJ, Watanabe C, Stults JT (2003) Protein identification: the origins of peptide mass fingerprinting. J Am Soc Mass Spectrom 14:931–942
150. Sadygov RG, Cociorva D, Yates JR 3rd (2004) Large-scale database searching using tandem mass spectra: looking up the answer in the back of the book. Nat Methods 1:195–202

151. Breuker K, Jin M, Han X et al (2008) Top-down identification and characterization of biomolecules by mass spectrometry. J Am Soc Mass Spectrom 19:1045–1053
152. Siuti N, Kelleher NL (2007) Decoding protein modifications using top-down mass spectrometry. Nat Methods 4:817–821
153. Ge Y, Lawhorn BG, ElNaggar M et al (2003) Detection of four oxidation sites in viral prolyl-4-hydroxylase by top-down mass spectrometry. Protein Sci 12:2320–2326
154. Zabrouskov V, Han X, Welker E et al (2006) Stepwise deamidation of ribonuclease A at five sites determined by top down mass spectrometry. Biochemistry 45:987–992
155. McLafferty FW, Horn DM, Breuker K et al (2001) Electron capture dissociation of gaseous multiply charged ions by Fourier-transform ion cyclotron resonance. J Am Soc Mass Spectrom 12:245–249
156. Meng F, Forbes AJ, Miller LM et al (2005) Detection and localization of protein modifications by high resolution tandem mass spectrometry. Mass Spectrom Rev 24:126–134
157. Zhai H, Han X, Breuker K et al (2005) Consecutive ion activation for top down mass spectrometry: improved protein sequencing by nozzle-skimmer dissociation. Anal Chem 77:5777–5784
158. Parks BA, Jiang L, Thomas PM et al (2007) Top-down proteomics on a chromatographic time scale using linear ion trap Fourier transform hybrid mass spectrometers. Anal Chem 79:7984–7991
159. Jebanathirajah JA, Pittman JL, Thomson BA et al (2005) Characterization of a new qQq-FTICR mass spectrometer for post-translational modification analysis and top-down tandem mass spectrometry of whole proteins. J Am Soc Mass Spectrom 16:1985–1999
160. Xie Y, Zhang J, Yin S et al (2006) Top-down ESI-ECD-FT-ICR mass spectrometry localizes noncovalent protein-ligand binding sites. J Am Chem Soc 128:14432–14433
161. Macek B, Waanders LF, Olsen JV et al (2006) Top-down protein sequencing and MS3 on a hybrid linear quadrupole ion trap-orbitrap mass spectrometer. Mol Cell Proteomics 5:949–958
162. Tran JC, Zamdborg L, Ahlf DR et al (2011) Mapping intact protein isoforms in discovery mode using top-down proteomics. Nature 480:254–258
163. Ge Y, Lawhorn BG, ElNaggar M et al (2002) Top down characterization of larger proteins (45 kDa) by electron capture dissociation mass spectrometry. J Am Chem Soc 124:672–678
164. Zhang Z, Shah B (2007) Characterization of variable regions of monoclonal antibodies by top-down mass spectrometry. Anal Chem 79:5723–5729
165. Toler KN, Johnson KA, Rouse JC (2005) Characterization of the fragmentation behavior of light and heavy chain ions from recombinant monoclonal antibodies. In: Proceedings of 53rd ASMS conference on mass spectrometry allied topics, San Antonio, TX
166. Garcia BA (2010) What does the future hold for top down mass spectrometry? J Am Soc Mass Spectrom 21:193–202
167. Karabacak NM, Li L, Tiwari A et al (2009) Sensitive and specific identification of wild type and variant proteins from 8 to 669 kDa using top-down mass spectrometry. Mol Cell Proteomics 8:846–856
168. Chait BT (2006) Mass spectrometry: bottom-up or top-down? Science 314:65–66
169. Antes B, Amon S, Rizzi A et al (2007) Analysis of lysine clipping of a humanized Lewis-Y specific IgG antibody and its relation to Fc-mediated effector function. J Chromatogr B Analyt Technol Biomed Life Sci 852:250–256
170. Johnson KA, Paisley-Flango K, Tangarone BS et al (2007) Cation exchange-HPLC and mass spectrometry reveal C-terminal amidation of an IgG1 heavy chain. Anal Biochem 360:75–83
171. Dick LW Jr, Qiu D, Mahon D et al (2008) C-terminal lysine variants in fully human monoclonal antibodies: investigation of test methods and possible causes. Biotechnol Bioeng 100:1132–1143
172. Chelius D, Jing K, Lueras A et al (2006) Formation of pyroglutamic acid from N-terminal glutamic acid in immunoglobulin gamma antibodies. Anal Chem 78:2370–2376
173. Wedemeyer WJ, Welker E, Narayan M et al (2000) Disulfide bonds and protein folding. Biochemistry 39:7032
174. Gorman JJ, Wallis TP, Pitt JJ (2002) Protein disulfide bond determination by mass spectrometry. Mass Spectrom Rev 21:183–216

175. Yen TY, Joshi RK, Yan H et al (2000) Characterization of cysteine residues and disulfide bonds in proteins by liquid chromatography/electrospray ionization tandem mass spectrometry. J Mass Spectrom 35:990–1002
176. Zhang W, Marzilli LA, Rouse JC et al (2002) Complete disulfide bond assignment of a recombinant immunoglobulin G4 monoclonal antibody. Anal Biochem 311:1–9
177. Pitt JJ, Da Silva E, Gorman JJ (2000) Determination of the disulfide bond arrangement of Newcastle disease virus hemagglutinin neuraminidase. Correlation with a beta-sheet propeller structural fold predicted for paramyxoviridae attachment proteins. J Biol Chem 275:6469–6478
178. Martinez T, Pace D, Brady L et al (2007) Characterization of a novel modification on IgG2 light chain. Evidence for the presence of O-linked mannosylation. J Chromatogr A 1156:183–187
179. Chrisman PA, Pitteri SJ, Hogan JM et al (2005) SO2-* electron transfer ion/ion reactions with disulfide linked polypeptide ions. J Am Soc Mass Spectrom 16:1020–1030
180. Wu SL, Jiang H, Lu Q et al (2009) Mass spectrometric determination of disulfide linkages in recombinant therapeutic proteins using online LC-MS with electron-transfer dissociation. Anal Chem 81:112–122
181. Wang Y, Lu Q, Wu SL et al (2011) Characterization and comparison of disulfide linkages and scrambling patterns in therapeutic monoclonal antibodies: using LC-MS with electron transfer dissociation. Anal Chem 83:3133–3140
182. Wu SL, Jiang H, Hancock WS et al (2010) Identification of the unpaired cysteine status and complete mapping of the 17 disulfides of recombinant tissue plasminogen activator using LC-MS with electron transfer dissociation/collision induced dissociation. Anal Chem 82:5296–5303
183. Swaney DL, McAlister GC, Wirtala M et al (2007) Supplemental activation method for high-efficiency electron-transfer dissociation of doubly protonated peptide precursors. Anal Chem 79:477–485
184. Li S, Schoneich C, Borchardt RT (1995) Chemical instability of protein pharmaceuticals: mechanisms of oxidation and strategies for stabilization. Biotechnol Bioeng 48:490–500
185. Ji JA, Zhang B, Cheng W et al (2009) Methionine, tryptophan, and histidine oxidation in a model protein, PTH: mechanisms and stabilization. J Pharm Sci 98:4485–4500
186. Stadtman ER (1993) Oxidation of free amino acids and amino acid residues in proteins by radiolysis and by metal-catalyzed reactions. Annu Rev Biochem 62:797–821
187. Reddie KG, Carroll KS (2008) Expanding the functional diversity of proteins through cysteine oxidation. Curr Opin Chem Biol 12:746–754
188. Yang J, Wang S, Liu J et al (2007) Determination of tryptophan oxidation of monoclonal antibody by reversed phase high performance liquid chromatography. J Chromatogr A 1156:174–182
189. Wei Z, Feng J, Lin HY et al (2007) Identification of a single tryptophan residue as critical for binding activity in a humanized monoclonal antibody against respiratory syncytial virus. Anal Chem 79:2797–2805
190. Qi P, Volkin DB, Zhao H et al (2009) Characterization of the photodegradation of a human IgG1 monoclonal antibody formulated as a high-concentration liquid dosage form. J Pharm Sci 98:3117–3130
191. Simat TJ, Steinhart H (1998) Oxidation of free tryptophan and tryptophan residues in peptides and proteins. J Agric Food Chem 46:490–498
192. Schoneich C (2000) Mechanisms of metal-catalyzed oxidation of histidine to 2-oxo-histidine in peptides and proteins. J Pharm Biomed Anal 21:1093–1097
193. Gao J, Yin DH, Yao Y et al (1998) Loss of conformational stability in calmodulin upon methionine oxidation. Biophys J 74:1115–1134
194. Van Patten SM, Hanson E, Bernasconi R et al (1999) Oxidation of methionine residues in antithrombin. Effects on biological activity and heparin binding. J Biol Chem 274:10268–10276
195. Kornfelt T, Persson E, Palm L (1999) Oxidation of methionine residues in coagulation factor VIIa. Arch Biochem Biophys 363:43–54
196. Lu HS, Fausset PR, Narhi LO et al (1999) Chemical modification and site-directed mutagenesis of methionine residues in recombinant human granulocyte colony-stimulating factor: effect on stability and biological activity. Arch Biochem Biophys 362:1–11

197. Brot N, Weissbach H (2000) Peptide methionine sulfoxide reductase: biochemistry and physiological role. Biopolymers 55:288–296
198. Taggart C, Cervantes-Laurean D, Kim G et al (2000) Oxidation of either methionine 351 or methionine 358 in alpha 1-antitrypsin causes loss of anti-neutrophil elastase activity. J Biol Chem 275:27258–27265
199. Kim YH, Berry AH, Spencer DS et al (2001) Comparing the effect on protein stability of methionine oxidation versus mutagenesis: steps toward engineering oxidative resistance in proteins. Protein Eng 14:343–347
200. Chugha P, Sage HJ, Oas TG (2006) Methionine oxidation of monomeric lambda repressor: the denatured state ensemble under nondenaturing conditions. Protein Sci 15:533–542
201. Shen JF, Kwong YM, Keck GR et al (1996) The application of tertbutylhydroperoxide oxidation to study sites of potential methionine oxidation in a recombinant antibody. In: Techniques in protein chemistry VII. Academic Press Inc., New York, pp 275–284
202. Lam XM, Yang JY, Cleland JL (1997) Antioxidants for prevention of methionine oxidation in recombinant monoclonal antibody HER2. J Pharm Sci 86:1250–1255
203. Chumsae C, Gaza-Bulseco G, Sun J et al (2007) Comparison of methionine oxidation in thermal stability and chemically stressed samples of a fully human monoclonal antibody. J Chromatogr B Analyt Technol Biomed Life Sci 850:285–294
204. Presta LG, Shields RL, Namenuk AK et al (2002) Engineering therapeutic antibodies for improved function. Biochem Soc Trans 30:487–490
205. Liu D, Ren D, Huang H et al (2008) Structure and stability changes of human IgG1 Fc as a consequence of methionine oxidation. Biochemistry 47:5088–5100
206. Liu H, Gaza-Bulseco G, Xiang T et al (2008) Structural effect of deglycosylation and methionine oxidation on a recombinant monoclonal antibody. Mol Immunol 45:701–708
207. Pan H, Chen K, Chu L et al (2009) Methionine oxidation in human IgG2 Fc decreases binding affinities to protein A and FcRn. Protein Sci 18:424–433
208. Bertolotti-Ciarlet A, Wang W, Lownes R et al (2009) Impact of methionine oxidation on the binding of human IgG1 to Fc Rn and Fc gamma receptors. Mol Immunol 46:1878–1882
209. Gaza-Bulseco G, Faldu S, Hurkmans K et al (2008) Effect of methionine oxidation of a recombinant monoclonal antibody on the binding affinity to protein A and protein G. J Chromatogr B Analyt Technol Biomed Life Sci 870:55–62
210. Nimmerjahn F, Ravetch JV (2008) Analyzing antibody-Fc-receptor interactions. Methods Mol Biol 415:151–162
211. Luo S, Levine RL (2009) Methionine in proteins defends against oxidative stress. FASEB J 23:464–472
212. Dolnik V, Hutterer KM (2001) Capillary electrophoresis of proteins 1999–2001. Electrophoresis 22:4163–4178
213. Dukan S, Farewell A, Ballesteros M et al (2000) Protein oxidation in response to increased transcriptional or translational errors. Proc Natl Acad Sci USA 97:5746–5749
214. Ballesteros M, Fredriksson A, Henriksson J et al (2001) Bacterial senescence: protein oxidation in non-proliferating cells is dictated by the accuracy of the ribosomes. EMBO J 20:5280–5289
215. Yoshida T (1996) Determination of reduced and oxidized glutathione in erythrocytes by high-performance liquid chromatography with ultraviolet absorbance detection. J Chromatogr B Biomed Appl 678:157–164
216. Duenas ET, Keck R, De Vos A et al (2001) Comparison between light induced and chemically induced oxidation of rhVEGF. Pharm Res 18:1455–1460
217. Houde D, Kauppinen P, Mhatre R et al (2006) Determination of protein oxidation by mass spectrometry and method transfer to quality control. J Chromatogr A 1123:189–198
218. Sharp JS, Becker JM, Hettich RL (2003) Protein surface mapping by chemical oxidation: structural analysis by mass spectrometry. Anal Biochem 313:216–225
219. Patel K, Borchardt RT (1990) Chemical pathways of peptide degradation. II. Kinetics of deamidation of an asparaginyl residue in a model hexapeptide. Pharm Res 7:703–711

220. Patel K, Borchardt RT (1990) Chemical pathways of peptide degradation. III. Effect of primary sequence on the pathways of deamidation of asparaginyl residues in hexapeptides. Pharm Res 7:787–793
221. Tyler-Cross R, Schirch V (1991) Effects of amino acid sequence, buffers, and ionic strength on the rate and mechanism of deamidation of asparagine residues in small peptides. J Biol Chem 266:22549–22556
222. Robinson AB, Scotchler JW, McKerrow JH (1973) Rates of nonenzymatic deamidation of glutaminyl and asparaginyl residues in pentapeptides. J Am Chem Soc 95:8156–8159
223. Robinson AB, Rudd CJ (1974) Deamidation of glutaminyl and asparaginyl residues in peptides and proteins. Curr Top Cell Regul 8:247–295
224. Geiger T, Clarke S (1987) Deamidation, isomerization, and racemization at asparaginyl and aspartyl residues in peptides. Succinimide-linked reactions that contribute to protein degradation. J Biol Chem 262:785–794
225. Liu YD, van Enk JZ, Flynn GC (2009) Human antibody Fc deamidation in vivo. Biologicals 37:313–322
226. Huang L, Lu J, Wroblewski VJ et al (2005) In vivo deamidation characterization of monoclonal antibody by LC/MS/MS. Anal Chem 77:1432–1439
227. Gupta R, Srivastava OP (2004) Deamidation affects structural and functional properties of human alphaA-crystallin and its oligomerization with alphaB-crystallin. J Biol Chem 279:44258–44269
228. Cacia J, Keck R, Presta LG et al (1996) Isomerization of an aspartic acid residue in the complementarity-determining regions of a recombinant antibody to human IgE: identification and effect on binding affinity. Biochemistry 35:1897–1903
229. Harris RJ, Kabakoff B, Macchi FD et al (2001) Identification of multiple sources of charge heterogeneity in a recombinant antibody. J Chromatogr B Biomed Sci Appl 752:233–245
230. Wakankar AA, Borchardt RT, Eigenbrot C et al (2007) Aspartate isomerization in the complementarity-determining regions of two closely related monoclonal antibodies. Biochemistry 46:1534–1544
231. Rehder DS, Chelius D, McAuley A et al (2008) Isomerization of a single aspartyl residue of anti-epidermal growth factor receptor immunoglobulin $\gamma 2$ antibody highlights the role avidity plays in antibody activity. Biochemistry 47:2518–2530
232. Yan B, Steen S, Hambly D et al (2009) Succinimide formation at Asn 55 in the complementarity determining region of a recombinant monoclonal antibody IgG1 heavy chain. J Pharm Sci 98:3509–3521
233. Vlasak J, Bussat MC, Wang S et al (2009) Identification and characterization of asparagine deamidation in the light chain CDR1 of a humanized IgG1 antibody. Anal Biochem 392:145–154
234. Hambly DM, Banks DD, Scavezze JL et al (2009) Detection and quantitation of IgG 1 hinge aspartate isomerization: a rapid degradation in stressed stability studies. Anal Chem 81:7454–7459
235. Dick LW Jr, Qiu D, Wong RB et al (2010) Isomerization in the CDR2 of a monoclonal antibody: binding analysis and factors that influence the isomerization rate. Biotechnol Bioeng 105:515–523
236. Oliyai C, Borchardt RT (1993) Chemical pathways of peptide degradation. IV. Pathways, kinetics, and mechanism of degradation of an aspartyl residue in a model hexapeptide. Pharm Res 10:95–102
237. Li B, Gorman EM, Moore KD et al (2005) Effects of acidic N + 1 residues on asparagine deamidation rates in solution and in the solid state. J Pharm Sci 94:666–675
238. Goolcharran C, Stauffer LL, Cleland JL et al (2000) The effects of a histidine residue on the C-terminal side of an asparaginyl residue on the rate of deamidation using model pentapeptides. J Pharm Sci 89:818–825
239. Robinson NE, Robinson ZW, Robinson BR et al (2004) Structure-dependent nonenzymatic deamidation of glutaminyl and asparaginyl pentapeptides. J Pept Res 63:426–436
240. Robinson NE, Robinson AB (2001) Molecular clocks. Proc Natl Acad Sci USA 98:944–949

241. Radkiewicz JL, Zipse H, Clarke S et al (2001) Neighboring side chain effects on asparaginyl and aspartyl degradation: an ab initio study of the relationship between peptide conformation and backbone NH acidity. J Am Chem Soc 123:3499–3506
242. Kosky AA, Razzaq UO, Treuheit MJ et al (1999) The effects of alpha-helix on the stability of Asn residues: deamidation rates in peptides of varying helicity. Protein Sci 8:2519–2523
243. Chelius D, Rehder DS, Bondarenko PV (2005) Identification and characterization of deamidation sites in the conserved regions of human immunoglobulin gamma antibodies. Anal Chem 77:6004–6011
244. Wearne SJ, Creighton TE (1989) Effect of protein conformation on rate of deamidation: ribonuclease A. Proteins 5:8–12
245. Wakankar AA, Borchardt RT (2006) Formulation considerations for proteins susceptible to asparagine deamidation and aspartate isomerization. J Pharm Sci 95:2321–2336
246. Wakankar AA, Liu J, Vandervelde D et al (2007) The effect of cosolutes on the isomerization of aspartic acid residues and conformational stability in a monoclonal antibody. J Pharm Sci 96:1708–1718
247. Promega (2011) ISOQUANT® Isoaspartate detection kit. http://www.promega.com
248. Kroon DJ, Baldwin-Ferro A, Lalan P (1992) Identification of sites of degradation in a therapeutic monoclonal antibody by peptide mapping. Pharm Res 9:1386–1393
249. Terashima I, Koga A, Nagai H (2007) Identification of deamidation and isomerization sites on pharmaceutical recombinant antibody using H(2)(18)O. Anal Biochem 368:49–60
250. Lehmann WD, Schlosser A, Erben G et al (2000) Analysis of isoaspartate in peptides by electrospray tandem mass spectrometry. Protein Sci 9:2260–2268
251. Cournoyer JJ, Pittman JL, Ivleva VB et al (2005) Deamidation: differentiation of aspartyl from isoaspartyl products in peptides by electron capture dissociation. Protein Sci 14:452–463
252. Cournoyer JJ, Lin C, O'Connor PB (2006) Detecting deamidation products in proteins by electron capture dissociation. Anal Chem 78:1264–1271
253. Cournoyer JJ, Lin C, Bowman MJ et al (2007) Quantitating the relative abundance of isoaspartyl residues in deamidated proteins by electron capture dissociation. J Am Soc Mass Spectrom 18:48–56
254. Sargaeva NP, Lin C, O'Connor PB (2009) Identification of aspartic and isoaspartic acid residues in amyloid beta peptides, including Abeta1-42, using electron-ion reactions. Anal Chem 81:9778–9786
255. O'Connor PB, Cournoyer JJ, Pitteri SJ et al (2006) Differentiation of aspartic and isoaspartic acids using electron transfer dissociation. J Am Soc Mass Spectrom 17:15–19
256. Chan WY, Chan TW, O'Connor PB (2010) Electron transfer dissociation with supplemental activation to differentiate aspartic and isoaspartic residues in doubly charged peptide cations. J Am Soc Mass Spectrom 21:1012–1015
257. Bobst CE, Abzalimov RR, Houde D et al (2008) Detection and characterization of altered conformations of protein pharmaceuticals using complementary mass spectrometry-based approaches. Anal Chem 80:7473–7481
258. Wales TE, Engen JR (2006) Hydrogen exchange mass spectrometry for the analysis of protein dynamics. Mass Spectrom Rev 25:158–170
259. Kaltashov IA, Bobst CE, Abzalimov RR et al (2010) Conformation and dynamics of biopharmaceuticals: transition of mass spectrometry-based tools from academe to industry. J Am Soc Mass Spectrom 21:323–337
260. Tsutsui Y, Wintrode PL (2007) Hydrogen/deuterium exchange-mass spectrometry: a powerful tool for probing protein structure, dynamics and interactions. Curr Med Chem 14:2344–2358
261. Maier CS, Deinzer ML (2005) Protein conformations, interactions, and H/D exchange. Methods Enzymol 402:312–360
262. Marcsisin SR, Engen JR (2010) Hydrogen exchange mass spectrometry: what is it and what can it tell us? Anal Bioanal Chem 397:967–972
263. Houde D, Arndt J, Domeier W et al (2009) Characterization of IgG1 conformation and conformational dynamics by hydrogen/deuterium exchange mass spectrometry. Anal Chem 81:2644–2651

264. Hamuro Y, Coales SJ, Southern MR et al (2003) Rapid analysis of protein structure and dynamics by hydrogen/deuterium exchange mass spectrometry. J Biomol Tech 14:171–182
265. Houde D, Peng Y, Berkowitz SA et al (2010) Post-translational modifications differentially affect IgG1 conformation and receptor binding. Mol Cell Proteomics 9:1716–1728
266. Burkitt W, Domann P, O'Connor G (2010) Conformational changes in oxidatively stressed monoclonal antibodies studied by hydrogen exchange mass spectrometry. Protein Sci 19:826–835
267. Kaveti S, Engen JR (2006) Protein interactions probed with mass spectrometry. Methods Mol Biol 316:179–197
268. Chalmers MJ, Busby SA, Pascal BD et al (2006) Probing protein ligand interactions by automated hydrogen/deuterium exchange mass spectrometry. Anal Chem 78:1005–1014
269. Garcia RA, Pantazatos D, Villarreal FJ (2004) Hydrogen/deuterium exchange mass spectrometry for investigating protein-ligand interactions. Assay Drug Dev Technol 2:81–91
270. Chalmers MJ, Busby SA, Pascal BD et al (2007) A two-stage differential hydrogen deuterium exchange method for the rapid characterization of protein/ligand interactions. J Biomol Tech 18:194–204
271. Fenn LS, McLean JA (2008) Biomolecular structural separations by ion mobility-mass spectrometry. Anal Bioanal Chem 391:905–909
272. Kanu AB, Dwivedi P, Tam M et al (2008) Ion mobility-mass spectrometry. J Mass Spectrom 43:1–22
273. Ruotolo BT, Hyung SJ, Robinson PM et al (2007) Ion mobility-mass spectrometry reveals long-lived, unfolded intermediates in the dissociation of protein complexes. Angew Chem Int Ed Engl 46:8001–8004
274. Ruotolo BT, Benesch JL, Sandercock AM et al (2008) Ion mobility-mass spectrometry analysis of large protein complexes. Nat Protoc 3:1139–1152
275. Olivova P, Chen W, Chakraborty AB et al (2008) Determination of N-glycosylation sites and site heterogeneity in a monoclonal antibody by electrospray quadrupole ion-mobility time-of-flight mass spectrometry. Rapid Commun Mass Spectrom 22:29–40
276. Bagal D, Valliere-Douglass JF, Balland A et al (2010) Resolving disulfide structural isoforms of IgG2 monoclonal antibodies by ion mobility mass spectrometry. Anal Chem 82:6751–6755
277. McLean JA (2009) The mass-mobility correlation redux: the conformational landscape of anhydrous biomolecules. J Am Soc Mass Spectrom 20:1775–1781

Chapter 6
Molecular Variants Characterization in Protein Therapeutics Development

Richard Ludwig, Jacob Bongers, Li Tao, Yunping Huang, Jinmei Fu, Wei Wu, Peiran Liu, Hangtian Song and Reb Russell

6.1 Development of Protein Therapeutics

The fervor for biologics in the pharmaceutical industry has been fueled by several successful launches in recent years and the potential of biologics for delivering novel therapeutics to patients and high financial returns for pharmaceutical companies. While more and more promising candidates are generated from different drug discovery platforms to fill the pipeline, many challenges remain in developing efficacious and safe products. Due to specific interactions between biomolecules, protein therapeutics are generally more specific toward therapeutic targets resulting in fewer side effects. However, the complex nature of biomolecules and the sophisticated manufacturing processes pose more challenges than those for small molecule drugs with regard to analytics for the control of quality, and ultimately efficacy and safety. Many protein molecules consist of hundreds of amino acid residues, change in any one of them results in a change in their combined identity as a protein molecule. Any variation in primary, secondary, tertiary, and quaternary structure can generate variants. It is fair to say that a protein therapeutic is very rarely composed of a single species at the time of production, let alone after additional variants are introduced due to degradation during storage. Therefore, the characterization of variants in therapeutic proteins present unique challenges that differ from those encountered when characterizing small molecule drugs.

In general, the molecular composition of protein therapeutics, excluding formulation components, can be classified into the following categories:

1. Product-related substances: molecular variants that have comparable properties to the desired form of the drug with regard to efficacy and safety.
2. Product-related impurities: molecular variants that do not have comparable properties to the desired form with regard to efficacy and safety.

R. Ludwig · J. Bongers · L. Tao (✉) · Y. Huang · J. Fu · W. Wu · P. Liu · H. Song
Bristol-Myers Squibb, Drop Code HPW 9-2, 311 Pennington-Rocky Hill Rd,
Pennington NJ 08534, USA
e-mail: li.tao@bms.com

R. Russell
Bristol-Myers Squibb, 519 State Route 173, Bloomsbury NJ 08804-4047, USA

G. Chen (ed.), *Characterization of Protein Therapeutics using Mass Spectrometry*,
DOI: 10.1007/978-1-4419-7862-2_6, © Springer Science+Business Media New York 2013

3. Process-related impurities: molecules that are derived from the manufacturing process.

Variants of protein therapeutics, depending on their therapeutic properties related to efficacy and safety, can be product-related substances or product-related impurities. Although some variants are benign in terms of toxicity, others may pose serious side effects if their contents reach certain levels. Unfortunately, most protein variants cannot be categorized directly, since it is often technically unfeasible to isolate individual variants and test their properties independently. Unlike small molecule drugs in which most covalent modifications result in change in a drug's properties and activities, many modifications of protein therapeutics do not affect their intended activities. On the other hand, conformational changes in which covalent bonds remain intact, such as denaturation and aggregation, can have a significant and detrimental impact on a protein drug's properties. The assurance of quality for a protein therapeutic is achieved in part by detailed characterization of the protein and development of a well-controlled manufacturing process that is capable of generating highly reproducible product over time. Ultimately, the efficacy and safety of the protein therapeutics and their associated manufacturing process is validated by the results from clinical trials, for which the characterization of protein therapeutics always plays an important role.

Categorically, molecular variants can also be divided into those that are generated during biosynthesis and those that are formed through degradation. Heterogeneity and variants can be introduced through biosynthetic error or incomplete processing at any point in the protein biosynthesis, starting from amino acids being first assembled into the polypeptide chain, through post-translation modification (PTMs) such as glycosylation, N- and C-terminal processing, and disulfide bond formation. After the protein synthesis is complete, degradation such as oxidation, deamidation, isomerization, and fragmentation can occur throughout the lifetimes of therapeutic proteins. The following sections will focus on each of the main biosynthesis steps and the major degradation pathways for proteins under typical manufacturing and storage conditions and the common variants and degradation products are generated.

6.2 Sequence Variants Generated by Mutation or Mistranslation

The synthesis of a protein starting from genetic code through complete assembly of polypeptide chain is a culmination of multistep processes involving deoxyribonucleic acid (DNA) replication, messenger ribonucleic acid (mRNA) transcription, polypeptide synthesis, etc. Errors within any stage of gene expression can result in an unintended amino acid being incorporated into the polypeptide chain, thus generating a sequence variant of the target protein. Known mechanisms of sequence variants include mutations at the DNA level [1], or mistranslation (through either tRNA misaminoacylation or misreading) at the protein level [2]. While the overall fidelity of protein synthesis relies on the combined accuracy of all processes involved, each process exhibits different probabilities of introducing a variant. While the misincorporation rates have been found to vary and are influenced by different conditions and among different species [3], the often-quoted error rates for the individual processes in typical cells are as follows: ~1 in 10^8 for DNA replication by bacteriophages, *Escherichia coli* (*E. coli*), and various eukaryotes [4]; ~1 in 10^5 bases for transcription in *E. coli* [5]; and ~1 in 10^4 codons translated in proteins produced by *E. coli* and mammalian expression systems. At the DNA level, although most variants involve a single-base mutation—for which a total of 75 pairs of possible amino acid substitution can be derived [6]—sequence variants involving two bases substitution have also been reported [7]. In addition, mutations at the DNA levels are usually found to occur at one or a few specific sites [7–10], whereas translational errors are often found to be randomly distributed in the protein. The error rates can be significantly higher when expression occurs under stressed conditions, such as the presence of mutagenic reagents [11], exposure to reactive oxygen species [12], amino acid starvation [10], and the use of high-yield expression systems [13]. It is known that overexpression can lead to nutritional stress on host cells which in turn can cause increased misincorporation in recombinant proteins, especially in heterologous systems [10, 13]. Nevertheless, overexpression of recombinant proteins is almost always desirable in the production of recombinant proteins intended for structural and functional studies, as well as in the manufacture of therapeutic proteins.

Misincorporation of amino acid occurs in both nature and recombinant proteins. Ala to Ser misincorporation has been reported to occur in nature through either tRNA ribosomal frameshift [14] or misaminoacylation [15, 16], as well as through mutation at the DNA level [17, 18]. Hemoglobinopathy is a well-known disorder resulting from mutation and subsequent amino acid misincorporation. To date, more than a thousand hemoglobin variants have been discovered [19]. Sequence variants have also been observed during the production of human monoclonal antibodies and other recombinant proteins [8–10, 20–23]. Multiple reports for a variety of types of amino acid misincorporations have been reported when mammalian (mouse or human) proteins were expressed in *E. coli* cells, for instance, norleucine for Met [24], norvaline for Leu [25], Gln for Arg [26], His for Gln [27], norvaline for Leu for Arg [28], Gln for His [29], and Cys for Tyr [30]. Typically, the sequence variants are at low levels, however, a Lys for Arg misincorporation which produced two functional proteins with different properties at close to 1:1 ratio has been reported [28]. Although mammalian expression system is generally considered of higher fidelity, misincorporation has also been observed in recombinant proteins expressed from CHO cells such as Gln for Tyr [31], Ser for Asn or Asn for Ser [7, 32], Arg for Ser [11], Thr for Pro, Arg for Met, Gln for Leu, and Gly for Ser [9, 21]. Furthermore, both codon-specific (likely caused by misreading) [20] and noncodon-specific (likely caused by mischarging of tRNA) [21] mistranslations have been observed in recombinant proteins expressed in Chinese hamster ovary (CHO) cells.

Conceptually, the consequence of amino acid misincorporation in recombinant proteins can be varied. Even though many of the misincorporations are benign and do not affect a protein's properties, some of them can significantly alter a protein's characteristics such as catalytic constants, specificity, and stability [33–36]. In proteins produced for therapeutic use, amino acid misincorporation can potentially induce undesired immune response or abnormal receptor-ligand interactions. Sequence variants of the desired product are generally considered product-related impurities by regulatory authorities, unless proved otherwise. Misincorporated populations in protein therapeutics are generally very difficult to remove during downstream purification. Therefore, it is essential to understand all mechanisms underlying cellular processes that contribute to the generation of sequence variants and to control the production process to reduce or eliminate sequence variants whenever possible.

Historically, high-level misincorporations have been detected using electrophoresis-based methodologies such as isoelectric focusing [13] or two-dimensional gels [10]. Edman sequencing [20], ion-exchange chromatography [9], and DNA sequencing have also been employed. The electrophoretic and chromatographic methods can only detect variants that cause observable changes in the electrophoretic or chromatographic properties. In many cases, the misincorporated protein would need to be isolated from the normal protein pool in order to facilitate identification. However, the variant proteins with only one or a few amino acid substitutions often behave similarly to the unaltered proteins which are typically orders of magnitude greater in quantity. This makes isolation of variant proteins extremely difficult. As a result, detection and identification of amino acid sequence variants remains a challenging task.

Thanks to ever-improving performance of mass spectrometers in sensitivity, resolution, and throughput, mass spectrometry has become the workhorse for protein structural analysis. Misincorporated proteins can be detected as whole molecules by intact mass analysis when the misincorporation is not isobaric to the unaltered protein. Furthermore, peptide mapping can be performed to confirm the presence of misincorporations and to locate the specific site of the misincorporated residues using liquid chromatography-tandem mass spectrometry (LC-MS/MS) analysis. A typical enzyme digest of monoclonal antibody can be analyzed within two to three hours by LC-MS/MS. Sequence variant as low as 0.01 % has been detected using LTQ-Orbitrap XL instrument [21]. Analytical approaches combining LC-MS/MS and automatic Mascot or SEQUEST database searches were developed to detect and identify protein sequence variants [37, 38]. However, it is worth noting that manual assessment of the results is generally required to confirm the automatic assignment of sequence variants, since false positive results can be generated by automatic database searches. An inherent weakness of MS analysis is that it cannot detect a variant with no net mass change, for example, substitution between Leu and Ile. In addition, only high-resolution Fourier transform (FT) MS can detect a substitution between Lys and Gln due to their small mass difference ($\Delta m = 0.0364$ Da) [39].

6.3 Variants Resulting from Post-Translational Modifications

Post-translational modification (PTMs) is an important step in protein biosynthesis in which protein is covalently modified post-translation by various functional groups. PTMs convey many functional and pharmacokinetic attributes to therapeutic proteins. The application of MS in the characterization of PTMs has been discussed in several recent reviews [40–48]. The most prominent and most complex form of PTM is glycosylation which occurs in proteins expressed mainly by mammalian cells [49, 50]. In addition to the characterization of protein glycosylation, MS has also been used successfully to direct the control of cell culture conditions to manipulate oligosaccharide profiles for glycoproteins in mammalian expression systems [51]. In bacterial fermentations, feeding strategies have been used to mitigate amino acid misincorporation or translation errors that result in unexpected protein sequences or unwanted variants [52]. LC-MS has also been used to monitor chemical modifications following storage of formulated bulk solutions or lyophilized products [53].

Most PTMs happen intracellularly during protein expression, others such as disulfide formation result from protein production processes. Many PTMs in protein therapeutics are not relevant to protein function and therefore constitute a source of product-related variants. However, it is usually challenging to determine whether or not a modification affects the pharmaceutical properties of a protein, due to the technical difficulty in isolating the modified molecules from the rest of the population.

Although many types of modifications to proteins are possible, the nature of modifications correlates with the physiological conditions of the host cells and the specific processes in protein production. Some variants are generated during protein expression such as glycosylation and gluconoylation. The latter is frequently observed in *E. coli* expressed recombinant proteins [54–56]. Information about the expression system and the manufacturing processes can significantly facilitate identification of the modifications. In addition to glycosylation, the common PTMs observed in therapeutic proteins include phosphorylation, acetylation, methylation, N-terminal pyroglutamation, carbamylation, gluconoylation/phosphoglu conoylation, S-thiolation, etc. If molecular heterogeneity is not overwhelmingly complicated, a simple intact mass analysis can reveal the extent and nature of the PTMs. If the locations of PTMs are to be determined, LC-MS/MS analysis on enzymatically digested proteins is usually the method of choice. If the heterogeneity is overwhelmingly complicated for intact mass analysis by ESI-MS, MALDI-TOF can be used to obtain a mass spectrum at the expense of fine molecular mass profiles due to limited resolution of MALDI on larger molecules.

6.3.1 Glycosylation in Protein Therapeutics

Protein glycosylation represents one of the most prominent and most complex forms of protein post-translational modification. It has been estimated that in excess of half of all human proteins are functionalized with glycans attached to one or more sites [57] and that glycosylation is present in about the same fraction of approved biopharmaceuticals in the United States [58]. The biosynthetic pathway for glycosylation is complex; it consumes a substantial amount of energy and employs a set of well-defined cellular machinery that spans from the endoplasmic reticulum to the trans-Golgi network. The collective glycosylation machinery used is similar across eukaryotic expression systems, however, differences that are characteristic of or exclusive to specific host expression systems can potentially influence immunogenic responses, pharmacokinetic profiles, or drug stability characteristics [59–64]. As a consequence of this, the Food and Drug Administration (FDA) in the United States and the Committee for Proprietary Medical Productions (CPMP) of the European Community have been requiring more and more sophisticated carbohydrate analysis for new glycoprotein therapeutics intended for use in human therapy [65]. Accordingly, the use of suitable analytical methods in the characterization and monitoring of glycosylation characteristics within manufactured biotherapeutics has become a critical aspect in the understanding of both drug function and determination of manufacturing process capabilities. The importance of demonstrating control over glycosylation has recently been illustrated when the application for a scaled up manufacturing process for alglucosidase α (rhGAA) was rejected by the FDA due to concerns about the differences in glycosylation at two differing manufacturing scales, resulting in a determined lack of comparability [66].

Glycosylation is depended on cell, tissue and site-specific and sensitive to environmental variations as well. Therefore, a given glycoprotein may be differentially glycosylated depending on the conditions under which it is produced. Three factors are commonly accepted as exerting significant influence on the glycosylation profile of a glycoprotein with respect to both the degree of site occupancy and the distribution of glycans at each occupied site. These are as follows: (1) the specific polypeptide structure; (2) cell-specific parameters of the cell type used in expression of the protein; and (3) environmental factors and stresses which affect cellular homeostasis, or in some way alter the cellular phenotype [67]. Given the breadth of potential influencing factors, it is not surprising that producing biotherapeutics with a consistent oligosaccharide profile can be a challenge. Extensive characterizations as well as profile trending are required in order to ensure manufacturing process robustness and product consistency.

Glycosylation introduces heterogeneity within a biotherapeutics product through the incorporation of varying glycans resulting in the generation of different glycoform profiles in which the contribution from individual oligosaccharide components vary in both their core structure and by differences in the addition of outer-arm sugars. The presence and nature of the glycoforms present can impact a wide range of protein characteristics such as functional activity, folding, stability, pharmacokinetics, and immunogenicity. The presence of glycan structures on a therapeutic protein which are foreign to the host system in which the therapeutic protein is to be used can, therefore, induce undesirable effects. Mammalian expression systems produce glycosylation comparable to that of produced in humans and have, at least in part, for this reason become the dominant platform for the production of therapeutic glycoproteins. Even so, they exhibit an inherent glycan heterogeneity that is sensitive to culture conditions [68, 69], and these platforms require tight process control in order to ensure consistent product glycosylation.

6.3.1.1 Glycosylation and Biological Activity

The impact of protein glycosylation on protein function can be varied. There are cases where complete glycosylation equates with full biological activity [70, 71], and there are cases where the extent of glycosylation has no significant impact on the biological effectiveness of a protein at all [72, 73]. Glycosylation can affect many properties of protein therapeutics [49], such as structural stability [74], potency [75], efficacy [68], immunogenicity [76–78], and pharmacokinetics [79, 80]. Glycosylation of therapeutic proteins itself [81–86] and the applications of MS in this area have been extensively reviewed in some recent publications [87–91].

Glycosylation is known to be involved in regulating biological processes, particularly in immune response, in which it can have a direct impact on immunogenicity, manifested as an IgE response or by inducing the formation of antidrug antibodies and in doing so render the drug ineffective or even harmful to the recipient [58]. CHO cells produce glycosylation patterns that fairly closely resemble that of humans. However, there are significant differences. For example, CHO and mouse cell lines produce N-acetylneuraminic acid (NANA) which is present in human IgGs, but in addition CHO cells also express N-glycolylneuraminic acid (NGNA), a form of sialic acid (SA) which is not found in humans and has been reported as immunogenic [68, 80, 92–94]. High-mannose glycan structures also potentially raise the risk of adverse immune response as they can be substrates for the mannose receptors commonly expressed on the surface of macrophages and antigen presenting dendritic cells [95]. The absence of SA terminal residue [96] or of galactose residues [97] on the IgG heavy chain often results in immune complex formation. Slight variations in glycosylation can result in significant nonimmunogenic changes in protein properties as well. For instance, biantennary oligosaccharide structures lacking galactosylation have been reported to have slightly faster clearance rates than other structures in IgG1 antibodies [98]. Glycosylation can also have an indirect effect on immunogenicity through its impact on protein folding, solubility, or stability.

6.3.1.2 Solubility and Stability

The attachment of oligosaccharide structures on proteins frequently contributes to improved solubility and reduced protein aggregation [99]. The presence of altered or incomplete glycan structures has been reported to lead to conformational changes [100] as well as significant increases in protein aggregation [101].

Glycosylation has also been reported to have an effect on local secondary structure which in turn can play a role in the assembly of tertiary structure. Altered or absent glycosylation can potentially alter or eliminate protein epitopes or create new ones. The presence of attached oligosaccharides can alter solubility by shielding hydrophobic regions of a protein's surface thereby reducing the tendency to aggregate thereby enhancing stability by participating in stabilizing intrachain interactions [80, 81]. One example of this is the physicochemical properties of human granulocyte colony-stimulating factor (hG-CSF) which depends on the presence of O-linked glycan structures to prevent self aggregation which would result in complete biological inactivation [102]. Thermal stability may also be improved by glycosylation as has been shown for IFN-β [103], interleukin (IL)-5, rhEPO, and IgG1-Fc [104, 105]. In the case of the IgG1-Fc, the native glycosylated from was the most stable followed by various truncated glycoforms with the fully deglycosylated version being the least stable. Since IgGs have been used extensively within the field of biologics, a large amount of data has been generated regarding the impact of minor glycosylation changes and their impact on the function of this protein class. Therefore, IgG serves as a valuable example to illustrate the importance of glycosylation. It has been shown that the majority of IgG effector functions are compromised by the removal of the N-linked oligosaccharides: FcγRI activation is reduced by 2 orders of magnitude; FcγRII and FcγRIII, are abolished by complement activation and mannose-binding lectin (MBL) activation. In fact, protein's characteristics can be significantly affected even without complete

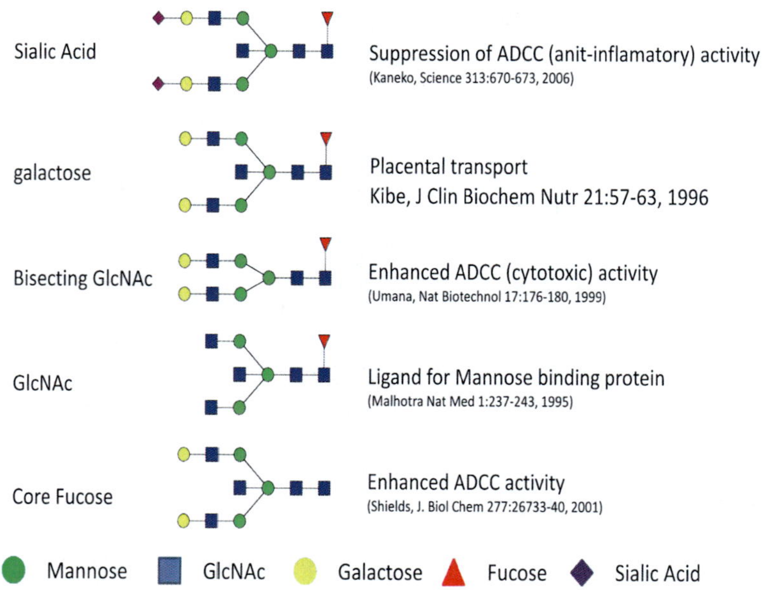

Fig. 6.1 Effect of terminal sugar residues of N-linked glycans on pharmacokinetic properties of IgG molecules

removal of glycosylation. Figure 6.1 illustrates some findings of the biological impact of individual glycoforms on biological activity of IgGs [106–110].

6.3.1.3 Glycoprotein Structure

The high degree of oligosaccharide complexity, resulting from the variable composition, linkage, branching points and configuration of monosaccharides, and the presence of various degrees of glycosylation at different glycosylation sites on glycoproteins are the main reasons for the diversity of analytical approaches that have been developed for the study of this post-translational modification [111]. To date, four types of protein-linked glycans are known, including (1) N-linked, (2) O-linked, (3) C-glycans, and (4) glycosylphosphatidylinositol anchors.

In N-linked glycosylation, the oligosaccharide is linked via a GlcNAc molecule in a β-N-glycosidic type bond to a nitrogen of the amide group of an asparagine (Asn) as illustrated in Fig. 6.2.

N-Linked glycosylation of proteins is a co-translational event occurring during protein synthesis and is initiated as the newly synthesized polypeptide chain enters the lumen of the endoplasmic reticulum (ER). N-linked oligosaccharide synthesis continues as the protein is transported from the ER to the Golgi apparatus and is completed by the time the glycoprotein leaves the trans-Golgi network. The presence of a consensus amino acid sequence within the protein amino acid sequence is a prerequisite for N-linked oligosaccharide incorporation. The consensus sequence for N-linked glycosylation is Asn-Xaa-Ser/Thr, in which Xaa may be any amino acid with the exception of proline. The consensus sequence allows recognition of the glycosylation site by the first enzyme involved in N-linked oligosaccharide production (oligosaccharyltransferase or OST) by providing a protein conformation which enables the enzyme to gain access to the glycosylation site.

The presence of the N-linked consensus sequence Asn-Xaa-Ser/Thr does not guarantee oligosaccharide attachment. The sequence may occur many times in a polypeptide chain with only a small number of the potential sites being glycosylated. Numerous factors influence whether a putative glycosylation site is coupled to an oligosaccharide. Occupied N-linked consensus sequences are frequently those

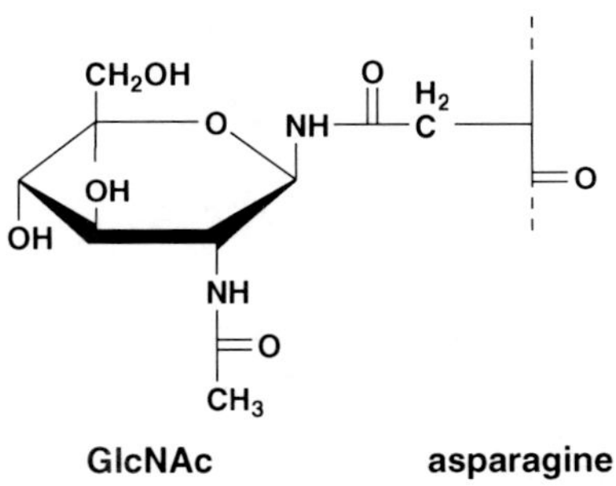

Fig. 6.2 N-Linked oligosaccharide structures linkage of GlcNAc to asparagine

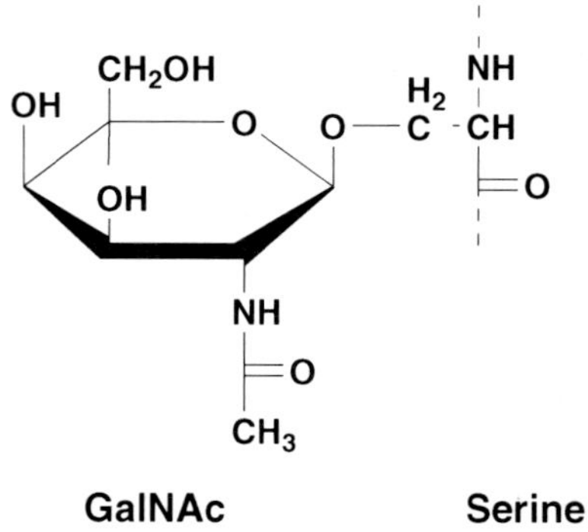

Fig. 6.3 O-Linked oligosaccharide structure linkage of GalNAc to serine

in which the consensus sequence is in a "turn" or "loop" in the polypeptide chain which enables access to the OST.

In O-linked glycoproteins, the first monosaccharide of the oligosaccharide chain, usually GalNAc, is attached through an α-O-glycosidic linkage to an oxygen atom of an amino acid residue, typically serine or threonine and to a lesser extent hydroxylysine or hydroxyproline on the polypeptide chain of a protein as illustrated in Fig. 6.3. Both N- and O-linked glycoproteins share common features and many proteins contain both types of glycan attachments within the same protein molecule.

The other two glycosylation structures C-glycans and glycosylphosphatidylinositol anchors are less common. In the former, the glycan (Mannose) is attached to the Trp residues by a C–C bond in a consensus sequence of Trp-Xxx-Xxx-Trp or Trp-Ser/Thr-Xxx-Cys; and in the latter, the glycan attaches to the carboxyl terminus of certain membrane-associated proteins by a phosphoethanolamine bridge with Mannose (Man) [112].

Within therapeutic proteins, the two most common forms of protein glycosylation are N- and O-linked glycosylation. Both N- and O-linked glycoforms are characterized by complex branched structures that vary greatly in form and size. Common core structures for both structures are provided in Figs. 6.4 and 6.5. N-linked glycans contain a common trimannosyl-chitobiose core ($Man_3GlcNAc_2$) with one or more antennae attached to each of the Man units [89]. Based on the location and nature of the additional monosaccharides added to the core, N-linked glycans are further classified into; (1) the "high Man" or "oligomannose" type ($Man_{5-9}GlcNAc_2$) N-glycans that have only Man residues added to the core; (2) N-glycans of the complex type that contain N-acetyllactosamine (Galb1-3/4GlcNAc) within their antennal region; and (3) the "hybrid type" N-glycans that contain both Man residues and N-acetyllactosamine attached to the trimannosyl-chitobiose core residues.

O-linked glycans, on the other hand, are characterized by the stepwise addition of sugar residues directly to a protein through a hydroxyl group. In mammals, the initiating step is typically the addition of N-acetylgalactosamine to Ser/Thr residues,

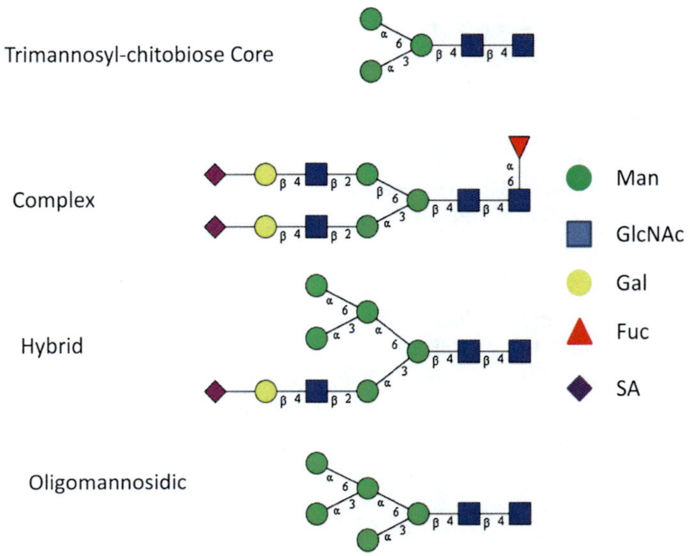

Fig. 6.4 N-Linked oligosaccharide structural classes

Fig. 6.5 O-Linked oligosaccharide core structures

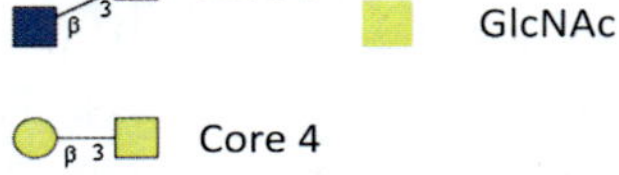

although other monosaccharide units, such as GlcNAc, or Man-linked oligosaccharides, have been reported to be involved in O-glycosidic linkages to hydroxyl amino acids [89]. Subsequent addition of Gal and/or GlcNAc leads to the formation of the common O-glycan core structures (Fig. 6.5). Biosynthesis of complex N- and O-linked glycans is completed by a variety of capping reactions, the most import in mammals being sialylation and fucosylation [113]. Because of the acidic nature of SA residues, primarily NANA (Neu5Ac) and N-glycolylneuraminic acid (Neu5Gc) impart a net negative charge to the otherwise neutral glycans. Glycans can be further modified by acetylation, methylation, phosphorylation, and sulfation, which can occur at internal or terminal positions in the glycan structure [83].

6.3.1.4 MS Detection of Carbohydrates: Ionization

The two most common ionization techniques use in the analysis of glycoproteins are electrospray ionization (ESI) and matrix-assisted laser desorption/ionization (MALDI). Both techniques posses their own strengths and weaknesses and one is often used as complementary approaches to the other. Conventional ESI-MS involves a flow of solution into the ion source of a MS instrument and has been observed to produce relatively weak ion signals for native oligosaccharides compared to those for peptides and proteins [114, 115]. It has also been observed that the ESI signal is increased when oligosaccharides are derivatized. This is consistent with the idea that the decrease of ESI signal for oligosaccharides relative to peptides is mainly due to decreased surface activity rather than decreased volatility [116]. In addition, nano-ESI has been reported to produce ion signals that are comparable in magnitude between the peptide and carbohydrate compound classes [117]. This increase in sensitivity in nano-ESI suggests that the hydrophilicity of oligosaccharides limits the surface activity in ESI droplets, and the barrier can be overcome by the more efficient droplet formation produced by the nanospray [116]. The fact that the ESI of carbohydrates appears to be more effective at the nanoscale could have important implications. Interfaces for online ESI LC/MS typically produce droplet sizes that are larger relative to those produced by spraying from a 1 to 2 μm-orifice nanospray emitter, and thus, the spray characteristics are typical of forced flow. As commercial nano-ESI systems continue to become more robust, the potential for routine use of nanospray ESI in the characterization and quantitation of underivatized oligosaccharides becomes an increasingly viable approach.

The Ionization mechanism for MALDI is different from that of ESI. For example, in contrast to that for ESI, where the ionization efficiency decreases with increasing molecular weight, MALDI-TOF ionization efficiency for neutral carbohydrate oligomers has been observed to be constant as the size of the molecule increases [118]. MALDI-TOF for neutral oligosaccharide analysis has advantages over ESI, particularly for applications that involve the profiling of mixtures released from glycoproteins. Quantitation of permethylated carbohydrate mixtures by MALDI-TOF has been shown to result in reproducible data with precision levels comparable to those obtained for the same oligosaccharide mixture after derivatization with a chromophore and chromatographic quantitation [119]. The MALDI-TOF approach has the added benefit of providing information on the structural composition. The commonly used MALDI matrices are listed in Table 6.1.

The advantages of MALDI in terms of ionization response must be balanced against the disadvantages of presence of meta-stable fragmentation that is caused by the high internal energies imparted to the ions in MALDI relative to those resulting from ESI. Although fragmentation is useful in the analysis of carbohydrate ion structure with post-source decay (PSD) on a MALDI-TOF instrument [120], it complicates MS profiling by introducing uncertainty as to whether or not an observed structure was generated in source. Nevertheless, the use of MALDI in carbohydrate and glycoconjugate analysis continues to advance and has been the subject of extensive review [121]. MALDI continues to be a major technique for the analysis of carbohydrates although

Table 6.1 Common MALDI matrices used in glycoprotein, glycopeptide, and oligosaccharide analysis

MALDI matrix and structure	Abbreviation	Application
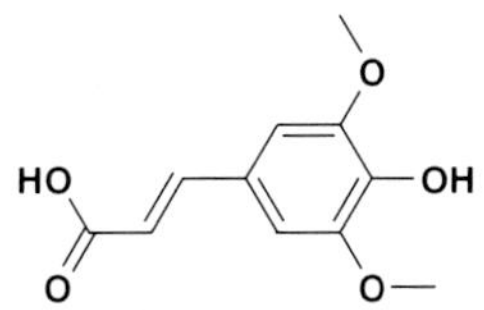 **sinapinic acid**	SA	Large peptides or proteins with molecular mass larger than 10 kDa
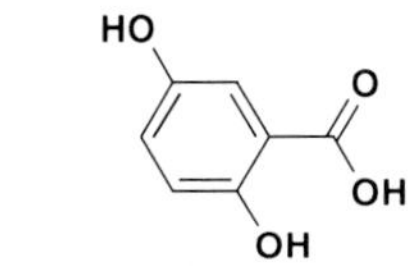 **2,5-dihydroxybenzoic acid**	DHB	Protein digests, carbohydrates, oligosaccharides, glycopeptides, and both proteins and peptides below 10 kDa. This matrix is also well suited for the negative ion MALDI-MS glycolipids
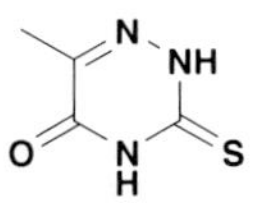 **6-aza-2-thiothymine**	ATT	Glycans, glycoproteins, and oligonucleotides
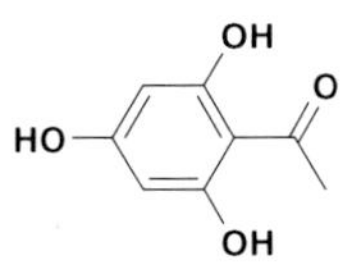 **2',4',6'-trihydroxyacetophenone**	THAP	Glycans, glycoproteins, and oligonucleotides work well in negative ion mode
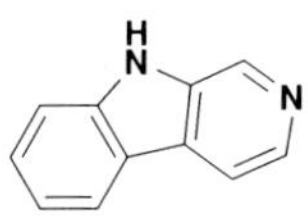 **norharmane**	--	Glycans and proteins
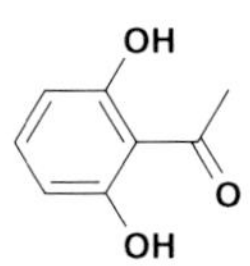 **2',6'-dihydroxyacetophenone**	DHAP	Glycoproteins and mixtures containing glycoproteins

(continued)

Table 6.1 Continued

MALDI matrix and structure	Abbreviation	Application
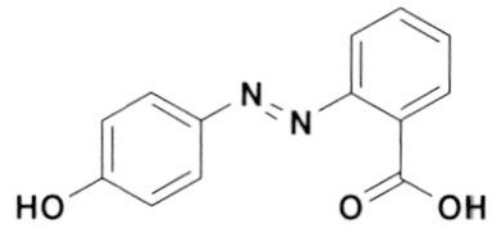 2-(4'-Hydroxybenzeneazo)benzoic acid	HABA	Oligosaccharides and glycoproteins (less discrimination against high-mass compounds)

electrospray, particularly with quadrupole–time-of-flight (Q-TOF) instruments, for the most part coupled with CID, has become increasingly popular. MALDI-TOF usually provides a superior profile of component glycan because of the predisposition of the technique to produce mainly singly charged ions. However, MALDI-TOF MS of native acidic glycans is less satisfactory due to problems arising from glycan fragmentation. ESI causes less fragmentation of these compounds but tends to produce ions in different charge states from glycans with several acidic groups, thus giving a profile that is not directly representative of the glycan content. This problem, and the instability of acidic carbohydrates under MALDI conditions, can be overcome by derivatization of the carboxylic acid group of SAs. Ultimately, both MALDI and ESI have advantages and disadvantages for carbohydrate work and the best technique to use will be dictated by the problem to be solved.

6.3.1.5 Nomenclature for the Fragmentation of Oligosaccharides and Glycoconjugates

Collision-induced dissociation (CID) of oligosaccharides and glycoconjugates results in the observation of ions that correspond to cleavage of the oligosaccharide portion of the molecule. Due to the labile nature of the glycan bonds under acidic conditions, ions are produced in greater abundances for the oligosaccharide portion than are those that occur in the aglycon (nonoligosaccharide) portion of the glycoconjugates. The nomenclature for oligosaccharide fragmentation used throughout the mass spectrometry field is shown in Fig. 6.6. Fragment ions that contain a nonreducing terminus are labeled with uppercase letters from the beginning of the alphabet (A, B, C) and those that contain the reducing end of the oligosaccharide or the glycoconjugate terminal end are labeled with letters from the end of the alphabet (X, Y, Z); subscripts indicate the cleaved ions location within the oligosaccharide structure. The A and X ion are produced by cleavage across the glycosidic ring and are labeled by assigning each ring bond a number and counting clockwise. Examples for two cross-ring cleavage ions are shown in Figs. 6.7, 6.8, and 6.9. Ions produced from cleavage of successive residues are labeled: A_m, B_m, C_m with $M = 1$

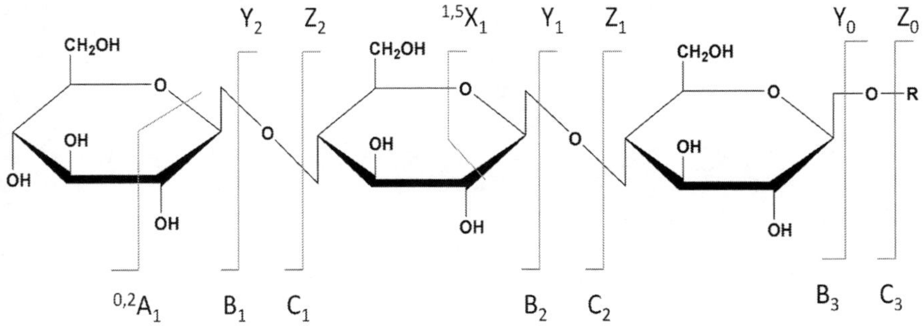

Fig. 6.6 Nomenclature for the fragmentation of oligosaccharides

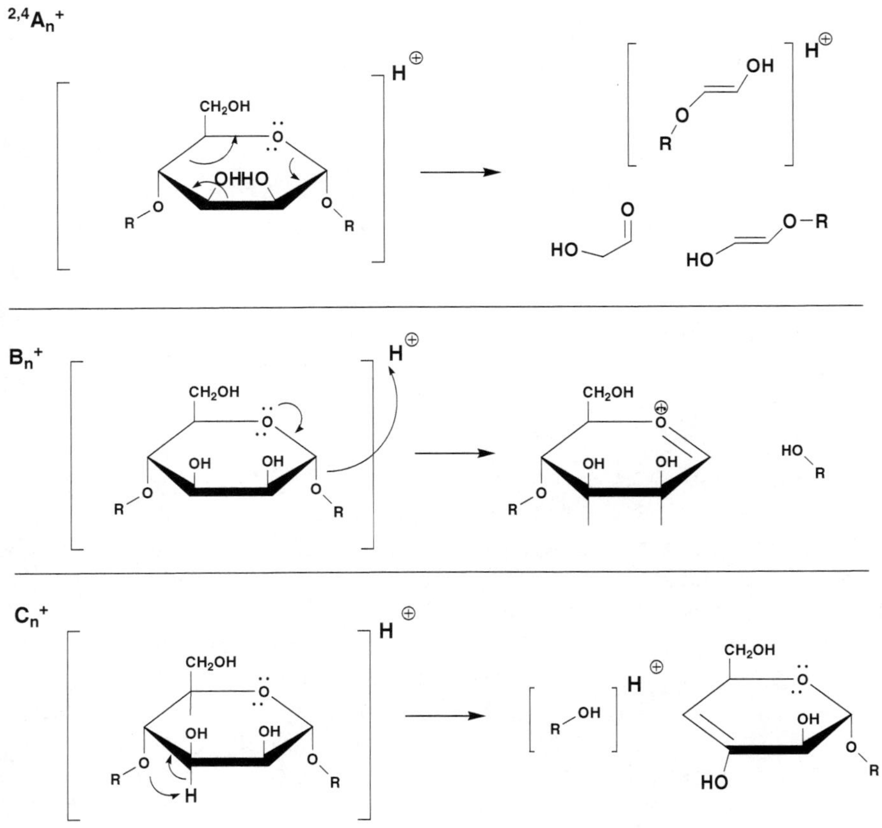

Fig. 6.7 Mass spectral fragmentation of hexose *A*, *B*, and *C* Ions

for the nonreducing end and X_n, Y_n, and Z_n with $n = 1$ for the reducing end residue. Note that Y_0 and Z_0 refer to the fragmentation of the bond to the aglycone.

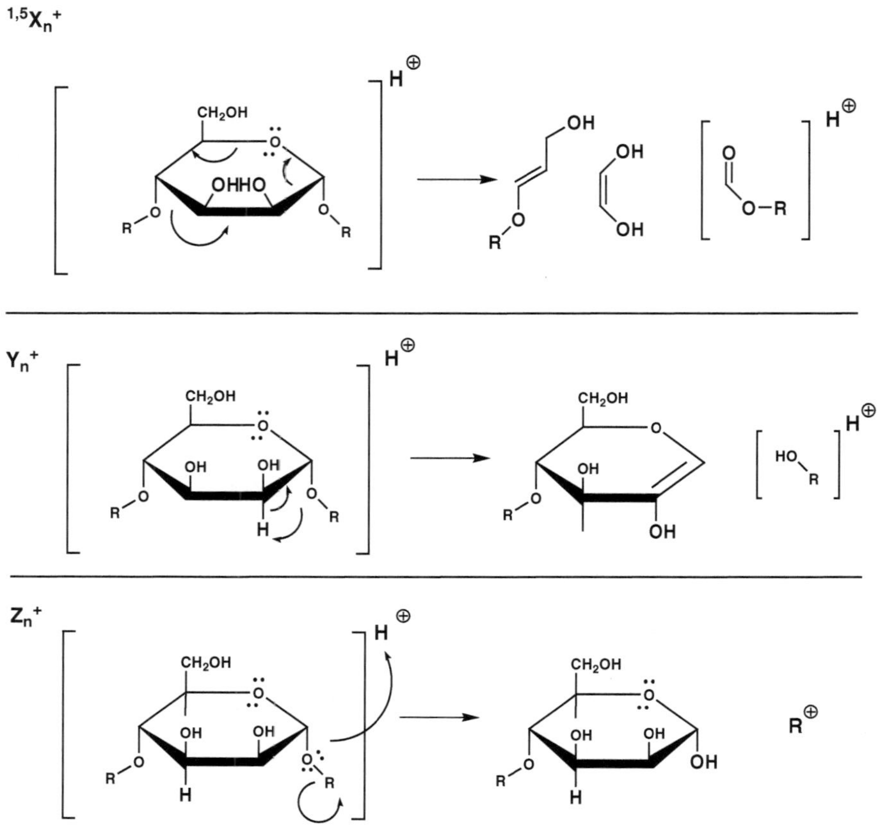

Fig. 6.8 Mass spectral fragmentation of hexose *X*, *Y*, and *Z* Ions

When performing tandem MS of oligosaccharides, there are multiple possibilities regarding the state of the precursor ion, the choice of which can significantly influence the product-ion pattern and the structural information obtained. For native oligosaccharides, common possibilities include protonated molecular ions $[M+nH]^{n+}$, deprotonated molecular ions $[M-nH]^{n-}$, and nitrated molecular ions, $[M+nNa]^{n+}$ as well as other metal- and salt-adducted ions.

Product-ion mass spectra of glycoconjugates are considerably more complicated than those of peptides because of the glycan branching structure. Within these branching structures, fragmentation occurs from the nonreducing end of each antenna and from the reducing end. Multiple cleavages are not uncommon, giving rise to a high level of spectral complexity. The input of energy into the molecule by collision most often breaks the single-bond glycosidic linkages. Using low-energy CID, fragmentation of glycosidic linkages is most likely, whereas fragmentation across the sugar rings is less likely because two covalent bonds must be cleaved. It does occur, however, and provides essential information on the location of substituents on branching monosaccharide residues.

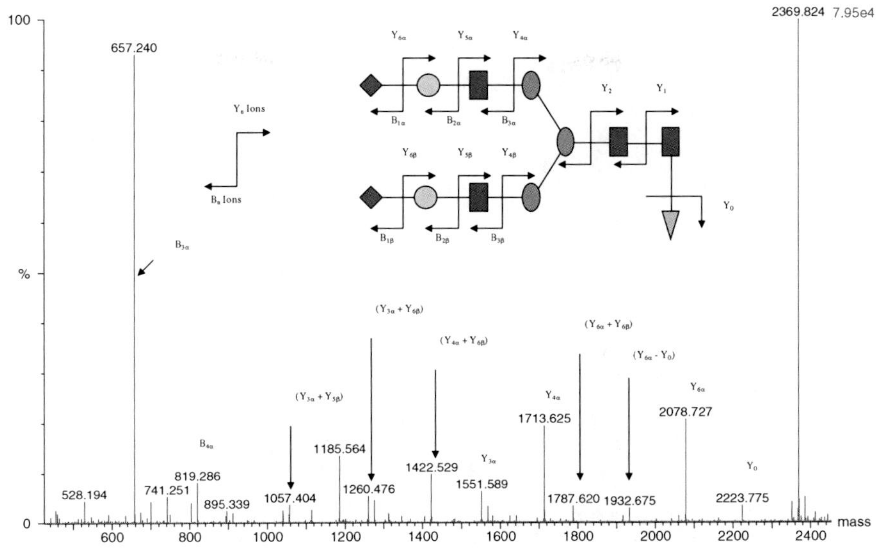

Fig. 6.9 Fragmentation of a biantennary N-linked oligosaccharide in CID experiments

It has been demonstrated that CID product-ion mass spectra can provide information on stereochemistry of individual sugar residues, the linkage position [122], and branching structure [123, 124]. Oligosaccharides that contain the same monosaccharides linked with a different branching structure often show distinct product-ion patterns because the local steric environments differ between the alternate isomers and result in different ion abundances in product-ion mass spectra. Due to the dependence of fragmentation patterns on the particular parameters and instruments used, it can be difficult to correlate the fragmentation patterns with confidence to data produced across multiple laboratories.

6.3.1.6 Analytical Approaches

The analysis of glycoproteins can be divided into three general approaches as shown in Fig. 6.10. Each approach provides information on different aspects of glycoprotein structure. Analysis of intact glycoproteins provides a global view of the glycan population or profile of the glycan content of the glycoprotein. It does not, however, provide information on the localized glycosylation sites within the protein, and in many cases, information on the oligosaccharide structures are confounded due to overlapping masses of glycoforms with multiple glycosylation sites. Glycopeptide analysis performed on proteolytic digest mixtures allows the identification of site-specific glycosylation and in many cases provides strong data for the determination of oligosaccharide structures when MS^n data are evaluated. Analysis of glycans after release from a glycoprotein enables the use of a wide

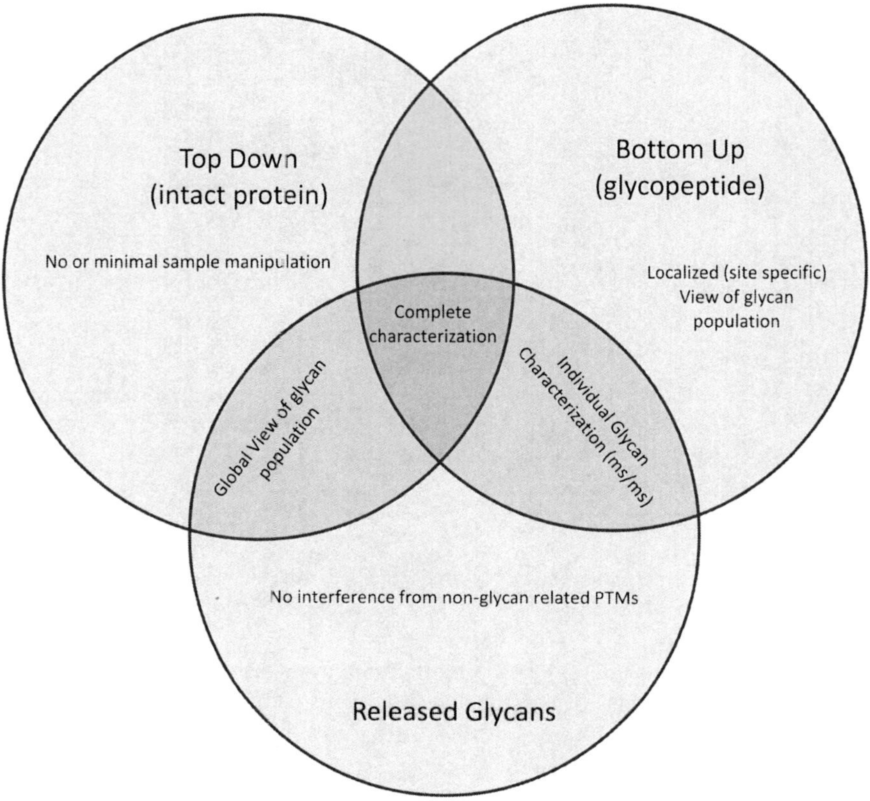

Fig. 6.10 Approaches used in glycosylation analysis

range of methods and analytical strategies which enable detailed structural investigation of individual glycan structures as well as the overall glycan profile without complication from additional protein modifications. Each of these approaches provide valuable information and taken together can produce a comprehensive view of the glycan structure of a given glycoprotein.

6.3.1.7 Intact Glycoprotein Analysis

MALDI-TOF and ESI have emerged as powerful tools for the analysis of large biomolecules. The characterization of glycoproteins by mass spectrometry is naturally more difficult than the mass spectrometric analysis of proteins, because glycoproteins exhibit extensive heterogeneity and because they are ionized less efficiently than nonglycosylated proteins. Nonetheless, to some degree, intact glycoproteins can be resolved to their individual glycoforms using both methodologies.

MALDI-TOF MS

MALDI-TOF MS has been extensively used for the exact molecular mass determination of peptides and proteins [125]. MALDI provides a rapid and simple method of obtaining molecular mass information at the picomole to femtomole sensitivity range. In fact, this analytical tool is routinely used to complement and in some instances even replace common protein biochemical techniques, such as sodium dodecyl sulfate polyacrylamide gel electrophoresis (SDS-PAGE) [126, 127]. MALDI is known to be well suited for detection of glycoproteins and high-mass proteins owing to its combination of high sensitivity and a theoretically unlimited m/z range for linear TOF analyzers, which has allowed the determination of singly charged molecular ions of up to 10^6 Da [128]. Due to the enormous heterogeneity arising from the macro- and microheterogeneity of glycosylation as well as other post-translational modifications (sulfation, phosphorylation, hydroxylation, carboxylation, etc.), MALDI-TOF analysis of intact glycoproteins has generally been unable to resolve individual protein glycoforms of larger proteins and typically is used to provide data on only the average carbohydrate content.

Instrumental resolution generally restricts the use of MALDI-TOF instruments to studies of glycoproteins with masses below 20 kDa. For example, MALDI spectra of ribonuclease B, a small glycoprotein (15 kDa) that contains five mannose N-linked glycans at a single glycosylation site, can be resolved well enough to determine the glycosylation pattern using linear instruments [129]. The careful selection and optimization of parameters which determines the desorption/ionization of glycoproteins can improve the quality of MALDI mass spectra as well as mass reproducibility and resolution. When parameters like the sample matrix, sample-matrix preparation technique, pH, and instrumental conditions are optimized, resolution of glycoforms can be achieved for both sialyl and nonsialyl oligosaccharide proteins [130, 131].

It has also been reported that the use of delayed ion extraction can result in improved glycoform resolution. This is because the delayed ion extraction minimizes the observed metastable fragmentation, which is known to adversely influence MALDI analysis of glycoproteins. By carefully controlling field strength and delay time, improvement of more than one order of magnitude in resolution has been reported relative to the nondelayed extraction case [132].

Given that existing TOF mass spectrometers can only achieve isotopic resolution for glycoproteins with masses below 10 kDa, and the fact that resolution of individual glycoforms becomes increasingly difficult due to the presence of non-glycosylation-related PTMs, salt adducts as well as fragment ions, the only viable solutions are analysis of glycans after release from the protein, or cleavage of the protein into smaller units. Even in the absence of glycoform resolution, the measurement of protein molecular mass before and after removal of the attached glycan provides information on the state of glycosylation. The difference between the glycoprotein molecular mass obtained by MALDI-MS and the predicted molecular mass from the amino acid sequence can also provide information about the carbohydrate content of the glycoprotein [133].

ESI Quadrupole and TOF

ESI coupled to a Q-TOF mass analyzer typically produces better results than MALDI-TOF in terms of resolution and mass accuracy for intact protein analysis. Advances in quadrupole technology continue to result in the maximum m/z of quadrupole analyzers moving higher. Given the higher mass ranges available and the computing power to rapidly deconvoluted the data generated, the mass of an intact mAb can be determined quickly and reproducibly on an ESI-quadrupole instrument with high mass accuracy and resolution.

Most modern orthogonal acceleration time-of-flight analyzers with reflection achieve a mass accuracy on the order of 2–10 ppm, a resolution of 5,000–30,000, and a maximum m/z of up to 10,000. When coupled with ESI, TOF instruments make an ideal system for mass determination of intact mAbs. For this reason, ESI with TOF analyzers, such as ESI-TOF or ESI-Q-TOF configurations, has become the method of choice for mass determination of intact mAbs. Using this methodology, ions from a given molecule produce an ion envelope in which a given ion differs by plus or minus one charge from adjacent ions in the series. In measuring the molecular mass, the charge on any one of the ions is first established by solving a series of simultaneous equations for any two consecutive ions in the series. From the combined data, the charge for all ions in the series can be deduced and the molecular mass calculated. As with any spectral data, overlapping of peaks may occur within electrospray data, particularly when several species are present, each giving rise to its own series of multiply charged ions. Under these conditions, accurate values for mass/charge ratios of the components in an unresolved multiplet may not be obtainable without some form of data deconvolution.

Maximum entropy (MaxEnt) techniques are particularly well suited for the effective deconvolution of mass spectral data. The MaxEnt deconvolution solution contains the minimum amount of spectral structure consistent with the data and is capable of providing a level of deconvolution which can enable overlapping peaks to be resolved and accurately centroided [134]. With external calibration on an ESI-TOF instrument, the mass of an intact mAb can be determined with a mass accuracy of less than 100 ppm. With carefully executed experiments, the mass of an intact mAb has been determined to an accuracy of 25–50 ppm [135, 136]. With experimental conditions optimized to minimize adduct formation, a calibrant analyzed immediately before sample analysis, and carefully controlled deconvolution parameters, mass accuracy that approached 10 ppm have been achieved [137]. This level of accuracy in mass measurement of intact mAbs by a TOF analyzer approaches the natural variation of the protein average mass. The mass range and mass accuracy of an ESI-TOF instrument, combined with an appropriate charge-deconvolution algorithm, make it an ideal match for the mass determination of intact mAbs [138].

Along with the high level of accuracy, there is a high level of mass resolution that is obtained from the combination of the instrumental mass resolution and the resolution enhancement afforded by deconvolution of the ion envelope data. The resulting level of resolution allows for resolved detection of individual glycoforms on glycoproteins of relatively large size. A typical example of an ESI-Q-TOF mass spectra of a mAb is shown in Fig. 6.11. For mAbs, typically in the 150-kDa mass range, the mass resolution after deconvolution easily allows for separation of glycoform masses.

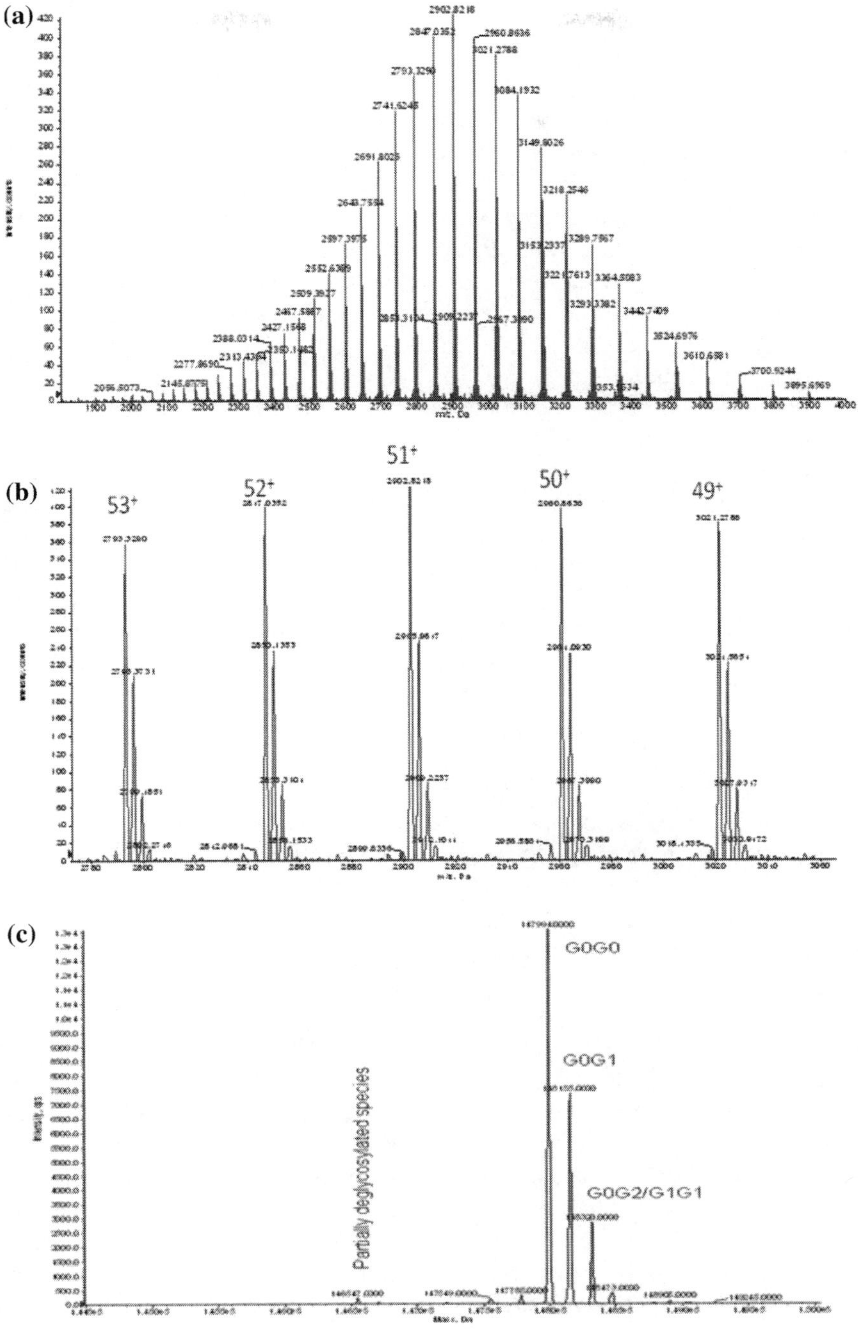

Fig. 6.11 Mass spectra of a monoclonal antibody

Typical raw and deconvoluted spectra of a mAb: (a) raw ESI-Q-TOF spectrum acquired on a Thermo Q-Star Elite. (b) Region of raw spectrum showing charge states 49–53. (c) Deconvoluted mass spectrum, in which clear resolution of glycoforms are identifiable.

It should be noted that isobaric glycoforms cannot be distinguished from one another from the intact analysis data. In the example presented in Fig. (6.11), a mAb with only two N-linked glycosylation sites and only three predominant oligosaccharide structures, the two glycoforms G0G2 and G1G1 are indistinguishable from one another. In glycoproteins with only moderate glycosylation, such as typical mAbs, the confounding of structural identification is minor. However, as a protein's glycosylation becomes more extensive, the number of isobaric glycoforms increases dramatically. In such cases, the intact mass data may be useful as a measure of glycan content and a means of monitoring profile consistency. However, complementary methods are needed in order to provide more detailed glycan characterization.

As with selection of the appropriate matrix and analyzer settings in MALDI-TOF, there are multiple options available when selecting the liquid chromatographic separation input for ESI MS. The two most commonly used include reversed-phase (RP) high-performance liquid chromatography (HPLC) and size-exclusion chromatography (SEC).

Chromatographic analysis of protein samples are usually performed on SEC, ion-exchange, or hydrophobic interaction chromatography. However, these chromatographic techniques are generally not compatible with MS detection due to the high salt content used in the mobile phases. The ideal combination of LC with MS is through RP-HPLC as the RP eluate is typically salt free and contains some levels of organic solvent. Historically, RP-HPLC analysis of proteins has suffered from poor chromatographic resolution and recovery. However, several laboratories have successfully developed reversed-phase methods for intact mAb analysis. A high column temperature of 60–80 °C is often required to minimize sample loss to the column and to minimize sample carryover [139, 140]. Other important factors include a careful choice of the column, and the usage of organic solvents with high eluotropic strength coefficients such as isopropyl or n-propyl alcohol. With an optimized reversed-phase method, separation of an intact mAb has been chromatographically separated from their C-terminal lysine variants, degradation products, cysteinylated form, and disulfide variants [141–145].

SEC has historically not been compatible with MS due to the use of high salt concentrations used in the mobile phase. However, newer column technology has allowed the modification the mobile phase to exclude salt, and the technique has become compatible with MS for LC/MS analysis of intact proteins. There is typically some sacrifice in chromatographic resolution upon the mobile-phase modification. A typical mobile phase appropriate for this application would contain some organic solvent under acidic conditions [146]. As an alternative, acidic aqueous mobile phase with post-column addition of organic solvent has also been reported [135]. Published results demonstrated that the SEC/MS analysis had advantages over the RP-LC/MS method in terms of sensitivity and overall quality of the mass spectra. Although most mAb isoforms are not resolved by SEC, the chromatography does not require elevated temperatures, minimizing concerns about method–induced glycoprotein degradation. SEC typically has less of a carryover problem than the RP method [146]. The selection of RP versus SEC LC methods will depend on the particular needs of the experimenter. RP-HPLC is often used when chromatographic resolution or sample throughput is important, whereas SEC is favored when better MS spectral quality is needed or when chromatographic separation is not a concern.

Generally, intact mass analysis gives a global view of the protein glycosylation, that is to say that specific information regarding local site glycosylation is not extractable from the intact view directly. However, if it were possible to fragment the intact protein within the spectrometer, it would be possible to gain some higher level of regiospecific information regarding the attached glycan location and local glycan populations. Top-down mass spectrometry refers to mass spectrometric evaluation of the instrument-induced fragmentation of biomolecular ions of any size [147–149]. This approach has the potential to be able to identify and characterize all types of post-translational modifications, including glycosylation, on the native protein of interest.

To date, the fragmentation of intact proteins the size of intact monoclonal antibodies in the gas phase has been attempted with CID, both through use of the dedicated CID cell as well as directly within the ion source itself [150, 151]. However, CID-generated sequence coverage, particularly from IgGs, has been limited. Electron-capture dissociation (ECD)- [152–154] and electron-transfer dissociation (ETD) [155]-induced fragmentation of large proteins provides an alternative, radical-based mechanism of fragmentation and creates additional opportunities in top-down MS analysis. The radical-induced fragmentation generally provides more extensive sequence coverage on large proteins and produces efficient rupture of disulfide bonds [156, 157] when compared to CID and Infrared Multiphoton Dissociation (IRMPD). However, possibly the most significant advantage of ECD/ETD over CID/IRMPD is in the characterization of labile PTMs on peptides [158], even though this advantage is less pronounced on proteins, where the distribution of vibrational energy can be more easily dispersed in the large protein structure which increases the chance for labile PTMs to remain intact during the fragmentation of the protein backbone bonds in CID/IRMPD [159, 160]. Although this approach presents great promise, it is still in its infancy. As the availability of ETD/ECD instruments becomes more accessible, the maturity of top-down analysis with respect to characterization of PTMs including glycosylation will continue to evolve.

Cases will always exist where intact mass analysis of a native protein will not be possible with available instrumentation or where more detailed structural information is desired. Under this circumstance, one must cleave the protein into smaller fragments before mass analysis. A variant of the top-down approach, named "middle-up," involves cleaving the protein into several large fragments before MS analysis. This approach is a convenient approach to use on mAbs as these structures are easily fractured into several medium-sized fragments.

A common strategy is to cleave a mAb into a few large fragments through reduction in the disulfide bonds to liberate the heavy chains and light chains. Mass analysis of the individual chains can be used to confirm the structure, or to locate variants or modifications in each individual chain. To illustrate how this simple variation in experimentation can increase structural information, we can use the intact mass example given previously. The inability to distinguish between the G1G1 and G0G2 glycoforms is due to the presence of multiple N-linked sites contained on the intact protein. By reducing the mAb, freeing the light and heavy chains to be analyzed individually without the interchain disulfide bonds, the number of N-linked sites per fragment is reduced to one and the elucidation of the contribution of the G1G1 and G0G2 glycoform contributions becomes evident. This level of detail was not possible from the native intact data in Fig. 6.11 and is obtained only after reduction and analysis of the heavy-chain data. Due in part to the prevalence of mAbs as biotherapeutics, a wide range of conditions have been reported which allow either complete reduction in disulfide bonds by employing denaturing conditions [161], or selective reduction in the interchain disulfide bonds in the absence of a denaturing agent [162]. This simple reduction method has been used to confirm mAb structures [163], to characterize the structure of antibody conjugates [164], to locate different modification of the light and heavy chains [165], and to examine carbohydrate structure on the heavy chain [166].

Another common middle-up approach is limited digestion of protein, under native condition, with a protease such as papain, pepsin, or Lys-C. Reduction in these enzymatically generated fragments yields even smaller fragments. Mass analysis of these smaller fragments allows for the elucidation of more detailed structural information than mass measurements of intact proteins alone. Similar to the reduction method, the limited-digestion method has been used to confirm mAb structures [167], to identify post-translational modifications [168], and to characterize the structure of antibody conjugates [169]. Although the middle-down approach provides the opportunity to gain additional regiospecific information, the top-down approaches currently do not, it has yet to be developed into a robust approach for detailed site-specific glycan characterization. Analysis of glycopeptides remains one of the most commonly used approach in obtaining detailed and localized characterization information regarding site connectivity and localized glycan population distributions.

6.3.1.8 Glycopeptide Analysis

Determining the glycosylation site specificity can be problematic in the top-down approach since the approach inherently gives a global, rather than local, view of the glycoprotein being investigated. Advances in fragmentation techniques and data analysis may ultimately allow the top-down approach to be used more routinely for this purpose, but currently the analysis of glycopeptide from a bottom-up approach is more common. This approach typically employs a combination of specific enzymatic proteolysis (usually with trypsin) followed by fractionation of glycopeptides by liquid chromatography or affinity chromatography and ultimately glycopeptide analysis by MS and MS/MS [170–174].

Due to its specificity and robust performance, the most common proteolytic enzyme employed in this type of analysis is trypsin. As the analysis of glycopeptides is typically part of a concurrent protein structural characterization, the same protocols for digestion are generally used and no additional optimization of the digestion protocols is performed to accommodate glycoproteins. Trypsin readily produces highly predictable peptide masses because of its high activity and specificity. In addition, tryptic glycopeptides guarantee a basic residue in every peptide which increases ionization efficiency during MS analysis. A significant drawback with this approach is that glycoproteins exhibit increased resistance to trypsin digestion. In addition, the resulting glycopeptides may often be too large for effective MS/MS analysis. This problem is often complicated by the presence of missed cleavages particularly near the sites of glycosylation. Despite these limitations, in the characterization of therapeutic proteins, where glycoprotein samples are typically relatively pure and available in significant quantity, this approach can give relatively comprehensive data on the location and micro-heterogeneity of glycosylation.

If needed, enrichment of the glycopeptides can be performed, but this introduces its own set of challenges. One enrichment strategy has been developed in which a cleavage of the carbon–carbon group between the diols of saccharide units produces aldehyde groups that can be captured by reaction with hydrazine which is immobilized on a solid support [175, 176]. Capture with immobilized boronic acid [177] and by hydrophilic interaction liquid chromatography (HILIC) have also been reported as well as lectin affinity chromatography using ConA, WGA, or a combination of lectins [172]. However, despite the wide variety of the methods, there is still no generally effective method for glycopeptide enrichment, as no single method is both comprehensive and highly specific [178].

6.3.1.9 Tandem MS of Glycopeptides

In theory, tandem MS can provide peptide and glycan sequence as well as the site of glycosylation. In practice, however, tandem MS analysis of glycopeptides can be problematic and far from routine. Studies of glycopeptide fragmentation reactions have focused almost exclusively on protonated tryptic glycopeptides. The typical glycopeptide fragments correspond to the loss of the fragments from the glycan moiety, while information on the peptide sequence and glycan attachment sites is often harder to obtain. Tandem MS is complicated by the sizes of tryptic peptides, which tend to be larger than the mass range which would allow for comprehensive sequence characterization using CID. This coupled with the labile nature of the glycan-peptide bond makes complete characterization using CID alone difficult. A common strategy is to determine the overall mass of the glycopeptide and perform tandem MS to yield the peptide mass. To sequence the peptide, the glycopeptide is first deglycosylated and evaluated separately in a second tandem MS experiment.

Multiple reviews on glycopeptide analysis by mass spectrometry [170, 179] conclude that the fragmentation behavior of glycopeptide ions under collision-induced dissociation tend to vary with the instrument, the instrumental parameters, the specific peptide composition, the charge carrier, and the charge state [173–182]. Upon instrumental and experimental optimization, detailed information on oligosaccharide structure can be obtained from the CID MS/MS data. Extensive fragmentation of the glycan can be obtained in the absence of peptide backbone fracture. MS/MS data on glycan stoichiometry and connectivity in combination with knowledge about the protein expression system used can then be useful in assignment of glycan structures. Ultimately detailed linkage information may require additional experimental data such as enzymatic glycan sequencing data. Glycopeptides when subjected to CID also yield low molecular weight ions such as m/z 163 (Hex + H), 204 (HexNAc + H), 292 (neuAc + H), and 366 (Hex-HexNAc) that are useful diagnostic peaks for the presence of glycosylation [171, 183]. In this way, glycopeptides can be readily identified by selective ion monitoring with ion trap MS [184] or Q-TOF mass analyzers [185, 186]. Additionally, neutral losses of saccharides such as hexose, n-acetylhexosamine, fucose, NANA can also be used to identify the presence of glycopeptides in mass spectra.

ECD and ETD have been applied to glycopeptides and show great promise as a complement to CID [187, 188]. These methods tend to cleave peptide bonds while leaving the attached glycan unaltered. The usefulness of the fragmentation technique has been demonstrated in a number of ECD and ETD studies on N-linked glycans where the glycosylation sites are predictable from the consensus sequence. The true power of ETD and ECD will become evident as application in the characterization of O-linked glycans becomes more prevalent. Therefore, ETD, ECD, and CID provide complementary information in a thorough characterization of glycopeptides, CID providing data for glycan characterization, and ETD/ECD providing peptide structural information including glycan site connectivity.

6.3.1.10 Free Glycan Analysis

When detailed glycan characterization of the global glycan population is required, it is often most practical to release the glycans from the glycoprotein and analyze the resulting free glycans directly or after derivatization. As native glycans typically do not contain strong chromophores, derivatization is an important consideration when high sensitivity is required. When characterizing a biotherapeutics, it is not uncommon to have significant enough quantities to alleviate the need for derivatization. However, derivatization is often required to detect and characterize glycan structures which may represent only a small fraction of the overall population or if quantification is needed. Many options are available for derivatization with fluorescent tags being the most common [189]. Prior to derivatization, however, glycans must be effectively released from the glycoprotein.

Release of Glycans from Glycoproteins

Different chemistries are used for the release of N- and O-linked glycans. This is due to the fundamental difference in the linkage connectivity. Intact N-linked glycans can be release effectively under relatively mild conditions because there exist readily available enzymes which will cleave the amide linkage between the protein and the glycan. PNGase F cleaves the linkage between the core GlcNAc and the Asn residue of all classes of N-linked glycans, with the exception of N-glycans that contain α(1,3)-linked fucose on the core GlcNAc directly attached to the protein. This type of structure is more likely to be found in plant- and insect-derived glycoproteins than those expressed in mammalian systems [190]. Also available are endoglycosidase D (endo D) which releases all classes of N-linked glycans through cleavage between the two GlcNAc residues within the chitobiose core, and endo H which cleaves at the same location and is selective for oligomannose and hybrid type structures. The ease of enzymatic release for N-linked glycans has made this approach the most commonly used, while release of N-linked glycans can also be performed by chemical methods which are more commonly used for O-linked glycan release.

Unfortunately, an enzyme of comparable activity and general effectiveness as PNGase F is for N-linked glycan release is not known for O-linked glycan release. An enzyme, endo-α-N-acetylgalactosaminidase (O-glycanase), which is specific for cleavage of core 1 O-glycan structures, has been reported [191]. However, given there are eight known O-linked cores and that in cases where the core 1 structure is present, the cores are often extended beyond the O-glycanase specificity, and O-glycanase cannot be used as a general solution for O-linked glycan cleavage. For this reason, chemical methods of O-linked glycan release are generally used. Two commonly used chemical methods are hydrazinolysis [192] and base-induced beta-elimination [193]. Care must be taken to ensure that the chemical means of removal does not alter the glycans being released. The conditions employed in alkaline beta-elimination can cause glycan degradation. An ammonia-based nonreductive beta-elimination has been reported which minimizes the potential for glycan degradation and is compatible with subsequent mass spectroscopic analysis [194]. Reductive beta-elimination with $NaBH_4$ results in the release of glycans and reduction in the resulting free reducing terminus to an alditol. This minimizes the degradation of the glycan but at the same time limits, the number of suitable reagents for derivatization if subsequent tagging and quantitation is desired.

6.3.1.11 Detailed Sequence and Linkage Analysis of Glycans

Complete structural analysis of glycans involves not only monosaccharide and sequence information but also the stereochemistry of each linkage and the level of branching. Regulatory agencies often require such detailed structural analysis of therapeutic glycoproteins (such as monoclonal antibodies) because the glycan structure can affect a biotherapeutics immunogenicity, stability, or pharmacokinetics [195]. Glycan analysis is important during glycoprotein production as a means of mapping process parameters and downstream processing impact on glycan structure [50]. Analysis of the primary structure of oligosaccharide is complicated by the number of parameters that must be determined. These include as follows: (1) the nature of the individual monosaccharides as well as their ring conformation; (2) the absolute stereochemistry of individual residues (D or L); (3) the anomericity of glycosidic bonds (a or b linkages); (4) substitution patterns and branch points; (5) the nature and location of any additional chemical modifications (such as acetylation, methylation, etc.) on a given monosaccharide. Although

Table 6.2 Commonly employed glycosidases for oligosaccharide sequencing

Enzyme	Source	EC number	Specificity
α-D-Sialidase	*Arthrobacter ureafaciens* sialidase	EC 3.2.1.18	Releases α-(2-6/3/8)-linked nonreducing terminal N-acetylneuraminic acid (NANA, Neu5Ac) and N-glycolylneuraminic acids (NGNA, Neu5Gc)
	Streptococcus pneumoniae sialidase	EC 3.2.1.18	Releases α(2-3)-linked nonreducing terminal sialic acids (NANA and NGNA)
β-D-Galactosidase	Bovine testes β-galactosidase	EC 3.2.1.23	Hydrolyzes nonreducing terminal galactose with β(1-3/4) linkages
	S. pneumoniae β-galactosidase	EC 3.2.1.23	Hydrolyzes nonreducing terminal galactose with β(1-3) linkages
α-D-Mannosidase	Helix Pomatio	EC 3.2.1.25	Hydrolysis of terminal, nonreducing β-D-mannose β(1-4) residues in β-D-mannosides
β-N-Acetyl-D-hexosaminidase	β-N-acetylglucosaminidase cloned from *S. pneumoniae* expressed in *Escherichia coli*	EC 3.2.1.30	Will digest β(1-4)-linked GlcNAc to mannose but not a bisecting GlcNAc β(1-4)-linked mannose
α-L-Fucosidase	Almond meal α-fucosidase	EC 3.2.1.51	Releases α(1-3/4)-linked nonreducing terminal fucose residues except core α(1-6) fucose
	Bovine kidney α-fucosidase	EC 3.2.1.51	Releases α(1-2/6) fucose-linked nonreducing terminal fucose residues more efficiently than α(1-3/4)-linked fucose

mass spectrometry can provide information on some of these parameters, by itself it cannot provide a complete characterization of all of the structural details. Although there are reports of using CID product-ion mass spectra to provide information on stereochemistry of individual sugar residues [196], the linkage position [122], and branching structure [123, 124] by evaluating the distinct product-ion patterns specific to oligosaccharide that contain the same monosaccharides linked with a different branching structure, use of these data is unlikely to provide strong enough evidence to make structural and stereochemical assignments with great confidence. Typically additional information is used to support the comprehensive assignment of oligosaccharide structure. Mass spectroscopic data coupled to enzymatic sequencing experiments are one method in which connectivity and stereochemistry is probed further. The principle of oligosaccharide sequencing is to take advantage of the ability of enzymes (endo- and exo-glycosidases) to remove terminal monosaccharides from the nonreducing end of oligosaccharides. The exo- and endo-glycosidases that are used in the structural analysis of oligosaccharides are very specific for the monosaccharide anomericity (α/β) of the glycosidic linkage, and the absolute stereoisomer (D/L) of the glycan. The individual specificities for some of the better defined exo-glycosidases are summarized in Table 6.2.

6.3.1.12 Future Considerations on Glycosylation

The importance of carbohydrates in biology, and therefore protein therapeutics, cannot be underestimated. The impact of glycosylation on mAbs, which have rather modest glycan diversity in comparison to other glycoprotein classes, has been well documented. The distinctive structural properties of oligosaccharides are particularly well suited for generating variability through primary structure (sequence and connectivity) but also in the spatial distribution. The variability introduced through branching and stereochemistry as well as the flexibility afforded within their cyclic structures produces an enormous degree of variability. Given this combination of impact on glycoprotein function and variability in structure, the ability to understand the causal relationships between the two will require technological advances to facilitate deeper structural characterization as well as greater analytical throughput. One of the main challenges of the glycoanalytics field will be the simplification and automation of the analysis to make it accessible to a wider range of researchers. The beginnings of this revolution in analytical development is evident in the development and application of microarray platforms, including lectin and glycan, the development of glycan-binding oligonucleotide ligands (aptamers) [197] and the availability of nanoflow HPLC-chip technologies for routine use [198]. More and more detailed glycan analysis will continue to be pushed for by regulatory agencies as glycosylation has been demonstrated to be a critical attribute of many therapeutic glycoproteins. This coupled with the continued discovery of the importance of glycosylation and its association with health and disease will drive technological advancements both in the short term and for the foreseeable future.

6.4 Variants Involving Cysteine, Cystine, and Disulfide Bridging

Cysteine is one of the rarest and most strongly sequence-conserved amino acid residues in proteins. The thiol (RSH) side chain makes cysteine the most nucleophilic and chemically reactive of all the common amino acids [199–205]. Cysteine thiols (sulfhydryls) in proteins participate in a many different chemical modifications, including oxidation, S-nitrosylation, and thiolation to form cysteinylated, glutathionylated, and other mixed-disulfides. Disulfide bond formation results from the oxidation of pairs of cysteine thiols and is critical for protein folding and maintaining three-dimensional structures. Additionally, disulfide heterogeneity, including formation of "nonnative" disulfide bonds from thiol–disulfide shuffling, mixed-disulfides, and various other transformations are of concern in the production and formulation of protein pharmaceuticals. Therefore, in addition to characterization of cysteine residues and disulfide bridging as part of an initial proof-of-structure, identifying and preventing unwanted cysteine modifications is also an important element of the comprehensive characterization and analytical testing of protein therapeutics.

The chemistry of cysteine disulfide bonding is often described as the chemistry of thiols. However, since the deprotonated thiolate anion (RS^-) form of cysteine is a much stronger nucleophile than the protonated thiol form (RSH), it is actually the chemistry of thiolates that predominates. Hence, the pKa values of cysteine residues are of critical importance to understand oxidation of cysteine to form disulfide links, thiol–disulfide exchange reactions, S-alkylations, metal ion coordination, and all the diverse chemical reactions of cysteine residues (Fig. 6.12). The pKa's of cysteine thiols are generally close to an average value of 8 for solvent-accessible cysteines. The exact thiol acid-ionization constants will vary depending on the local environment for the particular cysteine residue and pKa values can vary by as much as several units for certain buried cysteine residues, for instance.

Although the amino acid, cysteine, has a natural tendency to form cystine disulfide bridges via mild air-oxidation, only less than half of all cysteine residues in the natural proteins are actually present in the cystine disulfide-linked form. The remaining cysteines are involved in a variety of other interactions including reactive cysteine thiols in the active sites of enzymes, coordination to bound metal ions, covalent attachment to prosthetic groups such as the heme moiety in cytochrome c, S-lipolyation in lipoproteins, S-glutathionylated cysteines and other mixed-disulfides [206–211], and S-nitrosylated cysteines (RS-NO) in hemoglobin [212], and other nitrosylated proteins [213–217]. The strong affinity of cysteine thiols for zinc (Zn^{2+}) and other metal ions is exemplified by the metallothioneins which possess the highest known abundance of cysteine residues and bound metal ions in proteins. Mammalian metallothioneins [218–228] consist of a single chain of approximately 60 residues, of which no less than 20 are cysteines, all of which form stable thiolate metal ion clusters in the interior of the protein. Nitric oxide (NO) binding to thiolates is another interesting reversible modification of cysteine

property	serine	cysteine
side chain	alcohol (-**OH**)	thiol (-**SH**)
% frequency in proteins [a]	7.3%	1.8%
% SS-bridged in proteins [b]	--	42%
mean solvent-exposure in proteins [c]	44.2 A^2	13.9 A^2
water/octanol partition (π)[d]	-0.04	+1.54
metal ion binding	no	Zn^{2+}, Cu^{2+}, Fe^{2+} ,...
pK_a of side chain	13 (aprotic)	8
% ionized (thiolate) at pH 7.4	0%	25%

[a] NCBI database. Only Trp (1.3%) is rarer than Cys (1.8%).
[b] N. Nagano, et al., FEBS Lett., 458, 69-71, 1999.
[c] G.D. Rose, et al., Science, 22, 834-838, 1985.
[d] J.-L. Fauchere, V. Pliska, Eur. J. Med. Chem.-Chim. Ther., 18, 369-375, 1983. π scale is for acetyl-AA-amides and covers range from Arg (-1.01) to Trp (+2.25); Gly (0).

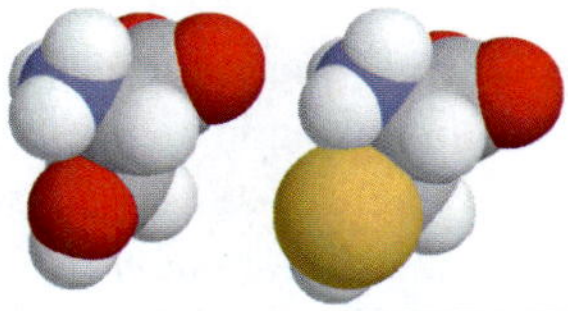

Fig. 6.12 The amino acid, cysteine (*Cys*), is one of the rarest and most strongly sequence-conserved amino acids. The table *above* compares physicochemical properties of cysteine and serine (*Ser*) which differ by a single atom in their side chains, that is, sulfur versus oxygen, respectively. The nucleophilic sulfur atom makes the cysteine side chain the most chemically reactive of all the twenty common amino acids. The amino acid, cystine, is the relatively stable disulfide-bridged form of cysteine. Individual disulfide-linked Cys residues in proteins are traditionally referred to as half-cystine residues although the term, cysteine, is very often used synonymously to refer to both cysteine and half-cystine residues in proteins

residues in proteins that remained unknown and unappreciated until fairly recently. One of the first reported examples was the discovery of the S-nitrosylation of the highly conserved cysteine 93 residue in the beta chain of hemoglobin [212] which led to the astounding realization of a previously unknown biological function in blood-pressure regulation for one of the most extensively characterized of all proteins. Cysteine residues are also key functional elements in certain intramolecular intein-splicing pathways. Clearly, cysteine is known to be a key component of a great many post-translational modifications of proteins in addition to its familiar role in disulfide bridging and the many known functional roles for cysteine residues in the natural proteins are beyond the scope of this brief review paper. While mass spectrometry has become an increasingly important tool for studying S-nitrosylation [213–215, 217], metal ion binding [227, 228], and many other cysteine modifications, this chapter focuses on MS studies of disulfide bridging in protein pharmaceuticals with a special emphasis on the IgG antibodies. Selected thiol modifications, labeling reagents, and degradative pathways of cysteine thiols and disulfides will also be reviewed briefly as well as an interesting interconversion of disulfides and trisulfides in antibody therapeutics via the reversible loss or gain of a sulfur atom.

Cysteine residues in proteins are generally found in either a stable static environment, that is, the disulfide-bridged cysteines buried in the hydrophobic interiors proteins or, alternatively, as dynamic elements for some type of chemically reactive site in an enzyme, for instance. The best characterized functional disulfides are those at the active sites of thiol–disulfide oxidoreductases. These dithiol/disulfides undergo a cycle of oxidation/reduction reactions with the thiols or disulfides in a protein substrate, resulting in net formation, reduction, or isomerization of disulfide bond(s) in the substrate. Most notably, the protein disulfide isomerase (PDI) family of enzymes are the major catalysts of thiol–disulfide exchange reactions in the ER [205]. PDI is a multidomain member of the thioredoxin superfamily and consists of four thioredoxin domains (a–b–b'–a') of which only the a and a' domains are active in catalyzing thiol–disulfide exchange reactions ("disulfide shuffling") by means of a Cys-Gly-His-Cys active-site CXXC motif. Interestingly, one of the PDI active-site cysteines has an unusually low pKa, such that the active thiolate state predominates at physiological pH, whereas the other active-site cysteine is reported to have an unusually high pKa, such that the inactive thiol state predominates at physiological pH.

Disulfide bonds are generally categorized as either structural or functional. Structural bonds stabilize a protein, while catalytic disulfide bonds in the active sites of enzymes catalyze thiol–disulfide interchange reactions in specific substrates. Even though disulfide bridges are generally perceived as primarily static structural elements in cytokines, monoclonal antibodies, and other protein therapeutics, it can be useful to keep an open mind to other more dynamic roles for cystine disulfides and/or cysteine thiols in natural systems and in vitro. Indeed, there is emerging evidence for a third type of disulfide bond that can control protein function [201–203, 205] by triggering a conformational change when it breaks and/or forms, that is, a type of "molecular switch." Furthermore, it is becoming apparent that some disulfide bonds regulate protein function in a nonenzymatic way by triggering changes in the intra- or intermolecular structure of proteins [201–203, 205] and these have been referred to as "allosteric disulfides." This section focuses on some interesting recent mass spectrometry and other structural studies of the interchain bridges in IgG4 and IgG2 antibodies which appear to exhibit dynamic thiol–disulfide exchange allosterism of structural and functional significance in molecular immunology and for protein pharmaceuticals.

6.4.1 Mass Spectrometry for Determining Cystine Disulfide Pairings and Unpaired Cysteine Thiols

Ellman's reagent (DTNB), 5,5′-dithiobis-(2-nitrobenzoic acid), is one of the best known classical methods for assaying cysteine thiols in protein and peptide samples, and this reagent and enhanced versions of it are still very useful. Traditionally, these are spectrophotometric assays which measure the absorbance of the yellow thionitrobenzoate anion (TNB) that is released upon thiol–disulfide exchange of DTNB with the thiols in the sample. In recent times, it has become quite feasible to also employ mass spectrometry to literally count the number of TNB-derivatized cysteine residues (mixed-disulfides) in the protein [229] which are the other products of the DTNB thiol–disulfide exchange chemistry aside from the TNB anion product. Mass spectrometry has thus added a new wrinkle to the venerable Ellman's assay for thiols.

Mass spectrometry has greatly augmented the experimental determination of cystine disulfide pairings and cysteine thiols in proteins [230–236] since the advent of MALDI and ESI in the 1980s and 1990s. Prior to wide availability of the now ubiquitous MALDI and ESI MS instruments, fast-atom bombardment (FAB) desorption/ionization was the only commonly available MS method for many years and this technique was usually limited to peptides smaller than about 6,000–8,000 Da. Of course, the FAB mode of ionization was not nearly as reliable or versatile as either MALDI or ESI. Many new possibilities for characterizing the chemical structures of peptides and proteins have opened up with the arrival in laboratories of these new readily available modes of ionization and new mass analyzers and hybrid MS instruments. One item to be aware from the outset, with regard to mass spectrometry of cysteine and other sulfur compounds, is the similarity in mass and isotopic distributions between a sulfur atom and two oxygen atoms. In some cases, this coincidental similarity in masses can lead to ambiguity in assigning accurate mass differences and interpreting results for cysteine residues in peptides and proteins. In most cases, this close similarity in mass between a sulfur atom and two oxygen atoms will, in all likelihood, not come into play nor cause any complications. Fortunately, the very high mass accuracy of the newer generation of FT-ICR and Orbitrap-type instruments can be called upon to resolve this ambiguity.

Much of the literature on mass spectrometric studies of cysteine and cystine disulfide bridging in protein therapeutics has been concerned with structural characterization and proof-of-structure determinations [234–236]. Recombinant cDNA-derived proteins expressed in *E. coli* microbial fermentation and other prokaryotic systems are typically recovered by lysis of the host cells to release reduced and denatured polypeptide chains which then undergo a "refolding step" prior to further downstream chromatographic purification steps. The refolding step generally consists of a simultaneous protein folding and mild oxidation of the cysteine residues to form a full set of correctly paired cystine disulfide bridges and allow the protein to assume its proper folded active structure. The determination of the disulfide-bridge pairings in the final product is of considerable importance because of this in vitro refolding step in the manufacturing process. On the other hand, recombinant monoclonal antibodies and other biologics manufactured by mammalian cell culture, that is, eukaryotic systems, are generally secreted from the host cells as fully folded, assembled, glycosylated, and correctly disulfide-bridged forms of the mature protein. However, it is generally still required that the disulfide pairings or other post-translational modifications of cysteines be confirmed as part of the Investigational New Drug (IND)-stage proof-of-structure for secreted proteins and other biologics manufactured in eukaryotic production systems such as CHO cell lines.

Ideally, one seeks to cleave between all the half-cystine residues with trypsin or some other reagent so as to generate a full set of disulfide-bridged pairs of peptides corresponding to the full complement of disulfide pairings in the protein sample [237–248]. Conventional methods for disulfide mapping include enzymatic digestion of nonreduced proteins by pepsin at pH 2–4 [239]. However, pepsin generates nonspecific cleavages, which increases the complexity of the digests. Alternatively, enzymes such as trypsin or Asp-N can provide specific cleavages; however, these enzymes prefer basic pH, which may lead to disulfide bond scrambling. In this case, it is advisable to perform an initial mild S-alkylation with N-ethylmaleimide (NEM) or similar reagent to label and block any small amounts of buried cysteine thiols prior to unfolding and fragmenting the sample in order to prevent unwanted method-induced artifacts arising from thiol–disulfide scrambling and/or air-oxidation of thiols.

6.4.2 Symmetrical Sulfur–Sulfur Cleavages of Cystine Residues and Nonsymmetric Carbon–Carbon Cleavages

MALDI of disulfide-bridged peptides [249–261] is known to include some degree of method-induced rupture of the sulfur–sulfur bond [249] which has been referred to as "prompt fragmentation." This type of S–S bond cleavage is quite useful for studying disulfide bridging and, more recently, several examples of gas-phase S–S bond cleavages have also been observed for ESI–MS by photolysis and other mechanisms. Katta et al. [252] and others have also reported even more useful cleavages at C–S bonds by MSMS that can be attained by MALDI with PSD [250, 251, 260].

To locate the exact linkage positions of disulfides, ESI-MSMS in the negative mode has been suggested [262–266]. Recently, ECD and ETD have been reported to facilitate the identification of disulfide linkages [267–274]. Traditional peptide mapping approaches rely on data generated with and without solution-phase reduction/alkylation whereas ETD can be used to create preferential gas-phase cleavage of the sulfur–sulfur bonds with little to no peptide backbone fragmentations and thereby lessen the need for solution-phase sample preparation with DTT and other reducing agents. Subsequent MSMS can then be employed for identification of individual peptides [267].

A new approach has been presented for the determination of the disulfide bond connectivity in proteins using negative ionization ESI mass spectrometry of nonreduced enzymatic digests [263]. Negative ion mode LC–MS peptide mapping was used to determine the disulfide structure of a human IgG2 antibody containing 18 unique cysteine residues linked via 11 unique disulfide bonds. The efficiency of the gas-phase dissociation of disulfide-linked peptides using negative ESI was evaluated for an ion trap mass spectrometer and an orthogonal acceleration TOF mass spectrometer and both techniques provided efficient in-source CID for the identification of the disulfide-linked peptides of the antibody. Seven of the 11 unique disulfide linkages have been determined in this way, including the H-to-L linkage. Only the precise disulfide connectivity of the H-to-H hinge peptide, CCVECPPCPAPPVAGPSVFLFPPKPK, could not be determined from these LC–MS peptide mapping data alone although this dimeric peptide was found in good yield as single species in the peptide map and an accurate MW was obtained [263].

A tandem MS approach employing ETD and CID was used to directly identify 5 out of a total of 9 disulfide linkages in an insulin-like growth factor binding protein (IGFBP-5), that is, Cys47-Cys60, Cys54-Cys80, Cys172-Cys199, Cys210-Cys221, and Cys223-Cys243 [267]. The combination of these MS results with ab initio molecular modeling was then used to predict the most likely arrangement of the remaining four disulfide pairs: Cys7-Cys33, Cys10-Cys35, Cys18-Cys36, and Cys25-Cys39. Taken together, these studies show that IGFBP-5 is composed of independent N- and C-terminal domains, containing six and three disulfide bonds, respectively [267]. Electron detachment dissociation (EDD) and IRMPD of peptide anions containing disulfide linkages result in preferential cleavage of S–S and C–S bonds and, therefore, both techniques can be used for probing disulfide bonds in peptide anions [248]. Factors such as precursor ion charge state and m/z value, peptide mass, and protease selection that may influence the dissociation outcome in ECD were investigated, aiming to improve peptide sequence coverage. It was shown that doubly protonated peptides did not fragment efficiently in ECD and that precursor ion m/z value is the main factor determining a successful ECD outcome. Highly charged precursor ions at m/z $<\sim$960 fragmented efficiently in ECD and yielded high peptide sequence coverage.

Often the limiting factor in characterizing proteins by MS methods is not the actual physical sample preparation or data acquisition. The real bottle-neck is processing and interpreting very large volumes of raw data generated by automated instruments and high-speed electronic computers [275–277]. And the fact that the MS and MSMS spectra of disulfide-bridged peptides are rather more complex to interpret than for the corresponding linear unbridged peptides certainly contributes to the challenge. Most automated analysis algorithms function based on the assumption that the preponderance of product ions observed during the dissociation of disulfide-bonded peptides result from the cleavage of just one peptide bond, and this assumption was tested recently for product ions generated when several disulfide-bonded peptides were subjected to CID on a QTOF instrument [276]. It was found that one of the most common types of product ions resulted from two peptide bond cleavages and for several of the disulfide-bonded peptides, the number of double cleavage product ions outnumbered those with single cleavages. The influence of charge state and precursor ion size was investigated to determine whether those parameters dictated the amount of double cleavage product ions formed. It was found that no strong correlation existed between the charge state or peptide size and the portion of product ions assigned as double cleavages. These ions could account for many of the product ions detected in CID data of disulfide-bonded peptides. The utility of double cleavage product ions was demonstrated for peptides connected by multiple cystine disulfide bridges, and this approach was able to fully characterize the bonding pattern of each half-cystine where typical single b/y cleavage products could not [276].

6.4.3 Labeling of Cysteine Thiols

The selective labeling or tagging of cysteine thiols [278–290] is an integral part of most strategies for characterizing disulfide pairings in proteins by mass spectrometry and peptide mapping. The other main component, generally, is the choice of proteolytic enzyme or other means of cleaving peptide bonds between individual half-cystine residues in nonreduced protein samples. Single labeling of thiols can be employed for either nonreduced or reduced samples or a double-labeling (differential labeling) scheme can be used with, for instance, an initial labeling of thiols for nonreduced sample, followed by a different label for the reduced protein sample or proteolytic digest sample. The reagents can range from simple protecting groups such as NEM and iodoacetamide (IAM) to specialized spectroscopic labels, affinity-handles, isotopic labels, and so on. The traditional Ellman's reagent (DTNB) described in the previous section can be used to good effect for covalently labeling cysteines for analysis by MS and LC–MS methods in favorable cases [278–281]. However, because the thiol–disulfide chemistry involved in derivatization with DTNB yields TNB mixed-disulfide products, the Ellman's reagent is usually not a good choice when highly stable derivatives are desired. The most widely used thiol-specific conjugation chemistries are either alkyl halides for direct S-alkylation, such as IAM and similar iodoacetyl-type reagents, or via Michael addition of the thiolate sulfur across the C=C double bonds in NEM and other reagents with maleimide and other vinyl functions. Both of these S-alkylation chemistries have their relative merits. The maleimide-type reagents are highly selective for thiols whereas some care should be taken with IAM and other alkyl halides so as not to "over-derivatize" the sample. In other words, if too high an excess of IAM is employed and/or the reaction is allowed to proceed for too long, complications will ensue when the less nucleophilic thioether of methionine begins to undergo a slow alkylation once all the more reactive cysteine thiols have been consumed. In the case of the maleimide reagents, things are complicated by the fact that the sulfur atom will generally add to both sides of the double-bond and generate two diastereomeric products which may or may not be resolved by HPLC into a "split peak" containing isobaric molecular ions. And to further complicate things, the succinimide ring in the maleimido ring can undergo hydrolysis (+18 Da) under certain conditions to yield two ring-opened carboxylic acid forms of the linker. The many diverse cysteine-specific tagging reagents available for proteomics and protein characterization have been extensively reviewed in the recent literature [281]; therefore, only selected examples will be discussed in this section.

Fluorescence-based tagging in protein characterization and proteomics is useful in tracking and quantifying target proteins during sample preparation or chromatographic processes. For example, the utility of thiol-reactive 5-iodoacetamido-fluorescein to target cysteinyl residues on the intact proteins has been demonstrated for ovalbumin, bovine serum albumin, and proteins in MCF-7 cells [282]. After trypsin digestion, samples were analyzed by nano-LC-ESI-Q-TOF or MALDI-TOF. The resulting MS spectra of tryptic fragments were similar to those of unlabeled or iodoacetamide-derivatized proteins, and the MSMS fragmentation of all fluorescein-tagged peptides was readily interpretable with intact label. Thus, fluorescein-derivatized proteins can be identified by automatic mass mapping or peptide sequencing with high confidence. It is notable that, in MSMS mode, a strong reporter ion (m/z 422) containing the fluorescein moiety was readily detected and was believed to derive from the immonium fragment of fluorescein-labeled cysteine residues (m/z 463), under CID conditions. Using a precursor scan of the reporter ion, a cysteinyl protein, ovomucoid, was identified to be present in the ovalbumin sample as an impurity. The fluorescein derivatives were further shown to have high affinities toward metal-chelating materials that have iminodiacetic acid functional groups either with or without the presence of bound metal ions. When combined with stable isotope labeling, fluorescein-tagged peptides could be selectively enriched, identified, and quantified [282]. Another useful reagent for labeling unpaired cysteine thiols is N-(Iodoacetyl)-N′-(5-sulfo-1-naphthyl) ethylenediamine (IA-EDANS) [283, 284]. EDANS derivatives can be detected by either UV absorbance at 340 nm or fluorescence (ex/em 360/500 nm) and are amenable to both MALDI and ESI–MS detection for intact proteins and LC–MS peptide mapping of proteins [284].

A low percentage of buried free thiol is a common feature of recombinant monoclonal antibodies although, in theory, all cysteine residues should be involved in disulfide bonds (Fig. 6.13). The 2002 report of Zhang and Czupryn [285] describes detection and quantification of free sulfhydryl in recombinant mAbs produced in CHO cells using fluorescent labeling with N-(1-pyrenyl) maleimide (NPM). Purified mAbs appear to be homogeneous under native conditions with approximately 0.02 mol of free sulfhydryl per mole of protein. Upon denaturation, minor species related to the mAbs are observed on SDS-PAGE, and the free sulfhydryl level is determined to be approximately 0.1 mol/mol of protein. These results suggest that a small portion of these recombinant mAbs lack in intermolecular disulfide bonds but remain noncovalently associated under native conditions [285]. The presence of free sulfhydryl groups in five recombinant monoclonal antibodies and their locations were investigated by labeling with 5-idoacetamidofluorescein (5-IAF), followed by reduction in disulfide bonds and alkylation with iodoacetic acid [286]. This double-labeling procedure allows differentiation of free cysteine residues from cysteine residues that are involved in disulfide bonding and a sensitive fluorescence detection of peptides with free sulfhydryl groups (thiols). The locations of the free sulfhydryl groups were determined using LC–MS peptide mapping for five different antibodies. Levels of free thiol varied for different antibodies, and residual thiol was detected commonly for intramolecular disulfides buried in the constant domains (Fig. 6.13). Furthermore, unpaired cysteine residues in the variable domains differed for particular antibodies and were rarely found for cysteines involved in interchain disulfide bonds [286].

A differential isotopic labeling method was developed to determine the percentage of free thiol at each cysteine residue of four recombinant monoclonal

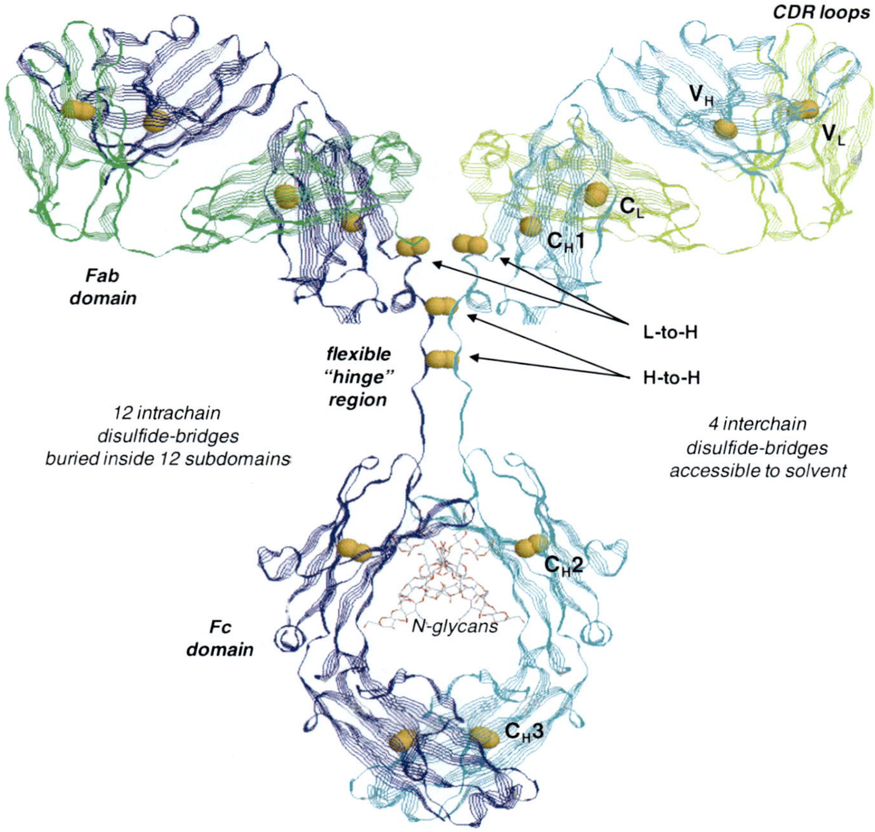

Fig. 6.13 The IgG1 antibody consists of two copies each of the light (*L*) and heavy (*H*) chains linked by 16 cystine disulfide-bridge pairings (*yellow spheres*) which consist of 12 intrachain bridges buried in immunoglobulin (*Ig*) folds, 2 partially buried interchain bridges linking L-to-H, and 2 solvent-accessible H-to-H interchain bridges. The 12 Ig subdomains consist of the H and L variable domains, (V_H and V_L) and constant domains (C_H and C_L) in the Fab antigen-binding domains and C_H2 and C_H3 subdomains which form the Fc domain. The six complementarity-determining region (*CDR*) hypervariable loops are located at the tips of the Fab antigen-binding arms as indicated

antibodies [287]. Free sulfhydryl was first alkylated with ^{12}C iodoacetic acid for a nonreduced antibody sample. Free sulfhydryl, resulting from the reduction in disulfide bonds for a DTT-reduced sample, was then alkylated with ^{13}C iodoacetic acid. Cysteine-containing peptides that were modified by ^{13}C iodoacetic acid showed a molecular weight that was 2 Da higher than the same peptide that was modified by ^{12}C iodoacetic acid. Peptides containing the same cysteine residues that had been modified with both alkylating reagents, coeluted on reversed-phase chromatography. Analysis by mass spectrometry resulted in two partially overlapped m/z series for each cysteine-containing peptide, corresponding to modification by iodoacetic acid with ^{12}C or ^{13}C. The percentage of free thiol was then calculated using the two m/z series at each cysteine site. A low percentage of free sulfhydryl was detected at every cysteine residue in the four antibodies studied. Although different antibodies contained different levels of free sulfhydryl, similar distributions of free thiol in the domain structures was observed in the four antibodies [287].

Traditionally, because of the high reactivity of thiol groups, alkylating reagents such as iodoacetic acid or 4-vinyl pyridine have been used to protect cysteine residues and prevent oxidation or thiol–disulfide exchange. However, these reagents require basic pH (~pH 8.0) for optimal reactivity and this can lead to unwanted disulfide mispairing (disulfide scrambling). Therefore, a reagent that would alkylate free sulfhydryl groups under acidic conditions would be desirable to obtain data for assigning free cysteine residues and maintaining the native disulfide linkages. An alternative way to protect thiol groups is cyanylation, which has been known and used for many years [237, 238]. New cyanylation reagents that show promise, such as 1-cyano-4-dimethylaminopyridinium tetrafluoroborate (CDAP), have been developed [288]. CDAP selectively cyanylates sulfhydryl groups of unpaired cysteine residues under acidic conditions resulting in a shift of +25 Da in molecular mass. Subsequent basic cleavage of the protein chain and analysis by RP-HPLC with online MS provides a simple method to identify the locations of unpaired as well as disulfide-paired cysteine residues. CDAP is reactive in a pH range of 2.0–8.0 with an optimal labeling efficiency at pH 5.0. CDAP labeling of cysteines under conditions of acidic pH diminishes thiol–disulfide exchange by reducing the concentration of reactive thiolate anion. The authors successfully employed CDAP as a thiol-directed probe to identify pH-dependent structural differences in recombinant methionyl G-CSF due to changes in the accessibility of the free thiol group of Cys-17. Cyanylation of Fc-OPG resulted in the characterization of an isoform where two unpaired cysteine residues were identified. In both G-CSF and Fc-OPG, the site specificity and acid compatibility of CDAP, as well as subsequent cleavages of the protein chains, afforded an accurate identification of the unpaired cysteines and served as a probe of their accessibility to solvent [288].

Monobromobimane (MBB) is a lipophilic reagent that selectively modifies free cysteine residues in proteins to yield highly fluorescent derivatives. A procedure has been described for the detection and relative quantitation of MBB-labeled cysteines using fluorescence and mass spectrometric analyses, which allow determination of free cysteine content and unambiguous identification of MBB-modified cysteine residues [289]. This approach was applied to the analysis of the redox-sensitive cysteine residues of a large membrane protein, the sarcoplasmic reticulum Ca^{2+} release channel with a molecular mass of 2.2 million Da. Labeling was performed for the active channel complex under physiological conditions followed by enzymatic digestion, and the resulting peptides were separated by RP-HPLC with fluorescence detection and identified by MALDI-TOF MS. Under MALDI conditions, partial photolytic fragmentation of the MBB–peptide bond occurred, thus allowing convenient screening for the MBB-modified peptides in the MS spectrum by detection of the specific mass increment of 190.07 Da for MBB-modified cysteine residues. Modification of the peptides was further confirmed by tandem mass spectrometric analysis, utilizing sequencing information and the presence of the specific immonium ion for the MBB-modified cysteine residues at m/z 266.6. Quantitative information was obtained by comparison of both fluorescence and MS signal intensities of MBB-modified peptides. Combination of fluorescence with MS detection and analysis of MBB-labeled peptides supported by a customized software program provides a convenient method for identifying and quantifying redox-sensitive cysteines in membrane proteins of native biological systems. Identification of one redox-sensitive cysteine in the native membrane-bound sarcoplasmic reticulum Ca^{2+} release channel was described [289].

Many proteomics studies make use of specialized fluorescent dyes to specifically stain the proteins either by adsorption after gel electrophoresis (in-gel staining) or by covalent coupling prior to gel electrophoresis (in-solution staining). Multiplex analysis of protein samples using maleimide-activated cyanine-based (Cy3 and Cy5) and rhodamine-based dyes (Dy505, Dy535, and Dy635) to permanently label all thiol groups of cysteine-containing proteins has been described [290]. The detection limits in SDS-PAGE were about 10 ng per band and even 2 ng for BSA due to its high content of cysteine residues.

6.4.4 Edman Chemistry as a Complementary Technique to LC–MS

Edman degradation chemistry is a useful microchemical technique for the N-terminal sequencing of picomolar amounts of samples when used in combination with LC–MS peptide mapping methods [291–297]. In addition to the commonly used approach of isolating peptides from an LC–MS peptide map for Edman microsequencing, it is also often feasible to do the converse, that is, recover samples after performing multiple Edman degradation cycles and then run MALDI-TOF MS or ESI–MS on the recovered sample. This tactic was nicely illustrated as part of a strategy for the elucidation of the disulfide structure of alfimeprase, a recombinant analog of fibrolase [293]. This protein contains closely spaced cysteine residues, 156 and 158, and the pairings, Cys-116/196, Cys-156/180, and Cys-158/163, were experimentally determined by a combination of peptide mapping, Edman degradation, and mass spectrometry. To determine the disulfide linkages among four cysteine residues within one proteolytic fragment in the nonreduced endoproteinase Asp-N digest, the peptide was first subjected to a specific number of cycles of Edman degradation, followed by MS analysis. Edman degradation was performed on an ABI Procise protein sequencer (PE Biosystems) for sample immobilized by adsorption to a PVDF membrane. After the specified number of Edman degradation cycles, the sample was eluted from the PVDF membrane with 0.1 % TFA in 50 % acetonitrile and analyzed on a Finnigan MAT LCQ quadrupole ion-trap mass spectrometer using a custom nanoelectrospray interface [293].

The formation of di-PTH-cysteine is a useful aspect to understand and be aware of for the Edman degradation of disulfide-bridged peptides. Edman degradation can be used to assign pairs of cysteine residues such that disulfide-bonded cysteine residues (i.e., Cys_X–Cys_Y) remain linked during the sequencing cycle. During the Edman derivatization with phenylisothiocyanate (PTC), cleavage, and cyclization to the phenylthiohydantoin (PTH) derivative, half-cystine residues (Cys_X) remain covalently linked to Cys_Y until the Edman cycle in which the latter cysteine is reacted. At that cycle, the disulfide-bonded cysteines are derivatized and released as di-PTH-cysteine. In favorable cases, multiple disulfide linkages between two or more peptides can be assigned using this strategy. This approach was used very nicely to dissect the "cystine-knot" and assign the Cys parings for a vascular endothelial growth factor (VEGF) [294]. Preparations of recombinant human vascular endothelial growth factor (VEGF165) expressed in CHO cells and *E. coli* were compared using a variety of analytical methods. These methods included determinations of the disulfide linkages for the eight cysteine residues in the carboxyl-terminal heparin-binding domain, which were assigned by amino-terminal sequencing of peptide fragments isolated from tryptic digests of each native molecule. The following closely spaced disulfide-knot pairings were identified for both CHO- and *E. coli*-derived VEGF165: Cys-117 and Cys-135, Cys-120 and Cys-137, Cys-139 and Cys-158, plus Cys-146 and Cys-160.

Gray [292] reported a partial reduction technique using TCEP under acidic conditions, followed by alkylation with a high concentration of iodoacetate or IAM at alkaline conditions. The conventional technique for disulfide determination is by proteolytic digestion and sequence analysis of the resulting peptides. The partial reduction technique releases some compactness of structure in the cystine-knot, therefore allowing disulfide analysis to reach enclosed regions where traditional methods fail. Although the low pH used in TCEP reduction prevented disulfide rearrangement, the higher pH required for IAM alkylation often resulted in disulfide exchange reactions during S-alkylation. More recently, alternative alkylation methods using acidic conditions (pH 5–6) have been developed which employ partial reduction with TCEP and S-alkylation with NEM. The disulfide bonds in the cystine knot structure of Md65 agouti-related protein (Md65-AGRP) were elucidated [297] by partial reduction with tris (2-carboxyethyl) phosphine (TCEP) under acidic conditions, followed by alkylation with NEM and Edman N-terminal sequencing. The procedure generated several isoforms with varying degrees of NEM alkylation. The multiple forms of Md65-AGRP generated by partial reduction and NEM modification were then completely reduced and carboxymethylated to identify unreactive disulfide bonds. Differentially labeled Md65-AGRP were then directly sequenced and analyzed by MALDI mass spectrometry [297].

6.4.5 *Cystine-Knot and Other Highly Disulfide-Bridged Proteins*

The "cystine knot" protein, VEGF, was given special mention in the previous section on Edman degradation as an example of a highly disulfide cross-linked protein which presented special challenges to common methods of disulfide mapping [294].

Similarly, the literature on disulfide bridging in conotoxins [298–304], a highly diverse superfamily of compact disulfide-knot peptide neurotoxins from cone snails, is well worth following for anyone seeking new strategies and tactics for disulfide mapping of highly bridged proteins and peptides. The critical and often quite difficult step in structure elucidation for the conotoxins is the determination of correct disulfide pairing between multiple closely spaced cysteine residues. A direct mass spectrometric analytical methodology for the determination of disulfide pairing of these highly bridged cystine-knot conotoxin peptides has been described by Gupta et al. [298]. CID of protonated ions of highly disulfide-bridged conotoxin peptides yielded fragmentation preferentially along the peptide backbone, with occasional fragmentation either by C–S bond cleavages or by S–S bond cleavages. Further MS^n fragmentation of the initial set of product ions yielded third- and fourth-generation fragment ions corresponding to various disulfide-bonded structures. This approach was illustrated by establishing cysteine pairing patterns in five conotoxins containing two disulfide bonds and was extended to the Conus araneosus peptides Ar1446 and Ar1430, two 14-residue sequences containing 3 disulfide bonds. The mass spectrometers used were a Bruker Ultraflex TOF/TOF MALDI MS and HCT-Ultra ETDII ion trap mass spectrometer. For all the intact fragmentation and MS^n experiments, purified HPLC fractions were injected directly into the ESI ion trap mass spectrometer using a syringe pump at a flow rate of 120 μL/h, and fragmentations were carried out inside the ion trap through the collision of helium gas with the protonated molecular ions of interest [298].

The examples of two and three disulfide-bonded conotoxins presented above establish that mass spectral fragmentation of intact disulfides can provide a means of establishing cysteine pairings. The question arose as to whether the generation of reactive thiol species can lead to disulfide scrambling in the gas phase. If such scrambling occurs, product ions must be observed which are diagnostic of the presence of dehydroalanine, cysteine persulfide, cysteine-thioaldehyde, and cysteine at specific positions along the sequence. These diagnostic product ions were not observed and so the authors concluded that thiol–disulfide interchange processes were not favored in the gas phase during MS experiments, although such reactions can occur readily in aqueous solution at alkaline pH [298].

De novo mass spectrometric sequencing using MALDI-MS can be used to determine disulfide connectivity in peptides, as in the case of two Conus peptides Vi1359 and Vi1361, from the vermivorous cone snail Conus virgo, found off the southern Indian coast [299]. The peptides, whose masses differ only by 2 Da, possess two disulfide bonds and an amidated C-terminus. Simple chemical modifications and enzymatic cleavage coupled with MALDI mass spectrometric analysis aided in establishing the sequences of Vi1359, ZCCITIPECCRI-NH_2, and Vi1361, ZCCPTMPECCRI-NH_2, which differ only at residues 4 and 6 (Z = pyroglutamic acid). The presence of the pyroglutamyl residue at the N-terminus was unambiguously identified by chemical hydrolysis of the cyclic amide, followed by esterification. The presence of isoleucine residues in both the peptides was confirmed from high-energy CID studies, using the observation of diagnostic $w(n)$ and $d(n)$

ions. Differential cysteine labeling, in conjunction with MALDI-MSMS, permitted establishment of disulfide connectivity in both peptides as Cys2-Cys9 and Cys3-Cys10. The cysteine pattern clearly reveals that the peptides belong to the class of T-superfamily conotoxins, in particular the T-1 superfamily [299].

6.4.6 Interchain Disulfide Bridging in IgG1 Antibodies and Isotypes

Modern molecular immunology and monoclonal antibody technology began to emerge in the 1960s with the pioneering work of Edelman and coworkers [305–307] on the first characterization of the full chemical structure of a human gamma-globulin (IgG), including all 16 disulfide pairings (Fig. 6.13). This comprehensive and accurate structural characterization of a protein molecule of unprecedented size and complexity was a monumental undertaking at the time. This work culminated in a detailed chemical structure of a human IgG1 and many important insights into their genetics and immunology. The entire project required years of painstaking sample preparation by classical column chromatography, manual dansyl-Edman sequencing, diagonal electrophoresis, and so on. It is even more impressive that this complete IgG structure was elucidated without the benefit of mass spectrometry or other powerful instruments and research tools routinely available today.

The disulfide bond structure is critical for antibody stability, antigen binding, and Fc effector functions [308–314]. IgG antibodies of various classes, IgG1, IgG2, IgG3, and IgG4, differ in the number and positions of the interchain disulfide bonds. The prototypical IgG1 molecule depicted in Fig. 6.13 is the most abundant subclass or isotype of the natural circulating blood serum IgG molecules as well as the most prevalent subclass or isotype of the IgG molecules in biomedical research, pharmaceutical development, and clinical diagnostics. However, the more rare and specialized IgG2 and IgG4 subclasses also occupy niches in the biopharmaceutical industry as immunotherapy drugs for selected clinical indications.

For *kappa* light chains, the light-chain cysteine residue that links the light chain to the heavy chain in the IgG1 molecule (L-to-H) is the C-terminal residue. For *lambda* light chains, on the other hand, the cysteine residue is the penultimate residue followed by a serine residue at the C-terminus. The effect of this added C-terminal serine residue for the *lambda* light chains on the susceptibility of disulfide bonds to reduction was investigated by reduction, differential alkylation using iodoacetic acid with either natural isotopes or enriched with carbon-13, and mass spectrometry analysis [315]. The effect of the serine residue on disulfide bond susceptibility was compared using three antibodies with differences only in the light chain last amino acid, which was either a serine residue, an alanine residue, or truncated. The results demonstrated that the presence of the amino acid (serine or alanine) increased the susceptibility to reduction in the interlight- to interheavy-chain disulfide bonds (L-to-H). On the other hand, susceptibility of the pair of IgG1 interheavy-chain disulfide bonds (H-to-H) was not changed significantly.

Several differences from the classical disulfide bond structures have been reported in the literature recently. Furthermore, low levels of free thiol are common and may even have a similar distribution pattern in different IgG1 molecules [286–288, 316–318]. For example, it has been well documented that the intrachain and interchain disulfide bonds in the hinge region are in equilibrium in IgG4 antibodies and IgG2 antibodies which exist as ensembles of several isoforms with different in disulfide pairings in the hinge region.

6.4.7 Interchain Disulfides in IgG4 Antibodies

IgG4 antibodies differ functionally from other IgG subclasses in their antiinflammatory activity, which includes a poor ability to induce complement and cell activation because of low affinity for C1q complement and Fc receptors [319–328]. Consequently, IgG4 has become the preferred subclass for immunotherapy in which recruitment of host effector functions is undesirable and thus considered as a "nondepleting" antibody for blocking a target antigen with little to no unwanted ADCC [320]. Another attribute of blood-derived IgG4 is its inability to cross-link identical antigens, which is referred to as "functional monovalency." IgG4 antibodies are considered to be less efficient in inducing inflammatory responses and may even inhibit the inflammatory effects of other antibodies. Prolonged stimulation with high doses of soluble protein antigen preferentially induces IgG4 antibodies. Human IgG4 antibody is unable to precipitate purified antigens as a "neutralizing antibody." The inability to cross-link and precipitate antigen is not due to a difference in intrinsic antigen-binding affinity but rather is caused by the inability of functionally monovalent IgG4 antibodies to cross-link two antigens. IgG4, often induced by chronic antigen stimulation, then may interfere with immune complex formation by other antibody isotypes and may dampen inflammatory reactions. In specific immunotherapy with allergen in allergic rhinitis, for example, increases in allergen-specific IgG4 levels indeed correlate with clinical responses. In contrast, bispecific cross-linking of nonidentical antigens has been observed under certain conditions and it was postulated that this apparent bispecificity might be explained by the exchange of half-molecules between distinct IgG4 molecules [321, 329]. The presence of serine in the canonical pair of H-to-H disulfides in the hinge region, CPXCP, explains the unique property of IgG4 molecules to form LH half-structures from the traditional LHHL configuration. Recent reports have shown that this property allows a dynamic exchange of the hinge disulfides in vivo resulting in bispecific IgG4 antibodies. This incomplete formation of interheavy-chain disulfide bonds (H-to-H) might make IgG4 antibodies susceptible to exchange of half-molecules. Exchange in a pool of polyclonal IgG4 would result in bispecific antibodies that behave as monovalent antibodies toward a single antigen. In contrast, monoclonal chimeric IgG4 antibodies would remain bivalent.

Bloom et al. [319] analyzed denatured samples of an IgG4 with 4-vinylpyridine labeling of free cysteines with and without 50 mM DTT (reduced vs. nonreduced) and compared them by LC–MS tryptic peptide mapping. In the nonreduced map of the CDP571 IgG4, a large peak at 118 min was found to contain a peptide with a measured mass of 2,827.5 Da. This mass matched that of the heavy-chain hinge-region peptide consisting of residues 223–249, YGPPCPSCPAPEFLGGPSVFLFPPKPK, with an intrachain disulfide bond having a calculated mass of 2,827.4 Da. In the 50 mM DTT map, the 118 min peak was not apparent and a new large peak was observed at 109 min that contained a peptide with a measured mass of 3,039.6. This mass matched that of the same 223–249 hinge peptide with 4-vinylpyridine modification of the two cysteines and calculated mass of 3,040.5 Da. The heights of the 118 and 109 min peaks in the two maps were nearly identical. These peptide mapping studies of CDP571 identified the factors preventing interheavy-chain disulfide bond formation between IgG4 half-molecules, that is, the two cysteines in the IgG4 and IgGl core hinge regions (CPSCP and CPPCP, respectively) are capable of forming an intrachain disulfide bond. Conformational modeling studies on cyclic disulfide-bonded CPSCP and CPPCP peptides suggested that the serine in the core hinge region of IgG4 allows more hinge-region flexibility than the proline of IgGl and thus more readily permits formation of a stable intrachain disulfide bond [319].

Since it was first suggested that IgG4 antibodies are bispecific as a result of half-molecule exchange, the role of the hinge cysteines has remained enigmatic [316, 319, 322, 324]. Although it was recognized that a noncovalent isomer of IgG4 exists without disulfide bonds between the heavy chains, its role in the exchange process could not be determined. More recently, it has been shown that IgG4 hinge disulfide isomerization results in Fab-arm exchange, with dissociation of the CH3 domains as the ultimate rate-determining step [325, 326, 329]. Under mild reducing conditions, the intrachain isomer can form dynamically from the interchain form, and this equilibrium determines the overall rate. Fab-arm exchange appears to be much slower in vivo compared to the rate of exchange observed for fully reduced IgG4. This implies that in vivo the process is under control of redox conditions. Blood levels of GSH seem too low to account for the observed exchange reaction. Thus, actual rates of Fab-arm exchange may vary considerably from site to site depending on local redox potentials and the presence of thiols to initiate the thiol–disulfide exchange chemistry.

To conclusively demonstrate that Fab-arm exchange is indeed the result of an intermolecular exchange reaction, intact antibody samples were analyzed using SEC and ESI-TOF MS [327, 330]. Samples containing 200 μg/mL of each antibody were deglycosylated overnight with peptidyl N-glycosidase F and then desalted with a BEH C8, 1.7 μm, 2.1 × 50 mm column at 60 °C. ESI-TOF MS was acquired online by a microTOF™ mass spectrometer operating in the positive ion mode. Plasma from mice injected with an IgG4 antibody mixture was analyzed and bispecific antibodies eluted at the expected position for monomeric IgG, which ruled out the possibility that the observed reactivity was due to aggregation. In a second experiment, mixtures of anti-CD20 IgG4 and anti-EGFR IgG4 were prepared in the absence or presence of reduced glutathione (GSH), followed by ESI-TOF MS. The molecular masses of anti-CD20 IgG4 (145.5 kDa) and anti-EGFR IgG4 (145.9 kDa) remained unchanged in the absence of GSH. In the presence of GSH, however, a new peak appeared with an intermediate mass (145.7 kDa) corresponded to the expected mass of the bispecific antibody against EGFR/CD20. Moreover, from the peak heights of the MS spectra, it was estimated that the bispecific antibody represented 50% of the total antibody mass in the mixture, which indicated a stochastic exchange [327, 330].

Dynamic Fab-arm exchange therefore represents a novel type of post-translational modification, which serves as an additional mechanism for generating anti-inflammatory activity. The mechanism by which IgG4 Fab-arm exchange occurs in vivo likely requires the reducing environment in blood or at cell surfaces to facilitate the breaking of interheavy-chain disulfide bonds located in the hinge region. Indeed, the addition of reducing compounds, such as GSH, to purified IgG4 alone was sufficient to induce in vitro Fab-arm exchange [327, 329, 330]. GSH, present in all cell types, may well perform this role in vivo, and so additional cofactors, chaperones, or receptors, as hypothesized for PDI and FcRn previously, may therefore not be essential. An important second antibody heavy-chain interface is located between the C_H3 domains, which has been shown to be critically involved in Fab-arm exchange, and the contributions of specific C_H3-domain amino acid contacts to the mechanism of this reaction have been investigated recently [328, 330].

In summary, antibodies of the IgG4 isotype are shown to be dynamic molecules, undergoing Fab-arm exchange in vivo and in vitro. The ability to engage in Fab-arm exchange appears to be an inherent feature of IgG4 that involves the third constant domain in addition to the hinge region and that only requires a reducing environment to be activated [325]. This novel protein modification challenges the commonly accepted one antibody–one antigen paradigm and redefines our thinking about the role of IgG4 in antibody-mediated immunity and the application of IgG4 monoclonal antibodies to immunotherapy. Fab-arm exchange is not limited to endogenous IgG4 but can also take place with therapeutic IgG4 antibodies. Natalizumab is an example of a therapeutic monoclonal antibody based on a wild-type IgG4 that was shown to participate in Fab-arm exchange in vivo [325]. Most IgG4 therapeutic antibodies in development now possess a hinge with a proline instead of a serine at position 228 in the hinge. Such antibodies were shown not to participate in Fab-arm exchange in animal models. However, at higher concentrations of reduced glutathione (GSH), an exchange reaction is observed for these "hinge-stabilized" IgG4 antibodies. Preliminary experiments with a hinge-stabilized IgG4 antibody indicated that at 5 mM GSH, equilibrium is reached after 2 days, whereas reducing the antibody with DTT resulted in similar kinetics as wild-type IgG4. Thus, slow Fab-arm exchange may still be observed using "hinge-stabilized" IgG4 antibodies depending on the redox conditions, and at present, it cannot be ruled out that such conditions are never met in vivo [325].

6.4.8 Interchain Disulfides in IgG2 Antibodies

Specific residues of the hinge region have been shown to have profound impact on the properties of immunoglobulins. For example, IgG2 is unique in presenting 4 cysteine residues in the hinge region, notably two consecutive residues, Cys-219 and Cys-220 that have no equivalent in any other immunoglobulin subclass (Fig. 6.14). Human IgG2s are also distinguished from other IgG isotypes by their preferential response toward polysaccharide antigens as well as their limited ability to engage in Fc-mediated effector functions [331]. In addition to the expected interchain disulfide pairings for the IgG2 structure, recent studies have revealed novel disulfide isoforms termed IgG2-B and IgG2-A/B [144, 334]. The A isoform is the familiar IgG2 structure with all four hinge cysteines engaged in symmetrical interchain disulfide bonding to the corresponding cysteines in the second heavy chain (Fig. 6.14). In the IgG-B isoform, two of the four hinge cysteines are

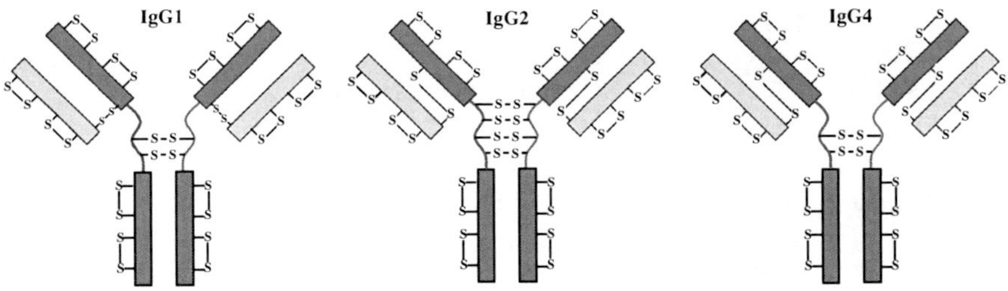

Eu #	128	129	130	131	132	133	134	135	136	137	138
IgG1	Leu	Ala	Pro	Ser	Ser	Lys	Ser	Thr	Ser	Gly	Gly
IgG2	Leu	Ala	Pro	**Cys**	Ser	Arg	Ser	Thr	Ser	Glu	Ser
IgG4	Leu	Ala	Pro	**Cys**	Ser	Arg	Ser	Thr	Ser	Glu	Ser
Eu #	219	220	221	222	223	224	225	226	227	228	229
IgG1	Ser	**Cys**	Asp	Lys	Thr	His	Thr	**Cys**	Pro	Pro	**Cys**
IgG2	**Cys**	**Cys**	-	-	-	Val	Glu	**Cys**	Pro	Pro	**Cys**
IgG4	-	-	-	Tyr	Gly	Pro	Pro	**Cys**	Pro	Ser	**Cys**

Fig. 6.14 IgG1, IgG2, and IgG4 general structure diagrams and sequence alignments of the C_H1 loop and *upper hinge* of the heavy chain. Residue numbering is listed in Eu format

bridged to the second heavy-chain hinge cysteines, the third is bridged to the light chain, and the fourth is bridged to a cysteine to the C_H1 domain which pairs with the light-chain cysteine in the A isoform. The A/B isoform is an intermediate containing both A- and B-isoform disulfide linkages. Moreover, IgG2 with a *lambda* light chain (IgG2λ) showed relatively smaller amounts of the IgG2-B disulfide isoform whereas IgG2 with *kappa* light chain (IgG2κ) generally populated this isoform in higher abundance. It was hypothesized that the additional C-terminal serine residue of the *lambda* light chain sterically interferes with the *lambda* light-chain cysteine forming a disulfide bond with the hinge cysteines. Although IgG2-A seems to be preferred for IgG2λ, IgG2-A/B is detected in approximately equal amounts for IgG2λ and IgG2κ [144, 332].

The report by Martinez et al. [333] presents arguments for a dynamic exchange of IgG2 disulfide bridges between hinge, C_H1 domain, and light chains. The authors also note that the presence of covalent IgG2 dimers in normal pooled human sera has been reported and attributed to cross-linking of the hinge cysteines. The two results appear complementary. It is possible that in vivo thiol–disulfide exchange could generate covalent dimers and this is quite interesting because it could represent a natural "allosteric disulfide" or shuffling mechanism for doubling the effective valency and thereby perhaps increasing antigen-binding avidity. On the other hand, for the biotechnology-derived recombinant IgG2 monoclonal antibodies manufactured by large-scale mammalian cell culture, the product was apparently secreted predominantly as monomer and, once purified, the IgG2 and isoforms were stable as monomers. Upon storage, however, both noncovalent and covalent dimers did gradually form at minor levels.

Evolution has favored emergence of a hinge region in mammalian immunoglobulins for increased flexibility and the ability to adopt a large number of conformations. Although it is often depicted as a static "Y-shaped" molecule in diagrams such as the one in Fig. 6.14, the IgG molecule is more accurately described as a dynamic "flexible-adaptor." In other words, it is rather inaccurate to describe the hinge region of an IgG as a "hinge." It certainly does not function as a rigid hinge, and it is more accurate to describe this region as a flexible polypeptide linker connecting compactly folded Fab and Fc domains. In this context, the arrangement of the disulfide bridges in IgG2 molecules would be appear to be less flexible and less dynamic than for an IgG1 or IgG4, for instance. These IgG2 structures could represent an early state of IgG development which is supported by the fact that the IgG2 subclass is, reportedly, the most evolutionarily distant of the human immunoglobulins [333]. Moreover, the Fc constant regions share >95 % homology, whereas the hinge regions are significantly divergent, illustrating a high evolutionary pressure on this region. Elimination of one of the two cysteine residues in the hinge region specific to the IgG2 subclass, Cys-221 or Cys-222, abolishes the structural isoforms and generated a homogeneous structure. Therefore, the authors propose that elimination of these two cysteine residues during human evolution may have resulted in an IgG1-like hinge, leading to simplification of the number of Ig structures, optimized flexibility within this family of molecules, and better adaptability of the immune response.

It is interesting to compare the disulfide shuffling of the hinge-region interchain disulfides in the IgG2 framework with the Fab-arm exchange in IgG4 antibodies which both occur via the very same thiol–disulfide exchange chemistry but give opposite net results. In vivo IgG2 thiol–disulfide shuffling may serve to form tetravalent dimers with stronger avidity [331, 332] whereas the thiol–disulfide shuffling of IgG4 antibodies, on the other hand, forms bispecific antibodies which are functionally monovalent and have weaker avidity. An intriguing picture of the "hinge" region–interchain disulfide bridges in IgG antibodies is emerging from these recent studies of the IgG4 and IgG2 isoforms in which the interchain disulfides are dynamic and functional rather than static and structural. At minimum, these recent MS and structural characterization studies show how these hinge-region disulfide bridges, in addition to having the known high solvent-accessibility to facilitate thiol–disulfide exchange chemistry, also possess a type of molecular dynamics characterized by conformational flexibility and thermal mobility. It seems that these interchain disulfides may be positioned in the hinge region of antibodies for more than simply cross-linking protein chains together. Indeed, there may be a closer relationship shared by the CXXC in the hinge region of IgG antibodies and the homologous redox-active CXXC in the active site of PDIs than previously thought.

6.4.9 Trisulfide Bridges

It has been known for some time that antibodies under long-term storage in solution can form a nonreducible entity that appears on SDS-PAGE arising from the expulsion of sulfur from a disulfide, resulting in a thioether bridge. ($-CH_2-S-CH_2-$). In addition, evidence has been found recently [335–339] for the insertion of a sulfur into the hinge to form a trisulfide bridge ($-CH_2-S-S-S-CH_2-$). One of the first reported trisulfide-containing proteins is the recombinant human growth hormone (rhGH) expressed in *E. coli* [336]. Although possible routes of trisulfide formation in rhGH have been postulated, the exact mechanism remains unknown. One possible mechanism of trisulfide formation in recombinant proteins involves reaction of dissolved hydrogen sulfide (H_2S) generated during the fermentation process, since treatment of rhGH with solutions that contained dissolved H_2S has been shown to lead to the formation of elevated levels of trisulfide forms of the protein. For monoclonal antibodies, the hinge-region interchain disulfides are more solvent-accessible than other disulfides and thus a possible target for reaction with H_2S and the formation of trisulfide. Pristatsky et al. [337] identified trisulfide modification in the hinge region of a human IgG2 mAb fractionated on an ion-exchange column. More recently, Gu et al. [338] have demonstrated that trisulfides can occur in all IgG frameworks of recombinant and natural antibodies as well as commercial therapeutics. Trisulfide levels were found to range from below the detection limit to more than 40 % although the presence of high levels of trisulfide had no observable effect on the function or stability of the antibody. Although very little has been published on the insertion of a sulfur atom into an interchain disulfide bond to form a trisulfide in proteins, these findings suggest that it may be a fairly commonplace modification.

Peptide mapping of nonreduced protein samples using an LC–MS system was found to be a sensitive and reliable method for identifying and quantifying trisulfides [338]. In all IgG antibodies evaluated, trisulfides were observed in the H-to-L and, to a somewhat lesser degree, in the H-to-H interchain linkages and not in any of the intrachain disulfides. The percentages of the trisulfide in the light–heavy-chain linkage estimated from LC–MS peptide map extracted ion currents (EICs) were similar to those from corresponding UV traces although the EIC method was found to be more sensitive with a detection limit of 0.1 % and RSD of 3.8 % estimated on more than 20 experiments. The Lys-C peptide mapping method is generally applicable to nearly all human IgG1 antibodies, and similar peptide mapping strategies were successfully developed for quantification of trisulfide linkages in IgG2, IgG3, and IgG4 antibodies.

Lys endopeptidase peptide mapping analysis of H-to-L bond trisulfide by LC–MS [150, 338] was performed at room temperature for 30 min with 4-vinylpyridine in the presence of 6 M guanidine hydrochloride, followed by ethanol precipitation. Alkylated mAb was digested with endoproteinase Lys-C in the presence of 2 M urea at room temperature for 16 h. The percentage of trisulfide–linked H-to-L peptides was calculated from LC–MS peak areas from the UV trace or on the extracted ion chromatogram of all three charged states, 1+ to 3+, with known peptides used as internal standards. The percent H-to-L trisulfide was calculated as (peak area of trisulfide) / (peak area of disulfide + peak area of trisulfide), and the lower limit of detection for the assay was <1.0 %. In the nearly 100 monoclonal antibody preparations that were purified and characterized, the level of trisulfide at the L-to-H linkage ranged from less than 1–40 %. Changes in culture conditions that seemed relatively minor and did not significantly affect growth and culture productivity resulted in large differences in trisulfide levels. In particular, culture duration and feeding strategy were important variables, and product with reproducible trisulfide levels was obtained by keeping cell culture conditions consistent [338].

The authors have shown that trisulfides can be incorporated into antibodies by exposure to hydrogen sulfide (H_2S), confirming the original finding with hGH [338]. Production of H_2S by mammalian cells and tissues through the enzymatic breakdown of cysteine and homocysteine could account for trisulfide formation during cell culture. The specificity of the chemical reaction is driven by solvent exposure of the interchain disulfides and by lack of exposure of intrachain disulfides, which are buried in the hydrophobic interior of antibodies. The trisulfide linkage was stable to prolonged storage at 4 °C and at room temperature and in rat serum in vitro but was rapidly converted to a disulfide within 24 h after systemic administration to rats. The rapid conversion of the trisulfide to a disulfide in rats is consistent with other reported reduction–oxidation-related changes that occur in vivo such as the rearrangement of human IgG2 antibody hinge disulfides and Fab-arm exchange of human IgG4 discussed in the previous section. In the absence of a reductant, the trisulfide linkage is very stable, but reversal of the trisulfide to a disulfide can occur both in vitro and in vivo by a mild reduction–oxidation process. Furthermore, it was reported that trisulfide linkages could be efficiently converted to disulfides by washing IgG antibody bound to a protein A affinity LC column with buffers containing millimolar concentrations of L-cysteine [339].

6.4.10 Cystine Desulfurization to Thioether and Dehydroalanine

The degradation of cystine disulfide residues via beta-elimination to generate dehydroalanine and lysinolalanine species at elevated temperatures has been studied fairly extensively for insulin and other proteins for many years [340–346]. More recently, a thioether bridge between the heavy and light chains of a IgG1 monoclonal antibody was observed as a 92 kDa species observed by SDS-PAGE analysis under reducing conditions [341]. LTQ ion-trap LC–MS endoproteinase Lys-C peptide mapping of an SEC fraction of the reduced/alkylated antibody showed that the heavy and light chains were cross-linked by a nonreducible thioether bond between Cys223 of the heavy chain and the C-terminal Cys213 residue of the light chain. The L-to-H thioether linkage was found for a "nonstressed" monoclonal antibody, and its content increased with the duration of incubation at 40°C. In addition to the disulfide-bridged pair of peptides, SFNRGEC/SCDK (m/z 1,260.4), approximately 0.4 % of an earlier eluting peptide with an m/z at 1,228.4 (32 Da less) was identified in the Lys-C digest of a control antibody sample and 13.6 % in a 40 °C heat-stressed antibody sample, in agreement with the presence of a thioether-linked SFNRGEC/SCDK pair. These structural assignments were further confirmed by MSMS and an accurate mass difference between the two Lys-C peptides of 31.97220 Da found by FT-ICR-MS which is consistent a theoretical value of 31.97207 Da for the loss of a sulfur atom.

The above partial H-to-L disulfide-to-thioether modification was also independently observed [342] for another IgG1 mAb formulated in phosphate buffered saline (pH adjusted between 4 and 10) stored for 17 days at 45 °C. The authors report LC–MS endoproteinase Asp-N peptide mapping results indicating the H-to-L disulfide-to-thioether formation through an initial beta-elimination mechanism to form a dehydroalanine intermediate with concomitant disruption of the L-to-H disulfide bridge followed by a Michael-like addition of thiocysteine across the dehydroalanine vinyl group (C=C) to form a stable thioether (lanthionine) linkage. In addition to the thioether formation, hydrolysis of the heavy-chain Ser219-Cys220 peptide bond in the hinge-region sequence, SCDKTHTCPPCPAP, to yield a Fab fragment was also observed and attributed to hydrolysis of the vinyl group in the dehydroalanine 219 intermediate. The rate of degradation was highly pH-dependent, minimizing at pH 6 and rapidly increasing at both pH extremes. It is also worth noting that, in addition to these liquid-phase cysteine desulfurizations, formation of dehydroalanine from cysteine in the gas phase has also been observed by MS and MSMS experiments under certain conditions.

Two method-induced modifications involving desulfurization of cysteine-containing peptides during sample preparation for protein characterization by MS have also been reported recently [343]. In addition to the above previously observed conversion of cysteine to dehydroalanine, a novel modification corresponding to conversion of cysteine to alanine was shown to occur by heating cysteine-containing peptides in the presence of the reducing agent, tris(carboxyethyl)phosphine (TCEP). Using model peptides, the conversion of half-cystine residues to dehydroalanine via beta-elimination of a disulfide bond was seen to result from the conditions of a typical tryptic digestion (37 °C, pH 7.0–9.0) done without disulfide reduction and alkylation. The conversion of cysteine to alanine was investigated by performing experiments in H_2O or D_2O, and the results suggested a radical-based desulfurization mechanism unrelated to beta-elimination. The formation of thioethers by beta-elimination of disulfides has been observed as a method-induced artifact in SDS-PAGE [343].

Mozziconacci et al. [245] have reported the conversion of disulfides to thioethers in an IgG1 following photolysis of the sulfur–sulfur bond with 254 nm ultraviolet radiation and subsequent formation of a thiohemiacetal. Unlike the above beta-elimination chemistry, this degradation of cystine proceeds by a free-radical mechanism to form a dithiohemiacetal and thioether products followed by the expulsion of hydrogen sulfide (H_2S) to form the stable thioether linkage. This photolytic sulfur–sulfur bond rupture is probably similar to the laser-induced "prompt fragmentation" of S–S bonds commonly observed by MALDI-TOF MS.

6.5 Variants Generated by Physical Degradation

Degradation of proteins can be divided into physical and chemical in nature. The former refers to denaturation, aggregation, precipitation, and other degradations without affecting the covalent bonds of the proteins. Because of protein molecule's polymeric nature and their ability to adopt various secondary, tertiary, and quaternary structures, their properties can be drastically different without change of chemical bonds [347, 348]. Among all physical degradation pathways, aggregation and precipitation are probably the most prominent which directly affect the functionality of therapeutic proteins and ultimately their efficacy and safety profiles. Consequently, aggregation and precipitation are among the critical quality attributes of therapeutic proteins. Aggregation and precipitation are characterized and analyzed by various chromatography and biophysical techniques [349, 350], such as SEC, field flow fractionation (FFF) [351], dynamic or static light scattering [352], and analytical ultracentrifugation (AUC) [353]. MS-based techniques have begun to be used successfully under certain circumstances to probe protein conformational changes [354–356]; among them, H/D exchange has proven to be a more dependable technique not only in the academic setting but also in industry as a robust technique to provide critical information for process and quality control, and as one of the critical methods for extensive characterizations [357].

6.6 Chemical Degradation

Degradation products in protein therapeutics can also be generated through various chemical reactions in which breaking or forming of covalent bonds occurs. Characterization and analysis of molecular variants resulting from chemical degradation is a daunting task due to multiple pathways of protein degradation, and the heterogeneous nature of protein molecules to begin with. However, since therapeutic proteins usually are not subjected to extreme conditions during their manufacturing processes and under storage conditions, degradation through regular pathways can be monitored by well-developed methods. The most challenging parts often involve detailed, site-specific, and accurate quantification of each degradation.

The major chemical degradation pathways for protein therapeutics include deamidation/isomerization, oxidation, fragmentation, disulfide scrambling, etc. Any of these degradation pathways can change therapeutic proteins' efficacy and safety. Therefore, it is critical to monitor and accurately measure chemical degradation during pharmaceutical development to optimize manufacturing processes and formulation conditions to reduce their impact. As deamidation/isomerization and oxidation are covered in Chap. 5, the following section will focus on fragmentation pathway.

6.6.1 Fragmentation

6.6.1.1 Asn- and Gln-related fragmentation

Besides being involved in protein deamidation/isomerization, Asn and Gln residues are also involved in spontaneous peptide bond cleavage at their carboxy sides [358, 359, 362 372]. When deamidation proceeds through direct hydrolysis or cyclic imide formation, it usually exhibits lower activation barriers than peptide bond cleavage and therefore is a more favorable pathway. But under certain circumstances, bond cleavage does happen. The fundamental distinction between the mechanisms leading to deamidation via a succinimide and backbone cleavage was found to be the difference in nucleophilic entities involved in the cyclization process (backbone vs. side-chain amide nitrogen). If deamidation is prevented by protein three-dimensional structure, cleavage may become a competing pathway. In addition, peptide bond cleavage at Asn residues is more likely to take place after it has deamidated into Asp [363]. Peptide bond cleavages at Asp and Glu residues have been reported also in peptides and proteins, and the occurrence rate is higher than Asn and Gln, respectively [364–366].

The mechanism of Asp/Glu-related cleavages is closely related to that of Asn/Gln deamidation process, in which it was proposed to involve nucleophilic attack of the ionized side-chain carboxylate on the protonated carbonyl carbon of the peptide bond to give a cyclic anhydride intermediate [365]. The cleavage rate is affected by the side-chain carboxylic acid group [366, 367]. Cleavage of

Asn-Pro residues turned out to be the fastest where succinimide formation could not occur [368, 369]. Needless to say, such peptide bond cleavage can inactivate proteins. For instance, one of the modifications observed in lens proteins is the progressive, age-dependent cleavage of specific peptide bonds in bovine A-crystallin [370]. In fact, peptide bond cleavages are observed in most peptide deamidation reactions, but the fragmentation is usually much slower than deamidation. Asn peptide deamidation half-time ranges from about 1 to 400 days, and Asn cleavage rate ranges from about 200 to >10,000 days [371].

6.6.1.2 Ab Fragmentation

Another type of fragmentation is observed in the hinge region of monoclonal antibodies [342, 361, 373–376]. This cleavage is not affected by protease inhibitors or EDTA which inhibits metal-mediated protease activities, indicating its spontaneous nature. This cleavage was observed on some antibody molecules after long-term storage for an extended period of time, even at 5 °C [377, 378]. The site of cleavage is usually in the heavy-chain hinge region near the papain cleavage site, generating Fab and Fab + Fc fragments which can be detected by MALDI-TOF MS [376]. Metal-mediated cleavage of the antibody molecules in the hinge region has also been reported [378].

Two mechanisms for antibody fragmentation in the hinge region have been proposed: beta-elimination and direct hydrolysis [342, 360]. Beta-elimination is more pronounced at pH 7 and above, which causes cleavage in between S/C in the SCDKTHTC region. Beta-elimination of the disulfide bond leads to the formation of a dehydroalanine residue, which hydrolyzes to form an amide group at the newly formed C-terminus and a pyruvyl group at the newly formed N-terminus [342]. Direct hydrolysis is accelerated by acidic and basic pH. While cleavage was found in every peptide bond in SCDKTHTC of the hinge region, the major cleavage sites have been identified to be in between S/C, C/D, D/K, and H/T [373]. It was also found that the major cleavage sites in the hinge region shift toward the C-terminus when pH changes from 9 to 5. At pH 4, the major cleavage site shifted to the CH2 domain. In addition, oligosaccharides only inhibit hinge-region fragmentation at pH 4. This shift was not observed at pH 9 to 5 [360].

6.7 Conclusions

The ultimate goal of drug development is to determine the safety and efficacy of drug candidates and to ensure drug candidates are produced by highly reproducible and well-controlled processes. The strengths of biologics and the associated challenges during drug development result from the complex biophysical properties of biomolecules. While the complex interaction between biomolecules allows specific inhibition or stimulation of therapeutic targets, thus with fewer side effects, it is quite challenging to characterize biomolecules extensively. In the past decade, MS has played a pivotal role in the characterization of biomolecules. Even though not all molecular variants can be tested for safety and efficacy, since it is almost impossible to isolate all components, MS made it possible to characterize and monitor these minor components to ensure that biomolecules are produced reproducibly with regard to the amounts of variants. Through this approach, the correlation of complex biomolecules to the corresponding therapeutic outcomes can be firmly established.

References

1. Lebkowski JS et al (1984) Transfected DNA is mutated in monkey, mouse, and human cells. Mol Cell Biol 4(10):1951–1960
2. Parker J (1989) Errors and alternatives in reading the universal genetic code. Microbiol Rev 53(3):273–298
3. Reynolds NM, Lazazzera BA, Ibba M (2010) Cellular mechanisms that control mistranslation. Nat Rev Microbiol 8(12):849–856
4. Kunkel TA (2004) DNA replication fidelity. J Biol Chem 279(17):16895–16898
5. Rosenberger RF, Foskett G (1981) An estimate of the frequency of in vivo transcriptional errors at a nonsense codon in Escherichia coli. Mol Gen Genet 183(3):561–563
6. Wada Y (1992) Mass spectrometry in the integrated strategy for the structural analysis of protein variants. Biol Mass Spectrom 21(12):617–624
7. Harris RJ et al (1993) Assessing genetic heterogeneity in production cell lines: detection by peptide mapping of a low level Tyr to Gln sequence variant in a recombinant antibody. Biotechnology 11(11):1293–1297
8. Wan M et al (1999) Variant antibody identification by peptide mapping. Biotechnol Bioeng 62(4):485–488
9. Dorai HST, Campbell A, Kyung YS, Goldstein J, Magill A, Lewis MJ, Tang Q, Jan D, Ganguly S, Moore G (2007) Investigation of product microheterogeneity: a case study in rapid detection of mutation in mammalian production cell lines. BioProcess Int 9:66–72
10. Guo D et al (2010) Mechanisms of unintended amino acid sequence changes in recombinant monoclonal antibodies expressed in Chinese Hamster Ovary (CHO) cells. Biotechnol Bioeng 107(1):163–171
11. Blank A et al (1986) An RNA polymerase mutant with reduced accuracy of chain elongation. Biochemistry 25(20):5920–5928
12. Kramer EB, Farabaugh PJ (2007) The frequency of translational misreading errors in E. coli is largely determined by tRNA competition. RNA 13(1):87–96
13. Ling J, Soll D (2010) Severe oxidative stress induces protein mistranslation through impairment of an aminoacyl-tRNA synthetase editing site. Proc Natl Acad Sci USA 107(9):4028–4033
14. Dayhuff TJ, Atkins JF, Gesteland RF (1986) Characterization of ribosomal frameshift events by protein sequence analysis. J Biol Chem 261(16):7491–7500
15. Guo M et al (2009) Paradox of mistranslation of serine for alanine caused by AlaRS recognition dilemma. Nature 462(7274):808–812
16. Beebe K et al (2008) Distinct domains of tRNA synthetase recognize the same base pair. Nature 451(7174):90–93
17. Fillingame RH, Oldenburg M, Fraga D (1991) Mutation of alanine 24 to serine in subunit c of the Escherichia coli F1F0-ATP synthase reduces reactivity of aspartyl 61 with dicyclohexylcarbodiimide. J Biol Chem 266(31):20934–20939
18. Saul B et al (1994) Point mutation of glycine receptor alpha 1 subunit in the spasmodic mouse affects agonist responses. FEBS Lett 350(1):71–76

19. Knowles JR (1987) Tinkering with enzymes: what are we learning? Science 236(4806):1252–1258
20. Yu XC et al (2009) Identification of codon-specific serine to asparagine mistranslation in recombinant monoclonal antibodies by high-resolution mass spectrometry. Anal Chem 81(22):9282–9290
21. Wen D et al (2009) Discovery and investigation of misincorporation of serine at asparagine positions in recombinant proteins expressed in Chinese hamster ovary cells. J Biol Chem 284(47):32686–32694
22. Ren D et al (2011) Detection and identification of a serine to arginine sequence variant in a therapeutic monoclonal antibody. J Chromatogr B Analyt Technol Biomed Life Sci 879(27):2877–2884
23. Que AH, Zhang B, Yang Y, Zhang J, Derfus M, Amanullah A (2010) Sequence variant analysis using peptide mapping by LC-MS/MS. BioProcess Int 8:52–60
24. Cupples CG, Miller JH (1988) Effects of amino acid substitutions at the active site in Escherichia coli beta-galactosidase. Genetics 120(3):637–644
25. Harris RJ (2005) Heterogeneity of recombinant antibodies: linking structure to function. Dev Biol (Basel) 122:117–127
26. Moura GR, Carreto LC, Santos MA (2009) Genetic code ambiguity: an unexpected source of proteome innovation and phenotypic diversity. Curr Opin Microbiol 12(6):631–637
27. Lu HS et al (1993) Isolation and characterization of three recombinant human granulocyte colony stimulating factor His→ Gln isoforms produced in Escherichia coli. Protein Expr Purif 4(5):465–472
28. Apostol I et al (1997) Incorporation of norvaline at leucine positions in recombinant human hemoglobin expressed in Escherichia coli. J Biol Chem 272(46):28980–28988
29. McNulty DE et al (2003) Mistranslational errors associated with the rare arginine codon CGG in Escherichia coli. Protein Expr Purif 27(2):365–374
30. Calderone TL, Stevens RD, Oas TG (1996) High-level misincorporation of lysine for arginine at AGA codons in a fusion protein expressed in Escherichia coli. J Mol Biol 262(4):407–412
31. Aguirre B et al (2011) A ribosomal misincorporation of Lys for Arg in human triosephosphate isomerase expressed in Escherichia coli gives rise to two protein populations. PLoS ONE 6(6):e21035
32. Masuda M et al (2006) Cysteine misincorporation in bacterially expressed human alpha-synuclein. FEBS Lett 580(7):1775–1779
33. Parker J et al (1978) Stuttering: high-level mistranslation in animal and bacterial cells. Proc Natl Acad Sci USA 75(3):1091–1095
34. Scorer CA, Carrier MJ, Rosenberger RF (1991) Amino acid misincorporation during high-level expression of mouse epidermal growth factor in Escherichia coli. Nucleic Acids Res 19(13):3511–3516
35. Langridge J (1974) Mutation spectra and the neutrality of mutations. Aust J Biol Sci 27(3):309–319
36. Nene V, Glass RE (1984) Genetic studies on the beta subunit of Escherichia coli RNA polymerase. IV. Structure-function correlates. Mol Gen Genet 194(1–2):166–172
37. Yang Y et al (2010) Detecting low level sequence variants in recombinant monoclonal antibodies. MAbs 2(3):285–298
38. Gatlin CL et al (2000) Automated identification of amino acid sequence variations in proteins by HPLC/microspray tandem mass spectrometry. Anal Chem 72(4):757–763
39. Tanaka K et al (2006) Determination of unique amino acid substitutions in protein variants by peptide mass mapping with FT-ICR MS. J Am Soc Mass Spectrom 17(4):508–513
40. Amoresano A et al (2009) Technical advances in proteomics mass spectrometry: identification of post-translational modifications. Clin Chem Lab Med 47(6):647–665
41. Sinha S et al (2008) Comparison of LC and LC/MS methods for quantifying N-glycosylation in recombinant IgGs. J Am Soc Mass Spectrom 19(11):1643–1654
42. Witze ES et al (2007) Mapping protein post-translational modifications with mass spectrometry. Nat Methods 4(10):798–806

43. Su X, Ren C, Freitas MA (2007) Mass spectrometry-based strategies for characterization of histones and their post-translational modifications. Expert Rev Proteomics 4(2):211–225
44. Garcia BA, Shabanowitz J, Hunt DF (2007) Characterization of histones and their post-translational modifications by mass spectrometry. Curr Opin Chem Biol 11(1):66–73
45. Salzano AM, Crescenzi M (2005) Mass spectrometry for protein identification and the study of post translational modifications. Ann Ist Super Sanita 41(4):443–450
46. Jensen ON (2004) Modification-specific proteomics: characterization of post-translational modifications by mass spectrometry. Curr Opin Chem Biol 8(1):33–41
47. Kuster B, Mann M (1998) Identifying proteins and post-translational modifications by mass spectrometry. Curr Opin Struct Biol 8(3):393–400
48. Barnes CS, Lim A (2007) Applications of mass spectrometry for the structural characterization of recombinant protein pharmaceuticals. Mass Spectr Rev 26:370–388
49. Hashimoto K et al (2006) KEGG as a glycome informatics resource. Glycobiology 16(5):63R–70R
50. Hossler P, Khattak SF, Li ZJ (2009) Optimal and consistent protein glycosylation in mammalian cell culture. Glycobiology 19(9):936–949
51. Yuk IH, Wang DI (2002) Changes in the overall extent of protein glycosylation by Chinese hamster ovary cells over the course of batch culture. Biotechnol Appl Biochem 36(Pt 2):133–140
52. Bogosian G et al (1989) Biosynthesis and incorporation into protein of norleucine by Escherichia coli. J Biol Chem 264(1):531–539
53. Eng M et al (1997) Formulation development and primary degradation pathways for recombinant human nerve growth factor. Anal Chem 69(20):4184–4190
54. Geoghegan KF et al (1999) Spontaneous alpha-N-6-phosphogluconoylation of a "His tag" in Escherichia coli: the cause of extra mass of 258 or 178 Da in fusion proteins. Anal Biochem 267(1):169–184
55. Du P et al (2005) Phosphorylation of serine residues in histidine-tag sequences attached to recombinant protein kinases: a cause of heterogeneity in mass and complications in function. Protein Expr Purif 44(2):121–129
56. Aon JC et al (2008) Suppressing posttranslational gluconoylation of heterologous proteins by metabolic engineering of Escherichia coli. Appl Environ Microbiol 74(4):950–958
57. Apweiler R, Hermjakob H, Sharon N (1999) On the frequency of protein glycosylation as deduced from analysis of the SWISS-PROT database. Biochim Biophys Acta 1473(1):4–8
58. Singh SK (2011) Impact of product-related factors on immunogenicity of biotherapeutics. J Pharm Sci 100(2):354–387
59. Schellekens H (2002) Bioequivalence and the immunogenicity of biopharmaceuticals. Nat Rev Drug Discov 1(6):457–462
60. Mahmood I, Green MD (2005) Pharmacokinetic and pharmacodynamic considerations in the development of therapeutic proteins. Clin Pharmacokinet 44(4):331–347
61. Butler M (2006) Optimisation of the cellular metabolism of glycosylation for recombinant proteins produced by Mammalian cell systems. Cytotechnology 50(1–3):57–76
62. Morrow KJ Jr (2007) Advances in antibody manufacturing using mammalian cells. Biotechnol Annu Rev 13:95–113
63. Dingermann T (2008) Recombinant therapeutic proteins: production platforms and challenges. Biotechnol J 3(1):90–97
64. Kawasaki N et al (2009) The significance of glycosylation analysis in development of biopharmaceuticals. Biol Pharm Bull 32(5):796–800
65. Liu DT (1992) Glycoprotein pharmaceuticals: scientific and regulatory considerations, and the US Orphan Drug Act. Trends Biotechnol 10(4):114–120
66. Mack G (2008) FDA balks at Myozyme scale-up. Nat Biotechnol 26(6):592
67. Goochee CF, Monica T (1990) Environmental effects on protein glycosylation. Biotechnology 8(5):421–427
68. Sethuraman N, Stadheim TA (2006) Challenges in therapeutic glycoprotein production. Curr Opin Biotechnol 17(4):341–346

69. Raju TS (2003) Glycosylation variations with expression systems and their impact on biological activity of therapeutic immunoglobulins. BioProcess Int 1:44–53
70. Fischer T et al (1990) Glycosylation of the human interferon-gamma receptor. N-linked carbohydrates contribute to structural heterogeneity and are required for ligand binding. J Biol Chem 265(3):1710–1717
71. Stockinger H et al (1992) Binding of recombinant variants of human tissue-type plasminogen activator (t-PA) to human umbilical vein endothelial cells. Thromb Res 67(5):589–599
72. Yamaguchi K et al (1991) Effects of site-directed removal of N-glycosylation sites in human erythropoietin on its production and biological properties. J Biol Chem 266(30):20434–20439
73. Imai N et al (1990) Physicochemical and biological characterization of asialoerythropoietin. Suppressive effects of sialic acid in the expression of biological activity of human erythropoietin in vitro. Eur J Biochem 194(2):457–462
74. Lis H, Sharon N (1993) Protein glycosylation. Structural and functional aspects. Eur J Biochem 218(1):1–27
75. Coloma MJ et al (1999) Position effects of variable region carbohydrate on the affinity and in vivo behavior of an anti-(1–>6) dextran antibody. J Immunol 162(4):2162–2170
76. Noguchi A et al (1995) Immunogenicity of N-glycolylneuraminic acid-containing carbohydrate chains of recombinant human erythropoietin expressed in Chinese hamster ovary cells. J Biochem 117(1):59–62
77. Jefferis R (2005) Glycosylation of recombinant antibody therapeutics. Biotechnol Prog 21(1):11–16
78. Gala FA, Morrison SL (2004) V region carbohydrate and antibody expression. J Immunol 172(9):5489–5494
79. Delorme E et al (1992) Role of glycosylation on the secretion and biological activity of erythropoietin. Biochemistry 31(41):9871–9876
80. Walsh G, Jefferis R (2006) Post-translational modifications in the context of therapeutic proteins. Nat Biotechnol 24(10):1241–1252
81. Sola RJ, Griebenow K (2009) Effects of glycosylation on the stability of protein pharmaceuticals. J Pharm Sci 98(4):1223–1245
82. Stallforth et al (2009) Claude S. Hudson Award in carbohydrate chemistry. Carbohydrates: a frontier in medicinal chemistry. J Med Chem 52(18):5561–5577
83. Brooks SA (2009) Strategies for analysis of the glycosylation of proteins: current status and future perspectives. Mol Biotechnol 43(1):76–88
84. Beck A et al (2008) Trends in glycosylation, glycoanalysis and glycoengineering of therapeutic antibodies and Fc-fusion proteins. Curr Pharm Biotechnol 9(6):482–501
85. Kamoda S, Kakehi K (2008) Evaluation of glycosylation for quality assurance of antibody pharmaceuticals by capillary electrophoresis. Electrophoresis 29(17):3595–3604
86. Ko K et al (2008) Glyco-engineering of biotherapeutic proteins in plants. Mol Cells 25(4):494–503
87. Amon S, Zamfir AD, Rizzi A (2008) Glycosylation analysis of glycoproteins and proteoglycans using capillary electrophoresis-mass spectrometry strategies. Electrophoresis 29(12):2485–2507
88. Budnik BA, Lee RS, Steen JA (2006) Global methods for protein glycosylation analysis by mass spectrometry. Biochim Biophys Acta 1764(12):1870–1880
89. Morelle W et al (2006) The use of mass spectrometry for the proteomic analysis of glycosylation. Proteomics 6(14):3993–4015
90. Wuhrer M, Deelder AM, Hokke CH (2005) Protein glycosylation analysis by liquid chromatography-mass spectrometry. J Chromatogr B Analyt Technol Biomed Life Sci 825(2):124–133
91. Harvey DJ (2005) Proteomic analysis of glycosylation: structural determination of N- and O-linked glycans by mass spectrometry. Expert Rev Proteomics 2(1):87–101
92. Sheeley DM, Merrill BM, Taylor LC (1997) Characterization of monoclonal antibody glycosylation: comparison of expression systems and identification of terminal alpha-linked galactose. Anal Biochem 247(1):102–110

93. Yuen CT et al (2003) Relationships between the N-glycan structures and biological activities of recombinant human erythropoietins produced using different culture conditions and purification procedures. Br J Haematol 121(3):511–526
94. Werner RG et al (1998) Appropriate mammalian expression systems for biopharmaceuticals. Arzneimittel-Forschung 48(8):870–880
95. Jefferis R (2009) Glycosylation as a strategy to improve antibody-based therapeutics. Nat Rev Drug Discov 8(3):226–234
96. Lawson EQ et al (1983) Effect of carbohydrate on protein solubility. Arch Biochem Biophys 220(2):572–575
97. Bond A, Cooke A, Hay FC (1990) Glycosylation of IgG immune complexes and IgG subclasses in the MRL-lpr/lpr mouse model of rheumatoid arthritis. Eur J Immunol 20(10):2229–2233
98. Huang L et al (2006) Impact of variable domain glycosylation on antibody clearance: an LC/MS characterization. Anal Biochem 349(2):197–207
99. Arakawa T et al (1991) Glycosylated and unglycosylated recombinant-derived human stem cell factors are dimeric and have extensive regular secondary structure. J Biol Chem 266(28):18942–18948
100. Krapp S et al (2003) Structural analysis of human IgG-Fc glycoforms reveals a correlation between glycosylation and structural integrity. J Mol Biol 325(5):979–989
101. Iwasaki H, Suzuki Y, Sinohara H (1996) Cloning and sequencing of cDNAs encoding plasma alpha-macroglobulin and murinoglobulin from guinea pig: implications for molecular evolution of alpha-macroglobulin family. J Biochem 120(6):1167–1175
102. Oh-eda M et al (1990) O-linked sugar chain of human granulocyte colony-stimulating factor protects it against polymerization and denaturation allowing it to retain its biological activity. J Biol Chem 265(20):11432–11435
103. Runkel L et al (1998) Structural and functional differences between glycosylated and non-glycosylated forms of human interferon-beta (IFN-beta). Pharm Res 15(4):641–649
104. Sinclair AM, Elliott S (2005) Glycoengineering: the effect of glycosylation on the properties of therapeutic proteins. J Pharm Sci 94(8):1626–1635
105. Mimura Y et al (2000) The influence of glycosylation on the thermal stability and effector function expression of human IgG1-Fc: properties of a series of truncated glycoforms. Mol Immunol 37(12–13):697–706
106. Kaneko Y, Nimmerjahn F, Ravetch JV (2006) Anti-inflammatory activity of immunoglobulin G resulting from Fc sialylation. Science 313(5787):670–673
107. Kibe T, Ishida C, Togari H, Wada Y, Okada S, Nakagawa H, Tsukamoto Y, Kawamura Y, Takahashi N (1996) Glycosylation and placental transport of immunoglobulin G. J Clin Biochem Nutr 21:57–63
108. Umana P et al (1999) Engineered glycoforms of an antineuroblastoma IgG1 with optimized antibody-dependent cellular cytotoxic activity. Nat Biotechnol 17(2):176–180
109. Malhotra R et al (1995) Glycosylation changes of IgG associated with rheumatoid arthritis can activate complement via the mannose-binding protein. Nat Med 1(3):237–243
110. Shields RL et al (2002) Lack of fucose on human IgG1 N-linked oligosaccharide improves binding to human Fcgamma RIII and antibody-dependent cellular toxicity. J Biol Chem 277(30):26733–26740
111. Temporini C et al (2008) Integrated analytical strategies for the study of phosphorylation and glycosylation in proteins. Mass Spectrom Rev 27(3):207–236
112. Morelle W (2009) Analysis of Glycosylation and Other Post-Translational Modifications by Mass Spectrometry. Curr Anal Chem 5:144–165
113. Gesslbauer B et al (2007) Proteoglycanomics: tools to unravel the biological function of glycosaminoglycans. Proteomics 7(16):2870–2880
114. Reinhold VN, Reinhold BB, Costello CE (1995) Carbohydrate molecular weight profiling sequence linkage and branching data: ES-MS and CID. Anal Chem 67(11):1772–1784
115. Burlingame AL, Boyd RK, Gaskell SJ (1994) Mass spectrometry. Anal Chem 66(12):634R–683R

116. Karas M, Bahr U, Dulcks T (2000) Nano-electrospray ionization mass spectrometry: addressing analytical problems beyond routine. Fresenius J Anal Chem 366(6–7):669–676
117. Bahr U et al (1997) High-sensitivity analysis of neutral underivatized oligosaccharides by nanoelectrospray mass spectrometry. Anal Chem 69(22):4530–4535
118. Harvey DJ (1993) Quantitative aspects of the matrix-assisted laser desorption mass spectrometry of complex oligosaccharides. Rapid Commun Mass Spectrom 7(7):614–619
119. Viseux N et al (2001) Qualitative and quantitative analysis of the glycosylation pattern of recombinant proteins. Anal Chem 73(20):4755–4762
120. Viseux N, Costello CE, Domon B (1999) Post-source decay mass spectrometry: optimized calibration procedure and structural characterization of permethylated oligosaccharides. J Mass Spectrom 34(4):364–376
121. Harvey DJ (2009) Analysis of carbohydrates and glycoconjugates by matrix-assisted laser desorption/ionization mass spectrometry: an update for 2003–2004. Mass Spectrom Rev 28(2):273–361
122. Laine RA, Pamidimukkala KM, French AL, Hall RW, Abbas SA, Jain RK, Matta KL (1988) Linkage Position in Oligosaccharides by Fast Atom Bombardment Ionization, Collision-Activated Dissociation, tandem Mass Spectrometry and Molecular Modeling. L-Fucosylp (β1_X)-D-N-acetyl-D-glucosaminylp-(β1_3)-D-galactosylp-(β1-> 0-methyl) where X = 3,4, or 6. J Am Chem Soc 110:6931–6939
123. Carr SA et al (1985) Enhancement of structural information in FAB ionized carbohydrate samples by neutral gas collision. Biomed Mass Spectrom 12(6):288–295
124. Domon B, Costello CE (1988) Structure elucidation of glycosphingolipids and gangliosides using high-performance tandem mass spectrometry. Biochemistry 27(5):1534–1543
125. Chait BT, Kent SB (1992) Weighing naked proteins: practical, high-accuracy mass measurement of peptides and proteins. Science 257(5078):1885–1894
126. Chait BT et al (1993) Protein ladder sequencing. Science 262(5130):89–92
127. Wada Y et al (1994) Diagnosis of carbohydrate-deficient glycoprotein syndrome by matrix-assisted laser desorption time-of-flight mass spectrometry. Biol Mass Spectrom 23(2):108–109
128. Nelson RW, Dogruel D, Williams P (1995) Detection of human IgM at m/z approximately 1 MDa. Rapid Commun Mass Spectrom 9(7):625
129. Bonfichi R et al (1995) Preliminary investigation of glycosylated proteins by capillary electrophoresis and capillary electrophoresis/mass spectrometry using electrospray ionization and by matrix-assisted laser desorption ionization/time-to-flight mass spectrometry. Rapid Commun Mass Spectrom Spec No:S95–106
130. Sottani C, Fiorentino M, Minoia C (1997) Matrix performance in matrix-assisted laser desorption/ionization for molecular weight determination in sialyl and non-sialyl oligosaccharide proteins. Rapid Commun Mass Spectrom 11(8):907–913
131. Gimenez E et al (2007) Towards a reliable molecular mass determination of intact glycoproteins by matrix-assisted laser desorption/ionization time-of-flight mass spectrometry. Rapid Commun Mass Spectrom 21(16):2555–2563
132. Bahr U et al (1997) Delayed extraction time-of-flight MALDI mass spectrometry of proteins above 25,000 Da. J Mass Spectrom 32(10):1111–1116
133. Morelle W, Michalski JC (2005) The mass spectrometric analysis of glycoproteins and their glycan structures. Curr Anal Chem 1(1):29–57
134. Ferrige AG, Seddon MJ (1991) Maximum entropy deconvolution in electrospray mass spectrometry. Rapid Comm Mass Spec 5:374–379
135. Brady LJ et al (2008) Molecular mass analysis of antibodies by on-line SEC-MS. J Am Soc Mass Spectrom 19(4):502–509
136. Gadgil HS et al (2007) Screening and sequencing of glycated proteins by neutral loss scan LC/MS/MS method. Anal Chem 79(15):5991–5999
137. Nichols AC, Bourell JH (2005) Accurate mass analysis of intact IgG1 and IgG2 antibodies using ESI-Q-Tof mass spectrometers. In: Proceedings of the 53rd ASMS Conference on Mass Spectrometry and Allied Topics, San Antonio, TX June 5–9
138. Zhang Z, Pan H, Chen X (2008) Mass spectrometry for structural characterization of therapeutic antibodies. Mass Spectrom Rev 28:147–176

139. Beck A et al (2005) Characterization by liquid chromatography combined with mass spectrometry of monoclonal anti-IGF-1 receptor antibodies produced in CHO and NS0 cells. J Chromatogr B Analyt Technol Biomed Life Sci 819(2):203–218
140. Dillon TM, Bondarenko PV (1053) Ricci M (2004) Development of an analytical reversed-phase high-performance liquid chromatography-electrospray ionization mass spectrometry method for characterization of recombinant antibodies. J Chromatogr A 1–2:299–305
141. Dillon TM et al (2008) Structural and functional characterization of disulfide isoforms of the human IgG2 subclass. J Biol Chem 283(23):16206–16215
142. Ren D et al (2007) Reversed-phase liquid chromatography of immunoglobulin G molecules and their fragments with the diphenyl column. J Chromatogr A 1175(1):63–68
143. Ren D et al (2008) Reversed-phase liquid chromatography-mass spectrometry of site-specific chemical modifications in intact immunoglobulin molecules and their fragments. J Chromatogr A 1179(2):198–204
144. Wypych J et al (2008) Human IgG2 antibodies display disulfide-mediated structural isoforms. J Biol Chem 283(23):16194–16205
145. Dillon TM et al (2006) Optimization of a reversed-phase high-performance liquid chromatography/mass spectrometry method for characterizing recombinant antibody heterogeneity and stability. J Chromatogr A 1120(1–2):112–120
146. Lazar AC et al (2005) Analysis of the composition of immunoconjugates using size-exclusion chromatography coupled to mass spectrometry. Rapid Commun Mass Spectrom 19(13):1806–1814
147. Garcia BA (2010) What does the future hold for top down mass spectrometry? J Am Soc Mass Spectrom 21:193–202
148. Kelleher NL (2004) Top-down proteomics. Anal Chem 76(11):197A–203A
149. Reid GE, McLuckey SA (2002) ‘Top down’ protein characterization via tandem mass spectrometry. J Mass Spectrom 37(7):663–675
150. Karabacak NM et al (2009) Sensitive and specific identification of wild type and variant proteins from 8 to 669 kDa using top-down mass spectrometry. Mol Cell Proteomics 8(4):846–856
151. Ryan CM et al (2010) Post-translational modifications of integral membrane proteins resolved by top-down Fourier transform mass spectrometry with collisionally activated dissociation. Mol Cell Proteomics 9(5):791–803
152. Zubarev RA (2004) Electron-capture dissociation tandem mass spectrometry. Curr Opin Biotechnol 15(1):12–16
153. Zubarev RA (2003) Reactions of polypeptide ions with electrons in the gas phase. Mass Spectrom Rev 22(1):57–77
154. Zubarev RAKNA, McLafferty FW (1998) Electron capture dissociation of multiply charged protein cations. A nonergodic process. J Am Chem Soc 120:3265–3266
155. Syka JE et al (2004) Peptide and protein sequence analysis by electron transfer dissociation mass spectrometry. Proc Natl Acad Sci U S A 101(26):9528–9533
156. Zubarev RA et al (2000) Electron capture dissociation for structural characterization of multiply charged protein cations. Anal Chem 72(3):563–573
157. Zubarev RAKNA, Fridriksson EK, Lewis MA, Horn DM, Carpenter BK, McLafferty FW (1999) Electron capture dissociation of gaseous multiply-charged proteins is favored at disulfide bonds and other sites of high hydrogen atom affinity. J Am Chem Soc 121:2857–2862
158. Mirgorodskaya E, Roepstorff P, Zubarev RA (1999) Localization of O-glycosylation sites in peptides by electron capture dissociation in a Fourier transform mass spectrometer. Anal Chem 71(20):4431–4436
159. Siuti N, Kelleher NL (2007) Decoding protein modifications using top-down mass spectrometry. Nat Methods 4(10):817–821
160. Mikhailov VA, Iniesta J, Cooper HJ (2010) Top-down mass analysis of protein tyrosine nitration: comparison of electron capture dissociation with “slow-heating” tandem mass spectrometry methods. Anal Chem 82(17):7283–7292
161. Yang J et al (2007) Determination of tryptophan oxidation of monoclonal antibody by reversed phase high performance liquid chromatography. J Chromatogr A 1156(1–2):174–182

162. Yu L, Remmele RL Jr, He B (2006) Identification of N-terminal modification for recombinant monoclonal antibody light chain using partial reduction and quadrupole time-of-flight mass spectrometry. Rapid Commun Mass Spectrom 20(24):3674–3680
163. Wang L et al (2005) Structural characterization of a recombinant monoclonal antibody by electrospray time-of-flight mass spectrometry. Pharm Res 22(8):1338–1349
164. Siegel MM et al (1991) Matrix-assisted UV-laser desorption/ionization mass spectrometric analysis of monoclonal antibodies for the determination of carbohydrate, conjugated chelator, and conjugated drug content. Anal Chem 63(21):2470–2481
165. Johnson KA et al (2007) Cation exchange-HPLC and mass spectrometry reveal C-terminal amidation of an IgG1 heavy chain. Anal Biochem 360(1):75–83
166. Rehder DS et al (2006) Reversed-phase liquid chromatography/mass spectrometry analysis of reduced monoclonal antibodies in pharmaceutics. J Chromatogr A 1102(1–2):164–175
167. Ashton T (1995) The purchaser-provider split in New Zealand: the story so far. Aust Health Rev 18(1):43–60
168. Chumsae C et al (2007) Comparison of methionine oxidation in thermal stability and chemically stressed samples of a fully human monoclonal antibody. J Chromatogr B Analyt Technol Biomed Life Sci 850(1–2):285–294
169. Adamczyk M et al (2000) Region-selective labeling of antibodies as determined by electrospray ionization-mass spectrometry (ESI-MS). Bioconjug Chem 11(4):557–563
170. Dalpathado DS, Desaire H (2008) Glycopeptide analysis by mass spectrometry. Analyst 133(6):731–738
171. Huddleston MJ, Bean MF, Carr SA (1993) Collisional fragmentation of glycopeptides by electrospray ionization LC/MS and LC/MS/MS: methods for selective detection of glycopeptides in protein digests. Anal Chem 65(7):877–884
172. Zhao J et al (2006) Comparative serum glycoproteomics using lectin selected sialic acid glycoproteins with mass spectrometric analysis: application to pancreatic cancer serum. J Proteome Res 5(7):1792–1802
173. Ito H et al (2006) Direct structural assignment of neutral and sialylated N-glycans of glycopeptides using collision-induced dissociation MSn spectral matching. Rapid Commun Mass Spectrom 20(23):3557–3565
174. Geyer H, Geyer R (2006) Strategies for analysis of glycoprotein glycosylation. Biochim Biophys Acta 1764(12):1853–1869
175. Zhang H, Aebersold R (2006) Isolation of glycoproteins and identification of their N-linked glycosylation sites. Methods Mol Biol 328:177–185
176. Zhang H et al (2003) Identification and quantification of N-linked glycoproteins using hydrazide chemistry, stable isotope labeling and mass spectrometry. Nat Biotechnol 21(6):660–666
177. Sparbier K et al (2005) Selective isolation of glycoproteins and glycopeptides for MALDI-TOF MS detection supported by magnetic particles. J Biomol Tech 16(4):407–413
178. Calvano CD, Zambonin CG, Jensen ON (2008) Assessment of lectin and HILIC based enrichment protocols for characterization of serum glycoproteins by mass spectrometry. J Proteomics 71(3):304–317
179. Wuhrer M et al (2007) Glycoproteomics based on tandem mass spectrometry of glycopeptides. J Chromatogr B Analyt Technol Biomed Life Sci 849(1–2):115–128
180. Seipert RR et al (2008) Factors that influence fragmentation behavior of N-linked glycopeptide ions. Anal Chem 80(10):3684–3692
181. Hakansson K et al (2003) Combined electron capture and infrared multiphoton dissociation for multistage MS/MS in a Fourier transform ion cyclotron resonance mass spectrometer. Anal Chem 75(13):3256–3262
182. Deguchi K et al (2007) Structural analysis of O-glycopeptides employing negative- and positive-ion multi-stage mass spectra obtained by collision-induced and electron-capture dissociations in linear ion trap time-of-flight mass spectrometry. Rapid Commun Mass Spectrom 21(5):691–698
183. Medzihradszky KF, Besman MJ, Burlingame AL (1997) Structural characterization of site-specific N-glycosylation of recombinant human factor VIII by reversed-phase high-performance liquid chromatography-electrospray ionization mass spectrometry. Anal Chem 69(19):3986–3994

184. Wuhrer M et al (2005) New features of site-specific horseradish peroxidase (HRP) glycosylation uncovered by nano-LC-MS with repeated ion-isolation/fragmentation cycles. Biochim Biophys Acta 1723(1–3):229–239
185. Jebanathirajah J, Steen H, Roepstorff P (2003) Using optimized collision energies and high resolution, high accuracy fragment ion selection to improve glycopeptide detection by precursor ion scanning. J Am Soc Mass Spectrom 14(7):777–784
186. Ritchie MA et al (2002) Precursor ion scanning for detection and structural characterization of heterogeneous glycopeptide mixtures. J Am Soc Mass Spectrom 13(9):1065–1077
187. Catalina MI et al (2007) Electron transfer dissociation of N-glycopeptides: loss of the entire N-glycosylated asparagine side chain. Rapid Commun Mass Spectrom 21(6):1053–1061
188. Hakansson K et al (2001) Electron capture dissociation and infrared multiphoton dissociation MS/MS of an N-glycosylated tryptic peptic to yield complementary sequence information. Anal Chem 73(18):4530–4536
189. Anumula KR (2006) Advances in fluorescence derivatization methods for high-performance liquid chromatographic analysis of glycoprotein carbohydrates. Anal Biochem 350(1):1–23
190. Royle L, Dwek RC RA, Rudd PM (2006) Detailed structural analysis of N-glycans released from glycoproteins in SDS-PAGE gel bands using HPLC combined with exoglycosidase array digestions. Glycobiology Protocols 124–143
191. Bhavanandan VP, Umemoto J, Davidson EA (1976) Characterization of an endo-alpha-N-acetyl galactosaminidase from Diplococcus pneumoniae. Biochem Biophys Res Commun 70(3):738–745
192. Merry AH et al (2002) Recovery of intact 2-aminobenzamide-labeled O-glycans released from glycoproteins by hydrazinolysis. Anal Biochem 304(1):91–99
193. Carlson DM (1966) Oligosaccharides isolated from pig submaxillary mucin. J Biol Chem 241(12):2984–2986
194. Huang Y, Mechref Y, Novotny MV (2001) Microscale nonreductive release of O-linked glycans for subsequent analysis through MALDI mass spectrometry and capillary electrophoresis. Anal Chem 73(24):6063–6069
195. Raju TS (2008) Terminal sugars of Fc glycans influence antibody effector functions of IgGs. Curr Opin Immunol 20(4):471–478
196. Muller DR, Domon BM, Blum W, Raschdorf F, Richter WJ (1988) Direct stereochemical assignment of sugar subunits in naturally occurring glycosides by low energy collision induce dissociation. Application to papulacandin antibiotics. Biomed Environ Mass Spectrom 15:441–446
197. Svarovsky SA, Joshi L (2008) Biocombinatorial selection of carbohydrate binding agents of therapeutic significance. Curr Drug Discov Technol 5(1):20–28
198. Alley WR Jr, Mechref Y, Novotny MV (2009) Use of activated graphitized carbon chips for liquid chromatography/mass spectrometric and tandem mass spectrometric analysis of tryptic glycopeptides. Rapid Commun Mass Spectrom 23(4):495–505
199. Trivedi MV, Laurence JS, Siahaan TJ (2009) The role of thiols and disulfides on protein stability. Curr Protein Pept Sci 10(6):614–625
200. Nagano N, Ota M, Nishikawa K (1999) Strong hydrophobic nature of cysteine residues in proteins. FEBS Lett 458(1):69–71
201. Schmidt B, Ho L, Hogg PJ (2006) Allosteric disulfide bonds. Biochemistry 45(24):7429–7433
202. Hogg PJ (2009) Contribution of allosteric disulfide bonds to regulation of hemostasis. J Thromb Haemost 7(Suppl 1):13–16
203. Hogg PJ (2003) Disulfide bonds as switches for protein function. Trends Biochem Sci 28(4):210–214
204. Wouters MA et al (2011) Thiol-based redox signalling: rust never sleeps. Int J Biochem Cell Biol 43(8):1079–1085
205. Karala AR, Lappi AK, Ruddock WL (2010) Modulation of an active-site cysteine pKa allows PDI to act as a catalyst of both disulfide bond formation and isomerization. J Mol Biol 396(4):883–892

206. Zee RS et al (2010) Redox regulation of sirtuin-1 by S-glutathiolation. Antioxid Redox Signal 13(7):1023–1032
207. Gadgil HS et al (2006) Identification of cysteinylation of a free cysteine in the Fab region of a recombinant monoclonal IgG1 antibody using Lys-C limited proteolysis coupled with LC/MS analysis. Anal Biochem 355(2):165–174
208. Chen XN et al (2009) Charge-based analysis of antibodies with engineered cysteines: from multiple peaks to a single main peak. MAbs 1(6):563–571
209. Liu P et al (2009) Characterization of S-thiolation on secreted proteins from E. coli by mass spectrometry. Rapid Commun Mass Spectrom 23(20):3343–3349
210. Banks DD et al (2008) Removal of cysteinylation from an unpaired sulfhydryl in the variable region of a recombinant monoclonal IgG1 antibody improves homogeneity, stability, and biological activity. J Pharm Sci 97(2):775–790
211. Kleinova M et al (2005) Characterization of cysteinylation of pharmaceutical-grade human serum albumin by electrospray ionization mass spectrometry and low-energy collision-induced dissociation tandem mass spectrometry. Rapid Commun Mass Spectrom 19(20):2965–2973
212. Jia L et al (1996) S-nitrosohaemoglobin: a dynamic activity of blood involved in vascular control. Nature 380(6571):221–226
213. Murray CI et al (2012) Identification and quantification of S-nitrosylation by cysteine reactive tandem mass tag switch assay. Mol Cell Proteomics 11(2):M111 013441
214. Raju K et al (2011) Strategies and tools to explore protein S-nitrosylation. Biochim Biophys Acta
215. Torta F, Elviri L, Bachi A (2010) Direct and indirect detection methods for the analysis of S-nitrosylated peptides and proteins. Methods Enzymol 473:265–280
216. Liu M et al (2010) Site-specific proteomics approach for study protein S-nitrosylation. Anal Chem 82(17):7160–7168
217. Cook SL, Jackson GP (2011) Characterization of tyrosine nitration and cysteine nitrosylation modifications by metastable atom-activation dissociation mass spectrometry. J Am Soc Mass Spectrom 22(2):221–232
218. Maret W (2006) Zinc coordination environments in proteins as redox sensors and signal transducers. Antioxid Redox Signal 8(9–10):1419–1441
219. Maret W, Li Y (2009) Coordination dynamics of zinc in proteins. Chem Rev 109(10):4682–4707
220. Maret W (2011) Redox biochemistry of mammalian metallothioneins. J Biol Inorg Chem 16(7):1079–1086
221. Maret W (2000) The function of zinc metallothionein: a link between cellular zinc and redox state. J Nutr 130(5S Suppl):1455S–1458S
222. Maret W, Vallee BL (1998) Thiolate ligands in metallothionein confer redox activity on zinc clusters. Proc Natl Acad Sci USA 95(7):3478–3482
223. Palacios O, Atrian S, Capdevila M (2011) Zn- and Cu-thioneins: a functional classification for metallothioneins? J Biol Inorg Chem 16(7):991–1009
224. Krezel A, Hao Q, Maret W (2007) The zinc/thiolate redox biochemistry of metallothionein and the control of zinc ion fluctuations in cell signaling. Arch Biochem Biophys 463(2):188–200
225. Bell SG, Vallee BL (2009) The metallothionein/thionein system: an oxidoreductive metabolic zinc link. ChemBioChem 10(1):55–62
226. Feng W et al (2006) Metallothionein disulfides are present in metallothionein-overexpressing transgenic mouse heart and increase under conditions of oxidative stress. J Biol Chem 281(2):681–687
227. Polec Pawlak K et al (2002) Monitoring of the metal displacement from the recombinant mouse liver metallothionein Zn(7)-complex by capillary zone electrophoresis with electrospray MS detection. Talanta 57(5):1011–1017
228. Perez-Rafael S et al (2011) Differential ESI-MS behaviour of highly similar metallothioneins. Talanta 83(3):1057–1061

229. Kirkpatrick RB et al (2003) A bicistronic expression system for bacterial production of authentic human interleukin-18. Protein Expr Purif 27(2):279–292
230. Wong JW, Hogg PJ (2010) Analysis of disulfide bonds in protein structures. J Thromb Haemost
231. Tang HY, Speicher DW (2004) Determination of disulfide-bond linkages in proteins. Curr Protoc Protein Sci Chap 11: Unit 11 11
232. Gorman JJ, Wallis TP, Pitt JJ (2002) Protein disulfide bond determination by mass spectrometry. Mass Spectrom Rev 21(3):183–216
233. Yen TY et al (2000) Characterization of cysteine residues and disulfide bonds in proteins by liquid chromatography/electrospray ionization tandem mass spectrometry. J Mass Spectrom 35(8):990–1002
234. Zhang Z Pan, Chen HX (2009) Mass spectrometry for structural characterization of therapeutic antibodies. Mass Spectrom Rev 28(1):147–176
235. Carr SA et al (1991) Integration of mass spectrometry in analytical biotechnology. Anal Chem 63(24):2802–2824
236. Bean MF, Carr SA (1992) Characterization of disulfide bond position in proteins and sequence analysis of cystine-bridged peptides by tandem mass spectrometry. Anal Biochem 201(2):216–226
237. Wu J, Watson JT (1997) A novel methodology for assignment of disulfide bond pairings in proteins. Protein Sci 6(2):391–398
238. Watson JT, Yang Y, Wu J (2001) Capture and identification of folding intermediates of cystinyl proteins by cyanylation and mass spectrometry. J Mol Graph Model 19(1):119–128
239. Wallis TP, Pitt JJ, Gorman JJ (2001) Identification of disulfide-linked peptides by isotope profiles produced by peptic digestion of proteins in 50 % (18)O water. Protein Sci 10(11):2251–2271
240. Legros C, Celerier ML, Guette C (2004) An unusual cleavage reaction of a peptide observed during dithiotreitol and tris(2-carboxyethyl)phosphine reduction: application to sequencing of HpTx2 spider toxin using nanospray tandem mass spectrometry. Rapid Commun Mass Spectrom 18(12):1317–1323
241. Fung YM et al (2005) Facile disulfide bond cleavage in gaseous peptide and protein cations by ultraviolet photodissociation at 157 nm. Angew Chem Int Ed Engl 44(39):6399–6403
242. Kleemann GR et al (2008) Characterization of IgG1 immunoglobulins and peptide-Fc fusion proteins by limited proteolysis in conjunction with LC-MS. Anal Chem 80(6):2001–2009
243. Kim HI, Beauchamp JL (2009) Mapping disulfide bonds in insulin with the Route 66 Method: selective cleavage of S-C bonds using alkali and alkaline earth metal enolate complexes. J Am Soc Mass Spectrom 20(1):157–166
244. Yamaguchi H, Miyazaki M, Maeda H (2010) Proteolysis approach without chemical modification for a simple and rapid analysis of disulfide bonds using thermostable protease-immobilized microreactors. Proteomics 10(16):2942–2949
245. Mozziconacci O, Kerwin BA, Schoneich C (2010) Photolysis of an intrachain peptide disulfide bond: primary and secondary processes, formation of H2S, and hydrogen transfer reactions. J Phys Chem B 114(10):3668–3688
246. Zhang Y, Dewald HD, Chen H (2011) Online mass spectrometric analysis of proteins/peptides following electrolytic cleavage of disulfide bonds. J Proteome Res 10(3):1293–1304
247. Yan B, Boyd D (2011) Breaking the light and heavy chain linkage of human immunoglobulin G1 (IgG1) by radical reactions. J Biol Chem 286(28):24674–24684
248. Kalli A (2009) Protein and Lantibiotic Sequencing by Gas Phase Dissociation Involving Vibrational Excitation and Ion Electron Reactions. Chap 2, Preferential cleavage of S–S and C-S bonds in electron detachment dissociation and infrared multiphoton dissociation of disulfide-linked peptide anions. Dissertation, University of Michigan
249. Patterson SD, Katta V (1994) Prompt fragmentation of disulfide-linked peptides during matrix-assisted laser desorption ionization mass spectrometry. Anal Chem 66(21):3727–3732

250. Gehrig PM, Biemann K (1996) Assignment of the disulfide bonds in napin, a seed storage protein from Brassica napus, using matrix-assisted laser desorption ionization mass spectrometry. Pept Res 9(6):308–314
251. Jones MD, Patterson SD, Lu HS (1998) Determination of disulfide bonds in highly bridged disulfide-linked peptides by matrix-assisted laser desorption/ionization mass spectrometry with postsource decay. Anal Chem 70(1):136–143
252. Katta V, Chow DT, Rohde MF (1998) Applications of in-source fragmentation of protein ions for direct sequence analysis by delayed extraction MALDI-TOF mass spectrometry. Anal Chem 70(20):4410–4416
253. Schnaible V et al (2002) Screening for disulfide bonds in proteins by MALDI in-source decay and LIFT-TOF/TOF-MS. Anal Chem 74(19):4980–4988
254. Huwiler KG, Mosher DF, Vestling MM (2003) Optimizing the MALDI-TOF-MS observation of peptides containing disulfide bonds. J Biomol Tech 14(4):289–297
255. Fukuyama Y, Iwamoto S, Tanaka K (2006) Rapid sequencing and disulfide mapping of peptides containing disulfide bonds by using 1, 5-diaminonaphthalene as a reductive matrix. J Mass Spectrom 41(2):191–201
256. Combes A et al (2009) Determination with matrix-assisted laser desorption/ionization tandem time-of-flight mass spectrometry of the extensive disulfide bonding in tarantula venom peptide Psalmopeotoxin I. Eur J Mass Spectrom (Chichester, Eng) 15(4):517-529
257. Kalkhof S et al (2008) Determination of disulfide bond patterns in laminin beta1 chain N-terminal domains by nano-high-performance liquid chromatography/matrix-assisted laser desorption/ionization time-of-flight/time-of-flight mass spectrometry. Rapid Commun Mass Spectrom 22(12):1933–1940
258. Okumura A et al (2009) Identification and assignment of three disulfide bonds in mammalian leukocyte cell-derived chemotaxin 2 by matrix-assisted laser desorption/ionization time-of-flight mass spectrometry. Biosci Trends 3(4):139–143
259. Yang H et al (2009) A new method for analysis of disulfide-containing proteins by matrix-assisted laser desorption ionization (MALDI) mass spectrometry. J Am Soc Mass Spectrom 20(12):2284–2293
260. Krasny L, Hynek R, Kodicek M (2011) Disulfide bond decay during matrix-assisted laser desorption/ionization time-of-flight mass spectrometry experiments. Rapid Commun Mass Spectrom 25(17):2468–2474
261. Janecki DJ, Nemeth JF (2011) Application of MALDI TOF/TOF mass spectrometry and collision-induced dissociation for the identification of disulfide-bonded peptides. J Mass Spectrom 46(7):677–688
262. Zhang M, Kaltashov IA (2006) Mapping of protein disulfide bonds using negative ion fragmentation with a broadband precursor selection. Anal Chem 78(14):4820–4829
263. Chelius D, Huff Wimer ME, Bondarenko PV (2006) Reversed-phase liquid chromatography in-line with negative ionization electrospray mass spectrometry for the characterization of the disulfide-linkages of an immunoglobulin gamma antibody. J Am Soc Mass Spectrom 17(11):1590–1598
264. Thakur SS, Balaram P (2008) Fragmentation of peptide disulfides under conditions of negative ion mass spectrometry: studies of oxidized glutathione and contryphan. J Am Soc Mass Spectrom 19(3):358–366
265. Bilusich D, Bowie JH (2009) Fragmentations of (M-H)- anions of underivatised peptides. Part 2: Characteristic cleavages of Ser and Cys and of disulfides and other post-translational modifications, together with some unusual internal processes. Mass Spectrom Rev 28(1):20–34
266. Andreazza HJ, Bowie JH (2010) The application of negative ion electrospray mass spectrometry for the sequencing of underivatized disulfide-containing proteins: insulin and lysozyme. Phys Chem Chem Phys 12(41):13400–13407
267. Nili M et al (2012) Defining the disulfide bonds of insulin-like growth factor-binding protein-5 by tandem mass spectrometry with electron transfer dissociation and collision-induced dissociation. J Biol Chem 287(2):1510–1519

268. Stephenson JL Jr, Cargile BJ, McLuckey SA (1999) Ion trap collisional activation of disulfide linkage intact and reduced multiply protonated polypeptides. Rapid Commun Mass Spectrom 13(20):2040–2048
269. Wu SL et al (2010) Identification of the unpaired cysteine status and complete mapping of the 17 disulfides of recombinant tissue plasminogen activator using LC-MS with electron transfer dissociation/collision-induced dissociation. Anal Chem 82(12):5296–5303
270. Wu SL et al (2009) Mass spectrometric determination of disulfide linkages in recombinant therapeutic proteins using online LC-MS with electron-transfer dissociation. Anal Chem 81(1):112–122
271. Wu SL et al (2007) On-line LC-MS approach combining collision-induced dissociation (CID), electron-transfer dissociation (ETD), and CID of an isolated charge-reduced species for the trace-level characterization of proteins with post-translational modifications. J Proteome Res 6(11):4230–4244
272. Mormann M et al (2008) Fragmentation of intra-peptide and inter-peptide disulfide bonds of proteolytic peptides by nanoESI collision-induced dissociation. Anal Bioanal Chem 392(5):831–838
273. Wang Y et al (2011) Characterization and comparison of disulfide linkages and scrambling patterns in therapeutic monoclonal antibodies: using LC-MS with electron transfer dissociation. Anal Chem 83(8):3133–3140
274. Agarwal A, Diedrich JK, Julian RR (2011) Direct elucidation of disulfide bond partners using ultraviolet photodissociation mass spectrometry. Anal Chem 83(17):6455–6458
275. Choi S et al (2010) New algorithm for the identification of intact disulfide linkages based on fragmentation characteristics in tandem mass spectra. J Proteome Res 9(1):626–635
276. Clark DF et al (2011) Collision induced dissociation products of disulfide-bonded peptides: ions result from the cleavage of more than one bond. J Am Soc Mass Spectrom 22(3):492–498
277. Lee T et al (2007) An algorithmic approach to automated high-throughput identification of disulfide connectivity in proteins using tandem mass spectrometry. Comput Syst Bioinformatics Conf 6:41–51
278. Le M, Means GE (1995) A procedure for the determination of monothiols in the presence of dithiothreitol–an improved assay for the reduction of disulfides. Anal Biochem 229(2):264–271
279. Wright SK, Viola RE (1998) Evaluation of methods for the quantitation of cysteines in proteins. Anal Biochem 265(1):8–14
280. Riener CK, Kada G, Gruber HJ (2002) Quick measurement of protein sulfhydryls with Ellman's reagent and with 4, 4′-dithiodipyridine. Anal Bioanal Chem 373(4–5):266–276
281. Giron P, Dayon L, Sanchez JC (2011) Cysteine tagging for MS-based proteomics. Mass Spectrom Rev 30(3):366–395
282. Chen SH, Hsu JL, Lin FS (2008) Fluorescein as a versatile tag for enhanced selectivity in analyzing cysteine-containing proteins/peptides using mass spectrometry. Anal Chem 80(13):5251–5259
283. Branchini BR et al (1997) Inactivation of firefly luciferase with N-(iodoacetyl)-N'- (5-sulfo-1-naphthyl)ethylenediamine (I-AEDANS). Arch Biochem Biophys 340(1):52–58
284. Clements A et al (2005) Fluorescence-based peptide labeling and fractionation strategies for analysis of cysteine-containing peptides. Anal Chem 77(14):4495–4502
285. Zhang W, Czupryn MJ (2002) Free sulfhydryl in recombinant monoclonal antibodies. Biotechnol Prog 18(3):509–513
286. Chumsae C, Gaza-Bulseco G, Liu H (2009) Identification and localization of unpaired cysteine residues in monoclonal antibodies by fluorescence labeling and mass spectrometry. Anal Chem 81(15):6449–6457
287. Xiang T, Chumsae C, Liu H (2009) Localization and quantitation of free sulfhydryl in recombinant monoclonal antibodies by differential labeling with 12C and 13C iodoacetic acid and LC-MS analysis. Anal Chem 81(19):8101–8108
288. Pipes GD et al (2005) Optimization and applications of CDAP labeling for the assignment of cysteines. Pharm Res 22(7):1059–1068

289. Petrotchenko EV et al (2006) Combining fluorescence detection and mass spectrometric analysis for comprehensive and quantitative analysis of redox-sensitive cysteines in native membrane proteins. Anal Chem 78(23):7959–7966
290. Volke D, Hoffmann R (2008) Quantitative proteomics by fluorescent labeling of cysteine residues using a set of two cyanine-based or three rhodamine-based dyes. Electrophoresis 29(22):4516–4526
291. Haniu M et al (1994) Direct assignment of disulfide bonds by Edman degradation of selected peptide fragments. Int J Pept Protein Res 43(1):81–86
292. Gray WR (1993) Disulfide structures of highly bridged peptides: a new strategy for analysis. Protein Sci 2(10):1732–1748
293. Jones G et al (2001) Disulfide structure of alfimeprase: a recombinant analog of fibrolase. Protein Sci 10(6):1264–1267
294. Keck RG et al (1997) Disulfide structure of the heparin binding domain in vascular endothelial growth factor: characterization of posttranslational modifications in VEGF. Arch Biochem Biophys 344(1):103–113
295. Shu Q, Huang R, Liang S (2001) Assignment of the disulfide bonds of huwentoxin-II by Edman degradation sequencing and stepwise thiol modification. Eur J Biochem 268(8):2301–2307
296. Trachsel C et al (2009) Elucidation of the disulfide bridge pattern of the recombinant human growth and differentiation factor 5 dimer and the interchain Cys/Ala mutant monomer. Anal Biochem 390(2):103–108
297. Young Y et al (1999) Disulfide assignment of the C-terminal cysteine knot of agouti-related protein (AGRP) by direct sequencing analysis. J Pept Res 54(10604596):514–521
298. Gupta K, Kumar M, Balaram P (2010) Disulfide bond assignments by mass spectrometry of native natural peptides: cysteine pairing in disulfide bonded conotoxins. Anal Chem 82(19):8313–8319
299. Mandal AK et al (2007) Sequencing of T-superfamily conotoxins from Conus Virgo: pyroglutamic acid identification and disulfide arrangement by MALDI mass spectrometry. J Am Soc Mass Spectrom 18(8):1396–1404
300. Qi J et al (2001) Determination of the disulfide structure of sillucin, a highly knotted, cysteine-rich peptide, by cyanylation/cleavage mass mapping. Biochemistry 40(15):4531–4538
301. Yen TY, Yan H, Macher BA (2002) Characterizing closely spaced, complex disulfide bond patterns in peptides and proteins by liquid chromatography/electrospray ionization tandem mass spectrometry. J Mass Spectrom 37(1):15–30
302. Iyer S, Acharya KR (2011) Tying the knot: the cystine signature and molecular-recognition processes of the vascular endothelial growth factor family of angiogenic cytokines. FEBS J 278(22):4304–4322
303. Chelius D et al (2001) Expression, purification and characterization of the structure and disulfide linkages of insulin-like growth factor binding protein-4. J Endocrinol 168(2):283–296
304. Neumann GM, Bach LA (1999) The N-terminal disulfide linkages of human insulin-like growth factor-binding protein-6 (hIGFBP-6) and hIGFBP-1 are different as determined by mass spectrometry. J Biol Chem 274(21):14587–14594
305. Edelman GM, Gall WE (1969) The antibody problem. Annu Rev Biochem 38:415–466
306. Gall WE et al (1968) The covalent structure of a human gamma G-immunoglobulin. IV The interchain disulfide bonds. Biochemistry 7(5):1973–1982
307. Gall WE, Edelman GM (1970) The covalent structure of a human gamma G-immunoglobulin. X. Intrachain disulfide bonds. Biochemistry 9(16):3188–3196
308. Padlan EA (1994) Anatomy of the antibody molecule. Mol Immunol 31(3):169–217
309. Correia IR (2010) Stability of IgG isotypes in serum. MAbs 2(3):221–232
310. Mhatre R, Woodard J, Zeng C (1999) Strategies for locating disulfide bonds in a monoclonal antibody via mass spectrometry. Rapid Commun Mass Spectrom 13(24):2503–2510
311. Dorai H, Wesolowski JS, Gillies SD (1992) Role of inter-heavy and light chain disulfide bonds in the effector functions of human immunoglobulin IgG1. Mol Immunol 29(12):1487–1491

312. McAuley A et al (2008) Contributions of a disulfide bond to the structure, stability, and dimerization of human IgG1 antibody CH3 domain. Protein Sci 17(1):95–106
313. Azuma T et al (1992) Kinetics of inter-heavy chain disulfide bond formation of liganded and unliganded human immunoglobulin G by radioimmunoassay. Mol Immunol 29(1):37–44
314. Thies MJ et al (2002) Folding and oxidation of the antibody domain C(H)3. J Mol Biol 319(5):1267–1277
315. Liu H et al (2011) Effect of the light chain C-terminal serine residue on disulfide bond susceptibility of human immunoglobulin G1lambda. Anal Biochem 408(2):277–283
316. Trexler-Schmidt M et al (2010) Identification and prevention of antibody disulfide bond reduction during cell culture manufacturing. Biotechnol Bioeng 106(3):452–461
317. Brych SR et al (2010) Characterization of antibody aggregation: role of buried, unpaired cysteines in particle formation. J Pharm Sci 99(2):764–781
318. Lacy ER, Baker M, Brigham-Burke M (2008) Free sulfhydryl measurement as an indicator of antibody stability. Anal Biochem 382(1):66–68
319. Bloom JW et al (1997) Intrachain disulfide bond in the core hinge region of human IgG4. Protein Sci 6(2):407–415
320. Aalberse RC, Schuurman J, van Ree R (1999) The apparent monovalency of human IgG4 is due to bispecificity. Int Arch Allergy Immunol 118(2–4):187–189
321. Schuurman J et al (1999) Normal human immunoglobulin G4 is bispecific: it has two different antigen-combining sites. Immunology 97(4):693–698
322. Schuurman J et al (2001) The inter-heavy chain disulfide bonds of IgG4 are in equilibrium with intra-chain disulfide bonds. Mol Immunol 38(1):1–8
323. Zhang W et al (2002) Complete disulfide bond assignment of a recombinant immunoglobulin G4 monoclonal antibody. Anal Biochem 311(1):1–9
324. Aalberse RC et al (2009) Immunoglobulin G4: an odd antibody. Clin Exp Allergy 39(4):469–477
325. Rispens T et al (2011) Mechanism of immunoglobulin G4 Fab-arm exchange. J Am Chem Soc 133(26):10302–10311
326. Rispens T, den Bleker TH, Aalberse RC (2010) Hybrid IgG4/IgG4 Fc antibodies form upon 'Fab-arm' exchange as demonstrated by SDS-PAGE or size-exclusion chromatography. Mol Immunol 47(7–8):1592–1594
327. Rispens T et al (2009) Human IgG4 binds to IgG4 and conformationally altered IgG1 via Fc-Fc interactions. J Immunol 182(7):4275–4281
328. Labrijn AF et al (2011) Species-specific determinants in the IgG CH3 domain enable Fab-arm exchange by affecting the noncovalent CH3-CH3 interaction strength. J Immunol 187(6):3238–3246
329. van der Neut Kolfschoten M et al (2007) Anti-inflammatory activity of human IgG4 antibodies by dynamic Fab arm exchange. Science 317(17872445):1554–1557
330. Rose RJ et al (2011) Quantitative analysis of the interaction strength and dynamics of human IgG4 half molecules by native mass spectrometry. Structure 19(9):1274–1282
331. Yoo EM et al (2003) Human IgG2 can form covalent dimers. J Immunol 170(12626570):3134–3138
332. Liu YD et al (2008) Human IgG2 antibody disulfide rearrangement in vivo. J Biol Chem 283(43):29266–29272
333. Martinez T et al (2008) Disulfide connectivity of human immunoglobulin G2 structural isoforms. Biochemistry 47(28):7496–7508
334. Van Buren N et al (2009) Elucidation of two major aggregation pathways in an IgG2 antibody. J Pharm Sci 98(9):3013–3030
335. Nielsen RW et al (2011) Trisulfides in proteins. Antioxid Redox Signal 15(1):67–75
336. Canova-Davis E et al (1996) Confirmation by mass spectrometry of a trisulfide variant in methionyl human growth hormone biosynthesized in *Escherichia coli*. Anal Chem 68(22):4044–4051
337. Pristatsky P et al (2009) Evidence for trisulfide bonds in a recombinant variant of a human IgG2 monoclonal antibody. Anal Chem 81(15):6148–6155

338. Gu S et al (2010) Characterization of trisulfide modification in antibodies. Anal Biochem 400(1):89–98
339. Aono H et al (2010) Efficient on-column conversion of IgG1 trisulfide linkages to native disulfides in tandem with Protein A affinity chromatography. J Chromatogr A 1217(32):5225–5232
340. Volkin DB, Klibanov AM (1987) Thermal destruction processes in proteins involving cystine residues. J Biol Chem 262(7):2945–2950
341. Tous GI et al (2005) Characterization of a novel modification to monoclonal antibodies: thioether cross-link of heavy and light chains. Anal Chem 77(9):2675–2682
342. Cohen SL, Price C, Vlasak J (2007) Beta-elimination and peptide bond hydrolysis: two distinct mechanisms of human IgG1 hinge fragmentation upon storage. J Am Chem Soc 129(22):6976–6977
343. Wang Z et al (2010) Desulfurization of cysteine-containing peptides resulting from sample preparation for protein characterization by mass spectrometry. Rapid Commun Mass Spectrom 24(3):267–275
344. Liu H et al (2007) Characterization of lower molecular weight artifact bands of recombinant monoclonal IgG1 antibodies on non-reducing SDS-PAGE. Biotechnol Lett 29(11):1611–1622
345. Herbert B et al (2003) Beta-elimination: an unexpected artefact in proteome analysis. Proteomics 3(6):826–831
346. Mozziconacci O, Kerwin BA, Schoneich C (2010) Exposure of a monoclonal antibody, IgG1, to UV-light leads to protein dithiohemiacetal and thioether cross-links: a role for thiyl radicals? Chem Res Toxicol 23(8):1310–1312
347. Philo JS, Arakawa T (2009) Mechanisms of protein aggregation. Curr Pharm Biotechnol 10(4):348–351
348. Chi EY et al (2003) Physical stability of proteins in aqueous solution: mechanism and driving forces in nonnative protein aggregation. Pharm Res 20(9):1325–1336
349. Philo JS (2009) A critical review of methods for size characterization of non-particulate protein aggregates. Curr Pharm Biotechnol 10(4):359–372
350. Philo JS (2006) Is any measurement method optimal for all aggregate sizes and types? AAPS J 8(3):E564–E571
351. Cao S, Pollastrini J, Jiang Y (2009) Separation and characterization of protein aggregates and particles by field flow fractionation. Curr Pharm Biotechnol 10(4):382–390
352. Kendrick BS et al (2001) Online size-exclusion high-performance liquid chromatography light scattering and differential refractometry methods to determine degree of polymer conjugation to proteins and protein-protein or protein-ligand association states. Anal Biochem 299(2):136–146
353. Philo JS (2001) Analytical ultracentrifugation. The BioPharm Guide to Bioanalytical Methods. Advanstar Communications, Cleveland, pp 52–54
354. Fowler JD et al (2009) Probing conformational changes of human DNA polymerase lambda using mass spectrometry-based protein footprinting. J Mol Biol 390(3):368–379
355. Kvaratskhelia M et al (2002) Identification of specific HIV-1 reverse transcriptase contacts to the viral RNA:tRNA complex by mass spectrometry and a primary amine selective reagent. Proc Natl Acad Sci USA 99(25):15988–15993
356. Wood TD et al (1998) Creatine kinase: essential arginine residues at the nucleotide binding site identified by chemical modification and high-resolution tandem mass spectrometry. Proc Natl Acad Sci USA 95(7):3362–3365
357. Engen JR (2009) Analysis of protein conformation and dynamics by hydrogen/deuterium exchange MS. Anal Chem 81(19):7870–7875
358. Geiger T, Clarke S (1987) Deamidation, isomerization, and racemization at asparaginyl and aspartyl residues in peptides. Succinimide-linked reactions that contribute to protein degradation. J Biol Chem 262(2):785–794
359. Tyler-Cross R, Schirch V (1991) Effects of amino acid sequence, buffers, and ionic strength on the rate and mechanism of deamidation of asparagine residues in small peptides. J Biol Chem 266(33):22549–22556

360. Gaza-Bulseco G, Liu H (2008) Fragmentation of a recombinant monoclonal antibody at various pH. Pharm Res 25(8):1881–1890
361. Sinha S et al (2009) Effect of protein structure on deamidation rate in the Fc fragment of an IgG1 monoclonal antibody. Protein Sci 18(8):1573–1584
362. Violand BN et al (1990) Formation of isoaspartate 99 in bovine and porcine somatotropins. J Protein Chem 9(1):109–117
363. Catak S et al (2008) Computational study on nonenzymatic peptide bond cleavage at asparagine and aspartic acid. J Phys Chem A 112(37):8752–8761
364. Joshi AB, Rus E, Kirsch LE (2000) The degradation pathways of glucagon in acidic solutions. Int J Pharm 203(1–2):115–125
365. Joshi AB et al (2005) Studies on the mechanism of aspartic acid cleavage and glutamine deamidation in the acidic degradation of glucagon. J Pharm Sci 94(9):1912–1927
366. Piszkiewicz D, Landon M, Smith EL (1970) Anomalous cleavage of aspartyl-proline peptide bonds during amino acid sequence determinations. Biochem Biophys Res Commun 40(5):1173–1178
367. Gerschler JJ, Wier KA, Hansen DE (2007) Amide bond cleavage: acceleration due to a 1, 3-diaxial interaction with a carboxylic acid. J Org Chem 72(2):654–657
368. Landon (1977) Cleavage at aspartyl-prolyl bonds. Methods Enzymol 47:145–149
369. Marcus F (1985) Preferential cleavage at aspartyl-prolyl peptide bonds in dilute acid. Int J Pept Protein Res 25(5):542–546
370. Voorter CE et al (1988) Spontaneous peptide bond cleavage in aging alpha-crystallin through a succinimide intermediate. J Biol Chem 263(35):19020–19023
371. Robinson NE (2002) Protein deamidation. Proc Natl Acad Sci USA 99(8):5283–5288
372. Liu H, Gaza-Bulseco G, Sun J (2006) Characterization of the stability of a fully human-monoclonal IgG after prolonged incubation at elevated temperature. J Chromatogr B Analyt Technol Biomed Life Sci 837(1–2):35–43
373. Cordoba AJ et al (2005) Non-enzymatic hinge region fragmentation of antibodies in solution. J Chromatogr B Analyt Technol Biomed Life Sci 818(2):115–121
374. Liu H, Gaza-Bulseco G, Lundell E (2008) Assessment of antibody fragmentation by reversed-phase liquid chromatography and mass spectrometry. J Chromatogr B Analyt Technol Biomed Life Sci 876(1):13–23
375. Jiskoot W et al (1990) Analytical approaches to the study of monoclonal antibody stability. Pharm Res 7(12):1234–1241
376. Alexander AJ, Hughes DE (1995) Monitoring of IgG antibody thermal stability by micellar electrokinetic capillary chromatography and matrix-assisted laser desorption/ionization mass spectrometry. Anal Chem 67(20):3626–3632
377. Rao PE, Kroon DJ (1993) Orthoclone OKT3. Chemical mechanisms and functional effects of degradation of a therapeutic monoclonal antibody. Pharm Biotechnol 5:135–158
378. Smith MA et al (1996) Specific cleavage of immunoglobulin G by copper ions. Int J Pept Protein Res 48(1):48–55

Chapter 7
Mass Spectrometry of Antibody–Drug Conjugates in Plasma and Tissue in Drug Development

Surinder Kaur, Keyang Xu, Ola Saad, Luna Liu, Tim Slattery and Randall Dere

7.1 Introduction

The discovery of "soft" ionization techniques in mass spectrometry (MS) such as electrospray ionization (ESI) [1] and matrix-assisted laser desorption/ionization (MALDI) [2] for measuring molecular masses of intact proteins was a significant breakthrough in the analysis of proteins. The work was recognized by a Nobel Prize in Chemistry in 2002 and led to the widespread use of ESI–MS to characterize intact protein molecular masses in protein therapeutics discovery and development in biotechnology. Previously, it was only possible to measure peptide molecular masses. In contrast to the analysis of intact purified proteins in simple buffers by ESI–MS, the ability to analyze biotherapeutic proteins in plasma or other tissues is significantly more challenging due to interference from the background plasma/tissue proteome and has only recently been reported for plasma [3]. The intact molecular mass measurement of biotherapeutic proteins in plasma by ESI–MS required isolation of the proteins from plasma using affinity capture followed by elution of the isolated intact biotherapeutics of interest and liquid chromatography (LC)–ESI–MS characterization to determine their intact molecular masses. The ability to obtain intact molecular masses and thereby characterize structural changes in biotherapeutics in plasma for in vivo studies provides key insights for large molecule drug development. Additional information can be obtained by enzymatic digestion followed by peptide analysis using LC-tandem MS (MS/MS) methodology. Understanding biotransformation and molecular changes of biotherapeutics in vivo is particularly valuable for the development of antibody–drug conjugates (ADCs) where efficacy and safety may be affected. It also provides essential structural characterization information for the ADCs in

S. Kaur (✉) · K. Xu · O. Saad · L. Liu · T. Slattery · R. Dere
Genentech, 1 DNA Way, South San Francisco, CA 94080–4990, USA
e-mail: kaur.surinder@gene.com

G. Chen (ed.), *Characterization of Protein Therapeutics using Mass Spectrometry*,
DOI: 10.1007/978-1-4419-7862-2_7, © Springer Science+Business Media New York 2013

vivo, necessary for designing appropriate quantitative assays for measuring pharmacokinetics (PK) and toxicokinetics (TK).

ADCs are monoclonal antibodies (mAbs) with covalently bound cytotoxic drugs. The drug is typically conjugated to the antibody via a chemical linker that reacts with either lysine or cysteine side chain residues in the antibody (Fig. 7.1) [4, 5]. The linker may be designed to be chemically or enzymatically cleavable or noncleavable [6–9]. The conjugation reaction results in a heterogeneous mixture of ADC molecules with a range of different drug-to-antibody ratios (DARs). Homogeneous ADCs with a defined DAR can also be produced by engineering reactive cysteine residues at specific sites in antibodies for the conjugation of drugs (Fig. 7.1) [10, 11]. The ADC molecules are designed to specifically bind to antigens that are overexpressed on the surface of tumor cells and minimally expressed on normal tissue. Upon binding, the ADCs are internalized and trafficked to lysosomes, where the cytotoxic drug or cytotoxic catabolites are subsequently designed to be released within the cell [6–9]. The targeted delivery and use of highly potent cytotoxic drugs are designed to enhance the antitumor effects of the molecule while reducing the systemic toxicity [12–14]. Highly potent cytotoxic agents that are otherwise too toxic to develop as therapeutics may be useful to develop as ADCs.

The concept of ADCs was first clinically validated by gemtuzumab ozogamicin, a conjugate of an anti-CD33 antibody and the cytotoxic agent calicheamicin. Gemtuzumab ozogamicin was approved by the Food and Drug

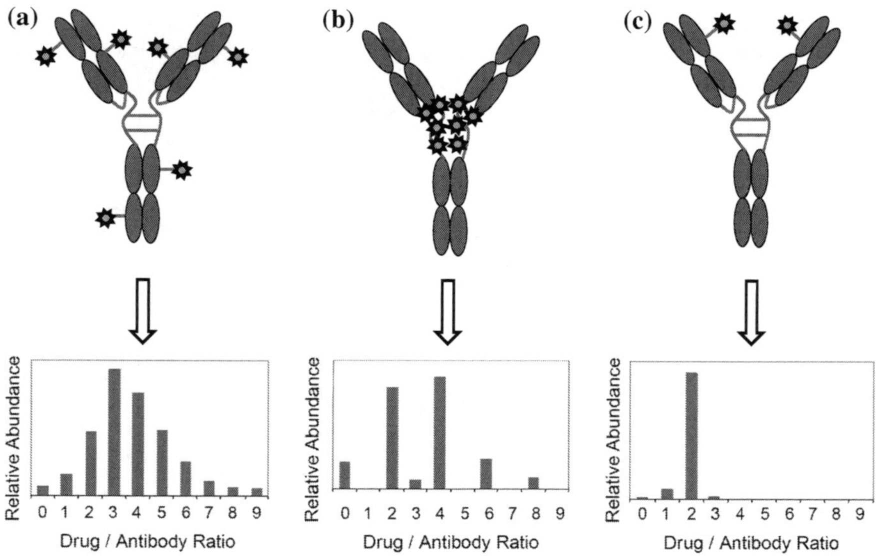

Fig. 7.1 ADC conjugation sites and drug/antibody ratio heterogeneity. **a** Conjugation through lysines. **b** Conjugation through reduced inter-chain disulfide bonds. **c** Conjugation through engineered cysteines

Administration (FDA) in 2000 for the treatment of patients with CD33-positive acute myeloid leukemia, a bone marrow cancer [15]. The product was recently withdrawn from the market after a later clinical trial raised new concerns about the product's safety and failed to demonstrate clinical benefit to patients [16]. A number of novel ADCs are currently in preclinical, early clinical, or late-stage clinical development for the treatment for solid and hematologic tumors [12–14, 17, 18]. In April 2011, the FDA approved Adcetris™ (brentuximab vedotin), an ADC, that targets CD30 on lymphoma cells, to treat Hodgkin lymphoma and a rare lymphoma known as systemic anaplastic large cell lymphoma under the accelerated approval program.

ADCs combine the molecular characteristics of small and large molecules; thus, methods designed for each therapeutic type have been used in drug development, in addition to new methods developed specifically for ADCs. Diverse bioanalytical methods are required for ADCs because of their structural complexity. Due to the tertiary structure and selective binding properties of large molecules, bioanalysis of biotherapeutics is predominantly performed by ligand-binding assays, for example, enzyme-linked immunosorbent assay (ELISA) [19–21]. Recently, MS-based methods have also been shown to provide quantitative data for large molecules in plasma and may provide an orthogonal bioanalytical method [22]. Chapters 2 and 3 in this book have detailed descriptions on the approaches. In general, these methods involve proteolytic digestion of the protein of interest in serum or plasma to generate a peptide specific for the protein; addition of an appropriate peptide stable isotope labeled internal standard and its quantification using LC–MS/MS. Bioanalysis of small molecule drugs is predominantly performed by plasma/serum extraction followed by LC–MS/MS quantification. It is challenging to develop ligand-binding assays for small molecules due to the lack of tertiary structure and potential steric hindrance issues of binding capture and detection reagents.

In addition to having complex molecular structures, ADCs are also typically complex mixtures comprising the antibody species with varying numbers of cytotoxic drugs attached resulting in a DAR distribution (Fig. 7.1). A variety of conventional large molecule and small molecule bioanalytical ELISA and LC–MS/MS methods have been employed to quantify ADCs and the cytotoxic drugs released from ADCs into circulation, respectively [23–25]. However, these methods have some limitations when used for the bioanalysis of ADCs. For instance, ELISA methods can measure the mAb concentration and conjugate concentration but cannot measure the ADC drug payload [25–27]. This is important to determine because the drug load or DARs can have a significant effect on in vitro and in vivo properties of ADCs [10, 28, 29]. On the other hand, current small molecule LC–MS/MS methods usually quantify a priori determined forms of the drug released from the ADC; however, this may not represent the major form of the drug released; for example, it is theoretically possible that the released drug contains part of the linker.

Bioanalytical methods developed specifically for ADCs include those that measure the total drug payload in circulation, including that conjugated to the antibody plus any released drug. For example, the total amount of monomethyl auristatin E (MMAE) in circulating ADC cAC10-MC-vc-PAB-MMAE, MMAE linked to antibody cAC10 by maleimido-caproyl-valine-citrulline-para-amino-benzyloxy-carbonyl, has been determined by incubating plasma samples with cathepsin B (to completely release the drug from the antibody) and then detecting the drug (MMAE) in a competition ELISA using an MMAE conjugated to horse-radish peroxidase (HRP) as a reporter and an anti-MMAE mAb coat [24]. Using a similar sample treatment strategy, LC–MS/MS could also provide an effective method to quantify the antibody-conjugated drug payload for ADCs in biological matrices. Another method developed specifically for ADCs is affinity capture capillary LC–ESI–MS to measure the intact molecular masses of ADCs in biological matrices [3]. This method involves isolation of the ADC from plasma or tissues by affinity capture followed by analysis of the molecular masses of the captured ADCs by capillary LC–ESI–MS. The molecular masses provide the DAR distribution and quantification of the relative amounts of the individual DAR ADC species. Thus, novel methods designed specifically for complex ADCs can provide additional information that cannot be obtained from conventional large and small molecule methods alone.

Additional complexity of ADCs can be generated in circulation in vivo due to biotransformations (Fig. 7.2). Early ADCs were based upon acid-cleavable hydrazone linkers that were relatively stable at neutral pH in the bloodstream (pH 7.3–7.5), while the more acidic environment within the cellular endosomes

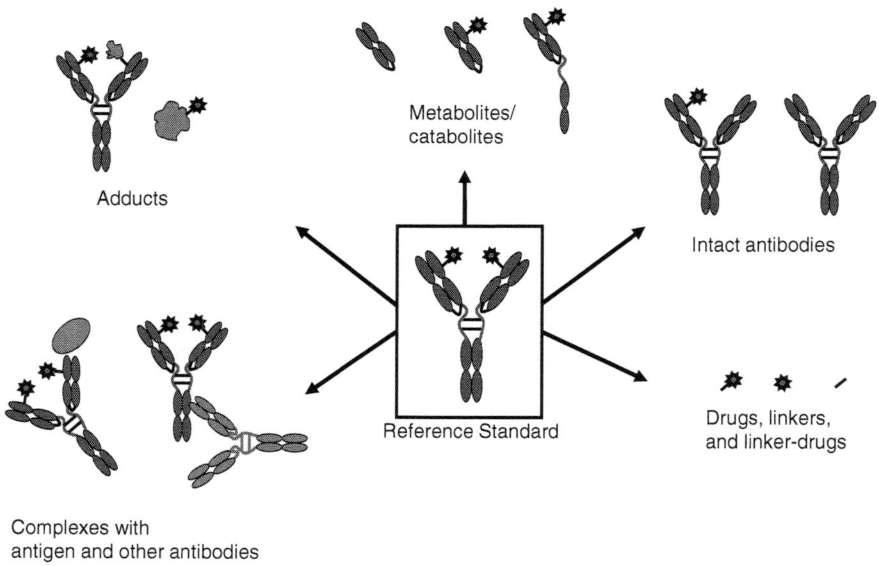

Fig. 7.2 ADC's complexity may increase in vivo due to catabolism or metabolism

(pH 5.0–6.5) and lysosomes (pH 4.5–5.0) resulted in hydrolysis of the linker after internalization of the ADCs and release of the drug payload [15, 30–32]. Enzymatically cleavable linkers, when the drug can be specifically released from the antibody by a lysosomal protease, such as cathepsin B, upon cleavage of the appropriate peptide bond in the linker have been reported [4, 33, 34] Linkers that are resistant to enzymatic or chemical cleavage have also been reported [35, 36].

Even for linkers designed to be entirely stable in plasma, unanticipated chemical or enzymatic activity in plasma may result in some drug release (deconjugation). Drug loss can result in the formation of ADC species containing lower DARs, thereby resulting in greater complexity in vivo. For a homogeneous ADC with a defined DAR, deconjugation in vivo may result in a mixture of DARs (Fig. 7.2) [3]. For example, loss of drug from a homogeneous ADC with a DAR of 2 (DAR2) in the dosing solution may result in a mixture of DAR0, DAR1, and DAR2. A reduction in the drug payload can affect the amount of drug being delivered to the site of action for efficacy and systemically released drug may pose a potential safety risk due to its high potency. For example, released drug may react with plasma proteins and peptides and could result in the formation of drug adducts that may have safety implications. Therefore, it is important to understand ADC structural changes (biotransformation) in vitro/in vivo. It is also important to understand biotransformations in plasma to ensure that assays developed for ADC quantification are suitable to measure the species that exist in circulation in addition to those present in the dosing solution.

In summary, ADCs present a new paradigm for bioanalysis to measure PK, TK, and catabolism/metabolism in drug development. It is important to use a variety of analytical methods to obtain both molecular characterization and quantification data. The molecular structures of ADCs are complex, they are typically mixtures, and there is the possibility of biotransformations in vivo that can increase the complexity further. Therefore, an integrated bioanalytical strategy incorporating a variety of large molecule ligand-binding assays, small molecule LC–MS/MS assays, novel quantitative approaches designed specifically for ADCs and protein mass spectrometric characterization methods are essential for ADC bioanalysis. Our bioanalytical strategy includes the molecular characterization of ADCs in biological matrices, for example, intact molecular mass measurement of ADCs in plasma by affinity capture LC–MS. This is important for understanding of the key analytes that circulate in vivo and must be measured appropriately by quantitative assays for PK and TK assessment. The quantitative assays include the use of ligand-binding and MS assays and are designed to measure specific molecular components in the mixture of ADC molecules in vivo, for example, the total amount of antibody, the total amount of conjugated antibody, the total amount of conjugated drug, released drug and released drug catabolites. A case-by-case strategy is used to select a subset of assays based on the ADC molecular structure. This chapter will focus on how MS can be incorporated into the bioanalytical strategies for ADCs and a case study including highlights of the ELISA and MS assays used.

7.2 Bioanalytical Methods

7.2.1 Affinity Capture Capillary LC–MS

Affinity capture LC–MS assay [3] was used to measure the relative intensities of ADCs with different DARs to obtain DAR distributions in plasma/serum. Typically, the biotinylated extracellular domain (ECD) of the receptor that is recognized by the antibody component of the ADC was immobilized onto streptavidin-coated paramagnetic beads. This affinity bead system was used to capture ADCs by incubating with the plasma/serum samples containing ADCs for approximately 2 h at room temperature. Following the affinity capture process, the bound ADCs were isolated, washed, and deglycosylated on the beads by incubating with PNGase F in HBS-EP buffer (0.01 M HEPES, pH 7.4, 0.15 M NaCl, 3 mM EDTA, 0.005 % surfactant P20) at 37 °C overnight. Subsequently, the beads were washed extensively with HBS-EP and water, and the ADC analytes were then eluted by 30 % acetonitrile in water for LC–MS analysis. A volume of 10 μL of the ADC elute was injected onto a PLRP-S column (50 × 0.3 mm, 5 μM, 4,000 Å) with a 15 μL/min flow rate. Typical mobile phase (acetonitrile and water containing 0.1 % formic acid) was used. Analytes were ionized by ESI and detected by a quadrupole time-of-flight (QTOF) mass spectrometer operated in the positive TOF–MS mode. Raw data of ADCs were deconvoluted, and peak area under curve was obtained for each ADC component of interest. Relative intensities for the ADC components were calculated.

7.2.2 LC–MS/MS Method for DM1 in Rat Plasma, Bile and Urine and Human Plasma

Since emtansine (DM1) has a thiol, it can dimerize and/or form disulfide bonds with thiol-containing molecules in plasma. Therefore, before extraction, lithium-heparin plasma samples were treated with 1 mM *tris*(2-carboxyethyl) phosphine (TCEP) a reducing agent to reduce any mixed disulfides with DM1, and then alkylated with n-ethyl maleimide (NEM) to block the resulting free thiol and prevent further reaction. The samples (30 μL) were extracted by protein precipitation using 80/20 acetonitrile and water (120 μL) containing 7.5 nM of maytansine (internal standard) and separated with a high-performance liquid chromatography (HPLC) system. A reverse phase analytical column (MAX-RP 80A, C12, 4 μ, 50 × 2.00 mm) heated to 50 °C was used for separation followed by analysis on a triple quadrupole mass spectrometer. The transition monitored for DM1–NEM was *m/z* 845.7/485.3, for DM1 (to ensure complete derivatization)–*m/z* 738.5/547.3, and for maytansine (IS)–*m/z* 692.6/547.2. For DM1–NEM, the lower limits of quantitation (LLOQ) in rat plasma were 0.24 nM (0.18 ng/mL), and 0.49 nM (0.36 ng/mL) in rat bile and urine. In human plasma samples, the LLOQ of DM1–NEM in the LC–MS/MS assay was 1.00 nM (0.737 ng/mL). DM1 was not detected in rat and human plasma, or in rat bile or urine.

7.2.3 LC–MS/MS Method for MCC-DM1, and Lys-MCC-DM1 in Rat Plasma, Bile and Urine and Human Plasma (Exploratory Catabolites)

The samples (30 μL) were extracted by protein precipitation using 80/20 acetonitrile and water (120 μL) containing 7.5 nM of maytansine (internal standard) and analyzed for MCC (4-(N-maleimidomethyl) cyclohexane-1-carboxylate)-DM1 and Lys-MCC-DM1 on a triple quadrupole mass spectrometer by LC–MS/MS (as described above). Multiple reaction monitoring scan mode was used for quantitation. Transition *m/z* 975.3/547.4 was monitored for MCC-DM1, *m/z* 1,103.9/485.5 for Lys-MCC-DM1 and *m/z* 692.3/547.1 for maytansine (IS). For MCC-DM1, the LLOQ of the assay was 3.91 nM (3.81 ng/mL) in rat plasma and bile and 0.98 nM (0.95 ng/mL) in rat urine. For Lys-MCC-DM1, the LLOQ in rat plasma, bile, and urine was 1.95 nM (2.15 ng/mL), 7.81 nM (8.61 ng/mL), and 3.91 nM (4.31 ng/mL), respectively. For patient samples, MCC-DM1 and Lys-MCC-DM1 were measured in a similar manner, and the LLOQs in human plasma were 1.95 nM (1.90 ng/mL) and 0.976 nM (1.08 ng/mL), respectively.

7.2.4 ELISA Methods for T-DM1 in Rat and Cynomolgus Monkey Serum

Two ELISAs, total trastuzumab ELISA and conjugated trastuzumab emtansine (T-DM1) ELISA, were developed and validated to measure T-DM1 (Figs. 7.3 and 7.4) concentrations in serum samples from Sprague-Dawley rats and cynomolgus monkeys. The total trastuzumab ELISA measures both conjugated and

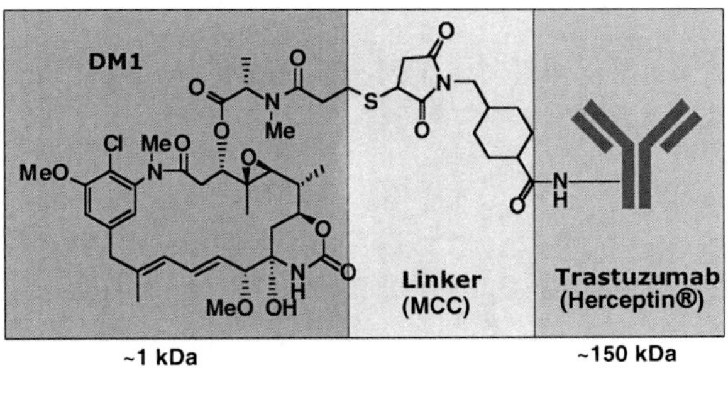

Fig. 7.3 Structure of trastuzumab-DM1 (T-DM1)

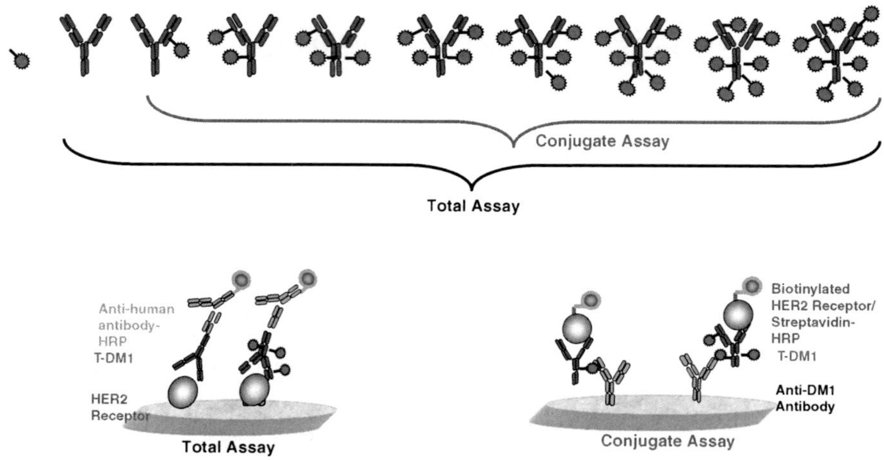

Fig. 7.4 Formats of T-DM1 ligand-binding assays

unconjugated trastuzumab and the conjugated T-DM1 ELISA measures trastuzumab conjugated with one or more DM1 [37]. The general formats of the assays are shown in Fig. 7.4. The total trastuzumab ELISA utilized recombinant human epidermal growth factor receptor 2 (HER2) ECD as capture reagent and peroxidase-conjugated F(ab′)2 goat anti–human IgG Fc for detection. The total trastuzumab ELISA was designed to measure T-DM1 with one or more covalently bound DM1 molecules and unconjugated trastuzumab. The minimum quantifiable concentration for total T-DM1 in both rat and cynomolgus monkey serum was 40 ng/mL. The conjugated T-DM1 ELISA utilized a murine anti-DM1 mAb as the capture reagent and biotinylated recombinant HER2 ECD and HRP-conjugated streptavidin for detection. The conjugated T-DM1 ELISA was designed to measure T-DM1 conjugate containing one or more covalently bound DM1. The minimum quantifiable concentrations for conjugated T-DM1 in rat and cynomolgus monkey serum were 30 and 40 ng/mL, respectively. For both validated assays, serum samples were quantified against T-DM1 calibrators with an average DAR of approximately 3.5.

7.2.5 Tissue Analysis by Affinity Capture Capillary LC–MS and LC–MS/MS

Exploratory tissue samples (lung, liver, and kidney) were collected during toxicity studies for ADC DAR distribution by affinity capture capillary LC–MS and for MMAE analysis by LC–MS/MS. At necropsies, appropriate tissue sections were collected, weighed, flash-frozen in liquid N_2 and stored at −80 °C. Tissues were then homogenized using a Mini-Beadbeater. To prevent any potential drug release from the ADC during tissue homogenization, a protease inhibitor cocktail consisting of 4-(2-aminoethyl) benzenesulfonyl fluoride (AEBSF), E-64, bestatin, leupeptin, aprotinin, and sodium EDTA was added to the tissues prior to homogenization. Tissue homogenate was analyzed by affinity capture LC–MS for the DAR distribution using a QTOF mass spectrometer coupled with a reversed phase capillary PolymerLab PLRP-S column (50 × 0.3 mm). Free MMAE drug was quantified by LC–MS/MS using a QTrap mass spectrometer coupled with a reversed phase Phenomenex Synergy Max-RP column (50 × 4.6 mm). In both cases, mobile phase A was 0.1 % formic acid in water and mobile phase B was 0.1 % formic acid in acetonitrile.

7.3 ADC Data Analysis and Discussion

The qualitative and quantitative MS methods for ADCs and how these methods can be integrated into the overall bioanalytical strategy during development of biotherapeutics are illustrated using a case study of T-DM1 during clinical development for HER2-positive metastatic breast cancer. Additional applications of MS to study ADC biotransformations in plasma and tissue and explore biotransformation mechanisms in vivo are illustrated using model trastuzumab thio-mAbs and thio-mAbs incorporating a variety of mAbs (thio-mAb1, thio-mAb2, and thio-mAb3) as models for ADC compounds.

7.3.1 T-DM1 Case Study

The HER2 is a transmembrane receptor tyrosine kinase that is part of a complex signal transduction network that plays an important role in cell differentiation, proliferation, and survival during morphogenesis [38]. In healthy adults, HER2 is expressed at relatively low levels in normal epithelial tissue [39] but is overexpressed in approximately 25–30 % of tumors from patients with breast cancer [40, 41]. These HER2-positive breast tumors are associated with aggressive growth and poor clinical outcomes [40, 41]. Trastuzumab, a humanized antibody directed against the extracellular region of HER2, is approved for the treatment of HER2-overexpressing breast cancer. However, some patients do not respond to trastuzumab or relapse following treatment. Therefore, there is a need for additional therapies [42, 43].

T-DM1 is an ADC that contains trastuzumab, a nonreducible thioether linker (MCC) and a maytansine derivative (DM1) that inhibits microtubule polymerization (Fig. 7.3) [6, 44–46]. T-DM1 is a mixture composed of trastuzumab with an approximately Poisson distribution of 0–8 DM1 molecules linked via MCC, primarily to lysine residues. The average DAR for T-DM1 achieved during conjugation is approximately 3.5 [47]. It has been shown that after T-DM1 binds to

HER2, the T-DM1/HER2 complex is internalized via endocytosis and degraded in lysosomes, ultimately leading to the intracellular release of lysine-MCC-DM1 [35, 36, 48]. Lys-MCC-DM1, MCC-DM1 and DM1 were identified as the major T-DM1 catabolites in rat plasma and were detected at low nanomolar levels in plasma collected from metastatic breast cancer patients during clinical development (data not shown).

A total of six assays were used to characterize the PK and catabolism of T-DM1 in serum/plasma for nonclinical and clinical studies (Figs. 7.4, 7.5, 7.6, 7.7). This included the use of three validated quantitative assays for all T-DM1 nonclinical and clinical studies: two large molecule ligand-binding assays (Fig. 7.4) and one small molecule LC–MS/MS assay (Fig. 7.5). In addition, two small molecule exploratory quantitative LC–MS/MS catabolite assays (Fig. 7.6) and one novel affinity capture LC–MS method specifically designed for ADC characterization in plasma and tissues (Fig. 7.7) were used in selected nonclinical and clinical studies.

The two large molecule PK ELISAs (Fig. 7.4) were designed to measure (1) total trastuzumab (including fully conjugated, partially deconjugated and fully deconjugated trastuzumab that are capable of binding to HER2 ECD) and (2) conjugated T-DM1 (trastuzumab conjugated to one or more DM1). The validated small molecule LC–MS/MS assay was designed to measure total DM1 released from T-DM1 (Fig. 7.5). Since DM1 contains a sulfhydryl moiety, it is possible that released DM1 could bind to plasma proteins or peptides or dimerize. Thus, the DM1 assay sample preparation included a reduction step to ensure measurement of these forms of DM1. Additional exploratory LC–MS/MS assays were used to

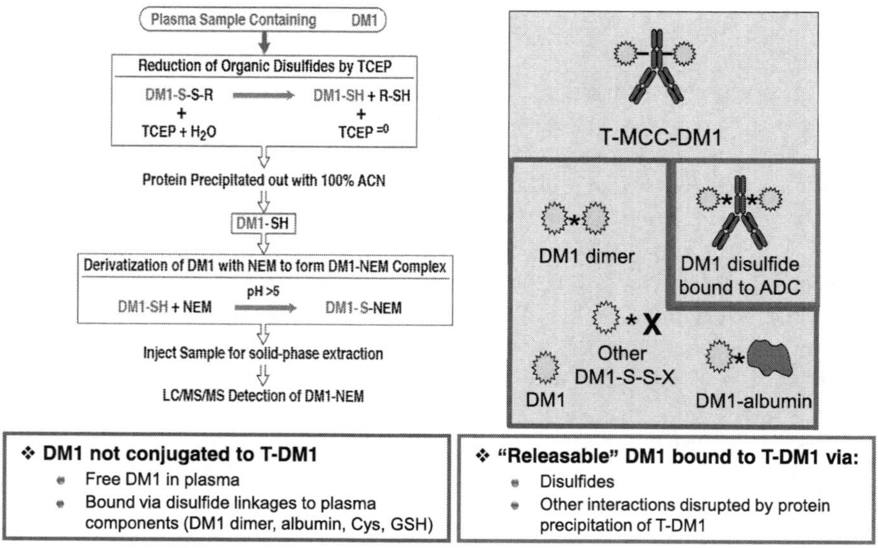

Fig. 7.5 General procedure for DM1 LC–MS/MS assay [53]. Reproduced with permission of Future Science Ltd

measure the catabolites, Lys-MCC-DM1 and MCC-DM1 for selected nonclinical and clinical studies (Fig. 7.6) (data not shown). In addition, a novel affinity capture LC–MS assay was used to measure intact T-DM1 isolated from plasma in selected samples collected from plasma stability, and rat and cynomolgus monkey studies (Fig. 7.7) [3].

ADCs present unique challenges for ligand-binding assays primarily due to the dynamically changing nature of the mixtures in vivo. The DAR distribution of the

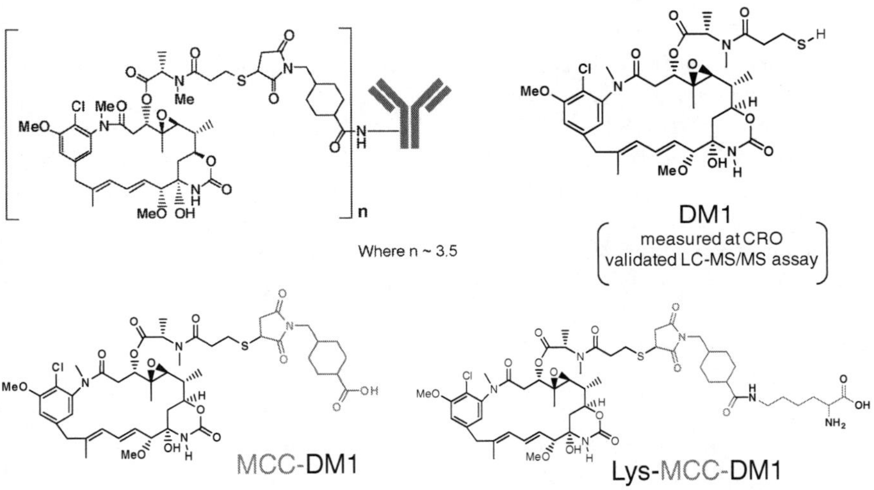

Fig. 7.6 T-DM1 analytes measured in biological matrices for nonclinical and clinical studies

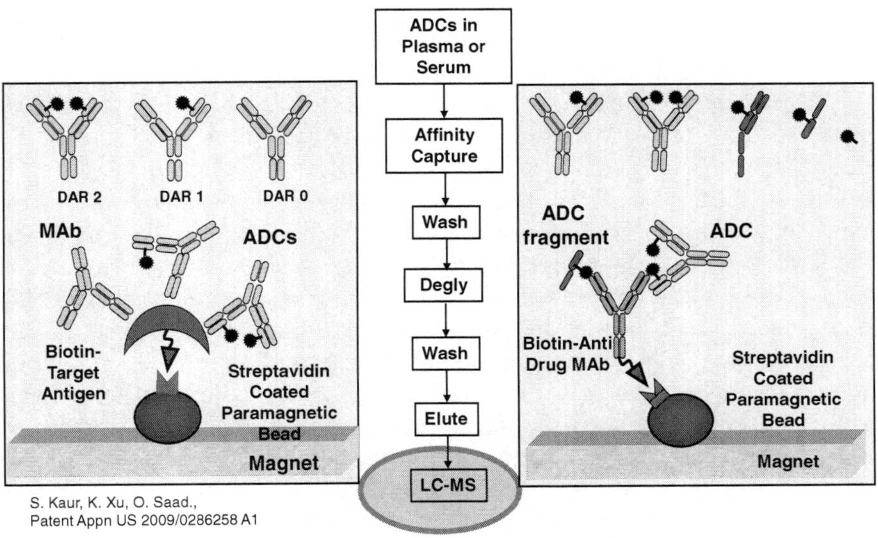

Fig. 7.7 General procedure for affinity capture capillary LC–MS

analyte mixture can change in vivo over time due to drug loss [3]. As the calibration curve of the assay is made up of the product reference material that represents the dosed starting ADC mixture, it is important to understand the impact of changing DARs on the assay accuracy. For example, the binding of the assay reagents to the analytes may vary with the individual DARs. For reagents that bind to the antibody portion of the ADC, a high drug load may interfere with binding. In the case of the anti-drug reagents, low avidity could result in under-quantification when the drug load is low, for example, DAR1. Ideally, it would be preferable to assess the assay recovery using individually isolated DARs for each component in the mixture, but in practice with existing technology, it is not often feasible to obtain individual DARs for a such complex mixture, particularly those present at low levels, for example, high DARs or DAR1.

A variety of T-DM1 DAR lots available for characterizing the ligand-binding assays were tested in the assays. This included data from testing T-DM1 mixtures with average DARs ranging from 2.6 to 4.1 in both of the T-DM1 ELISAs and confirmed that the assay had acceptable recovery across the range of DARs tested. Additional confirmation that T-DM1 reagents had acceptable recoveries across the range of DARs was obtained indirectly using affinity capture LC–MS (see below).

DM1 concentrations in nonclinical and clinical plasma samples were determined using validated LC–MS/MS assays. The assays were designed to measure free DM1 and any disulfide-bound forms of released DM1 (e.g., dimers, glutathione, cysteine, and albumin adducts) in plasma and to exclude DM1 that remained conjugated to trastuzumab via MCC-DM1 (Fig. 7.5). LC–MS/MS detection of DM1 was performed by detection of the alkylated form of DM1 (DM1–NEM). The three validated assays described above have been used for T-DM1 quantification in rat and cynomolgus monkey plasma samples collected during PK and TK studies (data not shown) and in human plasma samples collected during clinical studies.

Changes in T-DM1 DAR distribution in plasma in vitro and in vivo were measured using an affinity capture LC–MS method (Fig. 7.7) [3]. This method allowed the characterization of T-DM1 DAR distribution in vitro and in vivo directly by measuring the molecular masses of the ADC species isolated from plasma by affinity capture [3]. In addition to providing an understanding of the T-DM1 DAR species present in plasma, the data were important for appropriate characterization of the conjugate ELISA by confirming the ability of the anti-DM1 mAb reagent to recover all DAR species. Figure 7.7 shows the general procedure for the affinity capture LC–MS method for T-DM1. Streptavidin-coated paramagnetic beads coupled with biotinylated capture probes were used to capture intact T-DM1 in plasma. For ADCs, the capture probe can be the target antigen, or an antibody that specifically recognizes the ADC antibody. A probe that recognizes the antibody is important for measuring the DAR distribution as this will capture all DARs, including DAR0. Alternatively, the capture probe can be a mAb that recognizes the drug. A probe that recognizes the drug is important for understanding drug-related molecules that may be formed in vivo, for example, some drug may deconjugate and bind to other proteins. For T-DM1, two capture probes were used;

HER2 ECD to capture the ADC molecules via trastuzumab and an anti-DM1 mAb to capture DM1-containing molecules.

The method development included optimization of the paramagnetic bead type for analyte capture, the amount of beads, the amount of T-DM1, sample volumes, wash conditions for removal of background plasma proteins and elution buffers for recovery of T-DM1 analytes from the beads. For example, Fig. 7.8 shows the extracted mass spectra obtained using six different bead types. The optimal conditions are described in the methods section. The T-DM1 captured on the beads was deglycosylated, eluted from the beads, and analyzed by a Q-TOF MS coupled with a capillary flow LC. This method was able to resolve the molecular masses of individual T-DM1 molecules with different DARs to show the DAR distribution (DARs0–8) and allow determination of the relative abundance of each DAR. The DAR distribution accuracy and precision for the general affinity capture LC–MS method was previously published for more simple ADC mixtures made from individual known standards of DARs0–2 and found to be within 15 % [3].

To ensure that there was no bias during the affinity capture step, the DAR distribution data for T-DM1 spiked into human plasma and isolated by affinity capture was compared to the DAR distribution data for T-DM1 spiked into

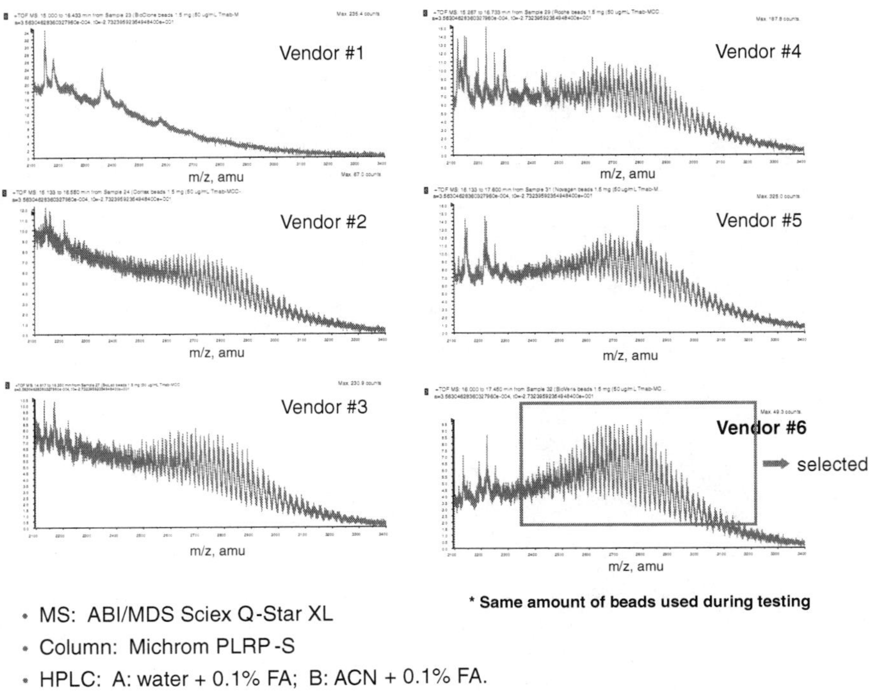

Fig. 7.8 Comparison of mass spectra from the affinity capture LC–MS of T-DM1 using six different bead types during assay development

buffer and analyzed directly without affinity capture. Figure 7.9 shows the deconvoluted mass spectrum of T-DM1 reference material in buffer overlaid with the corresponding affinity capture LC–MS mass spectrum for T-DM1 from plasma using an anti-DM1 mAb capture probe. The data showed comparable drug distributions with DARs ranging from 1 to 8, indicating that there were no selective losses or bias during the affinity capture LC–MS procedure. As expected, T-DM1 DAR0 (naked trastuzumab) was not observed in the mass spectrum after affinity capture with anti-DM1 antibody as the anti-DM1 affinity probe captures the analyte via DM1. These data (Fig. 7.9) also indicated that the anti-DM1 mAb affinity probe is capable of capturing low DAR species such as DAR1 effectively and provided support for characterization of the conjugate T-DM1 ELISA, where the anti-DM1 mAb is a critical reagent. Since drug development studies are conducted in a variety of animal species and in the clinic, affinity capture LC–MS data were compared for T-DM1 spiked into a variety of plasma matrices. Figure 7.10 shows the T-DM1 DAR distribution determined by LC–MS in buffer by direct injection and in plasma by affinity capture LC–MS using an anti-DM1 mAb capture probe in mouse, rat, cynomolgus monkey, and human plasma. For

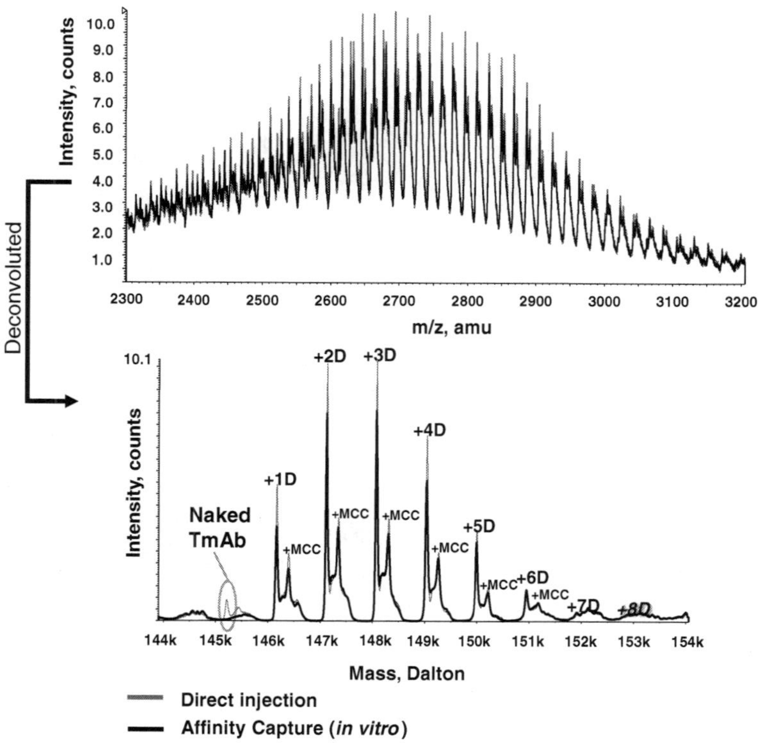

Fig. 7.9 Comparison of mass spectra from direct injection of T-DM1 and T-DM1 after affinity capture using an anti-DM1 mAb affinity probe

simplicity, only DARs1–7 are shown in Fig. 7.10 as the concentration of DAR8 was minimal and hence not meaningful to compare. In each plasma matrix, the DAR distribution for DARs1–7 was comparable to that obtained in buffer, indicating that there is no species-dependent bias. A comparison of the DAR distribution from affinity capture LC–MS using an anti-DM1 mAb capture probe and a target antigen HER2 ECD capture probe is shown in Fig. 7.11. As expected, only the HER2 ECD capture probe was capable of measuring DAR0, and this difference was observed in the DAR distribution. Also, the determined DAR distribution for the conjugated T-DM1 molecules was comparable for the two capture probes (Fig. 7.11).

Once the affinity capture LC–MS method was developed and optimized, it was used to gain insight into changes in the DAR distribution of T-DM1 for an in vitro plasma stability study (Fig. 7.12). The study involved incubation of 100 μg/mL of T-DM1 in human plasma at 37 °C for up to 96 h. Samples were frozen at −70 °C

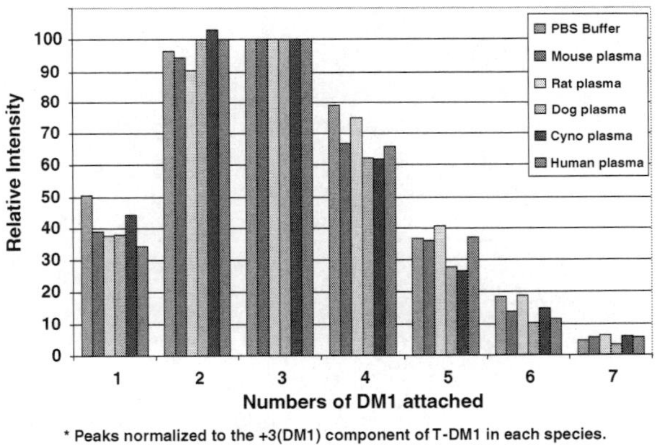

Fig. 7.10 Comparison of the DAR distributions from the affinity capture LC–MS of T-DM1 in PBS buffer, mouse, rat, dog, cynomolgus monkey and human plasma

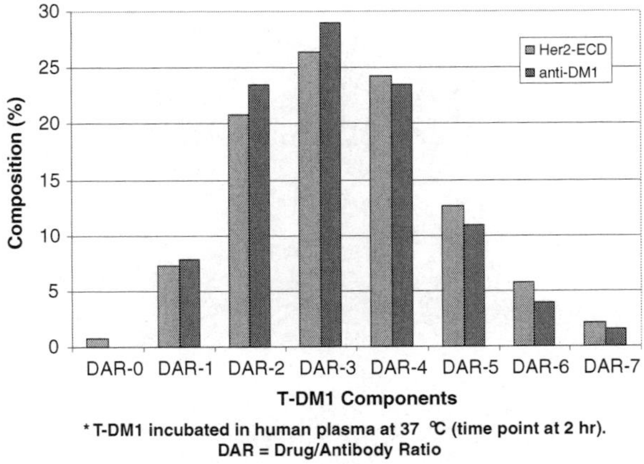

Fig. 7.11 Comparison of the DAR distributions from the affinity capture LC–MS of T-DM1 using an anti-DM1 mAb capture probe and an antigen capture probe

until analysis. An aliquot of 20 μL was used for affinity capture LC–MS analysis. A control experiment was conducted in the phosphate-buffered saline (PBS) buffer with 1 % bovine serum albumin (BSA) carrier protein under the same conditions. Figure 7.12 illustrates the affinity capture LC–MS data using an anti-DM1 mAb capture probe comparing T-DM1 DAR distribution in human plasma and the PBS buffer. The DAR in PBS buffer at 37 °C at all time points including the end of the 96 h study was virtually identical to the starting material, indicating that T-DM1 was stable in buffer under these conditions (Fig. 7.12). In contrast, changes in the DAR distribution were observed in plasma. For example, Fig. 7.12 shows the DAR distributions at 0 h and at 48 h in plasma. The relative amount of the higher DARs3–5 appeared to be decreasing, indicating losses of drug in plasma. The MCC linker is designed to be stable as observed in buffer; however, clearly in plasma some drug is lost with incubation at 37 °C. The mechanism of drug loss in plasma has not been studied in T-DM1 to date. However, insights into ADC drug deconjugation processes have been gained using less structurally complex engineered thio-mAb ADCs (DAR2) using the affinity capture LC–MS method [3]. It was observed that although maleimide is stable in buffer, in plasma, maleimide-containing linkers can exchange with sulfhydryl containing molecules such as albumin or cysteine in plasma by a nonenzymatic mechanism [49]. Therefore, it is possible that deconjugation of DM1 may also involve exchange of the maleimide in the MCC linker [50].

Affinity capture LC–MS using a HER2 ECD capture probe was used to determine the changes in T-DM1 DAR distribution in vivo in a cynomolgus monkey plasma samples collected during a PK study. Using a HER2 ECD capture probe allowed the measurement of fully deconjugated T-DM1 (DAR0) in addition to measuring changes in DARs1–8. Figure 7.13 shows the deconvoluted mass spectra of T-DM1 in cynomolgus monkey plasma up to 28 days. The spectra are normalized to the most abundant DAR to allow visualization of T-DM1 DAR distribution at later time points where the absolute amounts in plasma are relatively

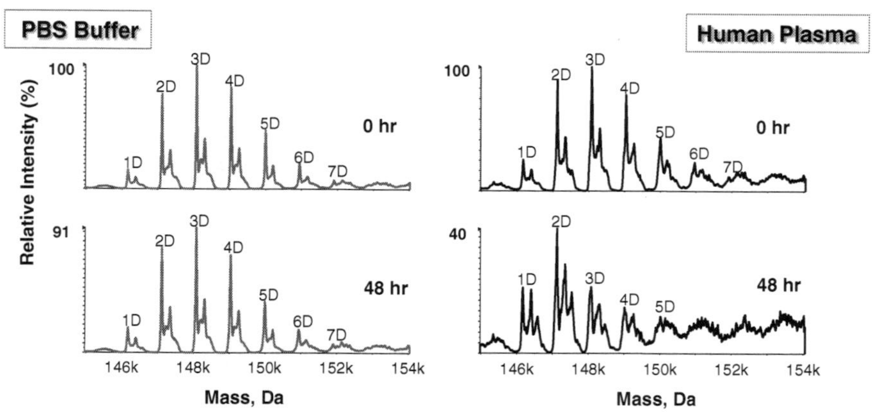

Fig. 7.12 Changes in the DAR distributions of T-DM1 by affinity capture LC–MS during in vitro plasma stability studies at 37 °C in PBS buffer and human plasma

low due to clearance. The DAR distribution at time 2 min (Fig. 7.13) is comparable to the DAR distribution in the dosing solution (Fig. 7.9). Up to day 7, the DAR distribution remains relatively unchanged. However, beyond day 7, the distribution is seen to change, where the relative abundances of higher DARs, for example, DAR4, DAR5, and DAR6 are seen to decrease. This is reflected in the average DAR for the distribution calculated from the peak areas that starts at an average DAR of 3.16 at 2 min and decreases over time to an average DAR of 0.66 at the end of the study (Fig. 7.13). The DAR0 relative abundance is low throughout the 28-day time course studied. Interestingly, at later time points, for example, day 10, new molecular masses are observed that are intermediate between the DAR molecules. The structures of these species are not yet known. Overall, the affinity capture LC–MS provided direct evidence that although some DM1 drug is lost from T-DM1, the majority of the trastuzumab molecules carry at least one DM1 throughout the 28-day duration of the PK study.

7.3.2 The Use of Structurally Less Complex Engineered Model ADCs (DAR2) to Elucidate the Mechanisms of ADC Deconjugation in Plasma

Understanding the stability and mechanisms of ADC deconjugation in vivo is important for designing and developing this new class of biotherapeutics. However, this is analytically challenging for biotherapeutic ADCs that are typically complex

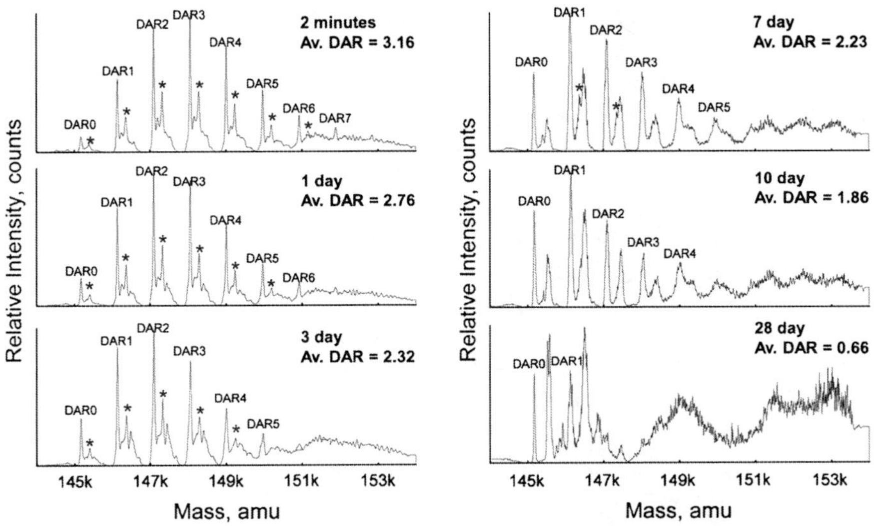

Fig. 7.13 Characterization of ADC (T-DM1) drug distribution in a cynomolgus PK study by HER2 ECD affinity capture LC–MS shows the DAR distribution shifts to lower values over time. Spectrum is normalized to the major component at each time point, * Extra MCC linker

heterogeneous mixtures, such as T-DM1 (DARs1–8). We have used engineered thio-mAb model ADCs that are structurally homogeneous with a DAR of 2 to gain insight into the mechanisms of deconjugation [49]. Three model ADCs trastuzumab-based thio-mAb structures with specific conjugation sites in the light chain (LC-V205C), heavy chain-Fab region (HC-A114C), or heavy chain-Fc region (Fc-S396C) were conjugated with MMAE via a maleimido-caproyl-valine-citrulline-p-amino-benzyloxy-carbonyl (MC-vc-PAB) linker or conjugated with monomethyl auristatin F (MMAF) via a maleimido-caproyl (MC) linker to obtain DAR2 model ADCs. The MC-vc-PAB (vc) linker contains an enzymatic cleavage site at valine-citrulline while the MC linker is enzymatically stable. These thio-mAb ADC molecules (Fig. 7.1), called thio-trastuzumab-vc-MMAE, thio-trastuzumab-MC-MMAF and analogous heavy chain thio-mAb ADCs with different mAbs (mAb1, mAb2, and mAb3), are structurally less complex than T-DM1 and allowed the accurate investigation of degree and rates of drug loss, the effect of the site of conjugation on drug loss, the structural characterization of the forms of the drug loss, that is, the component of the linker lost with the drug, and the molecular mechanism of the linker drug loss [49]. Mechanistic insights gained from these less complex ADC structures can be used to hypothesize analogous mechanisms for more complex biotherapeutic ADCs.

Model ADCs with cleavable linkers also allowed the development of a quantitative LC–MS/MS approach to understand drug deconjugation. As an example, the total antibody-conjugated drug payload was measured in vitro and in vivo. This process involved enrichment of the ADC in plasma using protein A affinity fractionation, enzymatic digestion to cleave the linker to release the drug and quantification of the drug using LC–MS/MS. This approach allowed assessment of the stability in plasma in vitro by quantifying the amount of antibody-conjugated drug in the ADC over time and allowed the direct comparison of changes in the antibody-conjugated drug payload for different model ADCs. Model ADCs were important to optimize and assess the performance of this method. Once optimized, the method was validated for more complex mixture ADCs, to ensure that the protein A affinity isolation and enzymatic digestion at the cleavage site in the linker is not compromised for mixtures. This provided a useful MS-based method for quantifying conjugated drug for in vitro stability and PK assessment in vivo.

In order to investigate the effect of the site of conjugation on the stability of ADCs, an in vitro plasma stability study was conducted by incubating LC-V205C, HC-A114C, and Fc-S396C thio-trastuzumab-vc-MMAE ADCs in human plasma at 37 °C for 96 h. They were also incubated with buffer for the same period of time as a control experiment. The DAR distribution was analyzed using the affinity capture capillary LC–MS method described earlier [3]. Prior to incubation with plasma, all three ADCs showed predominantly a DAR of 2. No changes were observed in any of the three ADCs in buffer after incubation for 96 h indicating these were stable in plasma [49]. In contrast, in plasma the three ADCs showed differing stability. The LC-V205C ADC showed virtually no change and retained its DAR of 2 throughout the 96 h incubation period, indicating it is stable in plasma. The HC-A114C ADC showed a significant formation of DAR1, indicating drug loss from DAR2, and by 96 h there were similar amounts of DARs1 and 2. The Fc-S396C ADC showed a rapid conversion of DAR2 to DAR1 and DAR0 species within the first 24 h indicating it was the least stable ADC. Clearly these changes in DAR distribution were plasma dependent as no changes in DAR distribution were observed in buffer over the same time period. These data indicate that the rate of drug loss from the antibody is dependent on the conjugation site and the three thio-trastuzumab-vc-MMAE ADCs can be ranked in the order of decreasing stability in plasma (from high to low): LC-V205C > HC-A114C > Fc-S396C. Thus, affinity capture LC–MS of model DAR2 thio-mAb ADCs provides valuable insights into ADC stability and structural considerations, such as site of conjugation, which impacts ADC stability [49].

MS was also used to study the mechanism of drug loss from ADCs in plasma during stability experiments [49]. The DAR distribution of the ADCs containing the enzyme cleavable MC-vc-PAB-MMAE linker drug was compared with DAR distribution data from the corresponding ADCs conjugated via the noncleavable MC-MMAF linker drug. Surprisingly, similar conjugation site–dependent DAR distribution changes in the order of decreasing stability in plasma (light chain (LC) > heavy chain (HC) > fragment, crystallizable (Fc)) were observed for both the enzymatically cleavable and the noncleavable linkers in plasma, while both linkers were stable in buffer [49]. This suggested that the cleavage mechanism in plasma was not dependent on the vc enzymatic cleavage site as similar deconjugation was observed in the ADC with the MC linker (noncleavable linker).

Additional insights into the deconjugation mechanism were gained by evaluation of the mass spectral data. Consistent molecular mass decreases were observed with deconjugation in plasma during in vitro stability and nonclinical PK studies for a variety of model DAR2 thio-mAbs: approximately 1,200 Da for MC-vc-PAB-MMAE conjugates and 810 Da for MC-MMAF conjugates [49]. The mass shifts suggested loss of the entire linker drug from the respective ADCs followed by a possible addition of cysteine to the antibody to form a disulfide bond. It was earlier reported that albumin-MC-MMAF adducts in plasma were observed in the case of an anti-CD30-MC-MMAF conjugate [51]. Therefore, it was of interest to determine whether albumin-linker drug could be observed for a variety of model ADCs using affinity capture LC–MS with an anti-drug antibody capture probe. Three thio-trastuzumab-MC-MMAF ADCs (LC-V205C, HC-A114C, and Fc-S396C) and heavy chain thio-mAb1 ADC with noncleavable MC-MMAF linker drug were incubated in plasma at 37 °C and analyzed by affinity capture LC–MS. In each case, as the relative abundance of the intact ADC DAR2 decreased, a lower molecular mass protein of approximately 67,370 Da was observed to increase [49]. The molecular mass of this new protein was consistent with albumin-MC-MMAF. Similarly, albumin-MC-vc-PAB-MMAE adducts were observed upon incubation of the corresponding ADCs with the enzyme cleavable MC-vc-PAB-MMAE linker drug in plasma. The Fc-S396C ADC showed the most albumin adduct formation, and the LC-V205C ADC showed very little albumin adduct formation indicating a good correlation with the rate of drug loss from ADC described above. This maleimide exchange phenomenon was also observed for ADCs with MC-vc-PAB-MMAE linker drugs, where linker drug was transferred to cysteine or glutathione if incubated in plasma with an excess of cysteine or glutathione. These results indicate that the maleimide exchange phenomenon for maleimide-containing linker drugs appears to be common for ADCs in plasma at 37 °C. It is noteworthy that maleimide exchange has only been observed in plasma and not in buffers under the same conditions.

Further insight into the mechanisms of deconjugation was gained from the intact molecular masses generated from model thio-mAb ADCs in vivo by affinity capture LC–MS [49]. Plasma samples collected from cynomolgus monkeys that received a single intravenous injection of thio-mAb2-MC-vc-PAB-MMAE or thio-mAb1-MC-MMAF ADCs (both HC DAR2 thio-mAb variants HC-A114C) were analyzed by affinity capture LC–MS using the corresponding target antigen affinity capture probes. Although both ADCs had different mAbs, linkers, and drugs, they showed similar DAR distribution profiles from day 1 to day 21, where the relative abundance of DAR2 decreased and that of DAR1 and DAR0 increased. This finding confirmed that the drug loss from these ADCs in vivo was independent of linker, drug, and antibody [49]. The same phenomenon was observed in in vitro plasma stability studies using the same ADCs [49]. Surprisingly, LC–MS analysis of plasma samples beyond day 35 from the PK studies showed minimal further loss of drug from DAR2 and DAR1 indicating the maleimide exchange process was no longer occurring after this time [49]. Careful examination of the mass spectra (day 1–day 35) revealed mass increases of 18 Da for DAR1 and 36 Da for DAR2 in mass spectra of samples collected on day 35 post-dose, indicating possible changes in their chemical structures [49]. The mass increase of 18 Da per linker drug was consistent with hydrolysis of the succinimide ring.

Further characterization of the proposed succinimide hydrolysis product was performed by tryptic digestion and peptide analysis. Plasma samples from the model thio-mAb2-vc-PAB-MMAE (HC-A114C variant) PK study above were affinity purified using an anti-antigen capture probe, digested with trypsin, and analyzed by LC–MS/MS. This further localized the 18 Da increase to the mass of tryptic peptide containing the MC-vc-PAB-MMAE linker drug. MS/MS sequencing data of this peptide identified that the site of the 18 Da mass increase was specifically at the Cys containing the MC-vc-PAB linker conjugation (likely on the succinimide ring) and not at any other amino acid in the peptide or at the MMAE portion of the linker drug (data not shown).

Thus, the MS data indicated that by day 35 in the PK studies of model DAR2 ADCs containing MC-vc-PAB or MC linkers, the succinimide ring in these linkers can hydrolyze resulting in an 18 Da mass increase. Once hydrolyzed, the succinimide no longer exchanges with other thiols in plasma and therefore helps stabilize the linker. Different rates of succinimide ring hydrolysis have been observed for model DAR2 thio-mAb ADCs in plasma in vitro and in vivo, depending on the site of linker drug conjugation [49]. This suggests that the linker stability of ADCs with site-specific conjugation on light chains may be higher due to faster succinimide ring hydrolysis in plasma [49].

The possibility of other structural alterations at the site of linker drug loss in model thio-mAb ADCs was explored, as the molecular mass change upon loss of the linker drug did not correspond exactly with the mass of the linker drug. For example, the 1,200 Da mass shift observed upon loss of linker drug in MC-vc-PAB-MMAE conjugates is greater than the molecular mass of the linker drug (1,319 Da). The hypothesis that a multi-step process giving rise to an overall mass shift of 1,200 Da was investigated by structural characterization of affinity-purified samples from a thio-mAb2-MC-vc-PAB-MMAE plasma stability study (data not shown). The DAR2, DAR1, and DAR0 affinity-purified mixture from capture with an anti-antigen probe in plasma was subjected to a partial proteolysis using endoproteinase Lys-C, without reduction. Analysis of the Lys-C digest by LC–MS showed components with molecular masses of 47,229 Da and 48,434 Da. The mass of 48,434 Da is consistent with the molecular mass of an expected Lys-C proteolytic product for the ADC, corresponding to the antibody fragment, antigen binding (Fab fragment) plus the linker drug. However, the mass of 47,229 Da is an unexpected mass, corresponding to the antibody Fab fragment plus 119 Da. Based on the molecular mass of cysteine being 119 Da, it was hypothesized that the (Fab+119) Da corresponds to Fab+cysteine. The loss of the entire linker drug combination (−1,319 Da) followed by the addition of cysteine (+119 Da) at the conjugation site would result in an overall mass loss of 1,200 Da from the intact MC-vc-PAB-MMAE ADCs. This corresponds to the mass loss observed in MC-vc-PAB-MMAE model DAR2 thio-mAbs in plasma upon conversion of DAR2 to DAR1 and an additional 1,200 Da loss for conversion from DAR1 to DAR0. Thus, the presence of (Fab+119) Da and the mass loss of 1,200 Da during loss of linker drug support the addition of 119 Da at the conjugation site in MC-vc-PAB-MMAE ADCs upon loss of linker drug in plasma. It is likely that this ADC structural modification at the site of linker drug loss in plasma is the addition of cysteine.

To confirm whether the 119 Da structural modification at the site (HC-A114C conjugate) of linker drug loss was cysteine, disulfide reduction was performed to assess if the 119 Da was linked via a disulfide bond and therefore reducible. Disulfide reduction of the partial Lys-C digest of thio-mAb2-MC-vc-PAB-MMAE mixture of DAR0, DAR1, and DAR2 (discussed above), containing the (Fab+119) Da species, was performed with TCEP. After the reduction, the LC–MS analysis showed unmodified light chain, unmodified HC Fd fragment (amino terminal half of the HC), HC Fd+(MC-vc-PAB) and HC Fd+(MC-vc-PAB) + 18 Da components (data not shown). The light chain and the conjugated HC (Fd-MC-vc-PAB-MMAE) components were expected products from reduction of the Lys-C digest of the thio-mAb2-MC-vc-PAB-MMAE DARs mixture. Interestingly, an unconjugated HC Fd fragment (23,632 Da) was observed, while an Fd+119 Da (23,751 Da) was not observed. This indicates that the 119 Da moiety in the Fab+119 Da fragment in the Lys-C digest of the thio-mAb2-MC-vc-PAB-MMAE thio-mAb DARs from plasma was reducible with TCEP. Therefore, these data support the possible addition of cysteine at the conjugation site after the loss of the entire MC-vc-PAB-MMAE, and this modification may also occur during deconjugation of other maleimide-containing linkers.

7.3.3 Measuring Intact ADCs and Released Drug in Tissues

Since the biodistribution of antibodies is not well understood, tissue analysis is important for ADCs as there may be nonspecific mechanisms for the biodistribution of ADCs. It is also possible that deconjugated drug-containing fragments or drug adducts could also traffic to tissues. Any drug-containing species trafficked to tissues could have the potential for toxicity. MS is commonly used to measure small molecule drugs in tissues using quantitative methods that are analogous to those used for plasma analysis, for example, organic extraction of the drug from tissue followed by LC–MS/MS. Similarly, qualitative MS methods developed for the analysis of proteins in plasma can be applied to tissues [52]. For ADCs, both the large molecule qualitative methods and small molecule quantitative methods developed in plasma can be extended to tissue analysis. Comparison of ADC analyses in plasma versus tissue, different tissues and subsections of the same tissue type can provide valuable insights into the fate of ADCs in vivo such as stability to proteolysis of the antibody in different tissue types and differences in ADC deconjugation processes across different tissues.

The following example illustrates the affinity capture LC–MS and released cytotoxic drug analysis in a variety of rat tissues using a model HC DAR2 thio-mAb3-MC-vc-PAB-MMAE ADC. Tissue was collected for analysis following the administration of four weekly intravenous doses. Plasma and tissue samples including lung and kidney were analyzed for the DAR distribution by affinity capture LC–MS and analyzed for the cytotoxic drug, MMAE by LC–MS/MS. It is possible that enzyme activity in the tissue may release some cytotoxic drug during tissue sample processing. To test this hypothesis, 15 μg of the thio-mAb3-MC-vc-PAB-MMAE was spiked into 0.3 g rat intact lung samples and the tissue was prepared using a variety of homogenization and room temperature storage conditions [52]. LC–MS/MS data showed an increase in MMAE with the homogenization time and the room temperature storage period [52]. Release of MMAE during homogenization was minimized by the addition of protease inhibitor cocktail. MMAE levels in rat tissues were measured by LC–MS/MS using an optimized tissue homogenization procedure with addition of a protease inhibitor cocktail. Tissue concentrations of MMAE appeared to be higher than that in plasma (Fig. 7.14a). The higher concentrations in tissues indicate that MMAE might be produced from the intact thio-mAb trafficked to the tissues. Or, it is possible that MMAE could partition there from plasma, as the octanol–water partition coefficient for MMAE suggests that it could accumulate in tissues. Figure 7.14b shows representative affinity capture LC–MS spectra from lung and kidney and indicates the formation of similar DAR distributions for thio-mAb3-MC-vc-PAC-MMAE in vivo in each of the tissues. The DAR distributions are also similar to those typically observed in plasma for HC thio-mAbs. Based on a limited sample set, this suggests that the deconjugation phenomenon in tissues is not significantly different from that in plasma.

The characterization of intact ADCs in tissues by affinity capture LC–MS and LC–MS/MS free drug quantification allows for the comparison of qualitative and

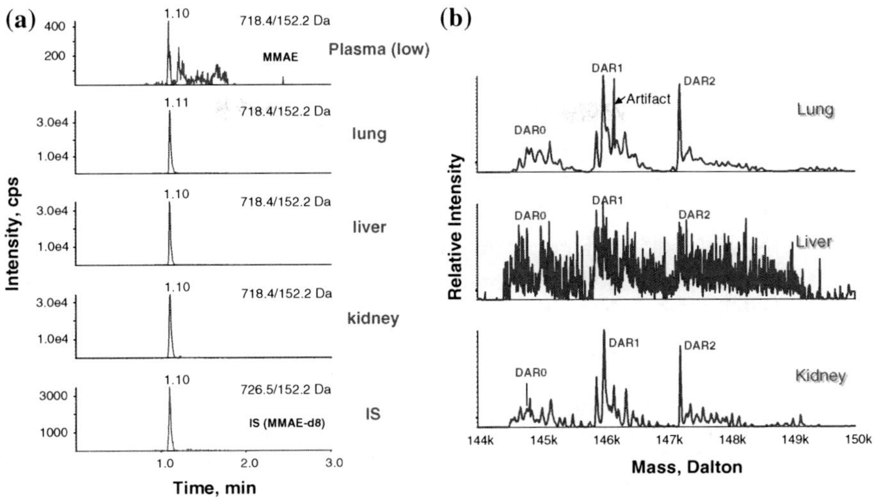

Fig. 7.14 Mass spectra of thio-mAb3-MC-vc-PAB-MMAE ADC from rat tissues **a** MMAE drug, **b** intact DAR distributions

quantitative concentrations of ADCs across different tissues and plasma. The quantification data for the rat study described above suggested that MMAE in tissues might be generated in situ upon tissue uptake of intact ADCs or MMAE could partition there from plasma. The qualitative characterization of the model heavy chain thio-mAb3-MC-vc-PAB-MMAE in a variety of tissues suggested that the deconjugation in tissues does not appear to be significantly different from that in plasma. Overall, both quantitative and qualitative MS measurements of ADCs in tissues may provide valuable insights into mechanisms of ADC trafficking and mechanisms of biotransformation in tissues.

7.4 Conclusions

ADCs are complex biotherapeutics where additional complexity can be generated by a variety of biotransformation mechanisms in vivo. MS is an essential tool to help understand the fate of ADCs in vivo to gain insights that can help evaluate safety and efficacy. This includes qualitative measurement of intact molecular masses of ADCs in plasma by affinity capture LC–MS using either target antigen affinity probes or anti-drug antibody affinity probes. Intact ADC molecular masses provide valuable DAR distribution data. Due to the high molecular mass of ADCs, in excess of 150 kDa, it may be challenging to detect all biotransformations at the intact molecular level. To assess structural changes that may not be apparent at the intact molecule level, the ADC can be reduced to light and heavy chains and then analyzed. With lower masses of the reduced form, additional resolution at the lower mass allows ready assignment of molecular mass shifts resulting from biotransformation. Confirmation of molecular structural changes, such as succinimide hydrolysis, can be obtained by enzymatic digestion of affinity captured ADCs and peptide-tandem MS. In addition to providing structural characterization information, MS can be used to quantify ADC components of interest. This includes quantification of catabolites and metabolites, for example, free drug, linker drug. Methods for protein quantification using proteomic approaches may also be applicable to quantify ADCs. Finally, MS is a critical component of the overall bioanalytical strategy for ADCs. Characterization of the biotransformation of ADCs in vivo provides important data for understanding analytes that should be quantified by ligand-binding assays. It is critical to integrate the qualitative and quantitative MS analytical strategies with ligand-binding strategies for ADC bioanalysis during drug development.

Acknowledgments The authors would like to thank the following Genentech colleagues for their support of mass spectrometric strategies for antibody drug conjugate bioanalysis, providing study samples and helpful discussions: Jagath Junutula, Ben-Quan Shen, Doug Leipold, Jay Tibbitts, Sandhya Girish, Kelly Flagella, Susan Spencer, Mark Sliwkowski, and Paul Polakis.

References

1. Fenn JB, Mann M, Meng CK et al (1989) Electrospray ionization for mass spectrometry of large biomolecules. Science 246:64–71
2. Karas M, Hillenkamp F (1988) Laser desorption ionization of proteins with molecular masses exceeding 10,000 daltons. Anal Chem 60:2299–2301
3. Xu K, Liu L, Saad OM et al (2011) Characterization of intact antibody-drug conjugates from plasma/serum in vivo by affinity capture capillary LC-MS. Anal Biochem 412:56–66
4. Carter PJ, Senter PD (2008) Antibody-drug conjugates for cancer therapy. Cancer J 14(3):154–169
5. Schrama D, Reisfeld RA, Becker JC (2006) Antibody targeted drug as cancer therapeutics. Nat Rev Drug Discov 5:147–159
6. Lewis Phillips GD, Li G, Dugger DL et al (2008) Targeting HER2-positive breast cancer with trastuzumab-DM1, an antibody-cytotoxic drug conjugate. Cancer Res 68:9280–9290
7. DiJoseph JF, Goad ME, Dougher MM et al (2004) Potent and specific antitumor efficacy of CMC-544, a CD22-targeted immunoconjugate of calicheamicin, against systemically disseminated B-cell lymphoma. Clin Cancer Res 10(24):8620–8629
8. Ojima I, Geng X, Wu X et al (2002) Tumor-specific novel taxoid-monoclonal antibody conjugates. J Med Chem 45(26):5620–5623
9. Wahl AF, Klussman K, Thompson JD et al (2002) The anti-CD30 monoclonal antibody SGN-30 promotes growth arrest and DNA fragmentation in vitro and affects antitumor activity in models of Hodgkin's disease. Cancer Res 62(13):3736–3742
10. Junutula JR, Raab H, Clark S et al (2008) Site-specific conjugation of a cytotoxic drug to an antibody improves the therapeutic index. Nat Biotechnol 26:925–932
11. Junutula JR, Bhakta S, Raab H et al (2008) Rapid identification of reactive cysteine residues for site-specific labeling of antibody-Fabs. J Immunol Methods 332:41–52
12. Lambert JM (2005) Drug-conjugated monoclonal antibodies for the treatment of cancer. Curr Opin Pharmacol 5:543–549
13. Wu AM, Senter PD (2005) Arming antibodies: prospects and challenges for immunoconjugates. Nat Biotechnol 23:1137–1146

14. Alley SC, Okeley NM, Senter PD (2010) Antibody-drug conjugates: targeted drug delivery for cancer. Curr Opin Chem Biol 14(4):529–537
15. Bross PF, Beitz J, Chen G et al (2001) Approval summary: gemtuzumab ozogamicin in relapsed acute myeloid leukemia. Clin Cancer Res 7:1490–1496
16. FDA (2010) Mylotarg withdrawal. http://www.fda.gov/Safety/MedWatch/SafetyInformation/SafetyAlertsforHumanMedicalProducts/ucm216458.htm
17. Payne G (2003) Progress in immunoconjugate cancer therapeutics. Cancer Cell 3:207–212
18. Polakis P (2005) Arming antibodies for cancer therapy. Curr Opin Pharmacol 5:382–387
19. Engvall E, Perlmann P (1971) Enzyme-linked immunosorbent assay (ELISA). Quantitative assay of immunoglobulin G. Immunochemistry 8:871–874
20. Lequin RM (2005) Enzyme immunoassay (EIA)/enzyme-linked immunosorbent assay (ELISA). Clin Chem 51:2415–2418
21. DeSilva B, Smith W, Weiner R et al (2003) Recommendations for the bioanalytical method validation of ligand-binding assays to support pharmacokinetic assessments of macromolecules. Pharm Res 20(11):1885–1900
22. Ezan E, Dubois M, Bscher F (2009) Bioanalysis of recombinant proteins and antibodies by mass spectrometry. Analyst 134:825–834
23. Francisco JA, Cerveny CG, Meyer DL et al (2003) cAC10-vcMMAE, an anti-CD30-monomethyl auristatin E conjugate with potent and selective antitumor activity. Blood 102:1458–1465
24. Sanderson RJ, Hering MA, James SF et al (2005) In vivo drug-linker stability of an anti-CD30 dipeptide-linked auristatin immunoconjugate. Clin Cancer Res 11:843–852
25. Tolcher AW, Ochoa L, Hammond LA et al (2003) Cantuzumab Mertansine, a maytansinoid immunoconjugate directed to the CanAg antigen: A phase I, pharmacokinetic, and biologic correlative study. J Clin Oncol 21:211–222
26. Stephan JP, Chan P, Lee C et al (2008) Anti-CD22-MCC-DM1 and MC-MMAF conjugates: impact of assay format on pharmacokinetic parameters determination. Bioconjug Chem 19:1673–1683
27. Stephan JP, Kozak KR, Wong WLT (2011) Challenges in developing bioanalytical assays for characterization of antibody–drug conjugates. Bioanalysis 3(6):677–700
28. King HD, Yurgaitis D, Willner D et al (1999) Monoclonal antibody conjugates of doxorubicin prepared with branched linkers: A novel method for increasing the potency of doxorubicin immunoconjugates. Bioconjug Chem 10:279–288
29. Hamblett KJ, Senter PD, Chace DF et al (2004) Effects of drug loading on the antitumor activity of a monoclonal antibody drug conjugate. Clin Cancer Res 10:7063–7070
30. Bross PF, Beitz J, Chen G et al (2001) Approval summary: gemtuzumab ozogamicin in relapsed acute myeloid leukemia. Clin Cancer Res 7:1490–1496
31. Hamann PR, Hinman LM, Hollander I et al (2002) Gemtuzumab ozogamicin, a potent and selective anti-CD33 antibody–calicheamicin conjugate for treatment of acute myeloid leukemia. Bioconjug Chem 13:47–58
32. Dijoseph JF, Armellino DC, Boghaert ER et al (2004) Antibody-targeted chemotherapy with CMC-544: a CD22-targeted immunoconjugate of calicheamicin for treatment of B-lymphoid malignancies. Blood 103:1807–1814
33. Doronina SO, Toki BE, Torgov MY et al (2003) Development of potent monoclonal antibody auristatin conjugates for cancer therapy. Nat Biotechnol 21:754–778
34. Doronina SO, Mendelsohn BA, Bovee TD et al (2006) Enhanced activity of monomethylauristatin F through monoclonal antibody delivery: effect of linker technology on efficacy and toxicity. Bioconjug Chem 17:114–124
35. Erickson HK, Park PU, Widdison WC et al (2006) Antibody-maytansinoid conjugates are activated in target cancer cells by lysosomal degradation and linker-dependent intracellular processing. Cancer Res 66:4426–4433
36. Chari RVJ (2008) Targeted cancer therapy: conferring specificity to cytotoxic drugs. Acc Chem Res 41:98–107

37. Kaur S (2008) Bioanalytical considerations for antibody-drug conjugates. In: Workshop proceedings of AAPS national biotechnology conference on overview of monoclonal antibody immunoconjugates in cancer therapy
38. Yarden Y, Sliwkowski MX (2001) Untangling the ErbB signalling network. Nat Rev Mol Cell Biol 2:127–137
39. Press MF, Cordon-Cardo C, Slamon DJ (1990) Expression of the HER-2/neu proto-oncogene in normal human adult and fetal tissues. Oncogene 5:953–962
40. Slamon DJ, Clark GM, Wong SG et al (1987) Human breast cancer: correlation of relapse and survival with amplification of the HER-2/neu oncogene. Science 235:177–182
41. Slamon DJ, Godolphin W, Jones LA et al (1989) Studies of the HER-2/neu proto-oncogene in human breast and ovarian cancer. Science 244:707–712
42. Nahta R, Yu D, Hung MC et al (2006) Mechanisms of disease: understanding resistance to HER2-targeted therapy in human breast cancer. Nat Clin Pract Oncol 3:269–280
43. Slamon DJ, Leyland-Jones B, Shak S et al (2001) Use of chemotherapy plus a monoclonal antibody against HER2 for metastatic breast cancer that over expresses HER2. N Engl J Med 344:783–792
44. Blattler WA, Chari RV (2001) Drugs to enhance the therapeutic potency of anticancer antibodies: antibody-drug conjugates as tumor-activated prodrugs. In: Ojima I, Vite G, Altmann K (eds) Anticancer agents—frontiers in cancer chemotherapy. American Chemical Society, Washington, DC, pp 317–338
45. Cassady JM, Chan KK, Floss HG, Leistner E (2004) Recent developments in the maytansinoid antitumor agents. Chem Pharm Bull (Tokyo) 52:1–26
46. Goldmacher VS, Blattler WA, Lambert JM, Chari RV (2002) Immunotoxins and antibody-drug conjugates for cancer treatment. In: Muzykantov VR, Torchilin V (eds), Biomedical aspects of drug targeting, pp 291–310
47. Krop IE, Beeram M, Modi S et al (2010) Phase I study of trastuzumab-DM1, an HER2 antibody-drug conjugate, given every 3 weeks to patients with HER2-positive metastatic breast cancer. J Clin Oncol 28(16):2698–2704
48. Erickson HK, Widdison WC, Mayo MF et al (2010) Tumor delivery and in vivo processing of disulfide-linked and thioether-linked antibody-maytansinoid conjugates. Bioconjug Chem 21:84–92
49. Shen BQ, Xu K, Liu L et al (2012) Conjugation site modulates the in vivo stability and therapeutic activity of antibody conjugates. Nat Biotechnol 30:184–189
50. Fishkin N, Maloney EK, Chari RVJ, Singh R (2011) A novel pathway for maytansinoid release from thioether linked antibody–drug conjugates (ADCs) under oxidative conditions. Chem Commun 47:10752–10754
51. Alley SC, Benjamin DR, Jeffrey SC et al (2008) Contribution of linker stability to the activities of anticancer immunoconjugates. Bioconjug Chem 19:759–765
52. Xu K, Liu L, Montserrat Carrasco-Triguero et al (2010) Monitoring tissue distributions of antibody drug conjugate biotherapeutics by LC–MS. Presented in the 58th American society for mass spectrometry (ASMS) conference proceedings, May 23–27, Salt lake City, Utah
53. Kaur S et al (2013) Bioanalysis, doi. 10.4155/BIO.13.72 in press

Chapter 8
Hydrogen/Deuterium Exchange Mass Spectrometry for Protein Higher-Order Structure Characterization

Hui Wei, Adrienne A. Tymiak and Guodong Chen

8.1 Introduction

As some of the most essential molecules of life, proteins fulfill a plethora of biochemical functions within every living organism. They are involved in virtually all cell functions, such as cell division, cell death, immune response, signal transduction, and ligand binding. In contrast to small molecules, proteins are significantly more complicated in structure. In biophysical conditions, proteins fold into unique three-dimensional structures in solution that are flexible and dynamic. The biochemical functions of a protein are directly related to such protein higher-order structures and structural dynamics. Proteins also interact with each other as well as with small ligands and other biopolymers to form complexes. Governed by the protein higher-order structure and dynamics, such interactions form the basis of signaling and regulatory processes and play a key role in drug action mechanisms [1]. Therefore, it is not only important to elucidate the chemical composition of a protein, but also to characterize its higher-order structure, or conformation, and the conformational dynamics in solution for a better understanding of the protein's functions.

With advances in mass spectrometry (MS), MS has become the standard method to identify cellular and therapeutic proteins, quantify expression levels, and characterize post-translational modifications, network relationships, and metabolism products, as described in the previous chapters of this book. Applications of MS methods are also being developed for determination of higher-order structures and dynamics of biomolecules. These methods successfully

H. Wei (✉) · A. A. Tymiak · G. Chen
Bristol-Myers Squibb, Route 206 and Province Line Road, Princeton, NJ 08543, USA
e-mail: hui.wei@bms.com

G. Chen (ed.), *Characterization of Protein Therapeutics using Mass Spectrometry*,
DOI: 10.1007/978-1-4419-7862-2_8, © Springer Science+Business Media New York 2013

complement other biophysical techniques, such as X-ray crystallography and NMR spectroscopy. X-ray crystallography and NMR can determine protein structures with high spatial resolution; however, the technologies require either successful crystallization or highly purified isotopically labeled proteins and are therefore unsuitable for rapid and routine sample analysis. NMR is further restricted by the size of the proteins to be analyzed. Structural MS, on the other hand, offers unique advantages over other techniques to study protein structures, including those of multi-protein complexes, regardless of the molecular size, amount of material available, solubility, or crystallization properties that often limit the utility of other techniques.

As an orthogonal biophysical technique, hydrogen/deuterium exchange mass spectrometry (HDX MS) represents one of the most widely used techniques for exploring protein higher-order structure in solution. As shown in Fig. 8.1, the protein backbone amide hydrogens can be exchanged with deuterium atoms from the surrounding solvent at different measurable exchange rates. The exchange rate is a function of the protein structure and solvent accessibility of the amide hydrogen, and it can be measured by MS as the heavier mass of deuterium gets incorporated into the protein. The hydrogen atoms at the surface of the proteins exchange more rapidly than the amide hydrogens buried in the interior or involved in stabilizing hydrogen bonds. Therefore, HDX MS can provide information on

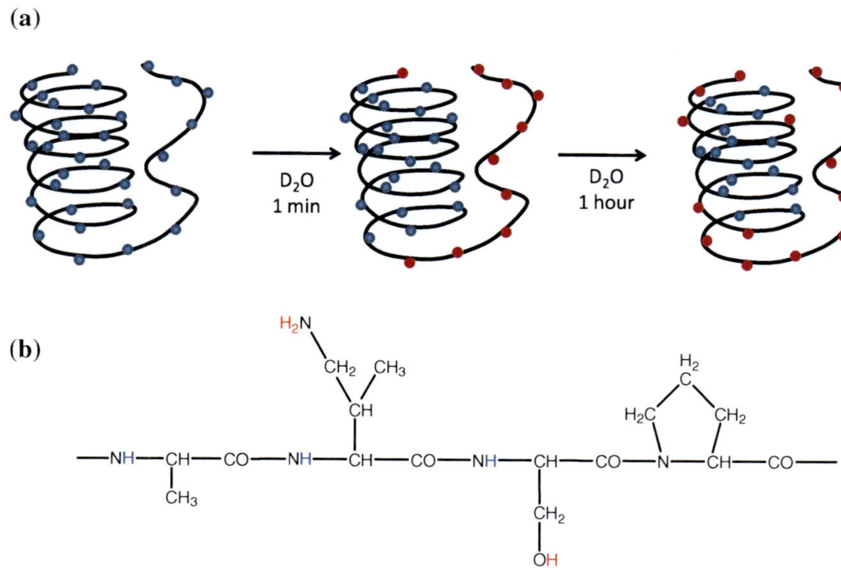

Fig. 8.1 **a** Schematic representation of protein backbone amide hydrogens (*blue dots*) exchange with deuterons in solution (*red dots*). **b** Types of hydrogens in a segment of a protein: backbone amide hydrogens (*blue*), hydrogens bonded to carbon (*black*) that do not exchange, and hydrogens on side chains containing –OH, –SH, $-NH_2$, –COOH, and $-CONH_2$ groups (*red*) that exchange too fast to be monitored

protein structure and solvent accessibility in solution. HDX was first introduced by Hvidt and Linderstrom-Lang in the mid-1950s, who realized that slower amide hydrogen exchange rates should reflect the presence of hydrogen bonding [2, 3]. Since then, HDX has been detected with various techniques such as scintillation counting [4], infrared and ultraviolet spectroscopies [5, 6], neutron diffraction [7], and NMR spectroscopy [8]. NMR methods remain particularly useful for measuring the exchange rates of individual amides in proteins [9, 10]. However, some amides exchange too rapidly to be measured with standard NMR techniques, making the study of most protein–protein interactions difficult by NMR [11]. The spatial resolution of HDX using other detectors was improved by combining limited proteolysis with separation of resulting peptides using the newly introduced high-pressure liquid chromatography (HPLC) techniques in the late 1970s and early 1980s [12, 13]. The coupling of MS for detection of HDX was introduced in 1991 [14] after it became possible to introduce protein molecules into a mass spectrometer for mass analysis, and the method was further enhanced by pairing the experiment with proteolytic digestion, enabling structural changes to be resolved at the peptide level [15]. The earliest experiments employed fast atom bombardment (FAB) for ionization of peptides, yielding a low number of peptides recovered from the HDX experiment (59 %). A year later, electrospray ionization mass spectrometry (ESI-MS) was coupled with HPLC peptide separation in HDX analyses of horse skeletal muscle myoglobin, and this combination demonstrated improved sequence coverage to 89 % [16]. Throughout the past two decades, the HDX MS method has developed rapidly with advances in both HDX methodology and MS instrumentation (hardware and software) and has emerged as an essential tool for the study of protein structure in solution including higher-order structures of protein therapeutics. It has been applied to study protein conformation, conformational dynamics, folding, binding, and aggregation [17–24]. The method requires only a small quantity of sample (picomoles) and can provide useful results with dilute solutions (submicromolar) [25]. It can be used to study proteins that are hard to purify and can reveal protein conformational dynamics on a wide timescale.

8.2 Theory

The kinetics of amide HDX has been measured by several methods, and the mechanism has been thoroughly reviewed [23, 24, 26–29]. In this section, a short summary of the basic HDX principles is described to lay the foundation for discussion of the HDX MS experimental design and data analysis.

There are three types of protons in protein that are grouped by their HDX behavior (Fig. 8.1b). Protons in side chains containing –OH, –SH, $-NH_2$, –COOH, and $-CONH_2$ groups and from the amino and carboxy termini exchange too rapidly to be measurable by isotope exchange methods; the carbon-bound aliphatic and aromatic protons have high covalent character, and the isotope exchange can only occur by catalyst activation, such as reaction with hydroxyl radicals [30]; protons on secondary amides comprising the backbone of a folded protein have measurable isotope exchange rates. The amide proton exchange rate is dependent on several factors including protein structure, pH, temperature, and adjacent amino acid side chains of the amide group that alter the pKa of the hydrogen atom [31]. Given a certain temperature and pH, the exchange rate for the amide hydrogen in an unstructured peptide can be affected by the primary structure, or sequence, by as much as 30-fold and can be predicted [31, 32]. In a folded protein, the exchange rate of each amide hydrogen can be decreased by its higher-order structure by up to eight orders of magnitude, ranging from milliseconds to many years [26]. The drastic difference in exchange rates allows the HDX method to serve as a sensitive probe for studying conformational changes in proteins. With the exception of proline and certain post-translationally modified residues, every non-N-terminal residue possesses amide hydrogen; therefore, HDX can probe structural features along the entire protein sequence.

In short peptides, amide hydrogen exchange involves proton abstraction described by a chemical exchange rate (k_{ch}). It depends on an "intrinsic" rate of exchange for that hydrogen (k_{int}) and the concentration of available catalyst, including OH^-, H_3O^+, water, and acidic or basic solutes:

$$k_{\text{ch}} = k_{\text{int}}\left[\text{catalyst}\right] \tag{8.1}$$

The exchange occurs in all aqueous solutions and can be catalyzed by acid or base. At lower pH, exchange occurs via proton addition, catalyzed by D_3O^+ (see acid-catalyzed reaction mechanism in Fig. 8.2a, [32–35]). Above pH 2.5, exchange occurs by proton abstraction, predominantly by OH^- (see base-catalyzed reaction mechanism in Fig. 8.2b, [35, 36]). The base-catalyzed exchange of backbone amide hydrogens is much more effective than acid-catalyzed or water-catalyzed exchanges by around eight or twelve orders of magnitude, respectively, at room temperature [31, 37]. Because chemical exchange rates for amide deuterium atoms are slower than for hydrogen atoms, with little solvent isotope effect, proton abstraction is rate-limiting for reactions occurring above pH 2.5 [38]. The rates of the base-catalyzed reactions increase at higher pH values. An increase of 1 pH unit increases the exchange rate constant by approximately 10-fold, with k_{ch} being minimum around pH ~ 2.5 (pH_{min}, sequence dependent) [37]. At pH 7.0, the chemical exchange occurs rapidly ($k_{\text{ch}} \sim 10^1\text{–}10^3\ \text{s}^{-1}$) [23, 31].

The backbone amide hydrogen exchange rates vary drastically after the protein has adopted a folded three-dimensional structure. To complicate matters even further, the rates are also affected by the protein flexibility and mobility, whereas the protein structure is altered briefly exposing regions that were previously inaccessible to solvent [39]. Below is a model describing HDX in native, folded proteins:

$$\text{Protein}\,(H)_{\text{folded}} \underset{k_f}{\overset{k_{\text{op}}}{\leftrightarrows}} \text{Protein}\,(H)_{\text{open}} \underset{D_2O}{\overset{k_{\text{ch}}}{\leftrightarrows}} \text{Protein}\,(D)_{\text{open}} \underset{k_{\text{op}}}{\overset{k_f}{\leftrightarrows}} \text{Protein}\,(D)_{\text{folded}} \tag{8.2}$$

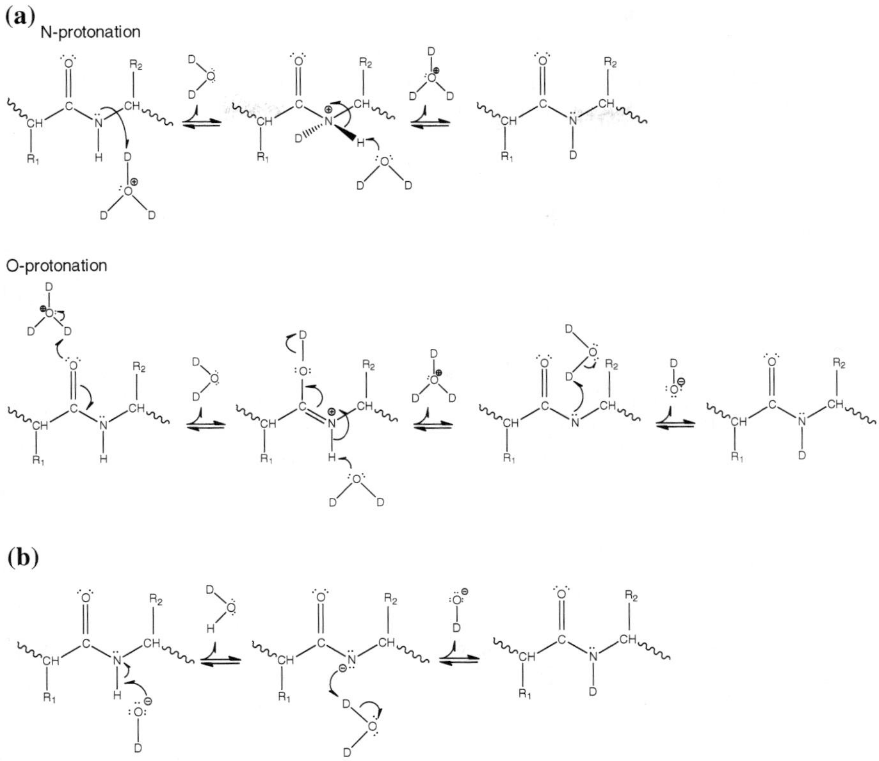

Fig. 8.2 **a** Acid-catalyzed HDX mechanism. **b** Base-catalyzed HDX mechanism

where k_{op} is the rate of unfolding, k_f is the rate of folding, *protein* (*H*) represents unlabeled protein, and *protein* (*D*) represents deuterated protein. Protein in the folded state contains inaccessible amide hydrogens, and protein in the open state contains accessible exchange-competent amide hydrogens. In this scheme, the HDX rate is given by Eq. (8.3) [40, 41]:

$$k_{\mathrm{HDX}} = \frac{k_{\mathrm{op}}\, k_{\mathrm{ch}}}{k_f + k_{\mathrm{ch}} + k_{\mathrm{op}}} \tag{8.3}$$

The assumption that stable proteins in the native state have a higher propensity for the folded form than for the open form implies that $k_{\mathrm{op}} \ll k_f$, which results in a simplified rate expression:

$$k_{\mathrm{HDX}} = \frac{k_{\mathrm{op}} k_{\mathrm{ch}}}{k_f + k_{\mathrm{ch}}} \tag{8.4}$$

This expression leads to the well-known limiting EX1 and EX2 regimes. When the small-amplitude fluctuations that convert solvent-protected hydrogens to solvent-exposed hydrogens are assumed to be completely reversible, the chemical exchange occurs quickly after conversion to the solvent-exposed form ($k_{ch} \gg k_f$). This represents a continuum of hydrogen exchange mechanisms which is termed as the EX1 regime or monomolecular exchange. The rate expression becomes

$$k_{HDX} = k_{op} \quad (8.5)$$

such that the HDX behavior of the protein reflects the kinetics of the opening event. The reaction rate is only proportional to the concentration of the single species unexchanged. The rate is independent of chemical exchange, and in most cases, complete pH independence will be observed if protein structure and the opening/folding rates are not affected by pH. Regions of proteins exhibiting EX1 kinetics can exchange all amide hydrogens during one unfolding event; hence, regions of proteins that undergo slow folding and refolding may display amide EX1 exchange kinetics with a short period of deuterium exposure [42]. Proteins can also be induced to exhibit EX1 kinetics with chemical denaturant [43] or by increasing pH [44].

A second type of bimolecular exchange, termed EX2, occurs when reconversion of the open form back to the folded, protected form occurs much faster than the rate of chemical exchange, or $k_f \gg k_{ch}$, then the equation changes into

$$k_{HDX} = \frac{k_{op}}{k_f} k_{ch} = K_{op} k_{ch} \quad (8.6)$$

Under conditions that obey EX2 kinetics, k_{HDX} depends on the equilibrium of folded and unfolded forms and on the chemical exchange rate. The unfolding–folding process may occur many times before the proton is exchanged. The exchange rate is pH dependent as the chemical exchange rate is pH dependent. For native-state proteins, experimental evidence based on the pH dependence of the exchange rate confirms that EX2 kinetics is the predominant mechanism of exchange for most proteins [39, 45]. However, some proteins may contain regions that undergo both mechanisms simultaneously [45]. Regions where either EX1 or EX2 mechanisms occur can also be distinguished by examining the characteristic isotope patterns in mass spectra [19].

In Eq. (8.6), the ratio of the rate constant for protein unfolding to folding, K_{op}, represents the unfolding equilibrium constant in the protein-opening reaction. HDX under EX2 conditions does not provide any information on the rate at which structural fluctuations occur. However, the unfolding rate constant (k_{op}) divided by the chemical exchange rate (k_{ch}) yields the unfolding equilibrium constant. Therefore, under EX2 conditions, the equilibrium free energy change upon unfolding (ΔG_{op}) can be obtained based on Eq. (8.6):

$$\Delta G_{op} = -\mathrm{RT} \ln \left(\frac{k_{HDX}}{k_{ch}} \right) = -\mathrm{RT} \ln K_{op} \quad (8.7)$$

where R and T correspond to the gas constant and the absolute temperature in Kelvin, respectively. The values of k_{ch} can be obtained from the literature [31] and calculated together with other values using spreadsheets downloadable from Walter Englander's website (http://hx2.med.upenn.edu). The values of k_{HDX} can be obtained from experimental data as it is the first-order rate constant that best fits the experimental deuterium uptake curves.

The ratio k_{ch}/k_{HDX} is referred to as the protection factor

$$P_f = \frac{k_{ch}}{k_{HDX}} \tag{8.8}$$

that provides a measure of the slowing of HDX rate induced by the structure of the protein or upon binding [46]. If a segment of a protein has an average protection factor of 1, then it indicates that the segment is essentially unstructured. If the protection factor is greater than 1, it indicates that the region is protected from HDX and vice versa.

In addition to pH, protein primary structure, higher-order structure and mobility, other factors such as temperature, solvent composition, and pressure can also affect the HDX rate. Changes in temperature alter the ionization constant of water, thus affecting the catalyst concentration in the exchange reaction [47]. The backbone amide exchange rate increases by about threefold for each 10 °C increment [35, 48]. The rate can be reduced by approximately tenfold by adjusting the temperature in the HDX experiment from room temperature to 0 °C. Solvent composition also has an impact on the exchange rates and has been reviewed comprehensively [13, 49]. Addition of organic solvent alters the equilibrium constant K_w for water which directly determines the OH^- ion concentration at a given pH value, thus affecting the base-catalyzed exchange rate. Miscible organic solvents used in reverse-phase HPLC generally slow down the exchange rate. For example, adding organic solvent such as dioxane lowers the concentration of OH^- ions, causing the shift of pH_{min} to progressively higher values [48]. Hence, reducing the temperature or changing solvent composition offers practical ways to minimize back exchange of the isotopic label.

8.3 Methodology

8.3.1 Labeling Approaches

There are two main approaches to perform the HDX experiments: continuous labeling and pulsed labeling (Fig. 8.3). In continuous labeling, the protein is exposed to the deuterated buffer under conditions where the native conformation of the protein is "stable." Incubation is carried out in aliquots of the protein sample for various amounts of times until quenched, allowing continuous deuterium incorporation into the protein structure to be monitored as a function of time (Fig. 8.3a). During the incubation, the populations of folded and unfolded states of the protein are constantly changing. Proteins or regions of the proteins that are

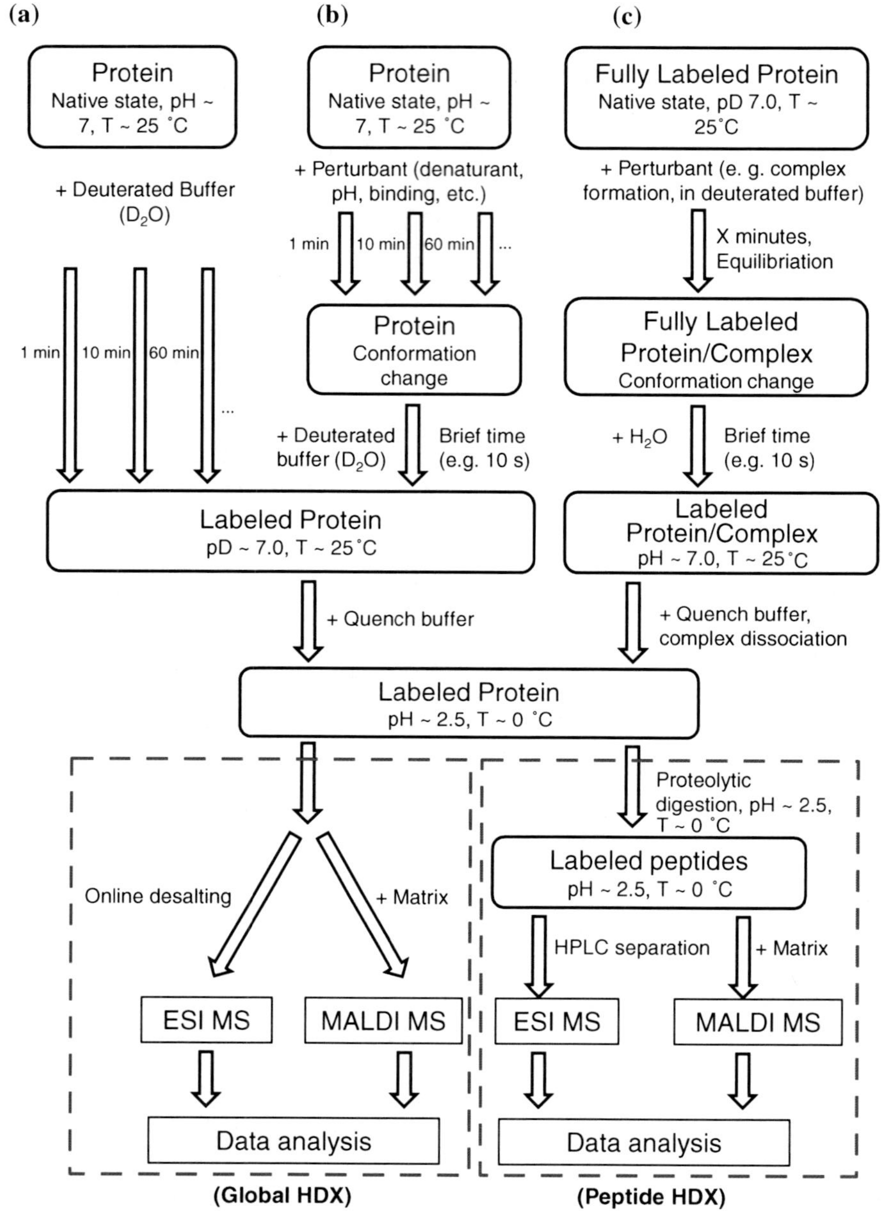

Fig. 8.3 Scheme for HDX MS experiments: **a** continuous labeling, **b** pulsed labeling, and **c** exchange-out experiments followed by either global (intact protein level) or local (peptide level) analysis. See text for detailed explanation for each workflow

or become unfolded during the incubation are labeled with deuterium, and the reaction is not reversible due to the high concentration of D_2O. Such information is readily measured by MS due to the heavier mass of deuterium. The deuterium incorporation rate monitored as a function of incubation time provides information on the conformational dynamics of a protein under equilibrium conditions. The continuous labeling method is particularly useful for monitoring slow unfolding transitions [50]. Due to the relatively simple experimental procedures, it has been widely adopted for a variety of studies, such as characterizing protein conformation and dynamics [51], protein/ligand binding [52, 53], and protein/protein interactions [54, 55].

As illustrated in Fig. 8.3b, in the pulsed labeling approach, a protein is usually perturbed in some manner, such as by adding chemical denaturant, forming complexes, changing pH, and changing temperature, in order to induce some structural changes [56, 57]. Then, the protein is exposed to deuterated buffer (typically with higher pH, e.g., pH 8–10) for a very brief period of time, typically 10 s or less [58]. The slightly basic solution serves to ensure a sufficient exchange rate during the brief labeling pulse. For example at pH 8.5, the exchange of unprotected amide hydrogens occurs on a timescale of 10 ms [31, 57]. Extensive HDX will only affect sites that are solvent exposed and not involved in hydrogen bonds. Binding to other moieties will decrease the HDX level as a result of the formation of intermolecular hydrogen bonds and steric protection of the exchangeable sites. After the brief labeling period, the deuterated protein or protein complex provides a snapshot of the protein (complex) that existed during the pulsed exposure to deuterium. The pulsed labeling approach is less commonly used as the technical requirements for performing such experiment are much higher than for the continuous labeling experiment. Nonetheless, pulsed labeling has unique advantages and applications since it only focuses on the rapidly exchanging amides, which typically occur on the surface of the protein and are the most relevant for binding events. Pulsed labeling has been adopted to study protein-folding kinetics and transient intermediate states [57, 59], as well as to characterize non-covalent protein complexes [58].

Instead of exchanging the deuterium atoms "into" the protein structure and measuring increased mass ("exchange in"), an alternative approach is to perform "exchange-out" reactions in which samples are fully deuterated before exposure to a pulsed labeling with H_2O [60] (Fig. 8.3c). Such an approach is also called "kinetic labeling." The deuterons in backbone amides with very fast exchange rates completely exchange back to hydrogens, and those with slow and intermediate exchange rates stay deuterated. This approach has often been used to probe the antigen epitopes of antibodies [55, 61]. The strategy involves complete deuteration of the free antigen, then loading the sample onto a column in which antibody has been covalently immobilized. The antigen then binds with antibody to form a complex, followed by a washing step with aqueous buffer for a predetermined period of time to exchange out the deuterium atoms. The antigen is then eluted out by a low-pH buffer, and the deuterium atoms retained by the epitope regions can be identified.

If the concentration of deuterium in the surrounding medium is low, then the deuterium in a protein will revert back to hydrogen form. The labeling experiments are almost always followed by a quenching step in order to stop the reaction and retain the specific isotope-labeled signature. As discussed in the previous section, the amide hydrogen exchange rate is minimal around pH 2.5 and is also temperature dependent. By changing the conditions from room temperature and physiological pH (pH 7.0–8.0) where most labeling experiments occur to 0 °C and pH 2.5, the exchange rate decreases by about five orders of magnitude [62, 63]. Under such conditions, if a protein is placed in a 100 % H_2O environment, the half-life for reversion of the deuterium label back to hydrogen is between 30 and 120 min, depending on the protein sequence [26, 31]. The subsequent steps to get the sample ready for MS analysis, such as enzyme digestion, desalting, and chromatographic separation, are usually performed in H_2O solution, and all the deuterium atoms that had been incorporated into terminal and side-chain positions should rapidly revert to the hydrogen atom forms, simplifying the data interpretation. However, these steps need to be performed rapidly under the quenched conditions in order to minimize the back exchange of secondary amide deuterium atoms. The quenching step also serves another purpose as it mildly denatures the protein and thereby facilitates the proteolytic digestion step.

8.3.2 Global Versus Peptide HDX MS

Isotopically labeled protein samples from the HDX experiments can be measured by mass spectrometers either at the intact protein level to reveal the global deuterium incorporation or at peptide level following enzyme digestion to reveal local structural information (see bottom part of Fig. 8.3). Global analysis was the first type of analysis reported for HDX MS [14] and is the most basic and simple application of the technique. As only the mass of the intact protein needs to be measured, there is no need for chromatographic separation. ESI is the preferred ionization mode for intact proteins because of its high sensitivity and suitability for detecting the low m/z values of the multiply charged species generated during ionization of intact proteins. Typically, a desalting column is used for removing salt that suppresses protein signals. The desalting is performed under quenched conditions to minimize back-exchange reactions. The desalting step also washes away N-terminal and side-chain deuterons, simplifying the subsequent data analysis as only the deuterons incorporated into the backbone amide positions are retained. The total number of deuterium atoms incorporated into the protein structure is then measured as a function of time. Though the global HDX analysis has no spatial resolution, it can quickly provide information on protein conformational stability and reflect the extent of overall folding of the protein. As the technical requirements for global HDX MS analysis are relatively simple, the experiments can be automated for high throughput with a platform programmed to perform labeling/quenching and online desalting automatically. This approach has the potential to be adopted in biopharmaceutical industries as a rapid structural screening tool for protein characterization and quality control studies.

Peptide-level HDX MS offers improved spatial resolution of protein conformations by measuring the deuterium incorporation into short stretches of peptide fragments that are produced after proteolytic digestion of the labeled protein. Therefore, more specific HDX information is provided, revealing both the rate and location of deuterium incorporation. The concept was first introduced by Rosa and Richards [12] and was later demonstrated in combination with MS analysis by Smith and coworkers [15]. Pepsin is the most widely used protease in HDX because it is stable and highly active under the acidic quenched conditions. Other acidic proteases such as *Aspergillus saitoi* (type XIII) and *Rhizopus* sp. (type XVIII) have also been used but with less digesting efficiency than obtained with pepsin under similar quenched conditions. Although pepsin exhibits preferential cleavage for hydrophobic, preferably aromatic, residues, it is still considered a non-specific protease and typically produces peptides 10–20 amino acids in length and with overlapping sequences [64]. The degree of overlap can be significant, reaching 3–4 overlapping peptides for a given region of the protein, which in turn helps improve the spatial resolution for assigning deuterium incorporation to short stretches of amino acids [28]. Though non-specific, pepsin cleavage is highly reproducible if the digestion conditions such as temperature, pH, concentration, and time are carefully controlled. Successive protease treatments can also be used to generate smaller fragments for mass spectrometric analysis, yielding even higher sequence coverage and spatial resolution [64].

Peptides generated from HDX experiments can be directly analyzed by matrix-assisted laser desorption/ionization mass spectrometry (MALDI MS) [65, 66] or separated chromatographically for analysis by ESI-MS [15], with the latter method being used more predominantly. Advantages for MALDI MS detection include easier interpretation of spectra as only singly charged ions are produced and a greater buffer salt tolerance than ESI therefore obviating the need for desalting. However, deuterium losses due to back exchange have been reported to be significantly higher in MALDI than in ESI [67]. Moreover, for a larger protein, for example, 50 kDa, directly measured spectra are complicated by more than one hundred peptides typically generated from pepsin digestion. With the alternative approach of coupling HPLC separation and ESI-MS, mass overlap of peptides is minimized and ionization suppression is reduced, although the separation time needs to be well controlled to minimize back exchange. During ionization in ESI, the cone temperature and desolvation gas temperature also need to be reduced to minimize back exchange. Proteolyzed peptides are then identified using a combination of exact mass and tandem MS analyses using control samples prepared under HDX conditions with the exception that the D_2O exchange buffer is replaced with an otherwise identical H_2O-based buffer. It is assumed that the same peptides (without isotopic labeling) are produced in the control experiment. Once identified, these peptides can be monitored for their deuterium uptake levels in subsequent labeling experiments.

8.3.3 Tandem MS

Further improvement in the spatial resolution of HDX MS can be achieved by dissociating peptic peptides into shorter fragment ions through collisions in the gas phase (MS/MS) in a tandem mass spectrometer. Ideally, deuteration of individual amides should then be revealed whenever a mass increment of 1 Da greater than the residue mass is observed. This concept was first tested using collision-induced dissociation (CID) as the method for peptide fragmentation [68, 69]. While it was possible to measure the extent of HDX at individual amide hydrogen, extensive scrambling of hydrogen or deuterium atoms attached to nitrogen and oxygen during the CID fragmentation process was observed [70–72]. Scrambling occurs when a peptide ion reaches the internal energy threshold that mobilizes protons before achieving backbone fragmentation.

Non-conventional fragmentation techniques such as electron capture dissociation (ECD) was demonstrated to proceed with a very low level of amide hydrogen scrambling [73]. ECD is a low energy, radical-driven fragmentation technique that cleaves the polypeptide backbone with minimal vibrational excitation [74] and is likely to limit the possibility of intramolecular amide hydrogen migration. ECD has been shown to fragment proteins into *c* and *z* ion series, but mainly at terminal regions of the peptide, yielding limited sequence information [75]. More efficient ECD has been demonstrated using the oscillating electric fields available in Fourier transform ion cyclotron resonance mass spectrometers (FT-ICR-MS) [74]; however, this technology is very expensive and not easily accessible to many laboratories. A more recently developed fragmentation technique, electron transfer dissociation (ETD), also yields *c* and *z* ion series [76]. It relies on the transfer of an electron to multiply charged peptide/protein ions using singly charged anions, generating odd-electron fragment ions via radical-based rearrangements. The resulting peptide backbone cleavage is assumed to proceed by the same mechanisms as for ECD. The ETD ion/ion reaction can be performed in any instrument that uses radio frequency oscillating electric fields such as in ion traps [76, 77]. Experiments conducted using peptides with unique selective labeling have demonstrated that very low levels of hydrogen scrambling can be observed when the collisional activation prior to the ETD event is carefully minimized [78]. Instrumental parameters that involve ion desolvation and transfer into the mass analyzer also need to be optimized to reduce the energy imparted to the precursor ion, hence minimizing scrambling. With careful experimental design, single-amide resolution for deuterium incorporation measurements has been achieved using model peptide systems [78] and has been applied to the characterization of conformational dynamics of proteins (the apo-form of amyloidogenic protein β_2-microglobulin) [79] and protein/ligand interactions (peroxisome proliferator-activated receptor bound with ligands) [80]. Figure 8.4 shows example data for the measurement of deuterium content of individual backbone amides for a section of β_2-microglobulin (residues 40–54) using the ETD approach [79]. An exchange-out experiment was performed on fully deuterated β_2-microglobulin for

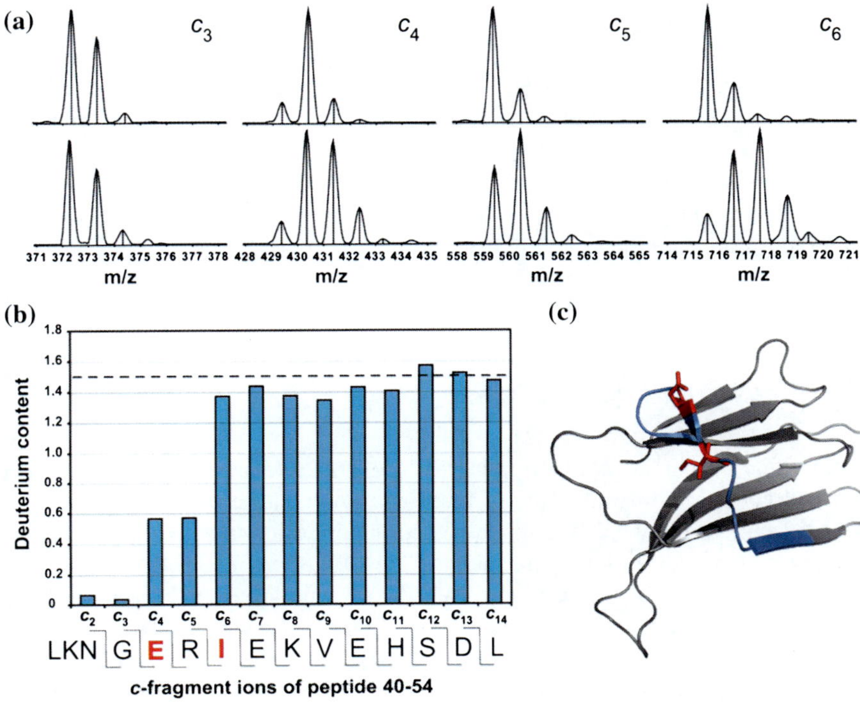

Fig. 8.4 Measuring the deuterium content of individual backbone amides of 2-microglobulin by ETD: **a** ETD mass spectra of the triply charged 2-microglobulin peptide KNGERIEKVEHSDL (residues 40–54) from an unlabeled (*upper panels*) and a labeled sample obtained after 40 min isotopic exchange (*lower panels*). The isotopic patterns of the fragment ions c_3–c_6 are displayed. **b** Bar chart showing the deuterium level of all *c* ions. The deuterium level of the intact peptide precursor ion is indicated by a *dotted line*. The protected residues E44 and I46 are *colored in red* in the sequence. Note that the amide hydrogen of residue *n* is contained in the c_{n-1} fragment ion. **c** Crystal structure of 2-microglobulin (PDB:1LDS) with the sequence LKNGERIEKVEHSDL *highlighted in blue*. Residues E44 and I46, which were identified as slow exchanging by recent NMR32 and present ETD data, are shown as *red sticks*. Figure was adapted with permission from Ref. [79]. Copyright (2009) American Chemical Society

40 min, followed by online pepsin digestion and LC/MS/MS with ETD fragmentation in an ion trap mass spectrometer. As shown in the figure, the exact location of retained deuterons is apparent from the masses of sequential fragment ions (*c* ions) of the peptic peptide. The exclusive retention of deuterium to E44 and I46 shown in the ETD data is in excellent agreement with the previously published HDX-NMR results obtained on the same molecule [81].

Studies examining full-length proteins by HDX combined with a top-down MS/MS approach with either ECD or ETD have also obtained single-amide resolution without requiring enzyme digestion or chromatographic separation [82, 83]. Another advantage of the top-down approach is that it provides a solution to the characterization of higher-order structure and dynamics in a conformer-specific fashion, while this level of discrimination is lost in the bottom-up approach [84]. However, the top-down HDX method is limited to small- and mid-sized proteins where relatively high sequence coverage can be obtained. For large proteins (>50 kDa), for example, monoclonal antibodies, and protein–protein complexes, the large number of fragment ions is confined to a narrow m/z region, leading to overlapping fragment ions and their isotopomers, thus complicating the data analysis and interpretation. Combining MS/MS fragmentation in the gas phase with proteolytic fragmentation in solution should help to yield better spatial resolution and higher sequence coverage for proteins.

8.3.4 Measuring Back Exchange

In HDX approaches coupled to either ESI or MALDI instruments, even under quenched conditions with low temperature and low pH, the occurrence of amide back exchange losing the deuterium labels during digestion, chromatography separation, or matrix preparation is unavoidable [85]. With well-controlled experimental conditions, much of the backbone amide deuteration can be preserved. Back exchange on side chain occurs very rapidly, and the conversion back to hydrogen substitution is almost complete, simplifying the data analysis by confining the mass increase to a unit dalton change reflecting a single deuterium label per residue. One exception is in the case of arginine, where the guanidine side chain has a minimum exchange rate similar to that of backbone secondary amides [35]. Proteins rich in arginine may complicate the analysis because not all side-chain deuterium atoms can be washed away prior to MS analysis.

The back-exchange process on backbone amides is significant, and corrections need to be made in order to achieve true values for HDX rates. One common approach for adjusting back exchange is to collect data on a pair of control peptides, one fully protonated and one fully deuterated, using identical conditions for digestion and HPLC separation [15, 62]. Assuming that the back-exchange rates are equal for different sites in the peptide, the measured mass shifts are used to correct measurements on digested aliquots using a simple scaling factor. Statistical analysis of 3,000 peptides with random sequences and 5–25 peptide linkages indicated that the average error in deuterium content would be 5.5 % of the adjustment [62]. Other methods have been described as well for correcting back exchange [23, 86, 87]. Some studies have demonstrated that backbone amide back exchange is more complex and dependent on the sequence [32, 88]. To address this, a set of parameters analogous to those developed by Englander and Bai for aqueous solution [31] could be derived for typical HPLC separation solvents and used to predict back-exchange rates [88]. In practice, however, an adjustment for back exchange is not always necessary. In the studies comparing proteins under various states, such as conformational changes, folding/unfolding and binding studies, determination of relative exchange levels may be sufficient assuming that the back-exchange rate for the same peptide under the same experimental conditions remains the same. Many studies have relied on relative exchange levels without the correction of back exchange [51, 89–91].

An alternative way to minimize back exchange is not to perform the analysis in proton-containing solvents. In one example, supercritical fluid chromatography (SFC) was used to replace HPLC as the desalting/separation technique prior to mass analysis [92]. The combined use of a non-exchanging CO_2 mobile phase, fast flow rates, and the short retention times afforded by SFC greatly reduced back-exchange rates comparing to fast, reduced temperature HPLC methods. However, some back exchange of deuterium still occurred in this study during the digestion step because it was conducted in a protonated aqueous buffer.

8.3.5 Data Interpretation and Instrumentation

When HDX is measured by MS, either MALDI or ESI instruments, each peptide is present in the mass spectrum as an envelope of peaks differing mainly by ^{2}H (D) in a deuterium-labeled sample. The most common method of determining the average deuterium uptake is to perform a centroid comparison, which takes the weighted mean of all points between user-defined upper and lower bounds for each mass envelope. The mass difference between unlabeled and labeled species determines the average amount of deuterium incorporated. The fractional deuterium uptake level of a protein or a protein segment is then determined as a function of labeling time t:

$$\text{fractional deuteration level } (t) = \frac{m\,(t) - m_0}{m_{100} - m_0} \tag{8.9}$$

where $m(t)$ is the centroid mass of the protein/peptide at labeling time t, m_{100} and m_0 are the centroid masses of completely labeled and unlabeled molecules, respectively. The deuteration level is the sum of deuterium uptake at all the backbone amide positions in the molecule. In MS/MS experiments with ECD or ETD fragmentation (see discussion in Sect. 8.3.3), the deuterium uptake level for each fragment ion can be determined from the centroid masses in the MS/MS spectra.

In addition to the deuterium incorporation, naturally occurring rare isotopes (^{13}C, ^{15}N, ^{18}O, ^{34}S, etc.) also contribute to the isotope distribution. One group applied a maximum entropy method (MEM) to determine the backbone deuteration levels [93]. The deconvolution of the natural abundance isotopic distribution pattern narrowed the mass spectral isotopic abundance envelope for greatly enhanced effective mass resolution. Another group developed a Fourier deconvolution method to explicitly determine the amount of backbone amide deuterium incorporated into protein regions or segments by HDX with high-resolution MS detection [94]. The method is based on a deconvolution calculation that removes both the natural isotopic abundance and fast-exchanging side-chain deuterons from the observed MALDI-TOF MS mass envelope for any desired peptide.

HDX data processing can be very time-consuming if performed manually. Automated data analysis for low- or medium-resolution [94, 95] or even high-resolution [96, 97] mass spectral data still remains challenging. There are several recently developed laboratory-built or commercial software solutions available to facilitate the processing of HDX MS data: DXMS, a software program that uses a curve-fitting approach to assign the centroid m/z values of the isotopic cluster envelopes of a set of peptides; an in-house analysis package was developed by Marshall's group that performs automatic mass assignment for HDX data collected on high-resolution, accurate mass FT-ICR-MS [98]; TOF2H, a downloadable laboratory-built toolbox for analyzing MALDI-based HDX data [99]; HD Desktop, an integrated platform for the analysis and visualization of HDX data [100]; HYDRA, a Web-based application for automatically extracting deuterium incorporation values from MS or MS/MS analysis and subsequent data visualization [101]; HX-Express, Excel-based software for mass assignment and plotting deuterium uptake curves [95]; and ExMS, a computer program that efficiently processes crowded mass spectra packed with sequentially and chromatographically overlapping peptide fragments to determine their HDX level [102]. These tools have substantially shorten the analysis time from weeks or even months of manual work down to days or hours, making the technique more practical for even analyzing more complex samples, such as large protein molecules and protein–protein interactions.

In addition to data interpretation, performing the HDX experiments, especially at the peptide levels, can be very laborious without automated instrumentation. Commercial and laboratory-built automated systems have been used to reduce human labor and improve the throughput for HDX experiments. For example, rapid quenched-flow techniques have been incorporated to facilitate the pulsed labeling HDX approach, and the method has been applied to examine the stabilization of secondary structure in refolding experiments [59, 103]. The Griffin Lab at Scripps introduced an automated platform for performing HDX experiments [53]. The platform contains an autosampler with cooled sample stacks which is connected to a three-valve unit contained in a cooling chamber that keeps the mobile phases at ~1 °C. An online pepsin digestion, peptide desalting, and HPLC separation are also integrated into the platform. The automated platform has been utilized for many HDX studies such as characterizing protein–ligand binding of PPARγ with full and partial agonist drugs. Another platform was based on a commercial nano-ACQUITY UPLC system that performs online pepsin digestion, as well as high-speed and high-resolution ultra-high-pressure LC (UHPLC) separation at 0 °C [104]. Integrated with an HDX manager for dual temperature control and automatic sampling, labeling, and quenching steps, the platform provides high-throughput HDX experiments [91]. Combining automation in instrumentation with software tools, HDX MS has become a more practical and attractive method in both academic and industrial research to provide insight into the behavior of macromolecular systems.

8.4 Protein Structure/Dynamics

The HDX method cannot determine the structure of a protein or define secondary structural elements. Because both solvent accessibility and hydrogen bonding contribute to the HDX rate, it is difficult to attribute a given exchange rate to a structural element without X-ray crystallography or NMR structural data. Nonetheless, conformational changes in proteins that result from post-translational modification, enzyme activation, drug binding, or other functional events can induce changes in HDX rate that are monitorable by HDX MS. This approach is particularly attractive for investigational studies because it employs an essentially continuous series of sensors along the entire length of the polypeptide backbone without a need to modify the protein.

8.4.1 Enzyme Conformational Dynamics

Measurement of protein dynamics at global or local scale is crucial to the understanding of protein functions. Especially in the case of enzymes, static structures are inadequate for describing the sometimes subtle, yet complex dynamic nature of enzyme processes that play an important role in catalytic efficiency. Several applications of HDX MS for examining dynamic and structural changes related to enzyme catalysis have been demonstrated and reviewed [29]. An example on how HDX MS was used in detecting the conformation mobility of extracellular regulated protein kinase-2 (ERK2) is discussed below [105].

Mitogen-activated protein (MAP) kinase ERK2 is an enzyme involved in a wide variety of cellular processes such as cell growth, differentiation, transcription regulation, and development in eukaryotic cells. Phosphorylation at Thr-183 or Tyr-185 leads to >1,000-fold enhanced specific kinase activity [106, 107]. In this study, Hoofnagle and coworkers applied a continuous HDX MS approach with pepsin digestion for peptide-level analysis to monitor HDX rates along the ERK2 sequence, and data obtained from unphosphorylated and dephosphorylated ERK2 (ppERK2) were compared. Thirty-nine peptic peptides were identified, providing 90 % sequence coverage with overlapping regions (Fig. 8.5a). The HDX rate for each peptide was obtained by fitting the kinetics of the exchange using corrected peptide masses (see Sect. 8.3.4) to sum exponentials by a nonlinear least-squared method [86]. The exchange rates were compared, and 20 peptides showed differences upon ERK2 activation by phosphorylation, with results mapped onto the sequence (Fig. 8.5a) and three-dimensional structures (Fig. 8.5b). The HDX data were also compared with the conformational changes in the corresponding regions of the known X-ray structures of ERK and ppERK [108]. In the activation lip and dimerization interface, as highlighted in Fig. 8.5b, the changes in exchange rates coincided with changes in conformation according to the X-ray structures. However, altered HDX upon activation in other regions, such as the ATP-binding

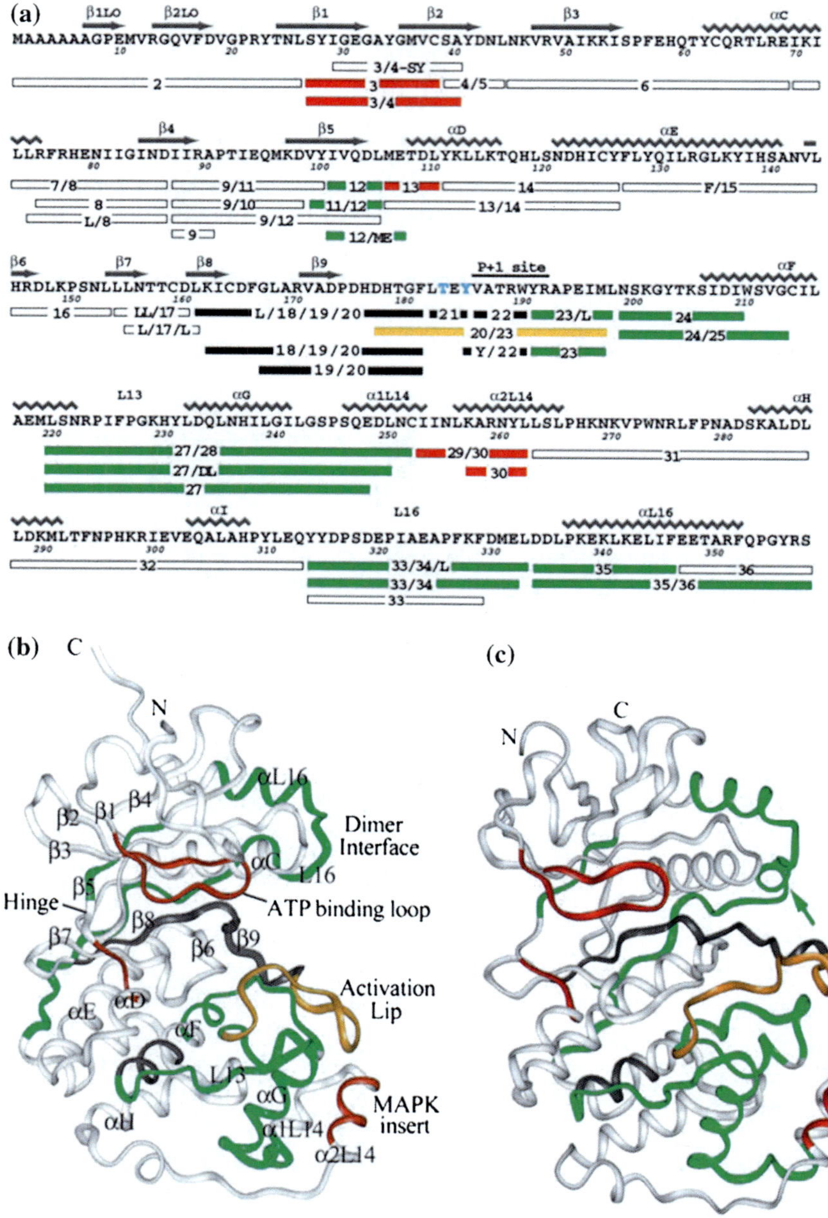

loop, the hinge region, the extended substrate binding groove, and the MAPK insert, could not be inferred from X-ray structures, but were consistent with expected changes in backbone flexibility upon activation. For instance, upon activation, the ATP-binding loop showed an increased HDX rate, indicating enhanced flexibility of this region which may also contribute to the increased specific activity by facilitating nucleotide binding or release. Such insights cannot be deduced

◀ **Fig. 8.5** **a** Sequence coverage of ERK2 indicating residue numbering and secondary structure [108]. Peptides *colored red*, *green*, *yellow*, or *white* showed HX rates that, respectively, increased, decreased, both increased and decreased, or did not change upon kinase phosphorylation. Peptides recovered only from unphosphorylated ERK but not from ppERK are shown in *black*. Residues phosphorylated in ppERK are shown in *blue*. **b** Structural representation of ERK [108]. Sequences not recovered from peptide digests are colored *gray*. **c** Structural representation of ppERK [107] as in (**b**), rotated slightly clockwise (as viewed down the vertical axis) to more clearly show the major differences with the inactive form, the reorganization of the activation lip (*yellow arrow*) and the formation of a 3/10-helix near the C-terminus (*green arrow*). Figure was adapted with permission from Ref. [105]. Copyright (2001) National Academy of Sciences, USA

from static crystallographic structures. Significant changes in protein mobility in regions important to catalysis upon enzyme activation are revealed by HDX MS. This work illustrated how the HDX technique can measure protein structural dynamics in solution and complement structural information obtained from X-ray crystallography by providing insight into internal motions of enzymes.

8.4.2 Folding/Unfolding

Protein-folding states can be readily resolved by HDX MS as it can identify bimodal populations of proteins or peptides as reflected by distinct patterns of deuteration. A single envelope of isotope peaks can be observed in HDX MS data if the sample is structurally homogeneous and if exchange is uncorrelated ($k_{ch} \ll k_f$, as in the EX2 regime). On the contrary, HDX can detect whether the molecules in a sample exist in two structurally different states that exchange through different mechanisms in some regions. For example, if the native form exchanges through equilibrium (EX2) mechanisms and a denatured or partially unfolded form exchanges through unfolding (EX1) mechanisms, then the mass spectra of peptides derived from these regions may display a bimodal isotope pattern [62]. Bimodal isotope patterns have been observed in the early folding studies of lysozyme [19], rabbit muscle aldolase destabilized by acid [62], and peptides derived from three segments of aldolase [22] with transient folding populations identified. In a folding/unfolding study of rabbit muscle triosephosphate isomerase, a pulsed labeling HDX approach was applied with the use of denaturant for the unfolding step, and a bimodal isotope pattern of labeled protein was observed (Fig. 8.6a), providing evidence for a two-state unfolding behavior [109]. During the renaturation step, three envelopes of isotope peaks were observed, suggesting an intermediate in the refolding pathway (Fig. 8.6b). The intermediate was then identified to have a folded C-terminal half and unfolded N-terminal half, by performing pepsin digestion immediately after the pulsed labeling.

These approaches for studying protein folding/unfolding lay the foundation for screening assays in the assessment of protein integrity and stability. The method,

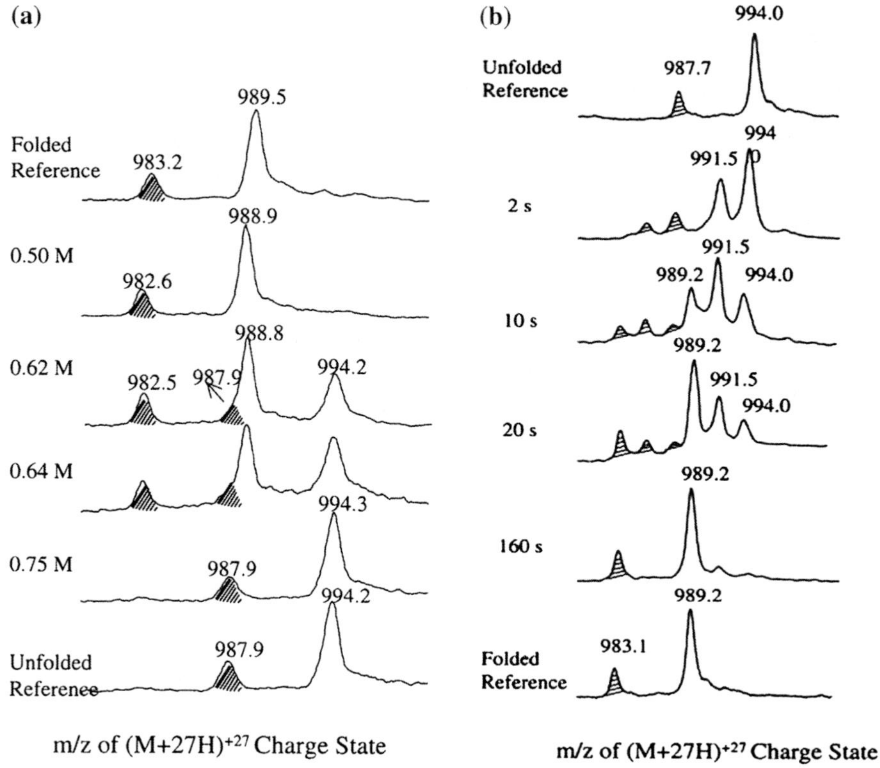

Fig. 8.6 **a** Mass spectra of the $(M+28H)^{+28}$ charge state of intact triosephosphate isomerase (TIM) unfolded in 1.5 M guanidine-HCl (GdHCl) for different times. Each sample was labeled for 10 s in GdHCl/D_2O prior to quenching isotope exchange. **b** Mass spectra of the $(M+27H)^{+27}$ charge state of intact TIM folded in 0.3 M GdHCl. The unfolded TIM was totally unfolded in 3.0 M GdHCl for 24 h and then diluted tenfold with phosphate buffer to initiate folding. Each sample was labeled for 3 s in D_2O (0.3 M GdHCl) prior to quenching isotope exchange. The peaks marked by hachure lines represent an impurity, which was identified as TIM truncated at the N-terminus. Figure was reprinted from Ref. [109]. Copyright (2004), with permission from Elsevier

named stability of unpurified proteins from rates of H/D exchange (SUPREX), uses HDX coupled with MALDI MS to estimate the stability of unpurified protein extracts, providing a high-throughput screen for assessing thermodynamic stability of proteins [25, 65]. In this approach, increasing concentration of denaturant is titrated into the protein solution, and protein stability is measured from the midpoint at which the HDX rate becomes dominated by cooperative opening events when a sharp increase in the total deuterium incorporation into full-length protein is observed. The method was also extended to the evaluation of protein–ligand binding as discussed in a later Sect. 8.6.2.

8.4.3 Comparability Studies for Protein Therapeutics

Higher-order structures and structural dynamics of protein therapeutics are critical factors in determining the function, efficacy, and safety of these drugs. Due to the complexity and limitations of NMR and X-ray crystallography techniques, which have been discussed earlier in this chapter, routine biopharmaceutical analyses of conformation and stability of protein drugs typically rely on classical biophysical methods, such as circular dichroism, FTIR spectroscopy, and calorimetry, which are more direct and less time-consuming methods [44, 110–114]. However, limited structural information can be revealed by these techniques. HDX MS has emerged as a more informative and sensitive method for higher-order structure analysis that can detect more subtle structural differences [115]. It can be used as an orthogonal measurement in structural comparison studies.

Changes in protein drug conformation and dynamics can be triggered by a variety of factors, such as chemical modification, mutation, surface binding, thermo denaturation, and pH denaturation caused by differences in manufacturing processes, cell media, or during storage. Covalent attachment of PEG (polyethylene glycol) to therapeutic proteins, or PEGylation, can be an effective strategy to improve the pharmacokinetic behavior of proteins [116, 117]. However, covalent attachment of PEG may cause possible conformational changes, steric interferences, and changes in electrostatic-binding properties for certain proteins [118, 119]. Aimed to determine the impact of PEGylation on protein conformation, a study was performed to compare the PEGylated and non-PEGylated forms of a protein drug, granulocyte colony-stimulating factor (G-CSF), using HDX MS [91].

In this study, peptide HDX MS experiments with a continuous "exchange-in" approach were carried out on both forms of G-CSF. Forty-six common peptides were identified in both proteins, constituting a linear sequence coverage of 91 % of the peptide backbone of G-CSF, with overlapping peptides in multiple regions. Deuterium uptake levels over a time span of 10 s to 4 h were obtained for each peptide. Some but not all peptides displayed differences in deuterium incorporation between the two forms of G-CSF. Figure 8.7a illustrates representative data of peptides in which deuterium uptake was the same or different. The location of each of the peptides was overlaid (Fig. 8.7b) onto the three-dimensional X-ray crystal structure of G-CSF (PDB code 2D9Q) [120]. Although small but significant differences in deuteration were found in multiple regions of G-CSF, for 65 % of G-CSF, there were no detectable differences in deuterium incorporation upon PEGylation. Many such peptides were located in the loop regions (e.g., residues 125–138, Fig. 8.7a, b), where exchange is predicted to be rapid. The peptide-covering residues 19–32 (part of an alpha helix) in which multiple amino acids are involved in receptor binding indicated no difference in deuteration rates and remained low in deuterium uptake for both forms. The HDX comparison data were visualized in a comparability butterfly chart (Fig. 8.7c) and a difference chart (Fig. 8.7d) to allow rapid qualitative and quantitative analysis of differences and the location of such differences. This plotting scheme was

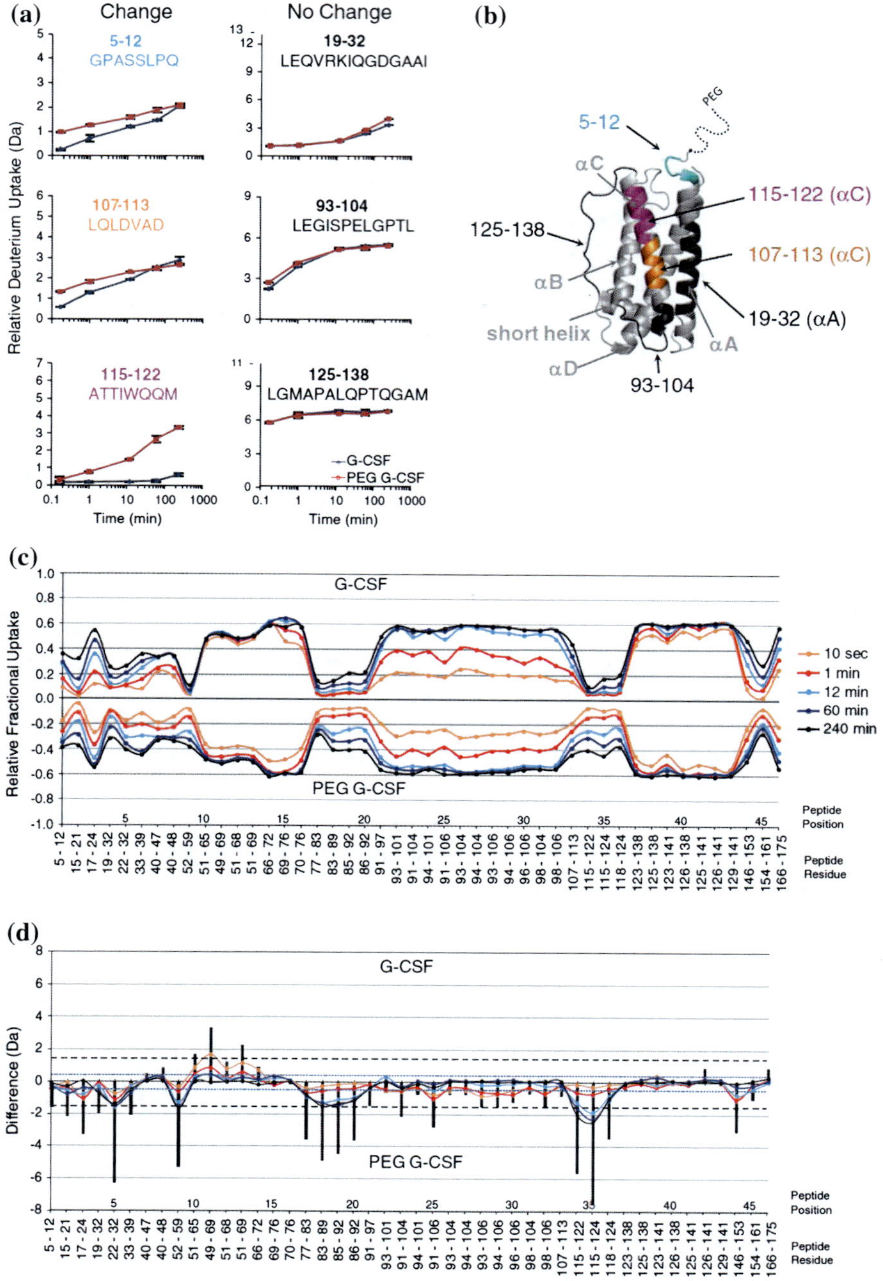

introduced by Houde et al. [51] and was used to compare a bound versus a free form of N-myristoyltransferase [121]. It is obvious from the charts that there are multiple regions along G-CSF with significant differences in deuterium incorporation between the two forms. Both the raw difference values (colored lines) and the

◀ **Fig. 8.7** **a** Deuterium uptake curves of six peptic peptides, three that had changes upon PEGylation (*left*) and three that did not (*right*). **b** The peptides are illustrated on the G-CSF crystal structure, with the color of the peptide matching the color of the residue labels in part A: *cyan*, residues 5–12; *orange*, residues 107–113; *pink*, residues 115–122; *black*, residues 19–32, 93–104, and 125–138. The three-dimensional structure of G-CSF is from PDB code 2D9Q [120]. **c** A butterfly chart of the raw deuterium levels and **d** a difference chart compare the forty-six common peptides (along the x-axis). Relative fractional uptake was calculated by dividing the deuterium level (in Da) by the total number of backbone amide hydrogens that could have become deuterated (equal to the number of amino acids, minus proline residues minus 1 for the N-terminal amide). In (**d**), the *blue dotted line* is set at 0.5 Da (both positive and negative differences), indicating the threshold for significant differences in raw fractional uptake. The *black vertical bars* indicate the summed differences in deuteration for each peptide, and the *black dotted line* is set at 1.5 Da (positive and negative) to indicate the threshold for a significant difference in summed deviations. Figure was adapted from Ref. [91] with kind permission from Springer Science + Business Media (Wei et al. [91] Figs. 2 and 3. Copyright 2012)

summed differences for each peptide (the vertical black bars) exceed significance lines for many peptides. However, no massive protein conformational changes occurred based on the extent of the difference. The conformation and conformational stability of the receptor-binding regions of the protein were not significantly affected by PEGylation either. The study indicated that the conformational differences detected may not have significant impact on the biological activity of G-CSF, which is consistent with other studies, showing that G-CSF maintains most of its biological activity after PEGlyation [122].

Comparability studies have also been conducted using similar peptide HDX MS approach on a protein drug interferon β-1a (IFN), a member of the type I interferon family (a group of homologous cytokines that display broad biological activity). One study compared the conformation of IFN versus a non-enzymatic version of the protein with alkylation at Cys-17 [115]. Dramatic destabilization of helix D by the modification was detected, and it affects a remote site in the amino acid sequence of this protein which is proximal to the Cys-17 side chain within the three-dimensional structure [123]. Another HDX MS study compared the HDX behavior of IFN reference material with a variety of other protein samples testing the effect of changed conditions such as freezing for 8 years, production from a different tissue culture growth medium, oxidation, and PEGylation [51]. The HDX MS data concluded that the only sample with significant differences from the reference material was the oxidized IFN. Upon oxidation, IFN showed widespread structural perturbation as reflected by changes in HDX.

These studies illustrate that changes in protein conformation as a result of some modification, process change, or other outside forces are readily assessed and can also be localized and quantified by HDX MS. Each of these experiments and data analysis were completed within a few days with automation in labeling, data collection, and data processing, making the technique potentially suitable for comparability studies, process monitoring, and protein therapeutic characterization in the biopharmaceutical industry.

8.5 Binding Studies

Protein binding studies are of crucial importance in biophysical studies and drug design. Non-covalent binding events, such as protein–ligand interaction (usually small molecule ligands or peptides) and protein–protein interaction, provide the foundation for virtually all drug action mechanisms [124]. HDX MS is an efficient analytical tool for analyzing these interactions because binding events usually enhance the thermodynamic stability of the protein, causing the protein complex to be more strongly protected from HDX than the free protein [125]. Spatially resolved HDX MS with the assistance of enzyme digestion and/or gas phase fragmentation is capable of measuring the deuterium incorporation rate at specific regions of a protein. A very important application of such an approach is to probe binding sites for molecular interactions. When a protein complex forms, the interface between the binding partners is likely to exclude solvent, hence reducing the exchange rate due to steric exclusion of solvent.

8.5.1 Protein–Protein Interactions

In order to probe binding sites, HDX experiments are usually performed on both the free protein as a control experiment and the protein–protein complex. The two experiments are compared, and regions in the protein that possess different exchange kinetics are highlighted. Both continuous labeling and pulsed labeling with "exchange-in" and "exchange-out" kinetics have been adopted to study binding regions [52, 54, 55, 63, 66, 126–130]. In several cases, the size and nature of the samples with complexity exclude the use of X-ray crystallography and NMR. One example from the continuous labeling approach for the study of a protein complex is briefly described [54].

In this study, HDX MS was used to map the structured and disordered regions of intrinsically disordered proteins (IDPs) and used to identify the disorder-to-order transitions. Two model IDPs were used in the study: an unstructured protein, ACTR (the unstructured activation domain of the activator of thyroid and retinoid receptors), and CBP (the molten globular nuclear coactivator-binding domain of CREB-binding protein), which together form a well-folded protein complex. Continuous "exchange-in" HDX MS experiments were performed followed by online pepsin digestion and LC separation on four independently prepared samples: free ACTR, free CBP, ACTR-CBP mixture (1:4.9 ACTR:CBP molar ratio) with 94 % of ACTR bound, and ACTR-CBP mixture (4/7:1 ACTR:CBP molar ratio) with 93 % of CBP bound. The labeling intervals ranged from 5 s to 12 h. The deuterium uptake for each peptic peptide was quantified via MS as a function of labeling time. Data obtained were used to calculate the protection factor for each peptide (see Eq. 8.8 in Method, and Ref. [46]). The protection factors were then compared between the free and bound states of each protein with and without the presence of a fivefold molar excess of its binding partner. Figure 8.8 shows the coverage maps for ACTR and CBP with peptide protection factors from both free and bound forms mapped onto the sequences. All the segments of the free ACTR exchange at rapid exchange rate, as predicted for an unstructured protein. Unlike ACTR, all segments in free CBP exchange more slowly than predicted by the intrinsic exchange calculation (k_{ch}). Similar HDX rates were observed across all segments, with protection factors ranging between 4.1 and 5.7. While in the bound complex, substantial changes were observed for both proteins. Many segments in ACTR were protected, especially in the middle regions of ACTR. For CBP, substantial changes were observed along almost the entire sequence, with the largest decrease in HDX rate in the Cα2 and Cα3 regions and in the linker connecting the two. While HDX data were informative for defining the complex interfaces, they also revealed dramatic conformational changes to each protein upon complex formation. The protein–protein interface or folding can be indicated by those peptides that retained more deuterons in the complex, or with increased protection factors, compared with control experiments in which only one protein was present. Segments of ACTR that had the largest decrease in HDX rate corresponded to the regions that make hydrophobic side chain contacts with hydrophobic residues lining a groove in CBP. While in CBP, though smaller changes were observed, the relatively largest decrease in HDX rate mapped to the core of the complex. The

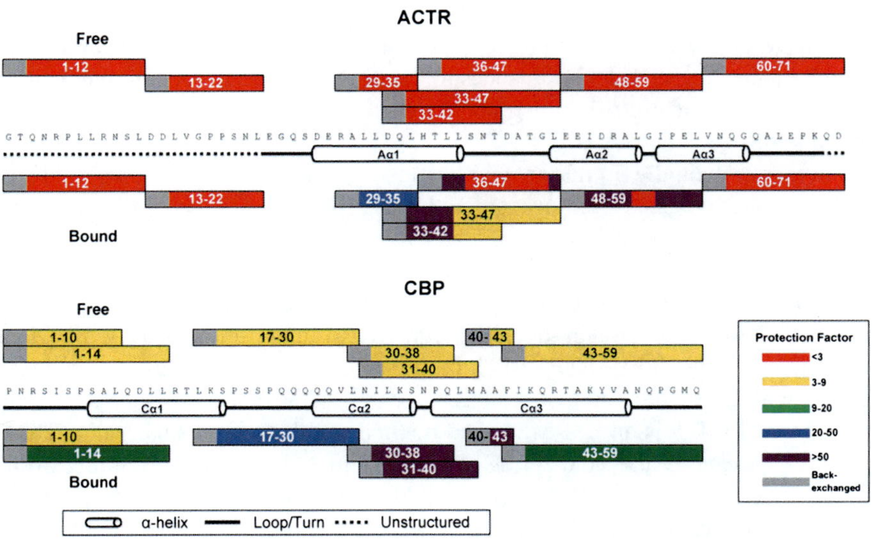

Fig. 8.8 Protection factors mapped onto the sequences of ACTR and CBP for each peptide for both free and bound forms. The assigned secondary structural elements from the CBP–/ACTR complex [152] are also indicated. The first two residues of each peptide are *colored gray* to indicate that rapid back exchange of the deuterium label at these two positions leads to no measurable deuterium uptake. Figure was adapted with permission from Ref. [54]. Copyright (2011) American Chemical Society

conclusions drawn from the HDX MS data were consistent with those obtained from other biophysical measurements such as NMR and X-ray crystallography.

Although proved very useful for probing interfaces for non-covalent protein–protein complexes, one needs to bear in mind that HDX cannot be used to locate binding interfaces for all interactions. Some protein–protein interactions driven by electrostatic forces of side chains are unlikely to change the exchange rate of backbone amide hydrogens, particularly if the amide hydrogens are located in stable structural elements such as alpha helices. In such circumstances, other biophysical methods are required.

8.5.2 Protein–Ligand Interactions

The characterization of protein binding to small ligands, such as small molecule drugs and peptides, is important for the development of small molecule therapeutics in pharmaceutical industries. Heme binding to myoglobin was one of the first published experiments using HDX MS to analyze small ligand binding to a protein molecule [16]. In the past two decades, the binding sites of small ligands on proteins have been identified in several instances: the binding region of the metalloproteinase inhibitor doxycycline on the active form of the matrilysin enzyme [131], the ligand-binding domain of the nuclear receptor PPARγ induced upon binding a full agonist and a partial agonist [53], and the two regions that are mainly involved in the interaction of insulin-like growth factor I binding to IGFBP-I [132], etc. Similar approaches to the study of protein–protein interactions can be used for identification of protein–ligand binding sites and will not be discussed in this section.

Another potentially important application of HDX MS is to quantitatively analyze protein–ligand binding interactions. Conventional methods for quantitative analysis of protein–ligand binding have limitations such as the relatively large quantity required for both proteins and ligands. High-throughput screening assays have been developed based on HDX MS technology that requires low amounts of samples. Protein–ligand interaction by mass spectrometry, titration, and H/D exchange (PLIMSTEX) is a high-throughput screening assay for quantitative analysis of protein–ligand binding that was developed by Gross and his group [133, 134]. It is an ESI-MS-based method which measures the global HDX kinetics of the apo- and holo-forms of the protein to quantify the binding affinity and stoichiometry of ligands. The method requires that a change, for example, conformational change and/or stability difference between the apo- and holo-form of the protein, occurs during a ligand titration. The approach is to conduct a titration where the ligand is added to aliquots of the protein solution with increasing ligand concentrations from zero (free protein) to excess (complex formation) and allowed to equilibrate. The aliquots are then diluted with excess deuterated buffer. Deuterium labeling is allowed for a predetermined period of time before quenching at which the global difference in deuterium (ΔD) between the apo- and holo-form is the greatest. Then, a plot of the mass difference between the deuterium uptake level and the ratio of total ligand concentration to the total protein concentration is constructed as the PLIMSTEX curve. The curve is fitted using a 1:n protein:ligand sequential binding model, where n is the number of binding sites for the same ligand, to extract the binding constants [134]. Figure 8.9 is a PLIMSTEX curve obtained from the titration of wild-type rat intestinal fatty-acid-binding protein (I-FABP) with potassium oleate fitted with a 1:1 binding model. The K_d derived from the model is $(2.6 \pm 0.2) \times 10^6\ \mathrm{M}^{-1}$, agrees with the literature value ($3.0 \times 10^6\ \mathrm{M}^{-1}$). Later on, the PLIMSTEX approach was applied to the characterization of human telomeric repeat-binding factor 2 (hTRF2) and DNA complex with the affinity constant determined [135]. By increasing the resolution of PLIMSTEX to the peptide level with pepsin digestion, the changes in deuterium uptake of hTRF2 as a function of varying amounts of a model oligodeoxynucleotide were also localized.

Another high-throughput screening assay for protein–ligand binding is SUPREX which was developed by Fitzgerald and coworkers [65, 136]. It is a MALDI-MS-based global HDX MS stability determination method and measures the increase in a protein's thermodynamic stability upon ligand binding in solution with the use of denaturants for inducing protein unfolding. The method was first used as a high-throughput stability screening approach (see Sect. 8.4.2). Higher concentration of denaturant is used in the presence of a bound ligand as the ligand usually induces stabilization of the protein structure [137]. The measured change in stability is used to calculate the dissociation constant (K_d) of the protein–ligand complex. The method has been applied to a range of different protein–ligand systems with obtained K_d values in good agreement with those reported on the same complexes using other techniques [138–141]. A single-point SUPREX protocol was also developed in a proof-of-concept study using the S-protein and a small test library of five peptides with known binding affinities for the S-protein [142]. The approach is suitable for performing high-throughput screening against large compound libraries in assays aimed at identifying high-affinity ligands to selected

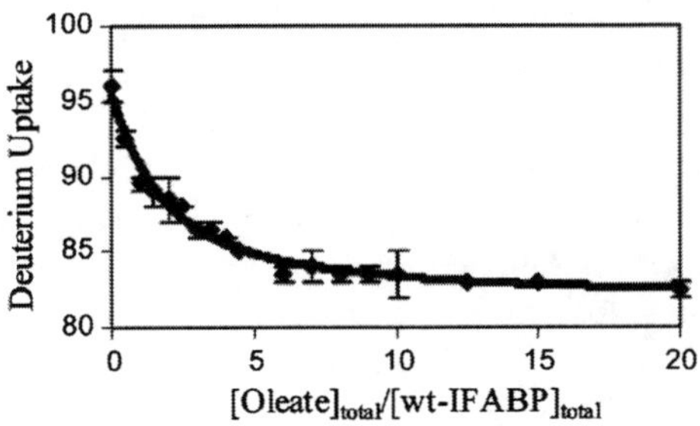

Fig. 8.9 PLIMSTEX data for 0.3 μM wild-type I-FABP titrated with K^+-oleate in 95 % D_2O, 20 mM pyrophosphate buffer, 135 mM KCl, and 10 mM NaCl (pH = 9.0), after 3 h of exchange. *Error bars* are from two independent runs. The *solid line* represents the fit by PLIMSTEX, using 1:1 binding and a three-parameter model (K_1, D_0, and ΔD_1). Figure was adapted with permission from Ref. [133]. Copyright (2003) American Chemical Society

protein targets. A protease digestion strategy was recently incorporated into the single-point SUPREX approach [143]. It has been demonstrated that the peptide readout was more efficient than the intact protein readout of the original single-point SUPREX protocol at discriminating hits from non-hits.

8.5.3 Allosteric Effects

In the binding studies, the observed information on the changes in HDX upon complex formation needs to be analyzed with extreme caution before drawing any conclusions regarding the binding interfaces, especially when the binding partner is a small molecule ligand. The reason for this caution is that upon ligand binding, other than through steric hindrance, protein conformational changes that are distant from the binding sites may be induced, leading to changes in HDX rates at sites remote from the binding interface. This effect is referred to as "allosteric effect." One example is the binding of cAMP to protein kinase A. In this case, binding causes the catalytic subunits to dissociate, releasing two active kinase molecules. The HDX MS experiments revealed the communication between the cAMP-binding site and the catalytic subunit, where upon cAMP binding, higher HDX rates were observed in the regulatory subunit that binds the catalytic subunit [144]. However, in cases where conformational changes reduced the local HDX rate, the interpretation of the HDX data becomes more complex, especially in the absence of three-dimensional structures. Recent HDX MS studies showed that the complex formation between cystatin (a thiol protease inhibitor) and its target enzyme, papain, reduced the flexibility throughout the sequence of cystatin [145]; binding of a peptide docking motif DEJL to a MAP kinase (ERK2) induced distal backbone flexibility changes in the activation lip [146]; upon the formation of a Dnase I:G-actin complex, conformational changes were observed to occur away from the binding site [147].

In order to distinguish the reduced HDX rate caused by ligand protection from the changes due to the allosteric effect, one effective approach is to combine site-directed mutagenesis with HDX MS experiments. This approach has been demonstrated by Brier and coworkers to identify the binding region of inhibitors targeting the human mitotic kinesin Eg5 [148]. By conducting mutagenesis on the two regions that showed decreases in deuterium incorporation rate: loop L5/helix alpha2 (region Tyr125-Glu145) and strand beta5/helix alpha3 (region Ile202-Leu227), the ligand-binding site of the protein was rapidly pinpointed. In another study, HDX MS combined with mutagenesis crystallography detected that the binding of GNF-5 to Abl via the myristate-binding region altered the conformation of the ATP-binding site of the protein [129]. GNF-5 is an analogue of a selective allosteric Bcr-Abl inhibitor (GNF-2). The changes in deuterium incorporation along the sequence of Abl in the presence versus absence of GNF-5 were observed in several peptides surrounding the myristate-binding cleft (e.g., residues 506–515), as well as in peptides near the ATP-binding site (e.g., residues 306–316).

The exchange rates were not altered in the non-binding Abl E505 K myristate mutant, indicating that the binding of GNF-5 is responsible for the allosteric conformational change.

Residues in binding regions can also be identified and distinguished from non-binding conformational changes by "exchange-out" experiments. In this approach, the apo-protein and its binding partner are first deuterated separately for a brief period of time, allowing surface amides to undergo rapid deuterium exchange. Then, the solutions are combined for complex formation to reach equilibrium, followed by dilution in H_2O-based buffer to exchange out the deuterium labels under conditions that preserve the non-covalent complex. The residues trapped in the binding regions will retain deuterium labels, while unprotected regions will lose the deuterium labels. After proteolysis, the segments of the protein that showed extensive deuterium incorporation are predicted to be the binding regions. Using this approach, the protein–protein interface between thrombin and a fragment of thrombomodulin, TMEGF45, was identified and differentiated from the regions of thrombin involved in allosteric changes [149]. Carefully designed control experiments are also critical to ensure the data accuracy for mapping binding regions using such an approach and to enhance the confidence in the conclusions drawn.

8.6 Future Perspectives

More than twenty years have passed since the initial protein studies were reported using the coupling of HDX to MS [14]. There have been significant advances in HDX MS instrumentation (hardware and software) and its expanded applications. With enhanced sensitivity, mass accuracy, and mass resolution of new MS instrumentation, HDX MS becomes more suitable to study proteins for comprehensive characterization. Advances in separation science will further improve the data quality for proteolysis-based HDX MS experiments, enabling the analysis of even larger proteins and complexes. The strategy to incorporate ETD/ECD into peptide-level HDX MS approach will greatly enhance the spatial resolution of the methodology/technology and is likely to be more widely adopted. With advancement in mass spectrometers, top-down HDX MS methods have the potential to spatially resolve deuteration patterns in larger proteins without proteolysis. This strategy can be applied to analyze heterogeneous proteins, such as protein conformers, protein conjugated to other molecules, which remained to be a challenge for the current HDX MS workflow. Further refinement in hardware automation and increasingly user-friendly software will pave the way for the technology being utilized for more sophisticated experiments, higher throughput, and more practical applications in higher-order structure characterization of protein therapeutics. With better data quality, accuracy, and throughput, another direction for HDX MS is likely to be coupled to computational protocols for structure–activity relation analysis, modeling, and docking studies [150, 151]. It is expected that HDX MS will continue to play an important role in protein structure characterization, help to gain new insight into the behavior of biomolecular systems, assist in drug design, and enhance quality control of biologics pharmaceuticals.

References

1. Kuntz ID, Chen K, Sharp KA, Kollman PA (1999) The maximal affinity of ligands. Proc Natl Acad Sci USA 96:9997–10002
2. Hvidt A, Linderstrom-Lang K (1954) Exchange of hydrogen atoms in insulin with deuterium atoms in aqueous solutions. Biochim Biophys Acta 14:574–575
3. Hvidt A, Linderstrom-Lang K (1955) The kinetics of the deuterium exchange of insulin with D_2O; an amendment. Biochim Biophys Acta 16:168–169
4. Englander SW (1963) A hydrogen exchange method using tritium and sephadex: its application to ribonuclease. Biochemistry 2:798–807
5. Haris PI, Chapman D (1995) The conformational analysis of peptides using fourier transform IR spectroscopy. Biopolymers 37:251–263
6. Englander JJ, Calhoun DB, Englander SW (1979) Measurement and calibration of peptide group hydrogen-deuterium exchange by ultraviolet spectrophotometry. Anal Biochem 92:517–524
7. Bentley GA, Delepierre M, Dobson CM, Wedin RE, Mason SA, Poulsen FM (1983) Exchange of individual hydrogens for a protein in a crystal and in solution. J Mol Biol 170:243–247
8. Englander SW, Mayne L (1992) Protein folding studied using hydrogen-exchange labeling and two-dimensional NMR. Annu Rev Biophys Biomol Struct 21:243–265
9. Jeng MF, Dyson HJ (1995) Comparison of the hydrogen-exchange behavior of reduced and oxidized Escherichia coli thioredoxin. Biochemistry 34:611–619
10. Dempsey CE (2001) Hydrogen exchange in peptides and proteins using NMR spectroscopy. Prog Nucl Magn Reson Spectrosc 39:135–170
11. Paterson Y, Englander SW, Roder H (1990) An antibody binding site on cytochrome c defined by hydrogen exchange and two-dimensional NMR. Science 249:755–759
12. Rosa JJ, Richards FM (1979) An experimental procedure for increasing the structural resolution of chemical hydrogen-exchange measurements on proteins: application to ribonuclease S peptide. J Mol Biol 133:399–416
13. Englander JJ, Rogero JR, Englander SW (1985) Protein hydrogen exchange studied by the fragment separation method. Anal Biochem 147:234–244
14. Katta V, Chait BT (1991) Conformational changes in proteins probed by hydrogen-exchange electrospray-ionization mass spectrometry. Rapid Commun Mass Spectrom 5:214–217
15. Zhang Z, Smith DL (1993) Determination of amide hydrogen exchange by mass spectrometry: a new tool for protein structure elucidation. Protein Sci 2:522–531
16. Johnson RS, Walsh KA (1994) Mass spectrometric measurement of protein amide hydrogen exchange rates of apo- and holo-myoglobin. Protein Sci 3:2411–2418
17. Pan J, Rintala-Dempsey AC, Li Y, Shaw GS, Konermann L (2006) Folding kinetics of the S100A11 protein dimer studied by time-resolved electrospray mass spectrometry and pulsed hydrogen-deuterium exchange. Biochemistry 45:3005–3013
18. Eyles SJ, Kaltashov IA (2004) Methods to study protein dynamics and folding by mass spectrometry. Methods 34:88–99
19. Miranker A, Robinson CV, Radford SE, Aplin RT, Dobson CM (1993) Detection of transient protein folding populations by mass spectrometry. Science 262:896–900
20. Zhu MM, Rempel DL, Zhao J, Giblin DE, Gross ML (2003) Probing Ca2+-induced conformational changes in porcine calmodulin by H/D exchange and ESI-MS: effect of cations and ionic strength. Biochemistry 42:15388–15397

21. Lanman J, Lam TT, Barnes S, Sakalian M, Emmett MR, Marshall AG, Prevelige PE Jr (2003) Identification of novel interactions in HIV-1 capsid protein assembly by high-resolution mass spectrometry. J Mol Biol 325:759–772
22. Zhang Z, Post CB, Smith DL (1996) Amide hydrogen exchange determined by mass spectrometry: application to rabbit muscle aldolase. Biochemistry 35:779–791
23. Hoofnagle AN, Resing KA, Ahn NG (2003) Protein analysis by hydrogen exchange mass spectrometry. Annu Rev Biophys Biomol Struct 32:1–25
24. Wales TE, Engen JR (2006) Hydrogen exchange mass spectrometry for the analysis of protein dynamics. Mass Spectrom Rev 25:158–170
25. Powell KD, Fitzgerald MC (2001) Measurements of protein stability by H/D exchange and matrix-assisted laser desorption/ionization mass spectrometry using picomoles of material. Anal Chem 73:3300–3304
26. Englander SW, Kallenbach NR (1983) Hydrogen exchange and structural dynamics of proteins and nucleic acids. Q Rev Biophys 16:521–655
27. Hamuro Y, Coales SJ, Southern MR, Nemeth-Cawley JF, Stranz DD, Griffin PR (2003) Rapid analysis of protein structure and dynamics by hydrogen/deuterium exchange mass spectrometry. J Biomol Tech 14:171–182
28. Garcia RA, Pantazatos D, Villarreal FJ (2004) Hydrogen/deuterium exchange mass spectrometry for investigating protein-ligand interactions. Assay Drug Dev Technol 2:81–91
29. Busenlehner LS, Armstrong RN (2005) Insights into enzyme structure and dynamics elucidated by amide H/D exchange mass spectrometry. Arch Biochem Biophys 433:34–46
30. Goshe MB, Anderson VE (1999) Hydroxyl radical-induced hydrogen/deuterium exchange in amino acid carbon-hydrogen bonds. Radiat Res 151:50–58
31. Bai Y, Milne JS, Mayne L, Englander SW (1993) Primary structure effects on peptide group hydrogen exchange. Proteins 17:75–86
32. Molday RS, Englander SW, Kallen RG (1972) Primary structure effects on peptide group hydrogen exchange. Biochemistry 11:150–158
33. Fersht AR (1971) Acyl-transfer reactions of amides and esters with alcohols and thiols. A reference system for the serine and cysteine proteinases. Concerning the N protonation of amides and amide-imidate equilibria. J Am Chem Soc 93:3504–3515
34. Eriksson MA, Hard T, Nilsson L (1995) On the pH dependence of amide proton exchange rates in proteins. Biophys J 69:329–339
35. Brier S, Engen JR (2008) Hydrogen exchange mass spectrometry: principles and capabilities. In: Chance M (ed) Mass spectrometry analysis for protein–protein interactions and dynamics. Wiley, New Jersey
36. Berger A, Loewenstein A, Meiboom S (1959) Nuclear magnetic resonance and the proteolysis of N-methylacetamide. J Am Chem Soc 81(1):62–67
37. Dempsey CE (2001) Hydrogen exchange in peptides and proteins using NMR spectroscopy. Prog Nucl Magn Reson Spectro 39:135–170
38. Connelly GP, Bai Y, Jeng MF, Englander SW (1993) Isotope effects in peptide group hydrogen exchange. Proteins 17:87–92
39. Barksdale AD, Rosenberg A (1982) Acquisition and interpretation of hydrogen exchange data from peptides, polymers, and proteins. Methods Biochem Anal 28:1–113
40. Hvidt A, Nielsen SO (1966) Hydrogen exchange in proteins. Adv Protein Chem 21:287–386
41. Konermann L, Tong X, Pan Y (2008) Protein structure and dynamics studied by mass spectrometry: H/D exchange, hydroxyl radical labeling, and related approaches. J Mass Spectrom 43:1021–1036
42. Maier CS, Schimerlik MI, Deinzer ML (1999) Thermal denaturation of escherichia coli thioredoxin studied by hydrogen/deuterium exchange and electrospray ionization mass spectrometry: monitoring a two-state protein unfolding transition. Biochemistry 38:1136–1143
43. Deng Y, Smith DL (1998) Identification of unfolding domains in large proteins by their unfolding rates. Biochemistry 37:6256–6262
44. Swint-Kruse L, Robertson AD (1996) Temperature and pH dependences of hydrogen exchange and global stability for ovomucoid third domain. Biochemistry 35:171–180

45. Clarke J, Itzhaki LS (1998) Hydrogen exchange and protein folding. Curr Opin Struct Biol 8:112–118
46. Chetty PS, Mayne L, Lund-Katz S, Stranz D, Englander SW, Phillips MC (2009) Helical structure and stability in human apolipoprotein A-I by hydrogen exchange and mass spectrometry. Proc Natl Acad Sci USA 106:19005–19010
47. Englander SW, Englander JJ (1972) Hydrogen-tritium exchange. Methods Enzymol, 26 PtC, 406–413
48. Maier CS, Deinzer ML (2005) Protein conformations, interactions, and H/D exchange. Methods Enzymol 402:312–360
49. Woodward CK, Ellis LM, Rosenberg A (1975) The solvent dependence of hydrogen exchange kinetics of folded proteins. J Biol Chem 250:440–444
50. Engen JR, Smithgall TE, Gmeiner WH, Smith DL (1997) Identification and localization of slow, natural, cooperative unfolding in the hematopoietic cell kinase SH3 domain by amide hydrogen exchange and mass spectrometry. Biochemistry 36:14384–14391
51. Houde D, Berkowitz SA, Engen JR (2011) The utility of hydrogen/deuterium exchange mass spectrometry in biopharmaceutical comparability studies. J Pharm Sci 100(6):2071–2086
52. Zhou B, Zhang ZY (2007) Application of hydrogen/deuterium exchange mass spectrometry to study protein tyrosine phosphatase dynamics, ligand binding, and substrate specificity. Methods 42:227–233
53. Chalmers MJ, Busby SA, Pascal BD, He Y, Hendrickson CL, Marshall AG, Griffin PR (2006) Probing protein ligand interactions by automated hydrogen/deuterium exchange mass spectrometry. Anal Chem 78:1005–1014
54. Keppel TR, Howard BA, Weis DD (2011) Mapping unstructured regions and synergistic folding in intrinsically disordered proteins with amide H/D exchange mass spectrometry. Biochemistry 50:8722–8732
55. Coales SJ, Tuske SJ, Tomasso JC, Hamuro Y (2009) Epitope mapping by amide hydrogen/deuterium exchange coupled with immobilization of antibody, on-line proteolysis, liquid chromatography and mass spectrometry. Rapid Commun Mass Spectrom 23:639–647
56. Deng Y, Zhang Z, Smith DL (1999) Comparison of continuous and pulsed labeling amide hydrogen exchange/mass spectrometry for studies of protein dynamics. J Am Soc Mass Spectrom 10:675–684
57. Konermann L, Simmons DA (2003) Protein-folding kinetics and mechanisms studied by pulse-labeling and mass spectrometry. Mass Spectrom Rev 22:1–26
58. Hossain BM, Konermann L (2006) Pulsed hydrogen/deuterium exchange MS/MS for studying the relationship between noncovalent protein complexes in solution and in the gas phase after electrospray ionization. Anal Chem 78:1613–1619
59. Yang H, Smith DL (1997) Kinetics of cytochrome c folding examined by hydrogen exchange and mass spectrometry. Biochemistry 36:14992–14999
60. Rogero JR, Englander JJ, Englander SW (1986) Individual breathing reactions measured by functional labeling and hydrogen exchange methods. Methods Enzymol 131:508–517
61. Baerga-Ortiz A, Hughes CA, Mandell JG, Komives EA (2002) Epitope mapping of a monoclonal antibody against human thrombin by H/D-exchange mass spectrometry reveals selection of a diverse sequence in a highly conserved protein. Protein Sci 11:1300–1308
62. Smith DL, Deng Y, Zhang Z (1997) Probing the non-covalent structure of proteins by amide hydrogen exchange and mass spectrometry. J Mass Spectrom 32:135–146
63. Engen JR (2003) Analysis of protein complexes with hydrogen exchange and mass spectrometry. Analyst 128:623–628
64. Jr Woods VL, Hamuro Y (2001) High resolution, high-throughput amide deuterium exchange-mass spectrometry (DXMS) determination of protein binding site structure and dynamics: utility in pharmaceutical design. J Cell Biochem Suppl 37:89–98
65. Ghaemmaghami S, Fitzgerald MC, Oas TG (2000) A quantitative, high-throughput screen for protein stability. Proc Natl Acad Sci USA 97:8296–8301

66. Mandell JG, Falick AM, Komives EA (1998) Identification of protein–protein interfaces by decreased amide proton solvent accessibility. Proc Natl Acad Sci USA 95:14705–14710
67. Kipping M, Schierhorn A (2003) Improving hydrogen/deuterium exchange mass spectrometry by reduction of the back-exchange effect. J Mass Spectrom 38:271–276
68. Anderegg RJ, Wagner DS, Stevenson CL (1994) The mass spectrometry of helical unfolding in peptides. J Am Chem Soc 5:425–433
69. Demmers JA, Haverkamp J, Heck AJ, Koeppe RE 2nd, Killian JA (2000) Electrospray ionization mass spectrometry as a tool to analyze hydrogen/deuterium exchange kinetics of transmembrane peptides in lipid bilayers. Proc Natl Acad Sci USA 97:3189–3194
70. Jorgensen TJ, Gardsvoll H, Ploug M, Roepstorff P (2005) Intramolecular migration of amide hydrogens in protonated peptides upon collisional activation. J Am Chem Soc 127:2785–2793
71. Kaltashov IA, Eyles SJ (2002) Crossing the phase boundary to study protein dynamics and function: combination of amide hydrogen exchange in solution and ion fragmentation in the gas phase. J Mass Spectrom 37:557–565
72. Ferguson PL, Pan J, Wilson DJ, Dempsey B, Lajoie G, Shilton B, Konermann L (2007) Hydrogen/deuterium scrambling during quadrupole time-of-flight MS/MS analysis of a zinc-binding protein domain. Anal Chem 79:153–160
73. Rand KD, Adams CM, Zubarev RA, Jorgensen TJ (2008) Electron capture dissociation proceeds with a low degree of intramolecular migration of peptide amide hydrogens. J Am Chem Soc 130:1341–1349
74. Zubarev RA (2003) Reactions of polypeptide ions with electrons in the gas phase. Mass Spectrom Rev 22:57–77
75. Horn DM, Breuker K, Frank AJ, McLafferty FW (2001) Kinetic intermediates in the folding of gaseous protein ions characterized by electron capture dissociation mass spectrometry. J Am Chem Soc 123:9792–9799
76. Syka JE, Coon JJ, Schroeder MJ, Shabanowitz J, Hunt DF (2004) Peptide and protein sequence analysis by electron transfer dissociation mass spectrometry. Proc Natl Acad Sci USA 101:9528–9533
77. Xia Y, Thomson BA, McLuckey SA (2007) Bidirectional ion transfer between quadrupole arrays: MSn ion/ion reaction experiments on a quadrupole/time-of-flight tandem mass spectrometer. Anal Chem 79:8199–8206
78. Zehl M, Rand KD, Jensen ON, Jorgensen TJ (2008) Electron transfer dissociation facilitates the measurement of deuterium incorporation into selectively labeled peptides with single residue resolution. J Am Chem Soc 130:17453–17459
79. Rand KD, Zehl M, Jensen ON, Jorgensen TJ (2009) Protein hydrogen exchange measured at single-residue resolution by electron transfer dissociation mass spectrometry. Anal Chem 81:5577–5584
80. Landgraf RR, Chalmers MJ, Griffin PR (2012) Automated hydrogen/deuterium exchange electron transfer dissociation high resolution mass spectrometry measured at single-amide resolution. J Am Soc Mass Spectrom 23:301–309
81. Villanueva J, Hoshino M, Katou H, Kardos J, Hasegawa K, Naiki H, Goto Y (2004) Increase in the conformational flexibility of beta 2-microglobulin upon copper binding: a possible role for copper in dialysis-related amyloidosis. Protein Sci 13:797–809
82. Pan J, Han J, Borchers CH, Konermann L (2008) Electron capture dissociation of electrosprayed protein ions for spatially resolved hydrogen exchange measurements. J Am Chem Soc 130:11574–11575
83. Abzalimov RR, Kaplan DA, Easterling ML, Kaltashov IA (2009) Protein conformations can be probed in top-down HDX MS experiments utilizing electron transfer dissociation of protein ions without hydrogen scrambling. J Am Soc Mass Spectrom 20:1514–1517
84. Kaltashov IA, Bobst CE, Abzalimov RR (2009) H/D exchange and mass spectrometry in the studies of protein conformation and dynamics: is there a need for a top–down approach? Anal Chem 81:7892–7899

85. Wang L, Pan H, Smith DL (2002) Hydrogen exchange-mass spectrometry: optimization of digestion conditions. Mol Cell Proteomics 1:132–138
86. Resing KA, Hoofnagle AN, Ahn NG (1999) Modeling deuterium exchange behavior of ERK2 using pepsin mapping to probe secondary structure. J Am Soc Mass Spectrom 10:685–702
87. Hoofnagle AN, Resing KA, Ahn NG (2004) Practical methods for deuterium exchange/mass spectrometry. Methods Mol Biol 250:283–298
88. Feng L, Orlando R, Prestegard JH (2006) Amide proton back-exchange in deuterated peptides: applications to MS and NMR analyses. Anal Chem 78:6885–6892
89. Weis DD, Wales TE, Engen JR, Hotchko M, Ten Eyck LF (2006) Identification and characterization of EX1 kinetics in H/D exchange mass spectrometry by peak width analysis. J Am Soc Mass Spectrom 17:1498–1509
90. Houde D, Arndt J, Domeier W, Berkowitz S, Engen JR (2009) Characterization of IgG1 conformation and conformational dynamics by hydrogen/deuterium exchange mass spectrometry. Anal Chem 81:5966
91. Wei H, Ahn J, Yu YQ, Tymiak A, Engen JR, Chen G (2012) Using hydrogen/deuterium exchange mass spectrometry to study conformational changes in granulocyte colony stimulating factor upon PEGylation. J Am Soc Mass Spectrom 23:498–504
92. Emmett MR, Kazazic S, Marshall AG, Chen W, Shi SD, Bolanos B, Greig MJ (2006) Supercritical fluid chromatography reduction of hydrogen/deuterium back exchange in solution-phase hydrogen/deuterium exchange with mass spectrometric analysis. Anal Chem 78:7058–7060
93. Zhang Z (1997) Enhancement of the effective resolution of mass spectra of high-mass biomolecules by maximum entropy-based deconvolution to eliminate the isotopic natural abundance distribution. J Am Chem Soc 8:659–670
94. Hotchko M, Anand GS, Komives EA, Ten Eyck LF (2006) Automated extraction of backbone deuteration levels from amide H/2H mass spectrometry experiments. Protein Sci 15:583–601
95. Weis DD, Engen JR, Kass IJ (2006) Semi-automated data processing of hydrogen exchange mass spectra using HX-Express. J Am Soc Mass Spectrom 17:1700–1703
96. Buijs J, Hakansson K, Hagman C, Hakansson P, Oscarsson S (2000) A new method for the accurate determination of the isotopic state of single amide hydrogens within peptides using Fourier transform ion cyclotron resonance mass spectrometry. Rapid Commun Mass Spectrom 14:1751–1756
97. Pascal BD, Chalmers MJ, Busby SA, Mader CC, Southern MR, Tsinoremas NF, Griffin PR (2007) The Deuterator: software for the determination of backbone amide deuterium levels from H/D exchange MS data. BMC Bioinformatics 8:156
98. Kazazic S, Zhang HM, Schaub TM, Emmett MR, Hendrickson CL, Blakney GT, Marshall AG (2010) Automated data reduction for hydrogen/deuterium exchange experiments, enabled by high-resolution Fourier transform ion cyclotron resonance mass spectrometry. J Am Soc Mass Spectrom 21:550–558
99. Nikamanon P, Pun E, Chou W, Koter MD, Gershon PD (2008) TOF2H: a precision toolbox for rapid, high density/high coverage hydrogen-deuterium exchange mass spectrometry via an LC-MALDI approach, covering the data pipeline from spectral acquisition to HDX rate analysis. BMC Bioinformatics 9:387
100. Pascal BD, Chalmers MJ, Busby SA, Griffin PR (2009) HD desktop: an integrated platform for the analysis and visualization of H/D exchange data. J Am Soc Mass Spectrom 20:601–610
101. Slysz GW, Baker CA, Bozsa BM, Dang A, Percy AJ, Bennett M, Schriemer DC (2009) Hydra: software for tailored processing of H/D exchange data from MS or tandem MS analyses. BMC Bioinformatics 10:162
102. Kan ZY, Mayne L, Chetty PS, Englander SW (2011) ExMS: data analysis for HX-MS experiments. J Am Soc Mass Spectrom 22:1906–1915

103. Tsui V, Garcia C, Cavagnero S, Siuzdak G, Dyson HJ, Wright PE (1999) Quench-flow experiments combined with mass spectrometry show apomyoglobin folds through and obligatory intermediate. Protein Sci 8:45–49
104. Wales TE, Fadgen KE, Gerhardt GC, Engen JR (2008) High-speed and high-resolution UPLC separation at zero degrees Celsius. Anal Chem 80:6815–6820
105. Hoofnagle AN, Resing KA, Goldsmith EJ, Ahn NG (2001) Changes in protein conformational mobility upon activation of extracellular regulated protein kinase-2 as detected by hydrogen exchange. Proc Natl Acad Sci USA 98:956–961
106. Robbins DJ, Zhen E, Owaki H, Vanderbilt CA, Ebert D, Geppert TD, Cobb MH (1993) Regulation and properties of extracellular signal-regulated protein kinases 1 and 2 in vitro. J Biol Chem 268:5097–5106
107. Canagarajah BJ, Khokhlatchev A, Cobb MH, Goldsmith EJ (1997) Activation mechanism of the MAP kinase ERK2 by dual phosphorylation. Cell 90:859–869
108. Zhang F, Strand A, Robbins D, Cobb MH, Goldsmith EJ (1994) Atomic structure of the MAP kinase ERK2 at 2.3 a resolution. Nature 367:704–711
109. Pan H, Raza AS, Smith DL (2004) Equilibrium and kinetic folding of rabbit muscle triosephosphate isomerase by hydrogen exchange mass spectrometry. J Mol Biol 336:1251–1263
110. Dong A, Matsuura J, Allison SD, Chrisman E, Manning MC, Carpenter JF (1996) Infrared and circular dichroism spectroscopic characterization of structural differences between beta-lactoglobulin A and B. Biochemistry 35:1450–1457
111. Sugeta H (1991) Study on conformation of biomolecules by infrared circular dichroism. Tanpakushitsu Kakusan Koso 36:1849–1858
112. Susi H, Byler DM (1986) Resolution-enhanced Fourier transform infrared spectroscopy of enzymes. Methods Enzymol 130:290–311
113. Nguyen LT, Wiencek JM, Kirsch LE (2003) Characterization methods for the physical stability of biopharmaceuticals. PDA J Pharm Sci Technol 57:429–445
114. Martin SR, Schilstra MJ (2008) Circular dichroism and its application to the study of biomolecules. Methods Cell Biol 84:263–293
115. Kaltashov IA, Bobst CE, Abzalimov RR, Berkowitz SA, Houde D (2010) Conformation and dynamics of biopharmaceuticals: transition of mass spectrometry-based tools from academe to industry. J Am Soc Mass Spectrom 21:323–337
116. Abuchowski A, McCoy JR, Palczuk NC, van Es T, Davis FF (1977) Effect of covalent attachment of polyethylene glycol on immunogenicity and circulating life of bovine liver catalase. J Biol Chem 252:3582–3586
117. Greenwald RB, Choe YH, McGuire J, Conover CD (2003) Effective drug delivery by PEGylated drug conjugates. Adv Drug Deliv Rev 55:217–250
118. Chapman AP (2002) PEGylated antibodies and antibody fragments for improved therapy: a review. Adv Drug Deliv Rev 54:531–545
119. Reddy KR (2000) Controlled-release, PEGylation, liposomal formulations: new mechanisms in the delivery of injectable drugs. Ann Pharmacother 34:915–923
120. Tamada T, Honjo E, Maeda Y, Okamoto T, Ishibashi M, Tokunaga M, Kuroki R (2006) Homodimeric cross-over structure of the human granulocyte colony-stimulating factor (GCSF) receptor signaling complex. Proc Natl Acad Sci USA 103:3135–3140
121. Morgan CR, Miglionico BV, Engen JR (2011) Effects of HIV-1 Nef on human N-myristoyltransferase 1. Biochemistry 50:3394–3403
122. Piedmonte DM, Treuheit MJ (2008) Formulation of Neulasta (pegfilgrastim). Adv Drug Deliv Rev 60:50–58
123. Bobst CE, Abzalimov RR, Houde D, Kloczewiak M, Mhatre R, Berkowitz SA, Kaltashov IA (2008) Detection and characterization of altered conformations of protein pharmaceuticals using complementary mass spectrometry-based approaches. Anal Chem 80:7473–7481
124. Bissantz C, Kuhn B, Stahl M (2010) A medicinal chemist's guide to molecular interactions. J Med Chem 53:5061–5084

125. Konermann L, Pan J, Liu YH (2011) Hydrogen exchange mass spectrometry for studying protein structure and dynamics. Chem Soc Rev 40:1224–1234
126. Yamada N, Suzuki E, Hirayama K (2002) Identification of the interface of a large protein–protein complex using H/D exchange and Fourier transform ion cyclotron resonance mass spectrometry. Rapid Commun Mass Spectrom 16:293–299
127. Lisal J, Kainov DE, Lam TT, Emmett MR, Wei H, Gottlieb P, Marshall AG, Tuma R (2006) Interaction of packaging motor with the polymerase complex of dsRNA bacteriophage. Virology 351:73–79
128. Derunes C, Burgess R, Iraheta E, Kellerer R, Becherer K, Gessner CR, Li S, Hewitt K, Vuori K, Pasquale EB, Woods VL Jr, Ely KR (2006) Molecular determinants for interaction of SHEP1 with Cas localize to a highly solvent-protected region in the complex. FEBS Lett 580:175–178
129. Zhang J, Adrian FJ, Jahnke W, Cowan-Jacob SW, Li AG, Iacob RE, Sim T, Powers J, Dierks C, Sun F, Guo GR, Ding Q, Okram B, Choi Y, Wojciechowski A, Deng X, Liu G, Fendrich G, Strauss A, Vajpai N, Grzesiek S, Tuntland T, Liu Y, Bursulaya B, Azam M, Manley PW, Engen JR, Daley GQ, Warmuth M, Gray NS (2010) Targeting Bcr-Abl by combining allosteric with ATP-binding-site inhibitors. Nature 463:501–506
130. Zhang Q, Willison LN, Tripathi P, Sathe SK, Roux KH, Emmett MR, Blakney GT, Zhang HM, Marshall AG (2011) Epitope mapping of a 95 kDa antigen in complex with antibody by solution-phase amide backbone hydrogen/deuterium exchange monitored by Fourier transform ion cyclotron resonance mass spectrometry. Anal Chem 83:7129–7136
131. Garcia RA, Pantazatos DP, Gessner CR, Go KV, Woods VL Jr, Villarreal FJ (2005) Molecular interactions between matrilysin and the matrix metalloproteinase inhibitor doxycycline investigated by deuterium exchange mass spectrometry. Mol Pharmacol 67:1128–1136
132. Ehring H (1999) Hydrogen exchange/electrospray ionization mass spectrometry studies of structural features of proteins and protein/protein interactions. Anal Biochem 267:252–259
133. Zhu MM, Rempel DL, Du Z, Gross ML (2003) Quantification of protein-ligand interactions by mass spectrometry, titration, and H/D exchange: PLIMSTEX. J Am Chem Soc 125:5252–5253
134. Zhu MM, Rempel DL, Gross ML (2004) Modeling data from titration, amide H/D exchange, and mass spectrometry to obtain protein-ligand binding constants. J Am Soc Mass Spectrom 15:388–397
135. Sperry JB, Shi X, Rempel DL, Nishimura Y, Akashi S, Gross ML (2008) A mass spectrometric approach to the study of DNA-binding proteins: interaction of human TRF2 with telomeric DNA. Biochemistry 47:1797–1807
136. Powell KD, Wales TE, Fitzgerald MC (2002) Thermodynamic stability measurements on multimeric proteins using a new H/D exchange- and matrix-assisted laser desorption/ionization (MALDI) mass spectrometry-based method. Protein Sci 11:841–851
137. Huang YJ, Montelione GT (2005) Structural biology: proteins flex to function. Nature 438:36–37
138. Tang L, Hopper ED, Tong Y, Sadowsky JD, Peterson KJ, Gellman SH, Fitzgerald MC (2007) H/D exchange- and mass spectrometry-based strategy for the thermodynamic analysis of protein-ligand binding. Anal Chem 79:5869–5877
139. Roulhac PL, Weaver KD, Adhikari P, Anderson DS, DeArmond PD, Mietzner TA, Crumbliss AL, Fitzgerald MC (2008) Ex vivo analysis of synergistic anion binding to FbpA in gram-negative bacteria. Biochemistry 47:4298–4305
140. Hopper ED, Roulhac PL, Campa MJ, Patz EF Jr, Fitzgerald MC (2008) Throughput and efficiency of a mass spectrometry-based screening assay for protein-ligand binding detection. J Am Soc Mass Spectrom 19:1303–1311
141. Dearmond PD, West GM, Anbalagan V, Campa MJ, Patz EF Jr, Fitzgerald MC (2010) Discovery of novel cyclophilin A ligands using an H/D exchange- and mass spectrometry-based strategy. J Biomol Screen 15:1051–1062
142. Powell KD, Fitzgerald MC (2004) High-throughput screening assay for the tunable selection of protein ligands. J Comb Chem 6:262–269

143. Hopper ED, Pittman AM, Tucker CL, Campa MJ, Patz EF Jr, Fitzgerald MC (2009) Hydrogen/deuterium exchange- and protease digestion-based screening assay for protein-ligand binding detection. Anal Chem 81:6860–6867
144. Anand GS, Hughes CA, Jones JM, Taylor SS, Komives EA (2002) Amide H/2H exchange reveals communication between the cAMP and catalytic subunit-binding sites in the R(I) alpha subunit of protein kinase A. J Mol Biol 323:377–386
145. Akashi S, Takio K (2000) Characterization of the interface structure of enzyme-inhibitor complex by using hydrogen-deuterium exchange and electrospray ionization Fourier transform ion cyclotron resonance mass spectrometry. Protein Sci 9:2497–2505
146. Lee T, Hoofnagle AN, Kabuyama Y, Stroud J, Min X, Goldsmith EJ, Chen L, Resing KA, Ahn NG (2004) Docking motif interactions in MAP kinases revealed by hydrogen exchange mass spectrometry. Mol Cell 14:43–55
147. Chik JK, Schriemer DC (2003) Hydrogen/deuterium exchange mass spectrometry of actin in various biochemical contexts. J Mol Biol 334:373–385
148. Brier S, Lemaire D, DeBonis S, Kozielski F, Forest E (2006) Use of hydrogen/deuterium exchange mass spectrometry and mutagenesis as a tool to identify the binding region of inhibitors targeting the human mitotic kinesin Eg5. Rapid Commun Mass Spectrom 20:456–462
149. Mandell JG, Baerga-Ortiz A, Akashi S, Takio K, Komives EA (2001) Solvent accessibility of the thrombin-thrombomodulin interface. J Mol Biol 306:575–589
150. Bailey-Kellogg C, Kelley JJ 3rd, Stein C, Donald BR (2001) Reducing mass degeneracy in SAR by MS by stable isotopic labeling. J Comput Biol 8:19–36
151. Bennett MJ, Barakat K, Huzil JT, Tuszynski J, Schriemer DC (2010) Discovery and characterization of the laulimalide-microtubule binding mode by mass shift perturbation mapping. Chem Biol 17:725–734
152. Demarest SJ, Martinez-Yamout M, Chung J, Chen H, Xu W, Dyson HJ, Evans RM, Wright PE (2002) Mutual synergistic folding in recruitment of CBP/p300 by p160 nuclear receptor coactivators. Nature 415:549–553

Chapter 9
Fast Photochemical Oxidation of Proteins for Structural Characterization

Lisa M. Jones

9.1 Oxidation as a Tool for Structural Biology

Protein footprinting coupled with mass spectrometry (MS) has become a powerful tool for studying protein interactions. There are many types of footprinting labels that can be reversible or irreversible, and in either case they can be either general or specific for a particular amino acid. Reversible labels used for footprinting include deuterium [1–3] and citraconic anhydride for lysine modification [4, 5]. Examples of irreversible labels include hydroxyl radicals [6, 7], N-ethylmaleimide (NEM) for cysteine modification [8], and glycine ethyl ester (GEE) for modification of carboxylic acids [9]. These footprinting methods are a good probe of solvent accessibility.

Hydrogen–deuterium exchange (HDX) is perhaps the most widely used footprinting method with a number of publications using this method in recent years [10–13]. HDX is advantageous because every amino acid except proline can be labeled with deuterium. This general labeling strategy can be used on many different protein systems regardless of the primary sequence. However, HDX utilizes a reversible label which requires rapid post-labeling sample handling to minimize back exchange. This could be unfavorable for complex protein systems that require post-labeling purification to reduce the complexity of the mass spectrum.

In recent years, another footprinting method, hydroxyl radical-mediated oxidative labeling, has emerged as a tool for studying protein structure [14]. A benefit of hydroxyl radicals (•OH) for footprinting is that their size is similar to water, thus making them an excellent probe of solvent accessibility. Oxidative labeling is considered a general label because •OH react with more than half of the amino acids. However, unlike deuterium, amino acids have different reactivities with •OH (Table 9.1). This reactivity–specificity may bias the experiment based on the primary sequence of the protein system being studied.

L. M. Jones (✉)
Department of Chemistry and Chemical Biology, Indiana University-Purdue University in Indianapolis, LD326, 402 N. Blackford St, Indianapolis IN 46202, USA
e-mail: joneslis@iupui.edu

G. Chen (ed.), *Characterization of Protein Therapeutics using Mass Spectrometry*,
DOI: 10.1007/978-1-4419-7862-2_9,

Table 9.1 Rate constants for reaction of amino acids with hydroxyl radicals

Amino acid	$k_{\bullet OH}(M^{-1}s^{-1})^a$
Cys	3.5×10^{10}
Trp	1.3×10^{10}
Tyr	1.3×10^{10}
Met	8.5×10^{9}
Phe	6.9×10^{9}
His	4.8×10^{9}
Arg	3.5×10^{9}
Ile	1.8×10^{9}
Leu	1.7×10^{9}
Val	8.5×10^{8}
Pro	6.5×10^{8}
Gln	5.4×10^{8}
Thr	5.1×10^{8}
Lys	3.5×10^{8}
Ser	3.2×10^{8}
Glu	2.3×10^{8}
Ala	7.7×10^{7}
Asp	7.5×10^{7}
Asn	4.9×10^{7}
Gly	1.7×10^{7}

[a]Buxton [76]

9.2 Oxidation Coupled with Mass Spectrometry

Tullius and Dombroski [15] first developed the use of hydroxyl radical footprinting to map DNA–protein interactions. They used Fenton chemistry to generate $^{\bullet}OH$ via the reduction of hydrogen peroxide by iron(II). The backbone of DNA is broken in the presence of $^{\bullet}OH$ in a sequence-independent manner, thus permitting analysis of the entire DNA molecule. When protein is bound, those regions of the DNA that interact with the protein are protected from hydroxyl radical-mediated backbone cleavage. In this study, the results of oxidative modification were examined by gel electrophoresis.

Hydroxyl-mediated oxidative labeling was first coupled with MS by Chance and coworkers [16]. Mass spectrometry provides higher-resolution data than gel electrophoresis allowing for site-specific information to be obtained from the footprinting experiment. Maleknia et al. [16] used synchrotron X-ray radiolysis of water to form $^{\bullet}OH$. This method eliminates the need for additional reagents to be added to the sample. Fenton chemistry requires the addition of hydrogen peroxide and Fe(II), but other transition metals such as Cu(II), Co(II), Ni(II), and Mn(II) can be used as the reductant in Fenton-like reactions [14]. Moreover, Fenton chemistry is slow with labeling performed on the minutes timescale [15]. Synchrotron X-rays generate $^{\bullet}OH$ much faster, on the millisecond timescale. The synchrotron method for oxidative labeling has been successful in studying the structure of nucleic acids [6, 17], identifying structural allostery and the binding interface in the transferrin–transferrin receptor complex [18], and characterizing conformational changes that occur in gating of potassium channels [19]. There are several other methods to generate •OH including radiolysis of water by γ-rays [20] and methods that photolyze hydrogen peroxide by UV light [21] or pulsed laser [22, 23]. These methods have been extensively reviewed by Xu and Chance [14].

To date, several groups utilize oxidative labeling to characterize protein structure, protein–ligand interactions, and protein–protein interactions. The Konermann group has mapped the structure of the integral membrane protein bacteriorhodopsin [24–26]. They have also used oxidative labeling to monitor protein unfolding [27]/folding [28]. The Sharp group has developed a pulsed electron beam water radiolysis method that footprints on the submicrosecond timescale [29]. The Fitzgerald group has developed a method, stability of proteins from rates of oxidation (SPROX), that utilizes oxidative labeling to perform thermodynamic analysis of protein–ligand complexes [30]. This chapter specifically focuses on the work done by the Gross group, who developed fast photochemical oxidation of proteins (FPOP), an oxidative labeling method that labels on the microsecond timescale [23].

9.3 Fast Photochemical Oxidation of Proteins

Previous methods used to generate hydroxyl radicals label on the millisecond or longer timescale. Long exposures to the radical could initiate protein unfolding [31–33]. To minimize oxidation-induced protein unfolding, Hambly and Gross [23] developed FPOP. This method utilizes an excimer laser to photolyze hydrogen peroxide to form hydroxyl radicals. The labeling is rapid, faster than protein unfolding [34], which is an advantage over other oxidative labeling methods.

For FPOP labeling, H_2O_2 is cleaved by using a pulsed 17 ns KrF excimer laser operating at 248 nm (Fig. 9.1). To ensure a bolus of sample receives only one laser shot, the sample is under constant flow in flow cell. The first test of FPOP labeling was performed on apomyoglobin where a large amount of oxidation is observed with many different oxidation states sampled (Fig. 9.2 a middle panel). Although the radicals are formed in nanoseconds, kinetic calculations indicate self-quenching takes more than 100 μs (Fig. 9.2b), a time that allows for radical-induced protein unfolding. Therefore, the large amount of oxidation observed for apomyoglobin may include residues that are only solvent exposed because of protein unfolding. To reduce the radical lifetime, a scavenger was added to the sample. Studies were done using two different amino acids as scavengers, the highly reactive phenylalanine residue and glutamine, a residue that has medium reactivity with •OH (Table 9.1). As expected, owing to its high reactivity with •OH, the use of phenylalanine as a scavenger produced almost no oxidation (Fig. 9.2a top panel). Before the radicals can label the protein, they are rapidly quenched, in ~70 ns (Fig. 9.2b), by the phenylalanine scavenger. Alternatively, the use of the moderately reactive glutamine as a scavenger leads to a satisfactory level of labeling, where many oxidation states are sampled (Fig. 9.2a bottom panel). In

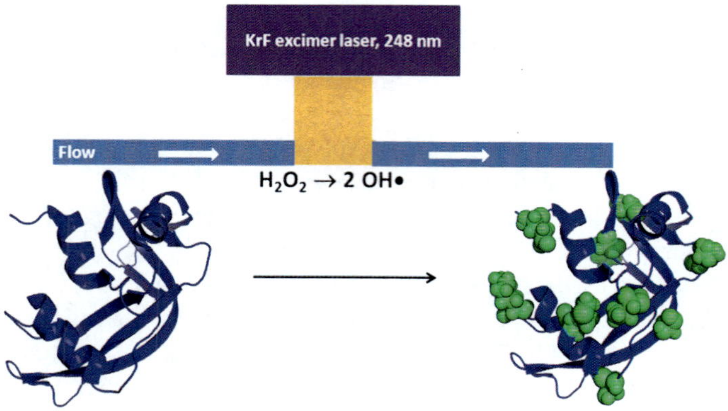

Fig. 9.1 Schematic of FPOP workflow. The sample, in a flow cell, is irradiated by an excimer laser at 248 nm. Hydrogen peroxide is photolyzed to form hydroxyl radicals that oxidize the protein (Oxidized side chains are highlighted in *green spheres*)

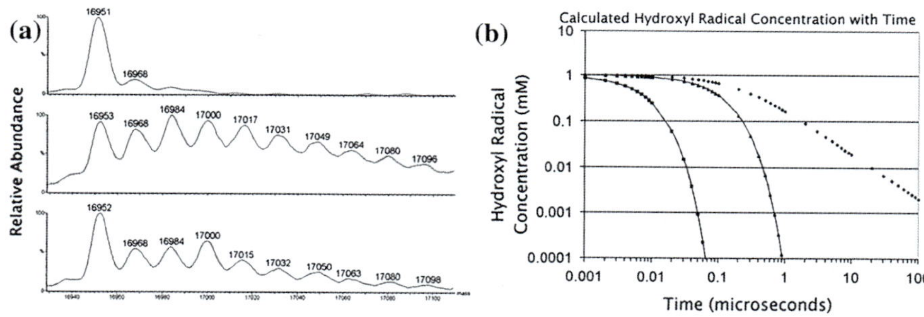

Fig. 9.2 **a** Oxidation of apomyoglobin with (*top*) phenylalanine as a scavenger, (*middle*) no scavenger present, and (*bottom*) with glutamine as a scavenger. **b** Disappearance of hydroxyl radical as a result of reaction with excess phenylalanine (*squares*), excess glutamine (*triangles*), and no scavenger (*diamonds*). Reprinted from Hambly and Gross [23] with permission from Elsevier

the presence of glutamine, the radical lifetime is limited to 1 μs (Fig. 9.2b), a time faster than protein folding. Presumably, the oxidation observed in the presence of glutamine is limited to the solvent-exposed regions of properly folded apomyoglobin.

An advantage of using hydroxyl radicals for protein footprinting is their size, which is similar to water, making them an excellent probe for solvent accessibility. With the proper experimental procedure, including using a scavenger molecule to limit the radical lifetime, FPOP labeling should be sampling solvent accessibility. Hambly and Gross [35] demonstrated the utility of FPOP as a measure of solvent accessibility by using apo- and holomyoglobin as model systems. The oxidation of residues in both states of myoglobin is consistent with their calculated solvent accessibility (Table 9.2). For example, the calculated solvent exposure of P43 is

Table 9.2 Solvent accessibility calculations and oxidations observed on myoglobin. Reprinted from Hambly [35] with permission from Elsevier

Amino acid		Solvent exposure (Å^2)		Observed oxidation	
Type	Number	Apo	Holo	Apo	Holo
MET	55	16	16	Y	Y
MET	131	0	0	Y	Y
TRP	7	19	19	Y	Y
TRP	14	7	7	Y	nd
TYR	103	28	12	Y	nd
TYR	146	18	18	Y	Y
PHE	43	56	6	Y	N
PHE	46	8	8	N	N
PHE	106	36	36	Y	nd
PHE	123	9	9	N	N
PHE	138	30	0	N	N
PHE	151	38	38	Y	Y
HIS	24	12	12	Y	Y
HIS	36	29	29	Y	Y
HIS	48	61	61	Y	Y
HIS	64	28	12	Y	Y
HIS	81	64	64	Y	Y
HIS	82	9	9	N	N
HIS	93	37	0	Y	N
HIS	97	38	17	nd	nd
HIS	113	48	48	Y	nd
HIS	116	47	47	Y	nd
HIS	119	24	24	Y	Y
ILE	21	54	54	Y	Y
ILE	30	14	14	N	N
ILE	75	14	11	N	N
LEU	72	27	0	N	N
LEU	76	0	0	N	N
LEU	86	10	10	Y	N
LYS	87	58	58	N	Y
PRO	88	46	46	Y	Y
LEU	89	56	22	Y	N
LEU	137	50	50	Y	N
LEU	149	55	55	N	Y

decreased by eightfold in holomyoglobin compared to the apo form. This residue is oxidized in the apo form; however, no oxidation is observed for this residue in holomyoglobin even though phenylalanine is highly reactive with $^{\bullet}OH$. This data suggest that FPOP is indeed monitoring solvent accessibility. Further, Phe43 is part of the heme-binding pocket. The absence of labeling of this residue in the ligand-bound form of the protein indicates the efficacy of FPOP in identifying protein–ligand interaction sites.

9.4 FPOP Labels Faster Than Protein Unfolding

A disadvantage of oxidative labeling is that, if not properly controlled, oxidative-induced conformational changes occur [31, 32, 36]. This would be detrimental for a footprinting experiment where you want to probe the native structure of a protein. Oxidative-induced structural changes can be minimized by carefully controlling modification levels [33]. FPOP does this by adding Gln as a scavenger to control the lifetime of the radical. Additionally, to limit post-labeling oxidation, the samples are collected from the flow cell in tubes that contain methionine and catalase to quench excess •OH and hydrogen peroxide, respectively. Hambly and Gross [23] proposed the speed of FPOP labeling circumvents the protein unfolding problem because the method is labeling faster than protein unfolding can occur. To test whether FPOP truly labels faster than protein unfolding, Gau et al. [34] assessed the FPOP modification patterns of three proteins, β-lactoglobulin, apo-calmodulin, and lysozyme. These proteins were chosen because they are sensitive to conformational changes induced by oxidation [32, 37].

Gau et al. [34] hypothesized that if oxidation occurred faster than protein unfolding, the resulting mass spectrum of the oxidized products should fit a Poisson distribution indicating a single conformation of the protein was present. Figure 9.3a

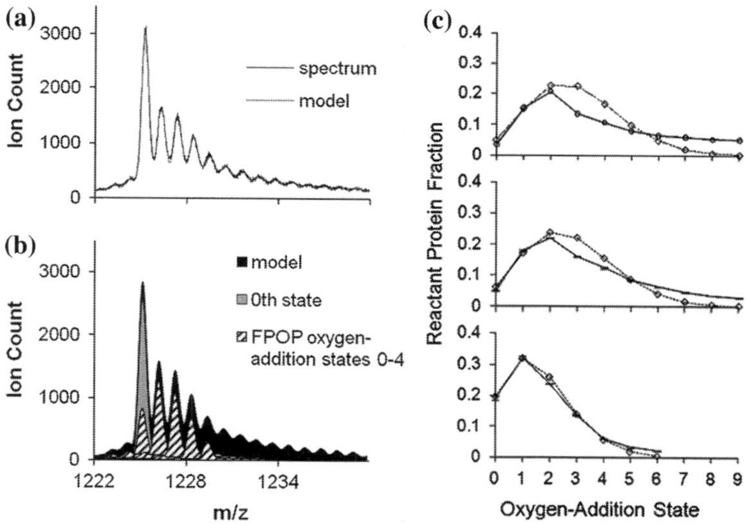

Fig. 9.3 **a** ESI spectrum of the 15th charge state of FPOP-treated β-lactoglobulin and its composite model. **b** Background-subtracted model with first five oxygen-addition states. **c** Oxygen-addition state ion counts modeled for β-lactoglobulin (*top*) treated without glutamine radical scavenger, post-FPOP catalase, or post-FPOP methionine, (*middle*) FPOP treated without glutamine scavenger, (*bottom*) FPOP treated with appropriate experimental controls. The *dotted line* with diamonds corresponds to the calculated Poisson distribution. The *solid line* corresponds to the averages of ion counts of replicate samples with error bars. Reprinted with permission from Gau et al. [34]. Copyright 2009 American Chemical Society

shows a mass spectrum of the 15th charge state of FPOP-treated β-lactoglobulin where the raw data correlate well with the Poisson model. All three proteins exhibited a pattern similar to the Poisson model of the unmodified with its first five oxygen additions shown in Fig. 9.3b. The nonlinear regression best-fit Poisson distribution fit of β-lactoglobulin labeled under varying FPOP conditions underscore the importance of experimental controls (Fig. 9.3c). When there is no Gln scavenger present, the data do not fit well to a Poisson distribution (Fig. 9.3c middle panel). The lack of the post-labeling quench solution of catalase and Met further reduces the fit of the data (Fig. 9.3c top panel). However, when the experiment is run with proper controls, including the presence of the Gln scavenger and the catalase-Met quench, the data fit very well to a Poisson distribution indicating FPOP is sampling a single conformation (Fig. 9.3c bottom panel). This data demonstrate that FPOP, done with the proper experimental controls, is indeed labeling faster than protein unfolding.

9.5 Data Acquisition and Processing

An advantage of FPOP is the ability to gain amino acid residue-level information with standard collision-induced dissociation (CID) methods. This is in contrast to hydrogen deuterium exchange, which owing to deuterium scrambling, requires electron transfer methods such as electron transfer dissociation (ETD) or electron capture dissociation (ECD) to provide residue-level information. ETD methods are not standard on many instruments and require specialized instrumentation. In order to capitalize on the use of residue-level information, rigorous data processing is required.

The inclusion of replicates is essential, not only for statistical information but also to expand the number of identified proteins in the experiment. The LC/MS/MS analysis is an important step in obtaining a large breadth of data. It is important to sample as many peaks as possible both in the MS and in the MS/MS domains. Good chromatography with 5–30 s wide peak widths, depending on the speed of the mass spectrometer that is used for analysis, is essential. In the case of a single peptide with multiple single modifications, it is best to get good enough chromatography to separate the single modifications within that peptide. Figure 9.4a shows an extracted ion chromatogram of a single peptide that has multiple modifications. The various modifications within this single peptide have been separated providing more detailed information on site-specific modification. This separation is especially important in peptides that contain a highly reactive residue such as methionine. Although Met has a high level of modification, the information found in the other less modified residues may be more informative. These data need to be examined to gain a more complete view of the system. In addition to good chromatography, it is also important to obtain high-quality MS/MS data (Fig. 9.4b). To date, all FPOP data have been recorded on a LTQ-Orbitrap (Thermo Fisher). This instrument provides high resolution and accurate mass on the peptide level and high sensitivity for MS/MS. All of the possible modifications

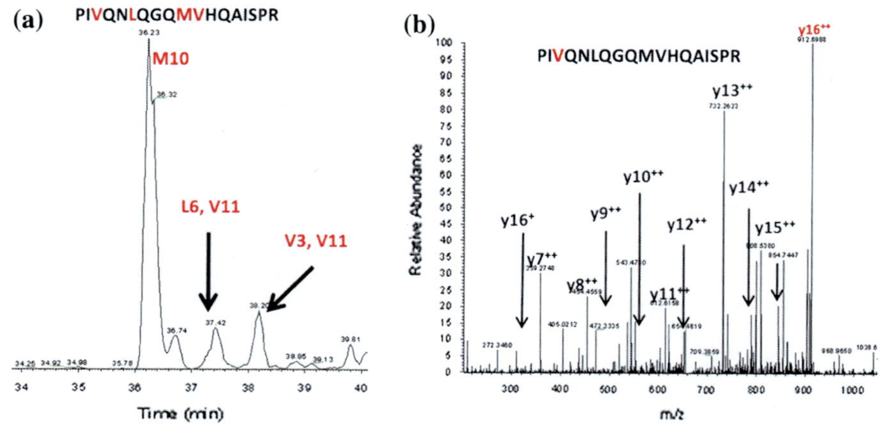

Fig. 9.4 **a** Extracted ion chromatogram of a single peptide with multiple modifications that are separated by liquid chromatography. **b** MS/MS spectrum identifying Val3 as the oxidized residue within the peptide. Oxidized residues are highlighted in *red* on the peptide sequence

Table 9.3 Possible amino acid modifications by hydroxyl radicals

	Amino acid	Possible modification
1	C	15.9949, 31.9898, 47.9847, −15.9772
2	M	15.9949, 31.9898, −32.008
3	W	15.9949, 31.9898, 47.9847
4	Y	15.9949, 31.9898, 47.9847
5	F	15.9949, 31.9898, 47.9847
6	H	15.9949, −23.0160, −22.0320, −10.0320, 4.9879
7	L	15.9949, 13.9793
8	I	15.9949, 13.9793
9	V	15.9949, 13.9793
10	P	15.9949, 13.9793
11	R	15.9949, 13.9793, −43.0534
12	K	15.9949, 13.9793
13	E	15.9949, 13.9793, −30.0106, −27.9949, −43.9898
14	Q	15.9949, 13.9793
15	D	15.9949, −30.0106, −27.9949, −43.9898
16	N	15.9949
17	A	15.9949
18	S	15.9949, −2.0157
19	T	15.9949, −2.0157
20	G	

for FPOP labeling are listed in Table 9.3 [14, 38–40]. Modifications other than +16 are possible on several amino acids. This leads to a complex analysis where high resolution and mass accuracy are very important.

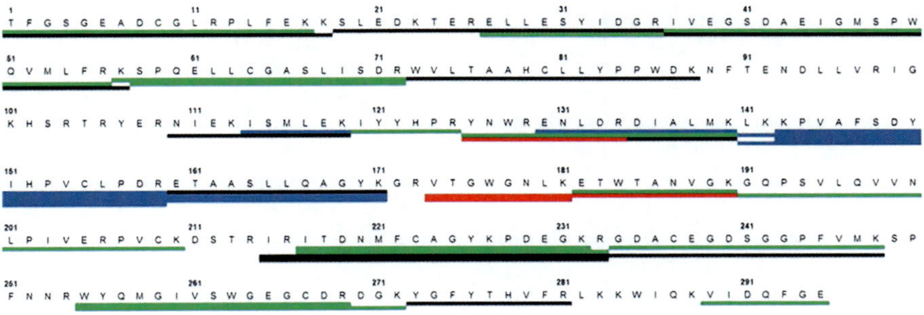

Fig. 9.5 Coverage map of the tryptic digest of thrombin. Peptides that were no oxidatively labeled are shown in *black*. Peptides that displayed no difference between the antibody-bound and antibody-free states are shown in *green*. Peptides that had protection in the antibody-bound form are shown in *blue*. Peptides that were de-protected in the antibody-bound form are in *red*. Reprinted with permission from Jones et al. [41]. Copyright 2011 American Chemical Society

Figure 9.5 shows the coverage map of thrombin after FPOP labeling of the protein bound to its antibody [41]. In total, 34 peptides were assigned for this protein. The high sequence coverage, 86 %, afforded a more detailed analysis of this protein system [41]. Additionally, several specific sites of modification were identified. The high sequence coverage was achieved even though the protein was analyzed in the presence of the antibody owing to good chromatography and high-resolution MS. As will be discussed later in this chapter, the residue-level data acquired for this protein were essential in identifying the antibody-binding site.

A first step in data processing is LC–MS feature alignment. In alignment, a sum of the extracted ion chromatogram peaks of ions with the same elution time and de-charged monoisotopic mass (±5 ppm) are organized. The list of features, each with a unique ID number, with their intensities from the LC–MS analysis is one of the outputs from alignment. Thus far, two alignment programs, Rosetta Elucidator version 3.3.0.0.220 (Rosetta Biosoftware, Seattle, WA) and Progenesis LC–MS (Nonlinear Dynamics, Durham, NC), have been used to process FPOP data. The alignment program also creates a data (dta) file for each product-ion spectrum. The dta files are merged into a single mascot generic format (mgf) file that is searched for modified and unmodified peptides by using MASCOT (Matrix Science, London, U.K.). All known side-chain reaction products of hydroxyl radical labeling were added to the modification database for search as variable modifications (Table 9.3) [14, 38–40]. In-house excel-based software is used to merge the MASCOT results with the feature list from the alignment program. An example of the data output for a particular peptide is shown in Table 9.4. Each entry in the table has a signal intensity for each replicate associated with it (data not shown). In some cases, database searching can lead to missed or incorrect assignments. Manual validation is recommended to check for errors in assignment.

Table 9.4 Annotated output of data merged from chromatography alignment and database searching

Peptide	Sequence	Theoretical mass	Mass centroid	Error (ppm)	Net mass change	MS2 ID'd modification	#Supporting MS2 spectra	Max score	Peak centroid time (min)	Unique ID
1–17	TFGSGEADCGLRPLFEK	1898.88330	1898.8788	−2.4	15.99488	F15	1	21.4	33.84	39330143_5
1–17	TFGSGEADCGLRPLFEK	1898.88330	1898.8788	−2.4	15.99488	R12	1	21.9	33.84	39330143_4
1–17	TFGSGEADCGLRPLFEK	1898.88330	1898.8788	−2.4	15.99488	L14	1	32.0	33.84	39330143_3
1–17	TFGSGEADCGLRPLFEK	1898.86210	1898.8788	8.8	15.99488	L14	2	27.7	33.84	39330143_2
1–17	TFGSGEADCGLRPLFEK	1898.88330	1898.8788	−2.4	15.99488	P13	2	43.2	33.84	39330143_1
1–17	TFGSGEADCGLRPLFEK	1898.88330	1898.8783	−2.6	15.99488	F2	8	43.3	36.02	39329648_1
1–17	TFGSGEADCGLRPLFEK	1898.88330	1898.8795	−2.0	15.99488	F2	2	33.2	36.54	39331727_1
1–17	TFGSGEADCGLRPLFEK	1882.88840	1882.8832	−2.8	0		55	44.8	37.80	39328770_1
1–17	TFGSGEADCGLRPLFEK	1882.88840	1882.8854	−1.6	0		105	89.9	37.80	39329325_1
1–17	TFGSGEADCGLRPLFEK	1825.86690	1825.8605	−3.5	0		72	36.9	39.42	39328313_1
1–17	TFGSGEADCGLRPLFEK	1825.86690	1825.8634	−1.9	0		48	69.4	39.43	39328446_1
1–17	TFGSGEADCGLRPLFEK	1873.85160	1873.8481	−1.9	47.98472	C9	19	108.4	40.67	39328972_1

The extent of modification can be calculated from the intensities of each peak using Eq. 9.1:

$$\text{Extent of Modification} = \frac{\sum_{i=1}^{n} I_{\text{OX1}}}{\sum_{i=1}^{n} I_{\text{OX1}} + \sum I} \tag{9.1}$$

where I_{ox1} is the signal intensity of each modification in the peptide, and I is the signal intensity of the unmodified peptide. This equation can be modified to calculate the extent of modification on individual amino acids to quantify site-specific information. It is important to note that quantitation of modification levels is done solely using the LC–MS data. MS/MS data are used exclusively for residue-level assignment.

9.6 Applications of FPOP

9.6.1 High-Throughput Analysis of Similarities in Ligand Binding

Many protein systems have multiple ligands that bind in a similar manner. The identification of the binding interactions of all of these ligands by high-resolution methods such as NMR and X-ray crystallography can be time consuming. Protein footprinting may be a useful tool for analyzing the binding interactions of multiple ligands to a single protein. Hambly and Gross [35] demonstrated the efficacy of FPOP in studying protein–ligand interactions using myoglobin as a model system. This ability of FPOP to identify interactions sites makes it a suitable method for screening interaction similarities in protein systems that bind multiple ligands.

To demonstrate the efficacy of FPOP for characterizing multiple ligand interactions, the calcium-binding protein calmodulin was studied [42]. Calmodulin binds multiple ligands including the skeletal muscle myosin light chain kinase (SK-MLCK), mastoparan (Mas), and melittin (Mel). An NMR structure of calmodulin bound to the M13 peptide, the binding domain of SK-MLCK, has been solved [43]. However, no high-resolution structures are available for calmodulin bound to Mas or Mel. FPOP analysis was done on M13-, Mas-, and Mel-bound calmodulin to evaluate the utility of FPOP as a method for comparing structures of protein–ligand complexes. A comparison of the oxidative labeling patterns could provide information on whether Mas and Mel bind calmodulin in a similar manner as the M13 peptide.

The similarity of binding between the three peptides was validated using a spectral-contrast angle θ that provides a confidence value that is related to similarity [44]. The θ values comparing the peptide labeling patterns of calmodulin bound to M13, Mas, and Mel were calculated. A comparison of calmodulin and M13-/Mas-/Mel-bound calmodulin gives a large θ indicating a significant change in conformation upon ligand binding. Conversely, the θ of pairwise comparisons of the three different peptides bound to calmodulin is similar to the smallest θ from replicate experiments indicating structural similarity when the three peptides are bound.

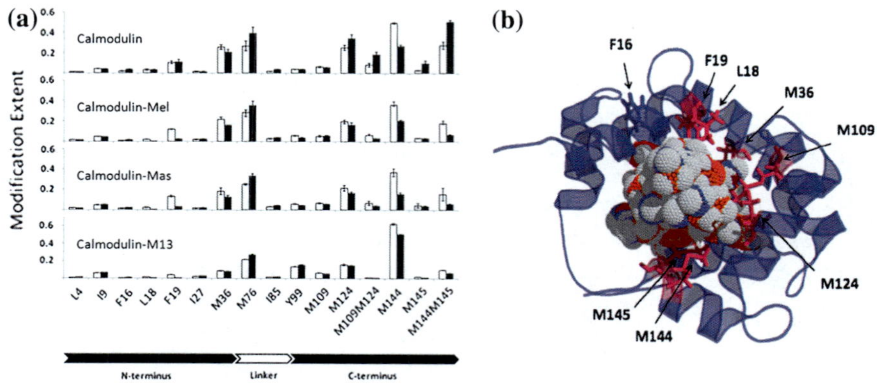

Fig. 9.6 **a** Extent of modification for Ca^{2+}-free (*white*) and Ca^{2+}-bound (*black*) states of each modified residue for various calmodulin complexes. **b** FPOP results mapped onto the structure of the calmodulin–M13 complex. Residues in *pink* are modified residues detected by LC–MS. Reprinted with permission from Zhang et al. [42]. Copyright 2011 American Chemical Society

To provide more detailed information, further analysis was done on the residue level. Figure 9.6a shows the labeling differences between calmodulin and calmodulin bound to M13, Mas, and Mel in the presence and absence of calcium. The data demonstrate that residues M109, M124, M144, M145, L18, and F19 are protected from labeling when any of the three peptides are bound to calcium-loaded calmodulin indicating these regions are part of the interaction site for the binding of all three peptides. These residues correlate well with interaction site that is highlighted in the NMR structure of M13-bound calmodulin (Fig. 9.6b). The data on both the peptide and residue levels validate the efficacy of FPOP as a method for comparing interactions of multiple ligands with a single protein. This method is more powerful when the structure of the protein with one of the ligands bound is known. This reference complex can be used to determine whether the other ligands bind in a similar manner. This relatively high-throughput screening of multiple ligands binding using FPOP may prove useful in the development of therapeutics.

9.6.2 Epitope Mapping

Antibodies are increasingly being used as therapeutics for a wide variety of diseases [45–47]. As is the case for all biologics, detailed analytical structural characterization is an important step in their use as therapeutics. One step in antibody characterization is the identification of the epitope. Multiple methods such as site-directed mutagenesis [48], X-ray crystallography [49], and epitope extraction [50] have been used in epitope mapping with success. FPOP could provide a high-throughput, specific method for epitope mapping.

The ability of FPOP to map an antibody epitope was tested using thrombin as a model system. The serine protease thrombin binds to its antibody via a discontinuous epitope, an epitope that consists of residues that are not together on the primary sequence but come together in the folded structure of the protein. Discontinuous epitopes are difficult to ascertain using site-directed mutagenesis. HDX was previously used to map the epitope of thrombin demonstrating the ability of protein footprinting to be a valuable tool in epitope mapping [51]. Since the epitope of thrombin is known, it is a good model system to determine whether FPOP will be useful for epitope mapping studies.

FPOP labeling was performed on apo-thrombin and antibody-bound thrombin. The peptide-level data indicated the oxidative modification of a six-residue peptide, 114–119, was decreased when the antibody was bound to thrombin, signifying it is a part of the epitope (Fig. 9.7a) [41]. This decreased level of modification was also observed in a second region of the antibody-bound protein. This region was represented by five peptides, including missed trypsin cleavages, which spanned 41 residues (Fig. 9.7b, e). It is highly unlikely the epitope spans such a large region, so further analysis needed to be done to determine the epitope.

To obtain a higher definition in the experiment, the residue-level data were analyzed. The acquisition of good MS/MS data allowed for the interrogation of residue-level information. Figure 9.7c displays the MS/MS spectrum of peptide 130–140, one of the peptides which displayed decreased oxidative modification in the antibody-bound form of thrombin. The modification in this peptide is a loss of carbon dioxide, a mass difference of −43.9898 Da, which is observed in modification of acidic residues [40]. The MS/MS spectrum indicates that Asp133 is the modified residue within this peptide. Analysis of the residue-level data indicates the regions of decreased oxidative modification are found specifically between residues 133–150 (Fig. 9.7d). A comparison of the interaction region identified by peptide-level data (Fig. 9.7e) and residue-level data (Fig. 9.7f) demonstrates the importance of obtaining residue-level information in an FPOP experiment. The epitope was identified at a higher resolution, when the residue-level data were analyzed.

The two regions that FPOP identified as the epitope of thrombin correlate well with the previously mapped hydrogen deuterium exchange data [51]. The HDX-identified epitope is represented by the peptic peptides 113–117 and 139–149, while the epitope mapped by FPOP is represented by the tryptic peptide 114–119 and residues 133–150. Additionally, when these regions are mapped onto the structure of thrombin, Fig. 9.7h, they are together in the folded protein even though they are far apart in the primary sequence, a characteristic of a discontinuous epitope. This data confirm that FPOP has utility for epitope mapping providing another structural tool for the characterization of antibodies.

The FPOP data also revealed regions where the antibody-bound thrombin had increased oxidative modification compared to apo-thrombin (Fig. 9.7g). Specific residues that displayed this labeling behavior were R134, P153, C155, and P157 (Fig. 9.7d). The regions of increased solvent accessibility are in the loop regions of the protein including the 99- and 148-loops (Fig. 9.7h). Thrombin is known

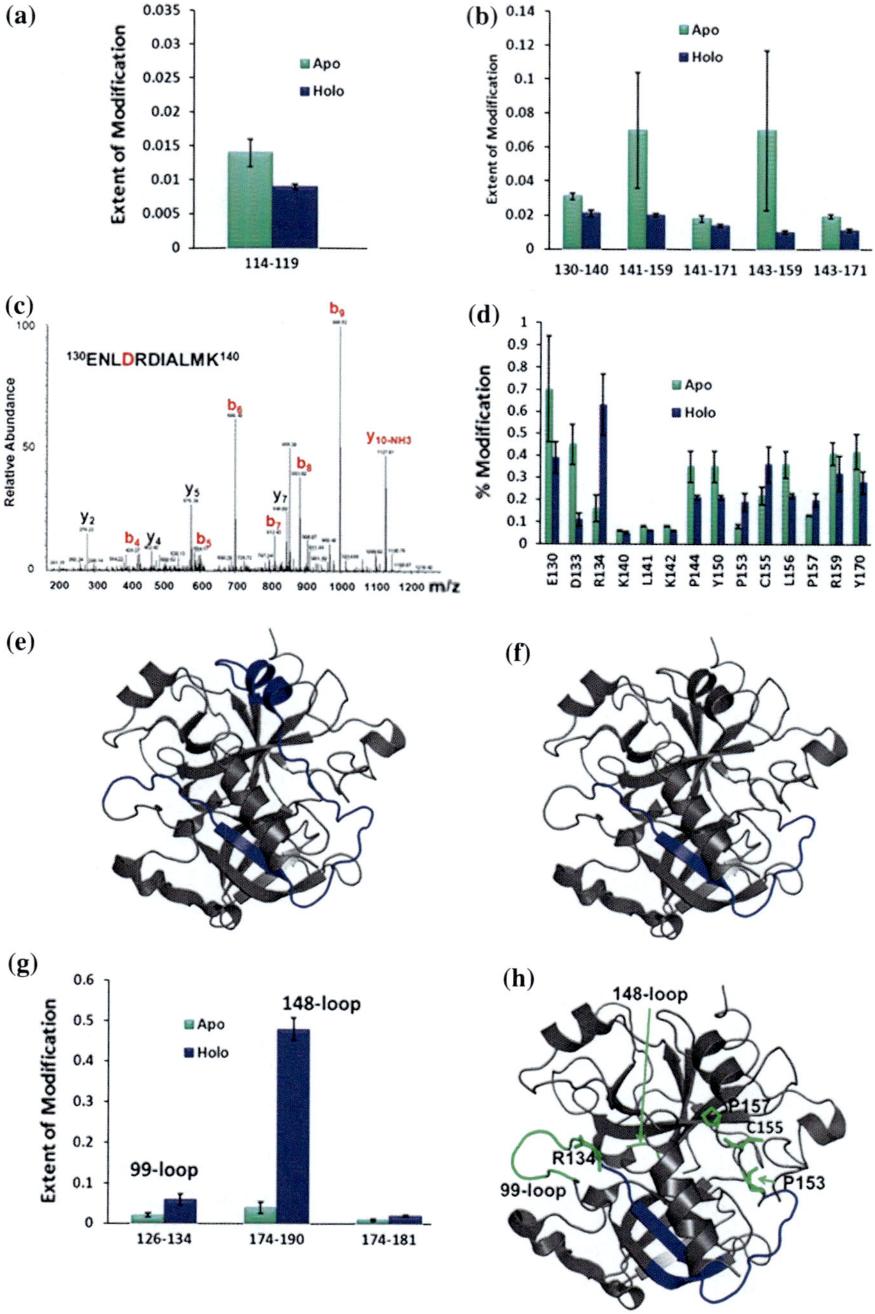
(a)
Extent of Modification
Apo
Holo
114-119
(b)
Extent of Modification
Apo
Holo
130-140
141-159
141-171
143-159
143-171
(c)
^{130}ENLDRDIALMK^{140}
Relative Abundance
b_9
b_6
$y_{10\text{-}NH3}$
b_8
y_5
y_7
b_7
y_2
b_4
y_4
b_5
m/z
(d)
% Modification
Apo
Holo
E130
D133
R134
K140
L141
K142
P144
Y150
P153
C155
L156
P157
R159
Y170
(e)
(f)
(g)
Extent of Modification
Apo
Holo
148-loop
99-loop
126-134
174-190
174-181
(h)
148-loop
P157
C155
R134
P153
99-loop

◀ **Fig. 9.7** **a** Extent of modification of peptide 114–119 in the antibody-bound (*holo*) and antibody-free (*apo*) states of thrombin. **b** Extent of modification for the five peptides that span the 130–171 region. **c** MS/MS spectrum of peptide 130–140 identifying D134 as the oxidized residue. **d** Extent of modification of thrombin on the residue level. Regions displaying modifications differences identified on the **e** peptide and **f** residue levels mapped onto the structure of thrombin (pdb: 2AFQ). **g** Extent of modification of thrombin that displayed increased solvent accessibility in the antibody-bound form. **h** Structural model of thrombin with the proposed epitope highlighted in *blue*. Regions of increased solvent accessibility in the antibody-bound form are highlighted in *green* with specific residues shown with sticks. Figure c reprinted, figures a, b, d, g, and h adapted with permission from Jones et al. [41]. Copyright 2011 American Chemical Society

to undergo allosteric conformational changes, particular in its loop regions, upon ligand binding [52, 53]. The FPOP data reveal these allosteric changes occur when thrombin is bound to the antibody. These antibody-bound allosteric changes have not been previously observed for thrombin. These changes were not observed in the HDX epitope mapping study presumably because the conformational changes are due to side-chain rotations. HDX monitors backbone amides and their role in hydrogen bonding and solvent accessibility. In this case, the antibody-induced allosteric conformational changes in the thrombin loops must not involve alterations in the hydrogen bonding network but rather side-chain rotations. FPOP monitors the side chains of amino acids and, therefore, would readily detect these side-chain rotations.

This data demonstrate the efficacy of FPOP as a tool for epitope mapping and for mapping protein conformational changes. Additionally, the method has a higher throughput than site-directed mutagenesis, X-ray crystallography, and NMR increasing its value as a reliable tool for antibody characterization [54].

9.6.3 Structural Analysis of Apolipoprotein E

Apolipoprotein E apoE is an important biological molecule that plays a role in Alzheimer's disease. apoE binds to multiple ligands to regulate lipid metabolism and control lipid redistribution in tissue and cells [55]. It is thought to interact with amyloid beta peptides, which possibly contributes to the progression of Alzheimer's disease [56, 57]. There are three apoE isoforms, apolipoprotein E2 (apoE2), apolipoprotein E3 (apoE3), and apolipoprotein E4 (apoE4) that differ slightly in their amino acid sequence at positions 112 and 158. apoE2 has cysteine residues at these two positions, but apoE4 has an arginine at these sites. apoE3 has a cysteine at position 112 and an arginine at position 158. Interestingly, the apoE4 isoform is strongly associated with Alzheimer's disease suggesting a relevancy to the residue differences between the isoforms [56, 58]. To examine the structural differences between the isoforms of apoE, Gau et al. [60] used FPOP. The labeling of the three isoforms was compared to determine whether the solvent accessibilities of specific residues are different.

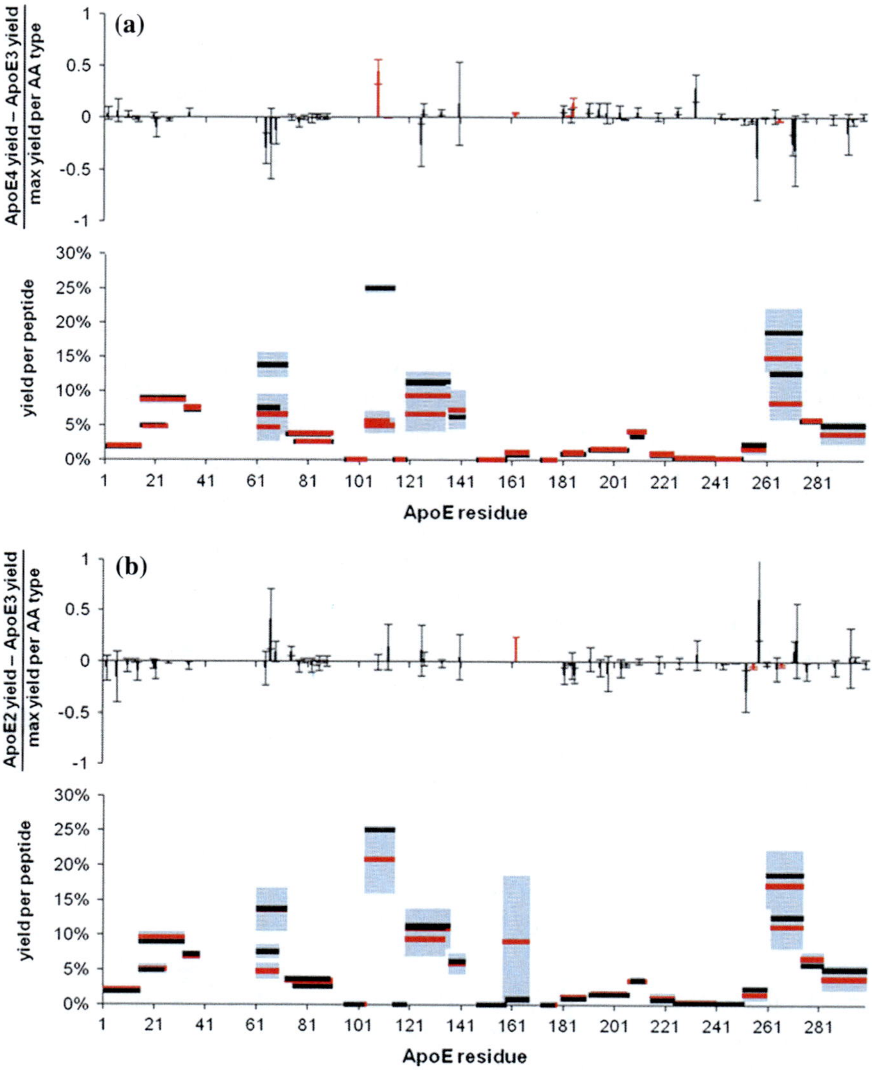

Fig. 9.8 Comparison of the tryptic peptide-level and residue-level FPOP labeling yields for **a** apoE3 and apoE4 and **b** apoE2 and apoE3. In the residue-level data, residues highlighted in *red* [M108, Y162, P183, V185, E266 in (**a**) Y162, E255, E266 in (**b**)] have significant difference between isoforms at 95 % confidence by the Student's t test. In the *bottom panels* of both **a** and **b**, the *light blue* area is the standard error. Reprinted with permission from Gau et al. [60]. Copyright 2011 American Chemical Society

Figure 9.8a shows the comparison of apoE4 and apoE3 on the residue (top) and peptide level (bottom). There are five specific residues, M108, Y162, P183, V185, and E266, that are modified significantly different between the isoforms at 95 % confidence by the Student's t test. Moreover, on the peptide level, three

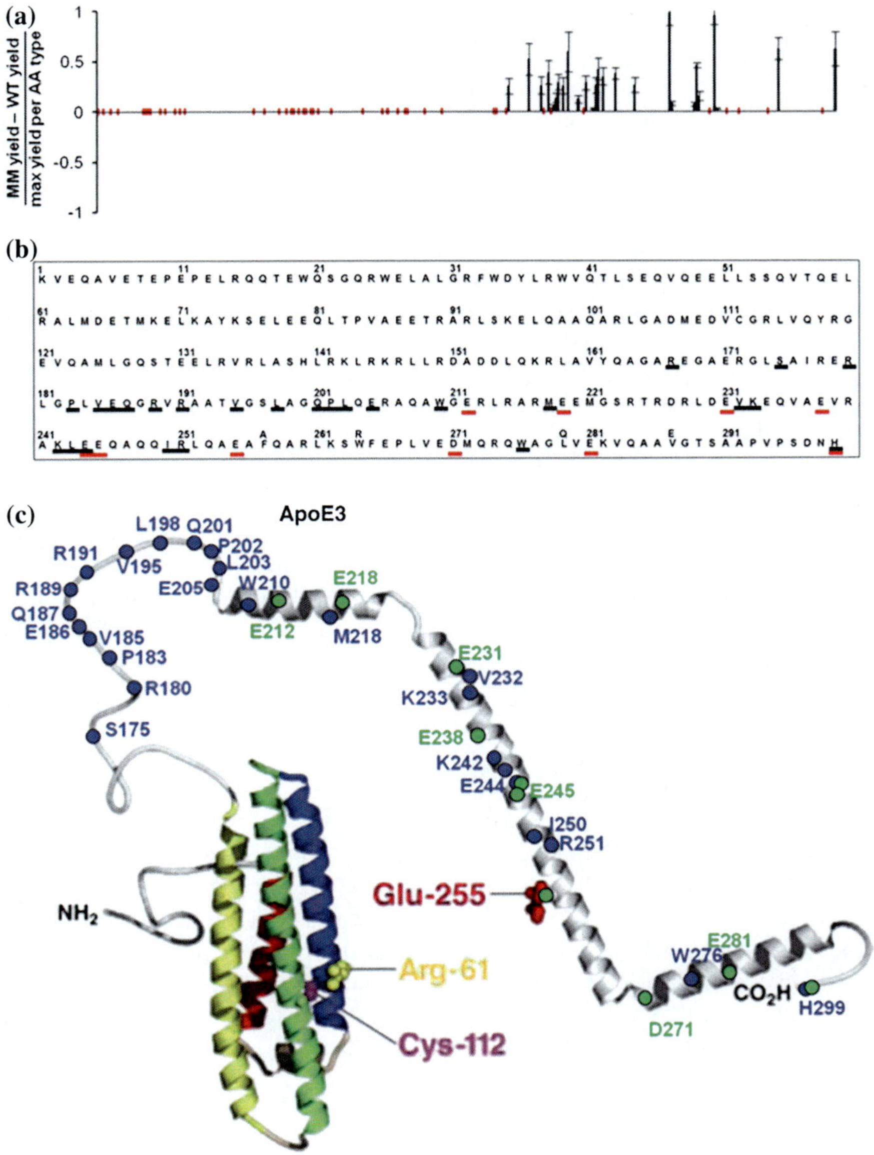

Fig. 9.9 **a** Comparison of the residue-level FPOP labeling yields for ApoE3 and ApoE3MM. **b** Residues along the ApoE3 sequence the display more labeling in ApoE3MM. Residues underlined in *black* are from the FPOP experiment. Residues underlined in *red* are from the GEE labeling experiment. Reprinted with permission from Gau et al. [60]. Copyright 2011 American Chemical Society

peptides, 62–72, 120–134, and 261–274, display increased oxidative modificationin apoE4 compared to apoE3, while peptide 104–112 has increased modification in apoE3. The differences between apoE2 and apoE3 are even smaller.

Three residues, Y162, E255, and E266, and one peptide, 62–68, display significant labeling differences (Fig. 9.8b). The largest difference between the three isoforms is M108 which is extensively modified in apoE4 but has only negligible modification in apoE2 and apoE3. This indicates a structural difference in this region of apoE4. This may provide insight into the source of the increased risk of Alzheimer's disease for apoE4.

It has been shown that apoE oligomerizes, forming dimers at nanomolar concentrations and tetramers at micromolar concentrations [59]. The self-association of apoE plays a major role in its lipid-binding ability [61]; thus, it is important to understand the characteristics of this self-association. The tendency of the protein to self-associate makes it difficult to study via NMR owing to the large size of the oligomeric states. Gau et al. [60] compared the oxidative modification of the tetrameric form of apoE3 to a monomeric mutant (apoE3MM). Four mutations, F257A, W264R, L279Q, and V287E, were engineered to produce the monomeric mutant [62], and its NMR structure has been solved [63]. A comparison of apoE3 and apoE3MM labeling shows the C-terminus of the protein has significantly greater oxidative modification for apoE3MM (Fig. 9.9a). In total, 26 residues in the C-terminus had increased modification for apoE3MM (Fig. 9.9b). This trend was observed on both the residue and peptide levels. This data correlate well with previous studies that indicated the C-terminus was involved in oligomerization [64–66]. The FPOP labeling data were further verified by using another footprinting method GEE footprinting. GEE specifically labels the carboxylic acids aspartate and glutamate. The GEE labeling also indicated that several Glu residues in the C-terminus had increased modification for apoE3MM compared to the apoE3 tetramer (Fig. 9.9b). The residues that demonstrated increased modification for apoE3MM for both FPOP and GEE labeling are highlighted on the apoE3MM structure in Fig. 9.9c. These regions of increased solvent accessibility are localized in the C-terminal portion of the protein. Since there are no primary sequence differences between the three isoforms in this region of apoE, it can be assumed that the same region is responsible for oligomerization of apoE2 and apoE4. This study reveals FPOP's proficiency in determining differences in protein isoforms that differ by a couple of amino acids as well as identifying oligomerization interfaces.

9.7 Method Development

FPOP has been successful as probe for solvent accessibility, elucidating protein–ligand and protein–protein interactions in a variety of systems. However, further development of the method can extend its use to other applications including protein folding. Moreover, the use of other radicals for footprinting may provide different information than $^{\bullet}$OH increasing the utility of FPOP for structural biology.

9.7.1 FPOP as a Tool for Studying Protein Folding

The high speed of FPOP labeling, 1 μs, makes its potential use to study protein folding dynamics interesting. Chen et al. [67] have developed a temperature jump (T jump) strategy to monitor protein folding using FPOP. The approach uses two lasers, an Nd:YAG laser to provide a T jump and an excimer laser to generate •OH (Fig. 9.10a). The delay time between the two lasers could be adjusted to provide kinetic information. This strategy allows for the analysis of protein folding on a submillisecond timescale.

Barstar was chosen as a test system for this approach. This protein is denatured at 0 °C and folds when its temperature is increased. First, the Nd:YAG laser provided a T jump (~20 °C) to the protein system. Then, the excimer laser was pulsed at various time intervals post-T jump to generate •OH for FPOP labeling. Figure 9.10b shows the differences in barstar modification at varying times between the heat pulse and FPOP compared to FPOP modification without a T jump (RT). As the time delay is increased, the amount of oxidative modification is decreased. This is consistent with the progression of protein folding where sites would become protected from labeling as the protein folds. Barstar has two intermediate states in its folding pathway. Based on the timescale of this experiment, it is presumed the first state is being monitored. This first transition state is more solvent exposed than the native state, even at 1 ms (Fig. 9.10b), compared to the native state control (RT control spectrum in Fig. 9.10b).

A rate constant for equilibrium of folding was calculated using the centroid of the 10+ charge state peak of the mass spectra. The centroid shift was plotted against the delay time of the excimer laser pulse (Fig. 9.10c). The data were fit using a single-exponential function, and a constant of 1.5 ms^{-1} was obtained [67].

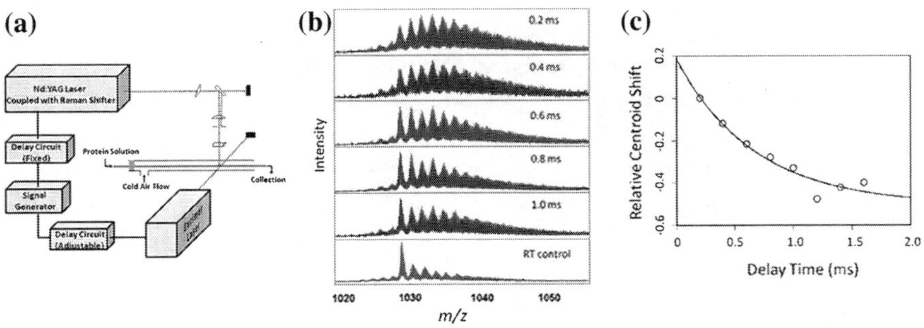

Fig. 9.10 **a** Schematic of the T jump system that shows flow system is intersected by two laser beams at a window in the tube. The delay circuit allows for time between the two laser pulses to be adjusted. **b** Mass spectra of FPOP-labeled barstar at different times between the heating pulse and the FPOP probe. The RT control spectrum is barstar FPOP-labeled at room temperature. **c** Plot of the centroid mass shift versus delay time with a curve fit obtained by fitting a single-exponential function (*solid curve*) to the data (O). Reprinted with permission from Chen et al. [67]. Copyright 2010 American Chemical Society

This agrees fairly well with the transition rate from the unfolded state to the first intermediate state measured by fluorescence which was 3.10 ms^{-1} [68].

These results indicate that a T jump coupled with FPOP is an effective approach for following protein folding kinetics on a submillisecond timescale. The use of MS as the detection method allows for more detailed information than global methods such as fluorescence. This experiment can be coupled with protein proteolysis to examine the changes in the local regions during folding.

9.7.2 New Reagents for FPOP-Based Labeling

Multiple laboratories have demonstrated the efficacy of $^{\bullet}OH$ for oxidative labeling of protein complexes [19, 29, 33]. However, other radicals could be as effective or more effective in footprinting proteins. Radicals have varying reduction potentials that determine its tendency to be either a strong oxidant or a reductant. For oxidative labeling, radicals with larger reduction potentials could be advantageous. Moreover, in certain protein systems, radicals that label more specifically than $^{\bullet}OH$ could be beneficial. An advantage of FPOP is that the method is tunable, so that other radicals can be used to label proteins.

9.7.2.1 The Sulfate Radical Anion

The sulfate radical anion, $SO_4^{-\bullet}$, has a reduction potential of 2,430 mV making it a slightly stronger oxidant than $^{\bullet}OH$ which has a reduction potential of 1,900 mV [69]. The efficacy of $SO_4^{-\bullet}$ as an oxidant has been previously described [70]. It has been shown to oxidize methionine, aromatic, carboxyl, and zwitterionic residues. The usefulness of $SO_4^{-\bullet}$ as a reagent for FPOP has been established by Gau et al. [40]. Sodium persulfate, $Na_2S_2O_8$, was used as the precursor molecule for the radical. A comparison of global labeling of β-lactoglobulin with $^{\bullet}OH$ and $SO_4^{-\bullet}$ with $Na_2S_2O_8$ and hydrogen peroxide in equimolar concentrations showed that $SO_4^{-\bullet}$ labeled with a higher yield (Fig. 9.11a). No oxidation was observed in the $Na_2S_2O_8$ control experiment where no laser was used (Fig. 9.11a top panel). Similar modification levels were only achieved once the $Na_2S_2O_8$ concentration was lowered from 15 to 5 mM (Fig. 9.11a). This correlates well with the reduction potential of the two radicals which indicate that $SO_4^{-\bullet}$ is a stronger oxidant than $^{\bullet}OH$. Further, the labeling data at 5 mM $Na_2S_2O_8$ fit well to a Poisson distribution indicating $SO_4^{-\bullet}$ labeling has sampled the native conformation, similar to $^{\bullet}OH$ labeling [34, 40].

For residue-level studies, the proteins apomyoglobin and calmodulin and peptides bradykinin and angiotensin II were labeled. The protein samples were proteolyzed with trypsin and analyzed by using LC/MS/MS. The residue-specific data indicate that similar to $^{\bullet}OH$, the sulfate radical anion is a promiscuous label. In total, 16 different amino acid residues types were labeled by $SO_4^{-\bullet}$. However, reagent reactivity differences were observed. A fraction modified metric was

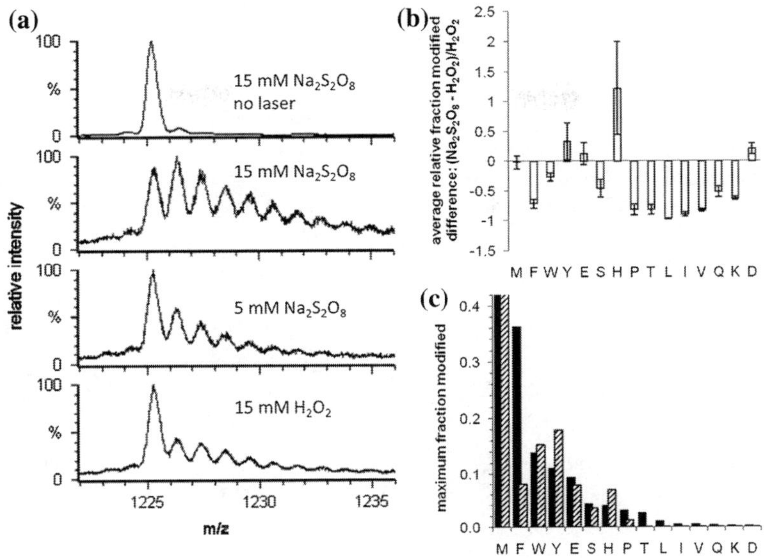

Fig. 9.11 **a** ESI mass spectra of the 15th charge state of β-lactoglobulin with varying labeling conditions. **b** The relative difference between persulfate and peroxide fraction modified of calmodulin, apomyoglobin, bradykinin, and angiotensin II residues (values averaged per amino acid type). **c** Maximum fraction modified of all amino acid residues of H_2O_2 FPOP labeling (*black bars*) and $Na_2S_2O_8$ FPOP labeling (*diagonal-pattern bars*). Reprinted with permission from Gau et al. [40]. Copyright 2010 American Chemical Society

calculated for each residue type for apomyoglobin, calmodulin, bradykinin, and angiotensin II [40]. Figure 9.11b shows a comparison of $^{\bullet}OH$ and $SO_4^{-\bullet}$ modification on a residue-by-residue basis. The data suggest that $^{\bullet}OH$ react more readily with certain residues such as Phe, Gln, and Lys as well as the aliphatic residues Pro, Thr, Leu, Ile, and Val. When the maximum fraction modified values for each residue are compared, a slightly different order of reactivity is observed for the $SO_4^{-\bullet}$ (Fig. 9.11 c). The reactivities for Met, Trp, and Glu are similar for the two radicals. Tyr and His have slightly higher reactivities with the $SO_4^{-\bullet}$, but its reactivity with Phe is greatly decreased in comparison to $^{\bullet}OH$ [40]. Gau et al. [40] attribute these differences to both the inherent reactivities of the two radicals and their different molecular sizes.

Lastly, the correlation between persulfate FPOP labeling and solvent accessibility was assessed. The per-residue fraction modified residue data for the histidines of myoglobin were compared to solvent accessible surface area calculated from an X-ray crystal structure. The reactivity does not correlate well when His64 is part of the fit ($R^2 = 0.63$, Fig. 9.12a). The reason for this is that His64 is an axial ligand of the heme iron and persulfate or $SO4^-$ may have a high affinity for the heme-binding pocket [40]. When His64 is omitted from the fit, the correlation is much better with an R^2 of 0.83 (Fig. 9.12b). The results indicate that similar to peroxide FPOP labeling, persulfate FPOP labeling is a suitable monitor of solvent accessibility.

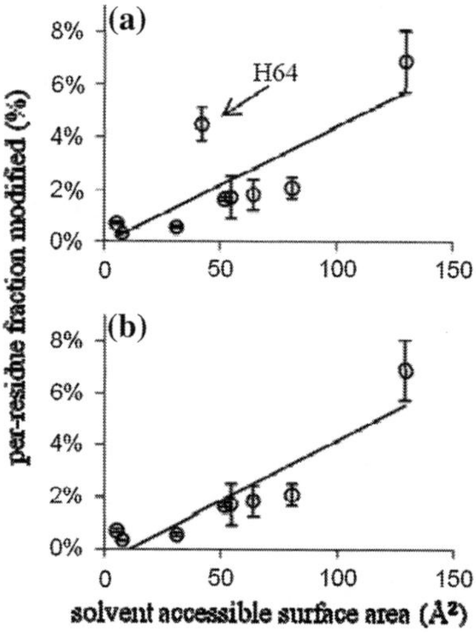

Fig. 9.12 Correlation of modification yields of apomyoglobin his residues with their calculated solvent accessible surface areas with least-squares fit. His64 is omitted from the fit in plot b. Reprinted with permission from Gau et al. [40]. Copyright 2010 American Chemical Society

9.7.2.2 The Iodine Radical

The sulfate radical anion, similar to hydroxyl radicals, is a nonspecific labeling method. Both radicals have the ability to oxidatively modify at least 14 amino acid residues. In certain protein systems, a radical that specifically modifies one or two residues is desirable. Iodination of proteins has been shown to be a very specific labeling strategy where tyrosine and to a lesser extent histidine residues are modified [71, 72]. The major product of iodination of tyrosine is 3,5-di-iodotyrosine. Three major products, 2-iodohistidine, 2,5-di-iodohistidine, and 1,2,5,-tri-iodohistidine, result from iodination of histidine. To take advantage of the specific labeling of iodine, an FPOP-based iodination method was developed [73].

To generate a radical species, iodobenzoic acid was used as the source. The photolysis of iodobenzoic acid by the excimer laser at 248 nm resulted in the formation of the iodide radical (I). Instead of glutamine, which is used in both peroxide and persulfate FPOP, free histidine was used as a scavenger to control radical lifetimes and reduce labeling-induced conformational changes. Histidine was chosen because its reaction with •I is 30–100 times slower than tyrosine [74]. It should consume radical at a rate that would still allow for adequate reaction with the protein. The higher reactivity of tyrosine may cause it to quench the radical prior to protein labeling. The modification of proteins with •I leads to a mass increase of 125.90 Da, in the case of a single modification.

Myoglobin was used as a model system to test FPOP-based iodination [73]. The labeling of myoglobin (Mb) and apomyoglobin (apoMb) with •I was compared on the global protein. MS analysis shows that modification levels in aMb (Fig. 9.13b) are much higher than those in Mb (Fig. 9.13a). In Mb, mono-iodinated species are present at 38 % of the unmodified species. In aMb, these species have increased to 130 %. Further, tri-iodination, which is present in aMb, is not detected in Mb.

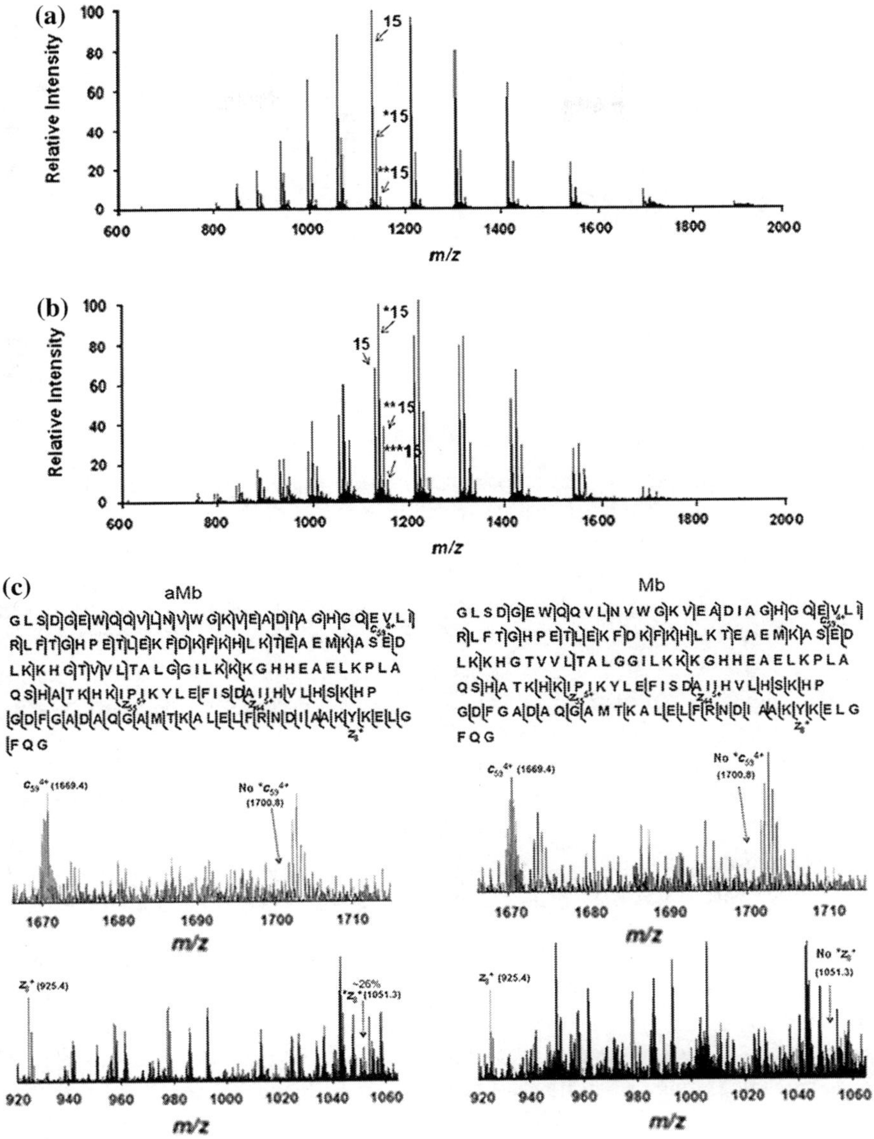

Fig. 9.13 Global protein MS analysis of iodinated Mb (**a**) and aMb (**b**). **c** Sequence coverage and ECD spectra of the 16th charge state of mono-iodinated aMb (*right panel*) and Mb (*left panel*). Top-down MS analysis of myoglobin. Reprinted from Chen et al. [73], with permission from Springer

The specificity of the •I makes iodinated proteins more suitable for top-down MS analysis than peroxide or persulfate FPOP-labeled proteins. Chen et al. [73] used top-down MS to identify the locations of modified sites. One state, the mono-iodinated species, was isolated and subjected to ECD fragmentation (Fig. 9.13c).

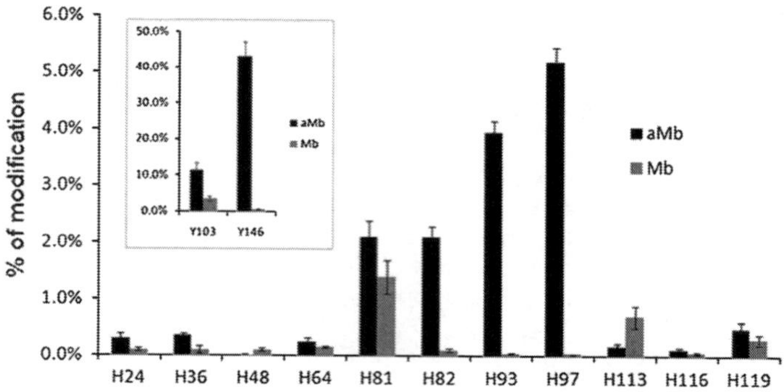

Fig. 9.14 Extent of iodination histidine and tyrosine (*inset*) residues of aMb (*dark bars*) and Mb (*light bars*). Reprinted from Chen et al. [73] with permission from Springer

Several modified sites were identified by top-down including Y146 and Y103. Results indicated that modified Y146 was observed in aMb (26 %) but not in Mb (Fig. 9.13c). Residue Y103 has similar levels of iodination in both states indicating the solvent accessibility of this residue does not change upon heme binding.

While top-down sequencing provided valuable information, it could not detect very low levels of labeling, thereby leaving large gaps in sequence coverage. Some modification sites could not be determined owing to their location in the middle of an unfragmented protein sequence. To increase the yield of modification information, bottom-up sequencing using trypsin digestion was applied. A significant modification difference was observed in residues H82, H93, and H97 (Fig. 9.14). The results agree well with previous NMR analysis of aMb and Mb that showed helix F is not formed in aMb but is well formed in Mb. Residues H82, H93, and H97 are all located in helix F. FPOP-based iodination was also successful in analyzing apo- and holo-carbonic anhydrase II as well as comparing lispro insulin, insulin-EDTA, and zinc-bound insulin.

The success of using persulfate and iodine as reagents in FPOP has opened the door for the use of a host of other radicals for protein labeling. It has been demonstrated that methylene carbene is a good reagent for footprinting of proteins via methylation [75]. Carbene radicals, unlike $^{\bullet}$OH, do not have an amino acid reactivity prejudice, thus providing a more broad-based labeling that is not biased by primary sequence. This and other radicals may be well suited for the photolysis strategy of FPOP.

9.8 Conclusions

To date, FPOP coupled with MS has been effective in the elucidation of protein–ligand and protein–protein interactions. The method reliably measures solvent accessibility, thus providing insight into protein structure. The irreversible nature of the label allows for post-labeling purification and long chromatography gradients for separation prior to MS analysis. Owing to this, the method is not limited by size and complex multi-protein systems can be analyzed. Further, the method has been shown to be tunable with other radicals such as $SO^{4-\bullet}$ and $^\bullet$I. FPOP has also been shown to be useful in studying protein folding. Further developments in methodology could further extend its use as a tool for structural biology and structural characterization of protein therapeutics.

References

1. Englander SW et al (1997) Hydrogen exchange: the modern legacy of Linderstrom-Lang. Protein Sci 6(5):1101–1109
2. Engen JR, Smith DL (2000) Investigating the higher order structure of proteins. Hydrogen exchange, proteolytic fragmentation, and mass spectrometry. Meth Mol Biol 146:95–112
3. Hoofnagle AN et al (2003) Protein analysis by hydrogen exchange mass spectrometry. Annu Rev Biophys Biomol Struct 32:1–25
4. Hanai R, Wang JC (1994) Protein footprinting by the combined use of reversible and irreversible lysine modifications. Proc Natl Acad Sci USA 91(25):11904–11908
5. Li W, Wang JC (1997) Footprinting of yeast DNA topoisomerase II lysyl side chains involved in substrate binding and interdomainal interactions. J Biol Chem 272(49):31190–31195
6. Brenowitz M et al (2002) Probing the structural dynamics of nucleic acids by quantitative time-resolved and equilibrium hydroxyl radical "footprinting". Curr Opin Struct Biol 12(5):648–653
7. Maleknia SD, Downard K (2001) Radical approaches to probe protein structure, folding, and interactions by mass spectrometry. Mass Spectrom Rev 20(6):388–401
8. Su DA et al (2011) Interactions of apurinic/apyrimidinic endonuclease with a redox inhibitor: evidence for an alternate conformation of the enzyme. Biochemistry 50(1):82–92
9. Wen J et al (2009) Membrane orientation of the FMO antenna protein from Chlorobaculum tepidum as determined by mass spectrometry-based footprinting. Proc Natl Acad Sci USA 106(15):6134–6139
10. Wei H et al (2012) Using hydrogen/deuterium exchange mass spectrometry to study conformational changes in granulocyte colony stimulating factor upon PEGylation. J Am Soc Mass Spectrom
11. Zhang J et al (2011) DNA binding alters coactivator interaction surfaces of the intact VDR-RXR complex. Nat Struct Mol Biol 18(5):556–563
12. Pan J et al (2009) Hydrogen/deuterium exchange mass spectrometry with top-down electron capture dissociation for characterizing structural transitions of a 17 kDa protein. J Am Chem Soc 131(35):12801–12808
13. Sperry JB et al (2008) Strong anion exchange for studying protein-DNA interactions by H/D exchange mass spectrometry. J Am Soc Mass Spectrom 19(6):887–890
14. Xu G, Chance MR (2007) Hydroxyl radical-mediated modification of proteins as probes for structural proteomics. Chem Rev 107(8):3514–3543
15. Tullius TD, Dombroski BA (1986) Hydroxyl radical "footprinting": high-resolution information about DNA-protein contacts and application to lambda repressor and Cro protein. Proc Natl Acad Sci USA 83(15):5469–5473
16. Maleknia SD et al (1999) Millisecond radiolytic modification of peptides by synchrotron X-rays identified by mass spectrometry. Anal Chem 71(18):3965–3973
17. Brenowitz M et al (2005) Catching RNA polymerase in the act of binding: intermediates in transcription illuminated by synchrotron footprinting. Proc Natl Acad Sci USA 102(13):4659–4660

18. Xu G et al (2005) Structural allostery and binding of the transferrin*receptor complex. Mol Cell Proteomics 4(12):1959–1967
19. Gupta S et al (2010) Conformational changes during the gating of a potassium channel revealed by structural mass spectrometry. Structure 18(7):839–846
20. Sharp JS et al (2006) Measurement of multisite oxidation kinetics reveals an active site conformational change in Spo0F as a result of protein oxidation. Biochemistry 45(20):6260–6266
21. Sharp JS et al (2004) Analysis of protein solvent accessible surfaces by photochemical oxidation and mass spectrometry. Anal Chem 76(3):672–683
22. Aye TT et al (2005) Nanosecond laser-induced photochemical oxidation method for protein surface mapping with mass spectrometry. Anal Chem 77(18):5814–5822
23. Hambly DM, Gross ML (2005) Laser flash photolysis of hydrogen peroxide to oxidize protein solvent-accessible residues on the microsecond timescale. J Am Soc Mass Spectrom 16(12):2057–2063
24. Pan Y et al (2009) Mapping the structure of an integral membrane protein under semi-denaturing conditions by laser-induced oxidative labeling and mass spectrometry. J Mol Biol 394(5):968–981
25. Pan Y et al (2009) Structural characterization of an integral membrane protein in its natural lipid environment by oxidative methionine labeling and mass spectrometry. Anal Chem 81(1):28–35
26. Pan Y et al (2010) Site-directed mutagenesis combined with oxidative methionine labeling for probing structural transitions of a membrane protein by mass spectrometry. J Am Soc Mass Spectrom 21(11):1947–1956
27. Stocks BB, Konermann L (2009) Structural characterization of short-lived protein unfolding intermediates by laser-induced oxidative labeling and mass spectrometry. Anal Chem 81(1):20–27
28. Stocks BB, Konermann L (2010) Time-dependent changes in side-chain solvent accessibility during cytochrome c folding probed by pulsed oxidative labeling and mass spectrometry. J Mol Biol 398(2):362–373
29. Watson C et al (2009) Pulsed electron beam water radiolysis for submicrosecond hydroxyl radical protein footprinting. Anal Chem 81(7):2496–2505
30. West GM et al (2008) Thermodynamic analysis of protein stability and ligand binding using a chemical modification- and mass-spectrometry based strategy. Anal Chem 80(11):4175–4185
31. Shum WK et al (2005) Onset of oxidative damage in alpha-crystallin by radical probe mass spectrometry. Anal Biochem 344(2):247–256
32. Sharp JS, Tomer KB (2007) Analysis of the oxidative damage-induced conformational changes of apo- and holocalmodulin by dose-dependent protein oxidative surface mapping. Biophys J 92(5):1682–1692
33. Konermann L et al (2010) Mass spectrometry combined with oxidative labeling for exploring protein structure and folding. Mass Spectrom Rev 29(4):651–667
34. Gau BC et al (2009) Fast photochemical oxidation of protein footprints faster than protein unfolding. Anal Chem 81(16):6563–6571
35. Hambly D, Gross M (2007) Laser flash photochemical oxidation to locate heme binding and conformational changes in myoglobin. Int J Mass Spectrom 259(1–3):124–129
36. Liu J, Konermann L (2009) Irreversible thermal denaturation of cytochrome C studied by electrospray mass spectrometry. J Am Soc Mass Spectrom 20(5):819–828
37. Venkatesh S et al (2007) Rapid identification of oxidation-induced conformational changes by kinetic analysis. Rapid Commun Mass Spectrom 21(23):3927–3936
38. Xu G et al (2003) Radiolytic modification of basic amino acid residues in peptides: probes for examining protein–protein interactions. Anal Chem 75(24):6995–7007
39. Xu G, Chance MR (2004) Radiolytic modification of acidic amino acid residues in peptides: probes for examining protein–protein interactions. Anal Chem 76(5):1213–1221
40. Gau BC et al (2010) Sulfate radical anion as a new reagent for fast photochemical oxidation of proteins. Anal Chem 82(18):7821–7827

41. Jones LM et al (2011) Fast photochemical oxidation of proteins for epitope mapping. Anal Chem 83(20):7657–7661
42. Zhang H et al (2010) Fast photochemical oxidation of proteins for comparing structures of protein-ligand complexes: the calmodulin-peptide model system. Anal Chem 83(1):311–318
43. Ikura M et al (1992) Solution structure of a calmodulin-target peptide complex by multidimensional NMR. Science 256(5057):632–638
44. Wan KX et al (2002) Comparing similar spectra: from similarity index to spectral contrast angle. J Am Soc Mass Spectrom 13(1):85–88
45. Cree B (2006) Emerging monoclonal antibody therapies for multiple sclerosis. Neurologist 12(4):171–178
46. Reichert JM, Valge-Archer VE (2007) Development trends for monoclonal antibody cancer therapeutics. Nat Rev Drug Discov 6(5):349–356
47. Aggarwal S (2007) What's fueling the biotech engine? Nat Biotechnol 25(10):1097–1104
48. Benjamin DC et al (1984) The antigenic structure of proteins: a reappraisal. Annu Rev Immunol 2:67–101
49. Mylvaganam SE et al (1998) Structural basis for the binding of an anti-cytochrome c antibody to its antigen: crystal structures of FabE8-cytochrome c complex to 1.8 A resolution and FabE8 to 2.26 A resolution. J Mol Biol 281(2):301–322
50. El-Kased RF et al (2011) A novel mass spectrometric epitope mapping approach without immobilization of the antibody. J Proteomics Bioinform 4(1):001–009
51. Baerga-Ortiz A et al (2002) Epitope mapping of a monoclonal antibody against human thrombin by H/D-exchange mass spectrometry reveals selection of a diverse sequence in a highly conserved protein. Protein Sci 11(6):1300–1308
52. van de Locht A et al (1997) The thrombin E192Q-BPTI complex reveals gross structural rearrangements: implications for the interaction with antithrombin and thrombomodulin. EMBO J 16(11):2977–2984
53. Rezaie AR et al (1998) Thrombomodulin increases the rate of thrombin inhibition by BPTI. Biochemistry 37(2):693–699
54. Herman C (2011) Fast, reliable too characterizes antibodies. Chem Eng News
55. Mahley RW (1988) Apolipoprotein E: cholesterol transport protein with expanding role in cell biology. Science 240(4852):622–630
56. Strittmatter WJ et al (1993) Apolipoprotein E: high-avidity binding to beta-amyloid and increased frequency of type 4 allele in late-onset familial Alzheimer disease. Proc Natl Acad Sci USA 90(5):1977–1981
57. Holtzman DM et al (2000) Apolipoprotein E isoform-dependent amyloid deposition and neuritic degeneration in a mouse model of Alzheimer's disease. Proc Natl Acad Sci USA 97(6):2892–2897
58. Corder EH et al (1993) Gene dose of apolipoprotein E type 4 allele and the risk of Alzheimer's disease in late onset families. Science 261(5123):921–923
59. Garai K, Frieden C (2010) The association-dissociation behavior of the ApoE proteins: kinetic and equilibrium studies. Biochemistry 49(44):9533–9541
60. Gau B et al (2011) Mass spectrometry-based protein footprinting characterizes the structures of oligomeric apolipoprotein E2, E3, and E4. Biochemistry 50(38):8117–8126
61. Garai K et al (2011) Dissociation of apolipoprotein E oligomers to monomer is required for high-affinity binding to phospholipid vesicles. Biochemistry 50(13):2550–2558
62. Zhang Y et al (2007) A monomeric, biologically active, full-length human apolipoprotein E. Biochemistry 46(37):10722–10732
63. Chen J et al (2011) Topology of human apolipoprotein E3 uniquely regulates its diverse biological functions. Proc Natl Acad Sci USA 108(36):14813–14818
64. Weisgraber KH (1994) Apolipoprotein E: structure-function relationships. Adv Protein Chem 45:249–302
65. Westerlund JA, Weisgraber KH (1993) Discrete carboxyl-terminal segments of apolipoprotein E mediate lipoprotein association and protein oligomerization. J Biol Chem 268(21):15745–15750

66. Aggerbeck LP et al (1988) Human apolipoprotein E3 in aqueous solution. II. Properties of the amino- and carboxyl-terminal domains. J Biol Chem 263(13):6249–6258
67. Chen J et al (2010) Temperature jump and fast photochemical oxidation probe submillisecond protein folding. J Am Chem Soc 132(44):15502–15504
68. Nolting B et al (1997) The folding pathway of a protein at high resolution from microseconds to seconds. Proc Natl Acad Sci USA 94(3):826–830
69. Davies MJ, Dean RT (1997) Radical-mediated protein oxidation: from chemistry to medicine. Oxford science publications. Oxford University Press, New York
70. Bridgewater JD et al (2006) Transition metal-peptide binding studied by metal-catalyzed oxidation reactions and mass spectrometry. Anal Chem 78(7):2432–2438
71. Kretsinger RH (1968) A crystallographic study of iodinated sperm whale metmyoglobin. J Mol Biol 31(2):315–318
72. Ghosh D et al (1999) Determination of a protein structure by iodination: the structure of iodinated acetylxylan esterase. Acta Crystallogr D Biol Crystallogr 55(Pt 4):779–784
73. Chen J et al (2012) New protein footprinting: fast photochemical iodination combined with top-down and bottom-up mass spectrometry. J Am Soc Mass Spectrom 23(8):1306–1318
74. Li CH (1944) Kinetics of reactions between iodine and histidine. J Am Chem Soc 66(2):225–227
75. Gomez GE et al (2012) Probing protein surface with a solvent mimetic carbene coupled to detection by mass spectrometry. J Am Soc Mass Spectrom 23(1):30–42
76. Buxton GV, Greenstock CL, Helman WP, Ross AB (1988) Critical-review of rate constants for reactions of hydrated electrons hydrogen-atoms and hydroxyl radicals (•OH/•O−) in aqueous- solution. J Phys Chem Ref Data 17:513–886

Chapter 10
Applications of Ion Mobility Mass Spectrometry for Characterization of Protein Therapeutics

Weibin Chen and Asish Chakraborty

10.1 Introduction

A major proportion of protein therapeutics marketed to date are produced by recombinant DNA technologies using well-chosen protein expression systems. Compared with small-molecule pharmaceuticals, recombinant therapeutic proteins are generally complex, heterogeneous, and subject to a variety of enzymatic or chemical modifications during expression, purification, and long-term storage. Because of unique structural features and production processes, the analytical strategies for characterization, quantitation, impurity profiling, and bioactivity evaluation of recombinant proteins represent a great challenge and a matter for debate.

The analysis of protein therapeutic products normally involves the appropriate combination of analytical tools in order to effectively address different aspects of the product's structural features. The need for many analytical technologies underlines the complexity of analytical support in biopharmaceutical development. It also calls for continued development of new tools, technologies, and assays to streamline the process, so that critical information regarding the biotherapeutic attributes can be reliably and seamlessly acquired.

Mass spectrometry (MS) has played an important role in the ensemble of analytical tools for in-depth characterization of biotherapeutic products [1]. MS is widely used for revealing the covalent structure and stability of protein therapeutics due to its analytical sensitivity, selectivity, and specificity. Information acquired from MS is particularly useful to demonstrate product quality and consistency and to identify the desired or undesired forms of protein products. Structural analysis is important to the production process as small changes in an optimized process may affect the structure of the product. On the other hand, assuring sufficient pharmaceutical stability is a

W. Chen (✉) · A. Chakraborty
Waters Corporation, 34 Maple Street, Milford MA 01757, USA
e-mail: weibin_chen@waters.com

G. Chen (ed.), *Characterization of Protein Therapeutics using Mass Spectrometry*,
DOI: 10.1007/978-1-4419-7862-2_10,

great challenge in protein drug formulation as a large number of functional groups in protein therapeutics are susceptible to chemical degradation.

Owing to the advantages that MS offers, it is of great interest to develop MS-based methods to characterize protein therapeutics not only by mass but also by other properties such as protein conformation (i.e., secondary/tertiary structures of proteins). Biophysical properties such as protein conformation, protein dynamics, and aggregation contribute either collectively or individually to the quality attributes of protein therapeutics. Reproducible safety and efficacy profiles for protein therapeutics require manufacturers to produce products with consistent higher-order structures. Methods for determining changes in higher-order structure are invaluable for such assessments. Despite the versatility and capability demonstrated by MS in protein characterization, most current applications of MS in biopharmaceutical industry are focused on characterizing primary structures of proteins [e.g., confirming protein sequence and identifying post-translational modifications (PTMs)]. Its role in studies of non-covalent features of proteins is less developed. Although methods that rely on direct electrospray ionization (ESI) MS analysis for higher-order structure study, such as probing non-covalent assemblies [2–4] or analysis of protein ion charge state distributions [5], have been reported, direct ESI MS measurements for protein conformation analysis have long been limited to academic laboratories. The embrace of these methodologies in industrial settings has been somewhat hindered due to the lack of robustness and direct connection between analytical data and protein structural information. Therefore, it is highly desirable to develop additional MS-based methods elucidating the three-dimensional molecular structure of proteins.

When coupled to spectroscopic techniques such as ion mobility spectrometry (IMS), or methods probing the conformation-dependent chemical reactivity such as hydrogen/deuterium (H/D) exchange (see Chap. 8) and fast photochemical oxidation (see Chap. 9), MS has demonstrated the potential of probing detailed molecular structures [6–9]. IMS is an electrophoretic technique that uses mobility rather than mass to separate gas-phase ions. Under the influence of an electric field and at either low vacuum or atmospheric pressure conditions, ions with different properties such as size, shape and charge show different characteristic behavior in the gas phase, which result in a specific drift velocity and thus ion mobility separation. Therefore, this gas-phase electrophoretic technique can be used analytically to deduce structural information regarding the ion of interest. As a stand-alone system, IMS has been in use for many decades for many analytical applications, ranging from the detection of chemical warfare agents [10, 11] to particle sizing [12, 13]. More recently, ion mobility has been coupled to MS providing a new dimension in the analysis of biomolecules with ion mobility offering direct molecular structural information for all resolved species. The unique information afforded by integrating the information from both the IM and the MS separation dimensions highlights the strength of ion mobility mass spectrometry (IMMS) techniques in life sciences research. With these characteristics, IMMS is viewed as a complementary tool in the context of structural biology, offering even greater insight into the properties of large and heterogeneous protein complexes by providing information on the stoichiometry, topology, and cross section of these assemblies and their composite sub-units [14–16].

The need for further advances in the ensemble of analytical tools to address different aspects of the quality assessment of protein therapeutics and in particular for conformation analysis and aggregates prompts much interest to explore the utility of IMMS. The aim of this chapter is to briefly introduce the development of IMS instrumentation and the major configurations of IMMS instruments. The focus of this chapter is to discuss the main applications of IMMS technique in the different aspects of the quality assessment of protein therapeutics.

10.2 Overview of Ion Mobility Spectrometry and Ion Mobility Mass Spectrometry

10.2.1 Brief History of Ion Mobility Spectrometry Development

The development of analytical IMS can date back to the mid- to late-nineteenth century when an early burst of scientific activities brought in a broad range and depth of inquiry about the formation and behavior of ions in gases at ambient pressure [17]. Through these activities, a wealth of data and experience was acquired regarding the identity of the ions formed in the gases, their mobilities as well as the factors that control the ion mobility (e.g., temperature, pressure, and gas purity). The development from these early years made important contribution to subsequent advances in mobility measurements and analytical IMS instrumentation. Langevin was one of the first to describe the theoretical treatment of mobility and practical experimentation. Two of his seminal papers [18, 19] properly recognized the interactions between ions and gaseous molecules and the influence of the gas on the mobility of the ion. Langevin's theory laid down the fundamental principles of modern analytical IMS; that is, ions can be separated by their characteristic velocity through a gas-filled electric field. It was also during this period of time, ion shutters technique for pulsed ion injection was developed by van de Graaf [20] and Cravath [21] (and later further modified by Bradbury [22]). This breakthrough solved one of the practical problems in IM measurement: how to couple continuous ion sources with IM drift tube for only discrete packet of ions is allowed to be injected into the ion mobility device (i.e., drift tube) at a time.

The next period of IMS progression started from 1948 when Lovelock [23] first reported his findings that a simple ionization detector [electron capture detector (ECD)] was able to detect ultralow concentrations of airborne industry–related organic vapors that were released into the atmosphere as pollutants. The significance of Lovelock's study lies in that it established for the first time that there is a direct link between the composition of a vapor sample, such as trace impurities in ambient air, and the ions created in an ionization source. Later on in the 1950s, due to research interest into pollution, warfare, and space exploration, extensive investigation of gas-phase ion-molecule reactions at elevated pressures was undertaken, and these studies formed a chemical foundation for the interpretation of the chemical events that occurred inside an IM drift tube. Experiments performed by

McDaniel [24] and Kebarle [25] with drift cells under a low electric field to study the separation and reactions of ions and small molecules in the gas phase led to the development of the predecessors of modern IM instrumentation.

On the basis of these early foundational work, the core component of modern analytical IMS, a drift tube, was created by McDaniel at Georgia Tech [26]. The drift tube consisted of a stack of electrically isolated rings, and a linear electric field was established with voltage dividers. In the mid-1960s, Martin Cohen and coworkers at Franklin GNO developed the first analytical IMS, so-called plasma chromatography [27], in an attempt to characterize vapors as ions in air using a drift tube. At a conceptual level, the instrument was a blend of Lovelock's ionization detector and the drift tube initially built at Georgia Tech. The instrumentation was soon commercialized with one of the configurations coupled with a mass spectrometer. The commercialization of IMS technology marked an end to the period of foundational discovery and introduced an era of extensive exploration of IMS as an analytical tool.

The introduction of commercial analytical IMS made it possible for researchers to explore its industrial applications and to transform the large laboratory instrument into portable or handheld rugged analyzers for detection of chemical warfare agents, explosives, and drugs. Although successful industrial application of IMS was rather limited in the era from 1970s to 1990s, miniaturization efforts put forth by industrial groups and US/UK military successfully brought in the deployment of IMS as a chemical agent monitor in combat and as explosives detectors in aviation security checks [17]. Since the 1950s, military-based applications had been a major driving force behind the development of IMS. The high acceptance and the trust placed by soldiers and the traveling public onto IMS demonstrated the unique applications of the technique and showed the strides that have been made over the years as an analytical instrument.

In the 1990s, with the advent of soft ionization methods, electrospray ionization (ESI) and matrix-assisted laser desorption/ionization (MALDI), the ionization of large intact macromolecules without fragmentation became a reality. Coupled with the soft ionization techniques, IMS was applied for analyzing biomolecules [28–31]. In the meantime, theoretical methods for elucidating ion structure were developed by correlating the ion mobility experimental values (cross sections) with calculated results for computer-generated trial structures/geometries [32, 33]. This development arguably represents one of the most important advances achieved with IMS technologies and paves the way to link the IMS technique with the ability to determine information about ion shape. In instrumentation, a number of MS techniques were successfully coupled to IMS to accelerate the study of biomolecules by IMS, including Fourier transform ion cyclotron resonance [34], linear quadrupoles [35], and trapping devices [36–38], as well as time-of-flight (TOF) [39] mass spectrometers.

10.2.2 General Principles of Ion Mobility Spectrometry

The basic principle behind IMS separations is straightforward: it measures how fast a given ion moves against an electrical field through an atmosphere composed of neutral

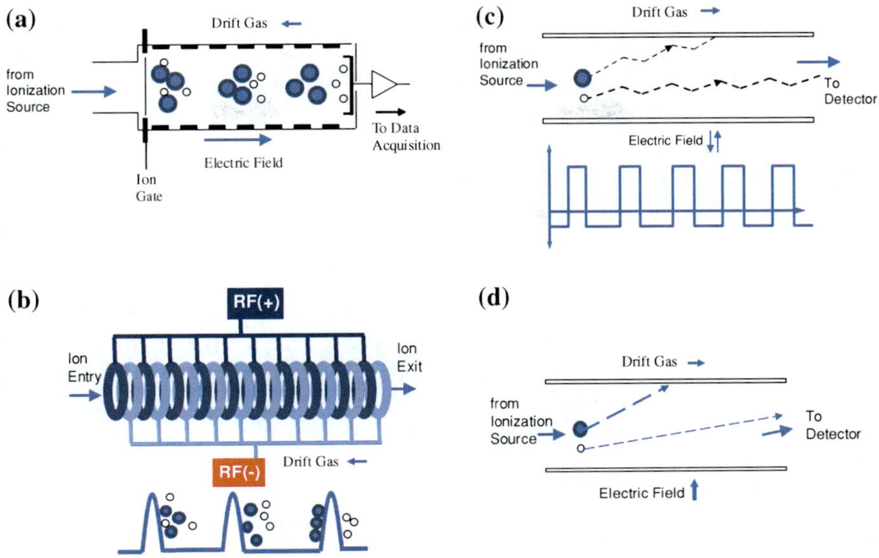

Fig. 10.1 Schematic illustration of separation principles of different ion mobility instrumentations. **a** Drift tube IM. **b** Traveling wave IM. **c** FAIMS. **d** GEMMA

drift molecules (e.g., N_2, He, etc.). Ion species with different properties, such as size, shape, and charge, travel with different velocities (i.e., ion mobilities) when they are pulled by an electric field through a drift cell filled with a buffer carrier gas (e.g., N_2). For example, an ion with a large average cross section undergoes more collisions with the buffer gas and travels more slowly than an ion with a small average collision cross section (CCS). Ion mobility separation is obtained when the ions disperse throughout the device. Through the measurement of the arrival time (distribution) at the detector, this information can be used analytically to deduce structural information regarding the ion of interest. The main advantages of IM measurement are the simplicity and speed of the measurement as well as the high sensitivity and selectivity of the analysis.

The principles of the three different ion mobility devices that are most frequently found in tandem with mass spectrometers will be described in this section (Fig. 10.1). Because the development of IM separation originated from linear drift tube (LDT) IMS, the principles employed in this type of instrument are discussed before the more recently developed traveling wave (TW) IMS and field asymmetric ion mobility spectrometry (FAIMS) are reviewed.

10.2.2.1 Linear Drift Tube IMS

The basics of LDT IMS have been known for a long time. The theoretical treatment of the technique is based on the primary conditions that the electric field employed for LDT IMS is weak and uniform throughout the device. The applied field is considered to be weak if the thermal energy obtained by the ions due to the collision with the drift gas molecules is greater than the energy the ions acquired from the electric field. Under those circumstances, the ions have the energies similar to that of the bulk buffer gas and diffusion process dominates over ion mobility. Under these weak-field conditions, according to Mason and McDaniel [40], the mobility of any ion, K, is defined as following (Eq. 10.1)

$$K = \frac{\vartheta_d}{E} = \frac{L}{Et_d} = \frac{L^2}{Vt_d} \tag{10.1}$$

where ϑ_d is the drift velocity of ions, $\boldsymbol{E}$ is the electric field (in units of V/cm and calculated as V/L), V is the voltage drop over the drift tube, L is the length of the drift cell, and t_d is the drift time.

The reduced ion mobility, K_0, normalized with respect to pressure, P, and temperature, T, can thus be defined as

$$K_0 = \frac{L}{t_d E} \times \frac{273.15}{T} \times \frac{P}{760} \tag{10.2}$$

K_0 is calculated using values in standard conditions to facilitate the comparison of ion mobility measurements between laboratories/settings.

When an ion is placed in the IM cell, it experiences an accelerated electric force, which is proportional to charge (ze), and a deceleration friction force caused by the collision with the buffer gas. This friction force is inversely proportional to the number density of buffer gas N and the CCS Ω of the ion. Therefore, this mobility constant can also be quantitatively described based on the characteristics of the ion and the conditions of the drift cell, namely the ion's charge (ze); average CCS Ω the number density of the drift gas (N).

$$K_0 = \frac{3ze}{16N_0}\sqrt{\frac{2\pi}{\mu k_B T}\frac{1}{\Omega}} \tag{10.3}$$

where $\mu = Mm\big/(M+m)$ is the reduced mass of the ion (with mass m) and buffer gas (with mass M), k_B is the Boltzmann's constant, T is the gas temperature, and N_0 is the number density of the drift gas at standard temperature and pressure.

When coupling IM with MS, the mass and charge of ions can be easily obtained. This information, combined with the measured ion arrival time, allows the Ω of an ion to be calculated using Eq. 10.3. This characteristic value represents the orientationally averaged area of the ion which is able to interact with the buffer gas. In comparison, the Ω for larger ions (e.g., proteins) can be approximated computationally using different theoretical models (e.g., relatively simple hard sphere scattering [32]) based on the coordinate data from either X-ray or NMR structural analysis.

During the transit process in the IM drift tube, the injected ion packet experiences a force exerted by an electric field and is accelerated along the field line until it collides with a gas molecule and scatters in random directions. This process is repeated throughout the entire navigation process until it exits out of the drift tube. Simultaneously, normal diffusion process will cause the ion pack to grow in size, while the ions are moved through the drift tube. These are opposing effects that limit the resolution obtainable with a drift tube.

Based on Eq. 10.1, the measurement of ion mobility, K, involves the determination of the arrival time t_d of an ion pack traveling in a weak field E over a given drift length L. The spread, Δx of a pack of identical ions due to diffusion, is given by

$$\Delta x = \sqrt{\frac{4k_B T L}{\pi E e}} = \sqrt{\frac{4k_B T}{\pi V e}} L \tag{10.4}$$

Therefore, the resolution of an IMS device, $t_d/\Delta t$, can be calculated as follows:

$$\frac{t_d}{\Delta t} = \frac{L}{\Delta x} = \sqrt{\frac{\pi V e}{4k_B T}} \tag{10.5}$$

Based on Eq. 10.5, the IM resolution is affected only by two experimental parameters: temperature, T, and drift voltage, V. To improve the resolution of IM separation, one can reduce T to decrease diffusion and thus reduce Δt. On the other hand, the resolution is proportional to $\sqrt{V/T}$, so increasing the voltage applied across the drift tube also improves the resolution. However, increasing V requires a simultaneous increase in the gas pressure (the buffer gas number density N) to maintain the ratio of E/N constant and to stay in the low-field regime desirable for ion mobility experiments.

IM separation based on drift tube is the simplest version among all the IM techniques available now. The IM device normally operates in one of two modes to obtain drift time ion mobility spectra: reduced pressure IMS and ambient pressure IMS. Historically, a field-deployable stand-alone IMS analytical instrument operates at ambient pressure for the separation and detection of trace quantities of explosives, drugs, and chemical warfare agents. Because of the higher gas pressure, ambient pressure IMS generally produces higher resolving power and greater separation selectivity but suffers from lower sensitivity due to inefficient transfer of ions from ambient pressure into the vacuum of the mass spectrometer [39, 41, 42]. In comparison with other variants of IM technologies (see below), the traditional LDT IMS provides the highest resolving powers, but has a decreased sensitivity due to its low duty cycle.

10.2.2.2 Traveling Wave IMS

Traveling wave ion mobility spectrometry (TWIMS) is a relatively new IMS technique developed by Giles et al. [43–45] and was implemented in the first integrated commercially available IMMS instrument, the Synapt™ HDMS system. TWIMS is operated at a reduced pressure (typically 3 mbar), and the TWIMS device [also called traveling wave ion guide (TWIG)] consists of a gas-filled cell that comprises a series of ring electrodes (similar to LDT device) arranged orthogonally to the direction of ion transmission. However, the operating principle of TWIMS is drastically different from the traditional LDT IMS device. Firstly, a radio frequency (RF) field is applied to a pair of consecutive electrodes in the stacked ring ion guide, providing an electric field that confines ions radially within the device. Secondly, instead of a low electrical field being applied uniformly across the cell, the TWIG uses a dynamically pulsed electrical field that is superimposed on top of the RF voltage to provide an electric field. This transient DC voltage is applied sequentially to pairs of ring electrodes to push the ions toward the exit of the ion guide. While this TW pushes ions through the gas, the friction from collisions between the ions and the buffer gas impedes the ion's travel, so that ions with a large CCS experience more drag force by the buffer gas and tumble backwards over the wave more frequently than small ions. Because the larger ions fall backwards more frequently, they remain in the TWIM cell longer, and separations are generated by the relative retention of different-sized ions within the device. The details of the ion separation process are yet to be completely understood although an attempt on theoretical treatment of the separation mechanism is reported in the literature [46].

The TWIM is a dispersive technology, and therefore, the measured mobility is the arrival time distribution (ATD) of ions at the detector. Since TWIMS is operated at reduced pressures, the resolution is generally lower than for a conventional LDT IMS. Resolving power will undoubtedly improve as more is learned about this complex but unique mobility cell. However, the TWIM technique has a high ion transmission efficiency compared with conventional LDT IMS. Unlike the LDT IMS, for which a quantitative relationship can be clearly defined between the mobility and the CCS of ions, the relationship between the measured drift time and the ions' CCS is complicated for the TWIM technique due to the complicated ion separation process, notwithstanding recent advances in theory and calibration methods that have made it possible to use TWIMS for cross-sectional measurements [46–48].

10.2.2.3 Field Asymmetric Ion Mobility Spectrometry

FAIMS, also known as differential mobility spectrometry (DMS), is a technique that separates gas-phase ions based on the IM difference in high and low electric field (opposite polarity) [49–52]. In a FAIMS device, ions are passed between two flat parallel electrodes and two concentric cylinder electrodes with a buffer gas flowing through the gap between the closely spaced electrodes. High drift electric field (far greater than the weak-field limit) is applied across the gap in a direction perpendicular to the gas flow. The field causes the ions to drift sideways, toward one or the other electrode. At high electric fields, the ion mobility is no longer constant but becomes dependent on the strength of the applied electric field. By applying a high frequency periodic asymmetric waveform across the two electrodes, ions alternately experience two distinct field strengths. The ions oscillate between the electrodes and, due to differential mobility in each field, will preferentially move toward one electrode. This process creates a net drift in the direction of the field but, unlike in low-field IMS, the net drift velocity is now a function of the difference in high- and low-field mobility of the ion. Because ions with different overall size and charge will acquire different net drift velocities, this effect causes the dispersion of ions according to their differential mobility. Separation of ions in FAIMS occurs under atmospheric pressure and room temperature conditions.

To counteract this sideways drift and enable the ions to transmit through the electrode region, a DC field is applied in the opposite direction to the ion drift. This is called the compensation field (CF). Because a given magnitude of CF only compensates for a specific drift velocity—and hence only ions that have the corresponding net drift velocity will be transmitted as a result of the applied CF—different compensation voltages are required for each ion to pass through the FAIMS cell. Therefore, the compensation voltage can be scanned to allow sequential detection of all ions present. Alternatively, a predetermined CF value is chosen to only allow specific ions of interest to pass through the FAIMS device for analysis (similar to a quadrupole mass analyzer). For a given species, it is the compensation voltage, rather than the drift time, which is generally reported.

In contrast to the LDT IMS setup described above, the principle of FAIMS is effectively based on the dependence of mobility on the field strength rather than an absolute value of mobility that is measured. Because ions are separated by the difference in their mobilities at high and low values of $\boldsymbol{E}$, the absolute value of $\boldsymbol{K}$ cannot be determined from FAIMS data. Therefore, the relationship between the mobility of an ion and its CCS is not as straightforward as that found in the LDT IM instruments. Additionally, a FAIMS spectrometer has lower resolving power than the LDT IMS instruments. When the instrument needs to operate in a voltage scan mode to monitor a range of mobilities, sensitivity is also reduced due to the duty cycle of the scan.

10.2.3 Ion Mobility Mass Spectrometry

The first ion mobility mass spectrometer was built in the 1960s. The instrument comprised of a LDT IMS and a magnetic sector analyzer and was used to study gas-phase ion-molecule reactions [24, 53, 54]. Before 1990, only a few IMMS instruments were employed in research and the roles of those instruments were mainly for supporting the development of IMS as an analytical method [17]. The next important development in the field of IMMS for biological applications occurred in the mid-1990s when Bowers et al. [33, 55] first reported the structural studies for peptides and Clemmer et al. [28, 56, 57] described the structural analysis of intact proteins by IMMS. The pioneering work led to a period of rapid and expanding applications of IMMS, with much activity focusing on characterizing biomolecular structures as represented by the work from Robinson et al. on the elucidation of protein complex quaternary structure [14, 58–60] using IMMS strategies. The application of IMMS in biological and biochemical research accelerated the IMMS instrumentation development, and instrumental designs of IMMS have been one of the most rapidly growing areas of MS as exemplified by the first commercially available integrated IMMS instrument in 2007 [44].

There have been a rich variety of IMMS instrument configurations reported for the combination of IM with MS. Among all the methods recording ion mobility, three principal types of ion mobility instrumentation have been successfully coupled with mass spectrometers. They are LDT IMS, TWIG IMS, and FAIMS. LDT IMS is operated under either reduced or ambient pressure conditions; FAIMS is operated under ambient pressure conditions, while TWIG IMS is operated at reduced pressure. Each of these ion mobility spectrometers can be interfaced to a variety of mass analyzers to create IMMS. TOF [39, 61], quadrupole [62], ion trap [29], or ion cyclotron [34], and magnetic sector analyzer [53, 54] have all been coupled with IMS. In addition to the various mass analyzers that can be coupled with ion mobility, an ion mobility cell can be placed in a number of positions relative to the mass analyzer. Among all the instrument designs, two main instrument configurations are frequently seen: the mass analyzer is either placed before the drift cell (MS/IMS) or positioned after the drift cell (IMS/MS). Details on the various IMMS configurations can be found in a comprehensive review by Kanu et al. [63].

There are several factors under consideration when coupling mass spectrometers with an IMS device. Firstly, from a typical performance parameter point of view, the mass resolution and mass spectral acquisition rate of mass spectrometers need to meet the analysis requirement. Since IMS operation uses a buffer gas, which potentially leaks into the mass spectrometer, vacuum requirement for various MS analyzers is another factor to consider. Not all types of mass spectrometers are readily coupled with IMS to the same degree due to the vacuum requirement. TOF mass spectrometers, although requiring high vacuum, have emerged as a powerful component in IMMS combination [39, 64, 65]. Some of the advantages of TOF MS are the wide mass range, the high sensitivity compared to scanning MS filters where unselected ions are lost, and the high resolution in a reflection arrangement. For mass spectral acquisition, the use of TOF [39] and in particular orthogonal acceleration (oa)-TOF technology provides full mass spectra on a timescale (microseconds) short enough to profile entire ion mobility peaks (usually milliseconds), a capability not possible with mass analyzers such as quadrupoles or ion traps. Because IMS spectra are obtained in milliseconds and TOF mass spectra are obtained in microseconds, thousands of mass spectra can be obtained for each ion mobility spectrum producing a two-dimensional array in which both mobility and mass of ions are recorded [65].

Perhaps the biggest challenge in the development of modern IMMS instruments is to address inherent sensitivity issues associated with the traditional drift tube IMS. There are two factors responsible for the poor sensitivity normally seen with the LDT IM device. For mobility measurements, ion pulses are intrinsically required. When combined with a continuous ion source such as ESI during the mobility experiment, ions have to be gated and only a narrow pulse of ions is introduced into the drift region. The majority of ion signal (up to 99–99.9 %) is discarded. The low duty cycle leads to severe undersampling of the ion signals and makes the method inherently insensitive. Another factor causing the low sensitivity is ion radial diffusion in the drift tubes. During the transportation process, due to the diffusion process, the sizes of ion could increase beyond the diameter of sampling apertures in the mass spectrometer.

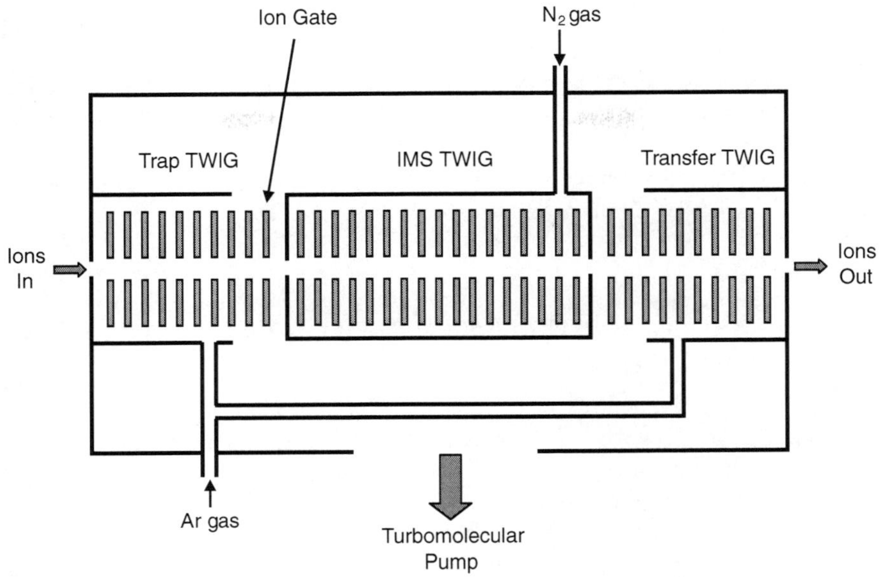

Fig. 10.2 Schematic diagram of TriWave technology implemented in the Synapt™ HDMS system. TriWave is comprised of three traveling wave ion guides (*TWIG*) labeled as *Trap TWIG*, *IMS TWIG*, and *Transfer TWIG* in the diagram. Detailed description about TriWave can be found in the literature [43]

Different approaches have been used during the development of IMMS to enhance the sensitivity. A Paul geometry ion trap was incorporated into a LDT IMMS to overcome the low duty cycle issue in sub-ambient pressure in linear drift cells. The ion trap accumulates ions, while the mobility separation is occurring for the previous packet of trapped ions, and then injects the trapped ions into the drift tube [61] for the next round of separation. Reports on the use of other ion storage geometries to improve duty cycle are also found [65, 66]. The issue of ion loss due to radial diffusion has been addressed through use of a periodic focusing DC drift tube design [67–69]. More recently, the use of duel ion funnels before and after the drift tube has been reported to provide essentially total ion transmission [70, 71]. The function of the first funnel is to accumulate ions and then pulse ions into the drift tube, whereas the second ion funnel to refocus and align the radially diffused mobility-separated ions.

In comparison with the traditional LDT IM setup, several technological features associated with the TWIM device make it very attractive to modern instrumentation. The TWIM technologies implemented in the commercial instrument consist of three stacked ring ion guides (called TriWave technology, see Fig. 10.2) with RF field applied to each of the ion guides [43–45]. The use of RF field creates a radial potential well to confine ions to the center of the guide. This essentially eliminates radial diffusion effects. In addition, the mobility cell is the center cell of three ion guides, so the first ion guide stores the ions, while the ion mobility is taking place and the final ion guide transfers the ions to the TOF analyzer. The primary advantage of this arrangement is that the sensitivity of the mass spectrometer is not compromised by the duty cycle of the IMS that is common with other mobility drift cells. As a result, this IMS/TOF instrument geometry provides mobility separation without sacrificing the base sensitivity of the mass spectrometer. In addition, because the TriWave technology is embedded in a high-performance oa-TOF tandem mass spectrometer, this instrument retains all the high-performance attributes of a TOF MS/MS instrument and yet adds a degree of orthogonality to the mass spectrometer.

10.3 Applications of IMMS for Characterization of Protein Therapeutics

Although the use of IMS in the battlefield and commercial aviation industry probably represents well-known applications of IMS, the focus of combined IMMS approach was historically used to perform basic research in ion-molecule reactions. In recent years, especially after the advent of ESI and MALDI soft ionization methods, the application of IMMS has been focused on biomolecule analysis. In particular, with commercial IMMS instruments available, applications expanded further into the field of analytical chemistry where the technique is used as a tool to address many challenges in biological studies. The studies range from identifying species in complex biological samples, simplifying complex datasets, to determining the conformations of large macromolecular complexes [58, 71]. Elaboration on these applications is out of scope of the chapter, and a book dedicated to IMMS theory and applications was recently published [72] to bring an update on the advances of IMMS and its routine applications. In this chapter, we focus on how IMMS is used as a tool in biopharmaceutical analysis to characterize the structures of protein therapeutics and help the development of therapeutic drugs. The selected applications cover a range of topics from the conformation analysis to the differentiation of isobaric species that are important to the attributes of therapeutic proteins.

10.3.1 Analysis of Conformation of Monomeric Proteins

The three-dimensional structure of a protein is critical to its functions and, therefore, its biological activity. Because the unique structural information that IMS offers, it is not surprising that the use of IMMS has gained attention as a tool for the analysis of macromolecules and in particular for its application in determining the conformations adopted by biological molecules in gas phases [55–60].

The analysis of protein conformation using IMMS generally requires a proper means to transfer protein ions from solution phase to the gas phase. This is normally accomplished by using ESI or nanoelectrospray ionization (nESI) methods [73]. Also important in the context of this measurement is that the ionization step should bring in little or no artificial perturbation, apart from the impact of natural desolvation processes, so the gas-phase ion structures can closely mimic the structural features in the solution states [14]. A common practice for this exercise is to electrospray the protein or protein complex samples from a fully aqueous environment (i.e., typically 10–500 mM ammonium acetate), where proteins can maintain its native structures. When proteins or protein complexes are analyzed under native (non-denaturing) aqueous conditions, the analytes under investigation do not carry a large number of charges and, therefore, appear in the high m/z range of the mass spectra. Hence, instruments with a wide mass range are needed to cope with the high m/z signals that are typically observed in such analyses. TOF MS fits particularly well to the analysis of high m/z ions [74, 75], because in theory it has an unlimited mass range.

Perhaps the most direct and convincing example to demonstrate the utility of IMMS for the characterization of therapeutic proteins is those effectively connecting the IM measurement with the quality attributes of protein therapeutics. Such connection is nicely illustrated by the work performed by Bagal et al. [76] for the rapid characterization of disulfide variants in intact IgG2 monoclonal antibodies (mAbs). Recombinant mAbs are an important class of therapeutic agent that has found widespread use for the treatment of many human diseases. It is known that intact human IgG2s have three distinct structural isoforms (IgG2-A, IgG2-B, and IgG2-A/B) which are caused by alternating disulfide connectivity from two of the four cysteine residues in the hinge region of the heavy chain (Cys-232 and Cys-233). For some IgG2 molecules, the disulfide-related structural isoforms have shown different activity against antigen targets. Therefore, the ability to rapidly detect and characterize IgG2 isoforms is of great interest to the development of mAbs.

In this study, Bagal et al. [76] developed a sensitive method using IMMS for quickly profiling the IgG2 isoforms with minimal sample preparation. For the IMMS analysis, the IgG2 samples were ionized from a solution of 160 mM ammonium acetate using nanoelectrospray. IMMS reveals 2–3 gas-phase conformer populations for IgG2s (Fig. 10.3), whereas a single gas-phase conformer was observed for both an IgG1 antibody and an IgG2 mutant (Cys-232 f Ser). Since both IgG1 and IgG2 mutant molecules are homogeneous with respect to disulfide bonding, the authors deduced the observed IgG2 gas-phase conformers are related to disulfide bond heterogeneity. Furthermore, the authors carried out IMMS analysis on the redox-enriched disulfide isoforms, so the identities of the mobility peaks can be assigned with respect to the established disulfide bonding patterns. Although analytical methods for structural characterization of disulfide isoforms of the human IgG2 subclass have been reported previously, the study clearly illustrates how IMMS can be used to quickly provide information on the higher-order structure of antibody therapeutics.

In a separate study, Atmanene et al. [77] showed how to use IMMS to probe protein conformation and provide the information for the lead optimization during the development of a therapeutic mAb. A murine mAb, mumAb 6F4, that has shown anti-proliferative and anti-tumoral properties was humanized to develop a new therapeutic drug. To determine whether the anti-tumoral properties observed for the murine mAb are conserved in the humanized version, it is necessary to study the formation of immune complexes involving between mumAb 6F4/hzmAb 6F4 and its antigen JAM-A.

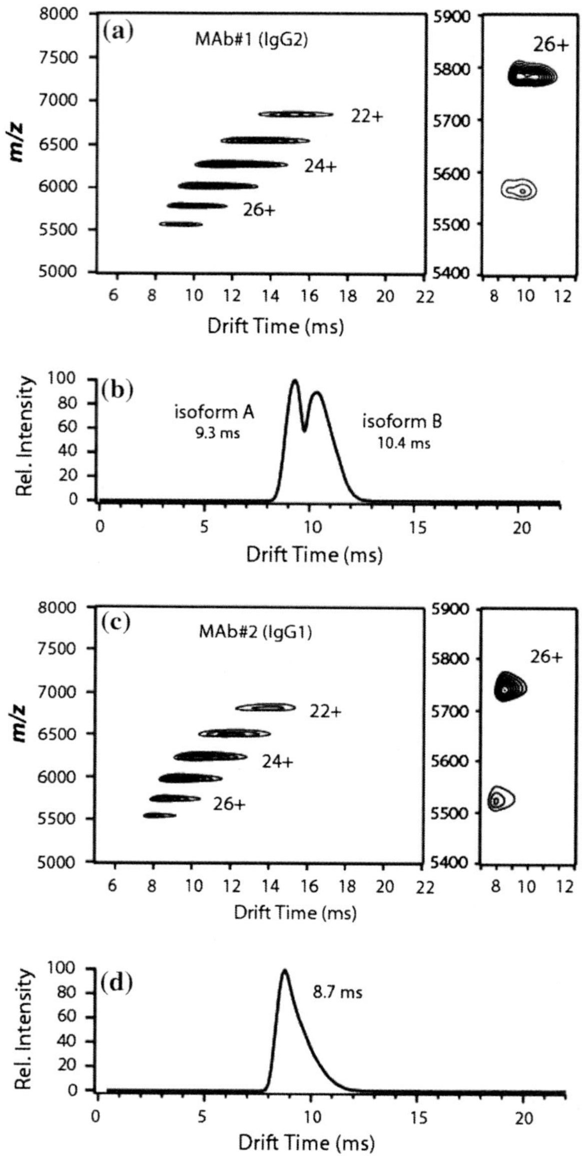

Fig. 10.3 Ion mobility separation of protein isoforms of **a** mAb#1 (*IgG2*) and **c** mAb#2 (*IgG1*) by TWIMS. The normalized ion mobility intensities are graphed as contour plots. Extracted ATDs for the 26+ charge states of **b** mAb#1 (*IgG2*) and **d** mAb#2 (*IgG1*) are also plotted to show the separation of *IgG2* isoforms by ion mobility. Reprinted with permission from Ref. [111]. Copyright 2010 American Chemical Society

In this study, several orthogonal analytical methods were developed to perform the structural assessment both for mAbs and for recombinant target antigen. Among them, IMMS was used to probe the structural features for several recombinant batches of human JAM-A and to ensure that the antigen displays the expected structural characteristics before forming the immune complexes with hzmAb 6F4. The goal of this study was to analyze recombinant antigen batches in order to verify their structural homogeneity, to determine their oligomerization state, and to check that folded conformation was maintained in the experimental conditions used for binding assays. The IMMS technique was initially used to check the purity and homogeneity of two antigen batches prepared either from the soluble fraction (JAM-A SF) or from inclusion bodies (JAM-A IB) of *E. coli* expression system. The nanoESI-TWIMS analysis in denaturing conditions revealed that multiply charged ions of JAM-A IB display lower m/z with longer drift times than those detected for JAM-A SF (Fig. 10.4). This observation indicates that heterogeneity existed between the two samples that were prepared differently. JAM-A IB contained more extended protein conformations which was believed to be caused by the formation of fewer disulfide bonds in the protein, conferring lower mobilities (i.e., longer drift times) to protein ions. The IM results correlated well with the information from the intact mass measurement in the same experiment for the same samples, which showed that the measured mass for JAM-A IB sample was 2 Da higher than that for JAM-A SF. Since the value of the measured mass for JAM-A SF sample (24540.3 $\pm$ 0.5 Da) was in good agreement with the theoretical value calculated from the amino acid sequence (24539.4 Da), the IMMS method provided direct evidence, based on different molecular properties, to demonstrate that absence of a disulfide bond in the preparation of JAM-A IB.

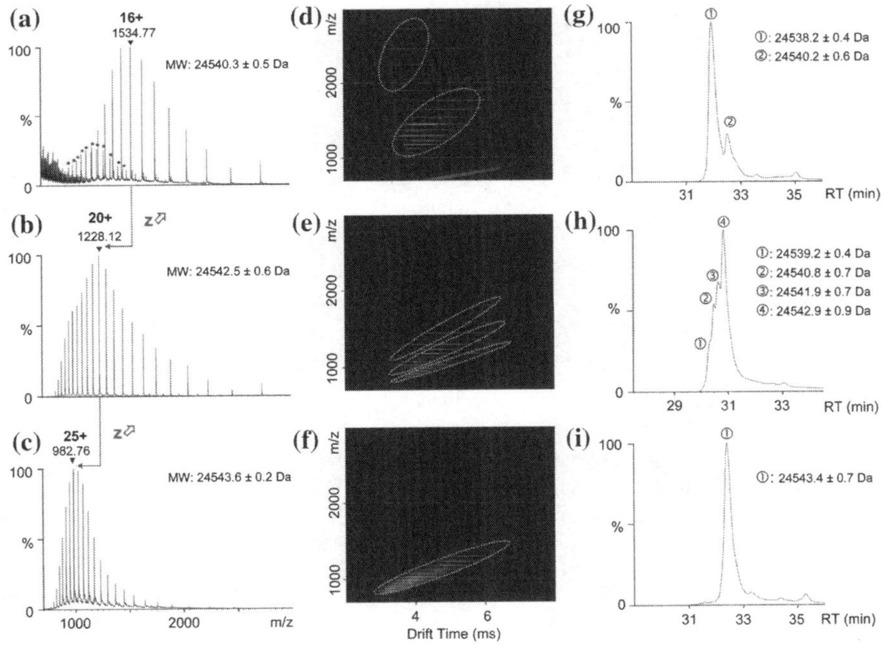

Fig. 10.4 Analysis of antigen JAM-A for sample purity and homogeneity using nanoESI-MS, nanoESI-IMS-MS and LC/MS under denaturing conditions. **a–c** nanoESI-MS analyses of JAM-A SF, JAM-A IB, and DTT-reduced JAM-A IB, respectively. **d–f** nanoESI-TWIMS analyses of JAM-A SF, JAM-A IB, and DTT-reduced JAM-A IB, respectively. **g–i** UV trace from LC–MS analyses of JAM-A SF, JAM-A IB, and DTT-reduced JAM-A IB, respectively. *Asterisk* (*) refers to a co-purified protein (26754.6 $\pm$ 0.9 Da). Circled ions populations correspond to species with different mobility properties. Reprinted with permission from Ref. [112]. Copyright 2009 American Chemical Society

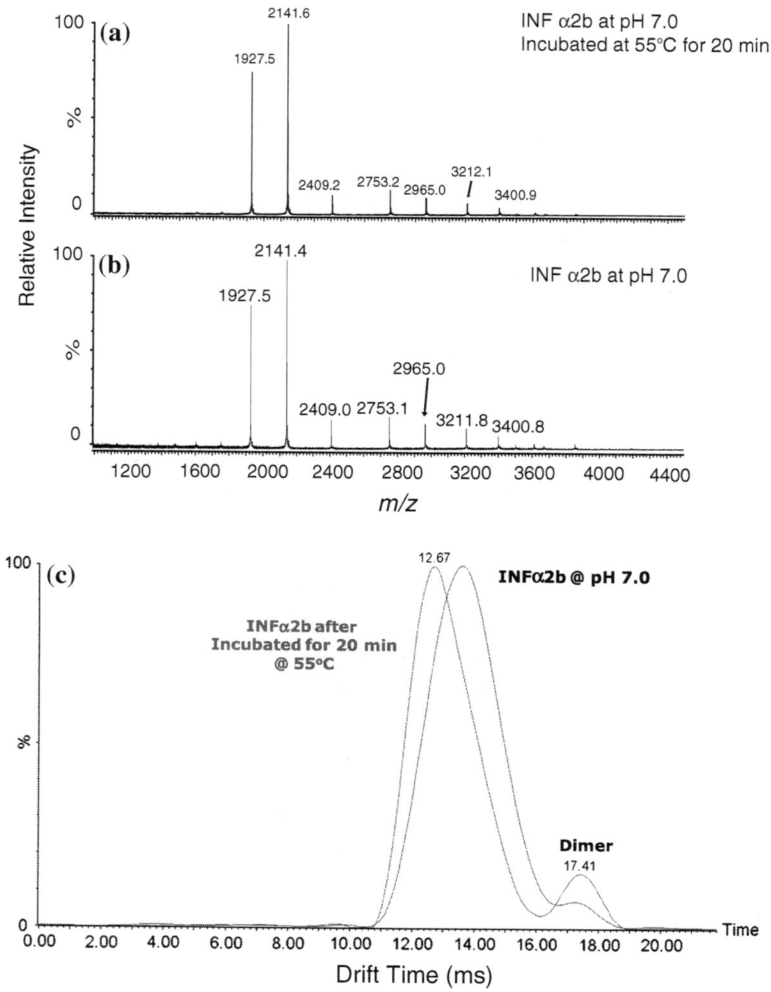

Fig. 10.5 Analysis of thermally stressed interferon α-2b by ion mobility mass spectrometry. **a** Mass spectrum of 10 pmol/μL interferon α-2b from a direct infusion experiment. The sample was dialyzed against 50 mM ammonium acetate, pH 7.0 for 2 h, and then incubated at 55 °C for 20 min before the experiment. **b** Mass spectrum of 10 pmol/μL interferon α-2b from a direct infusion experiment. The sample was dialyzed against 50 mM ammonium acetate, pH 7.0 for 2 h before the analysis. **c** Ion mobility arrival time distribution (*ADT*) of interferon α-2b ions (5+ charge state). The ions from the sample experiencing thermal stress (in *RED*) showed the shift of *ADT* as well as an increase in the dimer species (colour figure for online purpose)

The disulfide bridge pairing heterogeneity observed in denaturing conditions between JAM-A SF and JAM-A IB preparations is likely to induce differences in the native conformation of the protein. The difference, if it exists, would lead to incorrect conclusions about the binding assay between the hzmAb 6F4 and the antigens. nESI-TWIMS experiments were thus performed under non-denaturing conditions to get further insights into the native structure of the antigen batches under study. Their experimental results showed that homogeneous monomeric and dimeric ion populations were detected under non-denaturing conditions for both samples, suggesting both antigen batches were uniformly folded despite their heterogeneous disulfide bond connectivity. Furthermore, the authors estimated the CCS for the monomeric and dimeric species based on the IM measurement and found the CCS values were in good agreement with the CCS data obtained from a computational approximation approach using X-ray structural data. Therefore, the authors concluded the disulfide heterogeneities observed with the samples did not seem to affect the global protein conformation.

Interferon alpha-2 (IFN α-2) is widely prescribed for the treatment for hepatitis B and C as anti-viral drugs. Like many proteins, IFN has a problematic propensity to misfold, which leads to activity loss, aggregation, and increased immunogenic response [78, 79]. This structure loss can be accelerated by a variety of factors, such as chemical modifications, surface binding, exposure to elevated temperatures, lyophilization. Recently, we demonstrated the use of IMMS to detect the conformational changes of interferon protein from a thermal stress experiment [80]. A sample of IFN α-2b treated at 55 °C for 20 min was analyzed by nESI-IMMS from a buffer of 50 mM ammonium acetate, and the results were compared against the data acquired for the interferon samples without experiencing any thermal stress. The extracted ATDs (i.e., driftograms) for the lowest charge state (5+) interferon are shown in Fig. 10.5. Also shown are the mass spectra for the interferon samples with and without thermal stress. Clear shifts in the drift times are shown for the charge state, suggesting the changes of ion conformation after thermal treatment. In addition, an increase in the peak attributed to the IFN dimer (the peak at 17.41 ms) in the diagram is also seen. These results correlated well with the study on interferon protein aggregation under the influence of thermal treatment using orthogonal techniques such as size exclusion chromatography (SEC) and spectroscopy [81]. On the contrary, comparison of the MS spectra for the two samples shows no evidence that the native conformation of IFN was compromised as a result of thermal stress since similar charge state distribution was observed for both samples. Any attempt to deduce the conformational changes based on the direct mass analysis is not possible in this case.

10.3.2 Applications for Aggregation and Complexes Analysis of Protein Therapeutics

Protein aggregation is a common issue encountered during the preparation, formulation, or storage of biotherapeutics. Protein aggregation can be caused by many factors, such as protein misfolding during protein expression, or denaturation during protein purification, or high protein concentration [82]. Protein aggregation has been linked to potential loss of therapeutic efficacy or unwanted immune reactions [78, 79]. For these reasons, the biotechnology industry is under increasing pressure from regulatory bodies to provide detailed information about the quantity and nature of any aggregates present in a biopharmaceutical product.

As a generic term, protein aggregation includes many types of aggregation, from rapidly reversible aggregation caused by non-covalent bonds to irreversible aggregation in the form of covalent oligomers. As a result, no single analytical method can detect all types of aggregates, and an arsenal of methods and techniques are frequently employed for detecting and quantifying protein aggregation. SEC is a well-established method and has been the workhorse technique in aggregation analysis [81, 83–85]. However, the validity of SEC results is often challenged because aggregates can be lost through non-specific binding to the columns [86], and thus not all the aggregate species in a product are represented in the SEC results. For that reason, column (matrix)-free techniques, such as analytical ultracentrifugation (AUC) [87], dynamic light scattering (DLS) [87], and field-flow fractionation (FFF) [88], now find increasing applications in aggregation analysis.

Electrospray ionization differential mobility analysis (ESI-DMA), also known as gas-phase electrophoretic mobility molecular analysis (GEMMA), was recently explored as an alternative method for measuring low-order soluble aggregates of IgG antibodies in solution because of its low sample consumption and short analysis time [89]. The instrument essentially consists of a charge-reduced ESI source to generate macromolecular ions in the gas phase, a differential mobility analyzer (DMA) to measure gas-phase electrophoretic mobility diameter (EMD), and a condensation particle counter (CPC) as the detector [90, 91] to measure the number concentration of particles in the gas phase. The direct output of the instrument is a size distribution based on IMS separation, and the size spectrum can be converted to a mass spectrum, due to the strong correlation between mobility size and molecular weight [91].

When applied to the analysis of protein aggregates of IgGs, the method [89] can measure a size distribution of protein species present from 3 to 250 nm. These particles correspond to a mass range of 8 kDa (monomer of insulin) to 80 MDa (whole cell organelles). Distinct resolution for IgG aggregates from monomers to pentamer was achieved, and the sizes of the IgG and its aggregates measured by DMA were found to be in good agreement with those calculated from simple models, based on structural coordinates of IgG from protein crystallographic data (see the references cited therein). This finding corroborates the results from a separate study, where the protein sizes measured by ES-DMA method were proven to be comparable to the values from established bioanalytical techniques such as multi-angle laser light scattering (MALLS) and quasi-elastic light scattering (QELS). The benefits of ES-DMA method is that the method provides direct and quantitative measure of aggregate distributions quickly (<1 h) and is well suited for studying early stages of aggregation. However, the technique is limited by its narrow concentration range that it can measure and the inability to examine proteins in formulation buffers (because of the suppression caused by nonvolatile salts in the formulation buffer).

The same GEMMA technique was also applied to confirm the binding and bridging of antibodies to their antigens and to determine the formation of stable non-covalent complexes [92]. A single-chain variant of a monoclonal antibody (scFv), a monoclonal IgG, antigens as well as the complexes formed between the binding pairs was analyzed by IMS sequentially. When converted into a mass scale, the analysis covers a mass range from ~25 to ~500 kDa. The mass profile generated by the IMS technique provides detailed information of the nature of the binding in terms of stoichiometry. Although this information is not directly from a study with physiological buffer, which is preferred for determining binding kinetics, it can be used to quickly assess whether or not binding will occur and determine the stoichiometry of the binding interaction.

In an effort to assess the utility of IMMS as a sensitive analytical tool to probe the higher-order structures of therapeutic proteins, a number of human insulin analogs on the market were analyzed recently by the use of Synapt™ G2 HDMS [80]. The purpose of the experiment was to compare the CCS values measured for the insulin samples and identify the tertiary structure differences, if any, of the insulin produced by different manufacturers. The recombinant biotherapeutics were analyzed under non-denaturing conditions, and the CCS for individual charge states of both insulin monomer and dimer were determined. Measured CCS values were found in very good agreement with the calculated values for solid-state and solution-phase structures based on "projection approximation" model in MOBCAL [32]. In comparison with the CCS values obtained from traditional drift tube IM instruments, the CCS values from TWIMS exhibit a typical variance of approximately 2.5 %. The corroboration provided by the orthogonal analytical techniques (theoretical calculation and the CCS values from LDT IMS) suggests that the IMMS techniques offer an effective means to monitor the three-dimensional structures of insulin. Figure 10.6 shows the driftograms of six

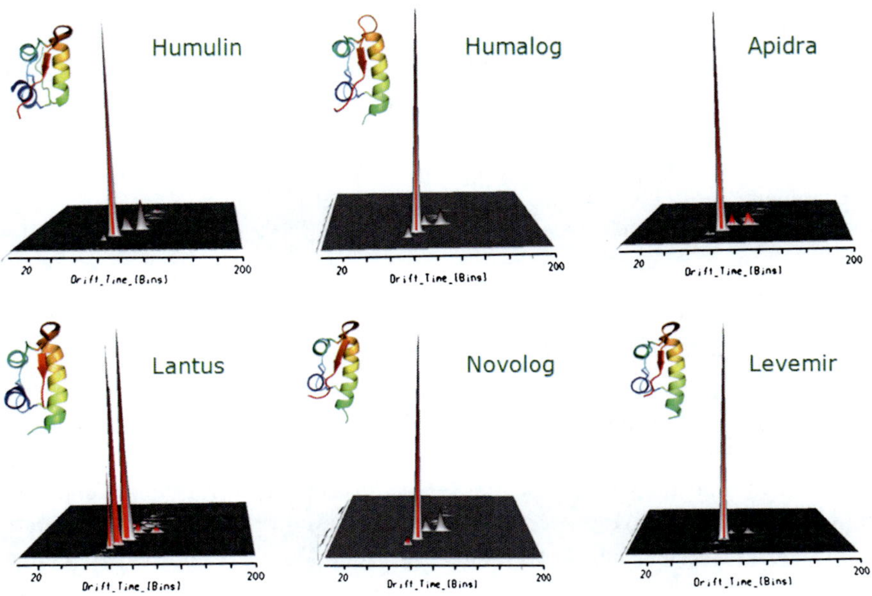

Fig. 10.6 Ion mobility analysis of insulin analogs using a Synapt G2 HDMS system. Six insulin analog products were analyzed by a direct infusion method after the samples were dialyzed against a solution of 50 mM ammonium acetate, pH 7.0. The concentration of the samples was estimated to be about 5 pmol/μL. The two small peaks behind the highest peak in each diagram are from either monomer (3+) or dimer (6+) of insulin

insulin analogs from different manufactures. The CCS values for monomeric and dimeric species for all the insulin samples were compared. Although little difference in measured CCS values among all the monomer species for insulin analogs were found, the data clearly demonstrate the power of the IMMS technique to profile the insulin oligomer populations that coexist in the solution and to examine each species at each individual state without any concern for the interferences caused by the presences of other species. For instance, closer examination of Fig. 10.6 showed that there is great variation in the concentration of dimers for the sample analyzed.

Because of the intrinsic nature, all measurements related to protein conformation by IMMS are performed in gas-phase environments. The CCS value from IMMS is modeled by the rotationally averaged CCS of proteins. Although strong evidence does exist, suggesting that ESI (to some extent MALDI) can produce ions that retain key aspects of solution structure, and thus the experimental CCSs of these molecules should have some relationship to crystal structure coordinates, care should be taken when interpreting results from IMMS measurements, and in particular correlation with measurements from crystallography or NMR spectroscopy. The influence of solvent molecules on the proteins structures should not be ignored when performing CCS calculations using theoretical modeling—the extent of correlations is difficult to gauge currently. Perhaps a sensible approach is first to adjust the crystal or NMR coordinates via molecular dynamics to factor out the impact imposed by the solvent molecules prior to calculation of the gas-phase CCS for comparison with gas-phase experimental CCS values. However, these challenges should not diminish the utility of IMMS as a technique for isolated proteins or protein complexes to provide both m/z values and information about the gas-phase conformation(s) for each m/z species, especially when the conformation information is sought for analytical support of biotherapeutic drug development.

10.3.3 Differentiation of Isobaric Species for Protein Therapeutics Characterization

Although the application of IMMS for rapid analysis of conformation adopted by biological molecules in gas phases has been an active research area for many years, the introduction of commercial IMMS instruments drives this technology from research into solving specific problems in routine analytical tasks. Many studies have focused as much on mixture separation as on structural measurement, reflecting the growing power of IMS as a unique separation technique. In this section, we present several examples to showcase the utility of IMS as an analytical tool for differentiating isobaric species that are relevant to biopharmaceutical characterization. Moreover, the use of IMS to separate complex mixtures of ions for mass spectral simplification is also discussed.

10.3.3.1 Differentiation of Isobaric Deamidated Peptides by IMMS

Deamidation of asparagine residues is one of the most common PTMs in recombinant therapeutic proteins [93]. It represents a non-enzymatic process in which an asparagine residue is converted into either aspartic acid (Asp) and/or isoaspartic acid (isoAsp). The deamidation process is generally regarded as a chemical degradation pathway and plays an important role in protein folding studies, including antibody-based therapeutics. Deamidation of therapeutic proteins may result in the loss of activity of the protein therapeutic or even trigger immunogenicity because of a change in tertiary structure. Accordingly, it is important to establish methods for characterizing the sites of deamidation as well as for evaluating the effect on biological activity and antigenicity.

With the employment of modern high-resolution mass spectrometers, a deamidation event (Asn $\rightarrow$ Asp) would be readily recognized, as it elicits a mass shift of +0.985 Da. However, since the resultant Asp and isoAsp residues share the identical molecular weight (isobaric isomers), structural differentiation of Asp and isoAsp products is inherently difficult. Mass measurement alone fails to resolve these two isomers, and even traditional CID experiments are not sufficient to differentiate these two isomers. Since the formation of isoAsp inserts a methylene group into the protein backbone, resulting in a beta-peptide linkage, and thus may significantly alter protein structure and function, there is strong interest in characterizing isoAsp in both biological research and pharmaceutical discovery [94, 95].

Currently, there are several existing methods for distinguishing isoaspartic acid from aspartic acid. These include NMR [96], Edman sequencing methods [97], antibody-based detection methods [98], and HPLC methods [99]. Due to the subtle differences in physical properties between Asp and isoAsp, effective separation and identification of aspartate-/isoaspartate-containing proteins and peptides by those methods are often tedious and laborious and require relatively large concentrations of peptides to generate differentiable signals. We recently introduced a rapid method for the differentiation of aspartate-/isoaspartate-containing peptides using IMMS. The peptide, FYPSDIAVEWESNGQPENNYK, also known as the "PENNYK" peptide, can be found in the tryptic digest of all humanized mAbs. It derives from the constant region (Fc) of humanized mAb and is known to be susceptible to deamidation in various experimental conditions. There are several potential deamidation sites in the peptide, and differentiation of these sites as well as identification of Asp from isoAsp for each site has always been an analytical challenge. In our experiment, deamidated PENNYK peptide isomers that have identical amino acid sequences but only differ at the deamidated amino acid residue (Asp vs. isoAsp) were synthesized and individually analyzed by a direct infusion experiment using a commercial IMMS system (Fig. 10.7a, e). The precursor ion with m/z 830.0 (MH_3^{3+}) was selected to undergo CID fragmentation. Fragment ions were subjected to IM separation to differentiate the presence of the Asp or isoAsp residue in the peptides. In an effort to make the method applicable for PENNYK peptides containing different deamidation sites, the IM separation was performed on the fragment ions instead of the precursors so the deamidation site could be simultaneously determined in a single experiment.

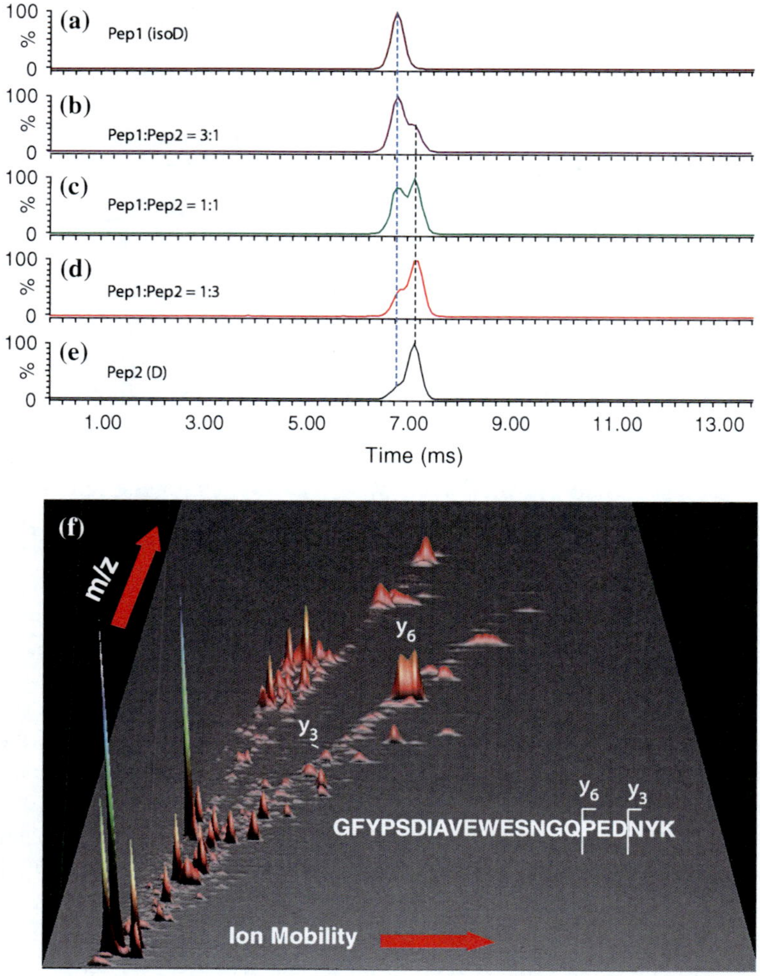

Fig. 10.7 Differentiation of deamidated PENNYK peptides that contain either isoaspartic (*Pep1*) or aspartic (*Pep2*) acid amino acid residue. **a** Synthetic PENNYK peptide containing only isoaspartic acid residue (*Pep1*). **b** A peptide mixture containing both *Pep1* and *Pep 2* at a ratio of 3:1. **c** A peptide mixture containing both *Pep1* and *Pep 2* at a ratio of 1:1. **d** A peptide mixture containing both *Pep1* and *Pep 2* at a ratio of 1:3. **e** Synthetic PENNYK peptide containing only aspartic acid residue (*Pep2*). **f** Ion mobility separation of fragment ions from the analysis of peptide mixture (*Pep1*:*Pep2* = 1:1). The resolving peaks of y_6 at the ion mobility dimension show the deamidated peptides containing both aspartic acid and isoaspartic acid residues

Figure 10.7b–d, f depict the IMMS results from an infusion experiment where binary mixtures of the peptides [GFYPSDIAVEWESNGQPEisoDNYK (Pep1) vs. GFYPSDIAVEWESNGQPEDNYK (Pep2)] at a ratio of 3:1, 1:1, and 1:3 were introduced into the system, respectively. The MS/MS spectrum contains a series of b,y-ion peaks from which the deamidation site can be deduced. In addition, it is of particular note that we observe two partially resolved peaks in the dimension of ion mobility ATDs for the fragment ion (y_6) for each of the binary peptide mixtures, and the intensity of each y_6 ion peaks (containing either Asp or isoAsp) varies according to the concentration of precursor peptides. Under identical instrumental conditions, IM arrival times are affected by ion structure (as indicated by CCS), mass, and charge [17]. Since both peptide isomers have identical mass and charge and were analyzed under identical conditions, these ATD differences are a clear indicator of their difference in molecular structure. In this case, the structural difference is caused by a single amino acid residue variation.

10.3.3.2 Analysis of Structural and Positional Isomers of Glycans

Glycosylation plays a vital role in stability, biodisposition, in vivo activity, solubility, serum half-life, and immunogenicity of therapeutic protein drugs and can affect efficacy, folding, target binding, and pharmacokinetic properties. In addition, glycosylation in recombinant proteins differs widely with cell culture parameters. As a result, it is important to accurately characterize the carbohydrate moieties of biotherapeutic proteins to demonstrate control of the manufacturing process.

A major challenge associated with the characterization of glycoproteins is the structural complexity and heterogeneity of their glycan moieties. Biological oligosaccharides frequently exist as sets of isomers that are identical in the number and type of carbohydrate monomers but differ in anomeric configurations, or glycosidic linkage, or connection sequence. Comprehensive characterization of the carbohydrate moiety of glycoproteins involves the resolution of the extensive presence of those isobaric species. Without prior chromatographic or electrophoretic separation, MS faces a number of distinct challenges in resolving isomeric structures, and typically extra steps such as sample derivatization are required for confident analysis.

The analysis of isobaric oligosaccharide structures by IMMS has been demonstrated by different IMMS combinations. In a broad sense, the published work can be generally grouped into two categories based on the nature of the carbohydrate structures: studies for differentiating isomers with different constitutional types and studies for analyzing spatial isomers comprised of the same number and type of carbohydrate monomers. The latter can be further divided between structural isomers and positional isomers. Positional isomers refer to the carbohydrate structures with the same number, type, sequence, and anomeric carbons of monomer units but differ in the specific glycosidic linkage arrangement (e.g., $1 \rightarrow 3$ vs. $1 \rightarrow 6$). On the other hand, structural isomers comprise identical carbohydrate monomeric units, but are arranged in different branching patterns.

The separation of positional isomers is probably more challenging than the differentiation of other carbohydrate isomers simply because the structure difference is rather small for this type of carbohydrate isomer. Using electrospray ionization-atmospheric pressure ion mobility time-of-flight mass spectrometry (a LTD IMS device), Clowers et al. [100] separated a series of disaccharide alditol positional isomers, some of which are derived from O-linked glycoproteins. Unfortunately, the sensitivity of the analysis is poor, requiring hundreds of pmol of material, mainly due to the low duty cycles of the instrument caused by the ion gating mechanism of LDT device. Similarly, the possibility of separating positional isomers of disaccharides was also shown using a FAIMS-type ion mobility separator [101]. However, the time needed to achieve the separation using FAIMS was relatively long (about 1 h) in comparison with LDT IMS approach (~5 min). In addition, the signal-to-noise ratio was low and multiple peaks corresponding to cluster dissociation were also seen.

Several studies have been reported to use IMMS to analyze positional carbohydrates isomers ranging from tri- to hexa-saccharides [102–104]. The final resolving power depends on the magnitude of differences in the glycosidic connectivity between isomers and the type of metal ions used for ionizing oligosaccharides. When only partial separation is achieved, the unique dual collision cell design associated with the instrument enhances the confidence in carbohydrate identification by fragmenting the carbohydrates either pre- and/or post-IM. The studies also suggest that the capability of the instrument to differentiate positional isomers is limited by the mass (size) with an upper limit of ~1,000 m/z [103] at the current IMS resolution.

The possibility of distinguishing structural isomers of protein-derived glycan was recently explored [104–106]. In favorable circumstances, structural isomeric glycan can be separated and characterized using IMMS approaches. For example, using TWIMS-MS, Williams et al. [105] successfully separated two structural isomers of $GlcNAc_1Man_3GlcNAc_2$, and proved that the two ATD peaks observed in driftogram indeed originated from the structural differences due to the capping GlcNAc group bound to the different branching arms of the glycan structures (1–3 branch vs. 1–6 branch, Fig. 10.8). In a separate report, Plasencia et al. [106] performed the analysis of N-linked glycans enzymatically released from ovalbumin using an LDT IMS-TOF/MS approach. The use of LDT IMS allowed them to achieve a higher resolution in IMS separation and obtain three sharp peaks in the ATD for the glycan ion with a composition of $[H_5N_4{+}2Na^+]^{2+}$. Through molecular modeling, the glycan structures corresponding to each of the peaks were assigned, and the relative ratios of the three structural isomers were determined. Together with the analysis for carbohydrate structural isomers from natural sources (e.g., milk) [103], these studies demonstrate that IMMS offers a number of advantages for the analysis of a complex mixture of released glycans, including short analysis time, minimum sample preparation, and the high information content of the experiment. However, while promising, the methodology for detailed characterization of glycans is clearly at an early stage of development. Care should be taken when assigning IMS peaks to glycan structures because a single isomer may exist as multiple conformations [107] and IMS separation may show multiple features for individual isomers.

The use of IMMS as a multi-dimensional separations tool to extract the spectra of N-glycans released for recombinant glycoprotein, gp120 (from the human immunodeficiency virus), was recently demonstrated by Harvey et al. [108]. The study took advantage of the IMS separation to extract glycan spectra that are obscured by the dominant signals from other abundant compounds such as detergents and residual buffer salts. Because the N-glycans signals fell into a unique region on the driftogram (ion mobility vs. m/z plot) after IMS separation, signals from other high-abundance

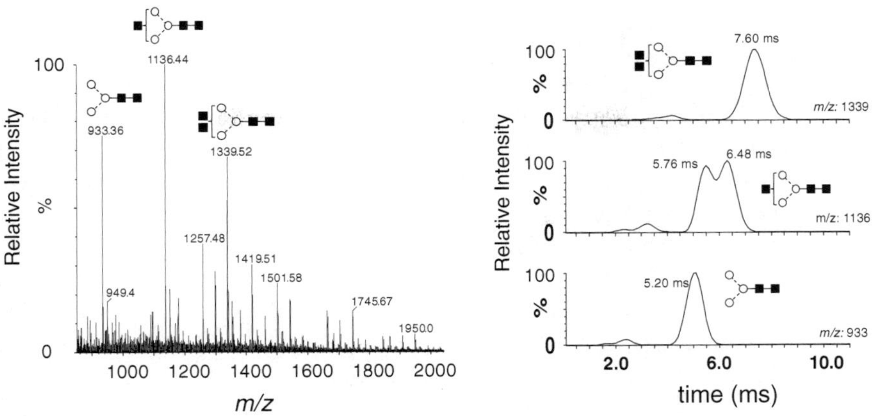

Fig. 10.8 Analysis of free glycans released from chicken ovalbumin using the Synapt HDMS system. The peaks for sodiated $GlcNAc_nMan_3GlcNAc_2$ (n = 0, 1, 2) are labeled with their respective structures. The corresponding TWIMS ATDs for these mass spectrum peaks are shown at *right*. Separation of structural isomers of $GlcNAc_1Man_3GlcNAc_2$ is shown in the *middle* of the *right panel*. Reprinted from Ref. [113]. Copyright (2009) with permission from Elsevier

interfering species can be filtered out and the glycan profiles are extracted with good signal-to-noise ratios. With this method, the glycans released from minute amount (sub-microgram) of glycoprotein material can be directly analyzed with little sample preparation, and a clear glycan profile can be obtained. In addition, structural analysis could be accomplished by MS/MS experiments after IMS separation.

10.4 Summary and Perspective

The inherent structural complexity and heterogeneity associated with therapeutic proteins require the deployment of many different analytical techniques to provide multi-faceted detailed characterization. The coupling of ion mobility separation with MS has created a powerful analytical technique that leads to important and revealing insights into protein analysis. More specifically, from the plethora of studies described in this chapter, IMMS is evidently a valuable addition to the toolbox for therapeutic protein characterization. Data obtained from IMMS experiments allow the identification and analysis of coexisting conformational states of proteins, observation of protein aggregates or conformational change in the absence of solvent, and the analysis of isobaric species that are difficult to handle by MS alone. Its ability to complement other structural methods in specific applications makes IMMS a powerful new tool in therapeutic protein characterization.

The development of this powerful hybrid technique and its application for biotherapeutic protein analysis has come a long way, but has much further to go in order to be fully adapted and incorporated into the analytical workflow for protein therapeutics development. Since IMMS essentially measures the protein conformation in gas phases, the protein conformation information derived from IMMS measurement has long been under critical assessment when effort is made to correlate the measurement to protein conformation in solution. Although gas-phase studies can eliminate the effects of solvent molecules, thus enabling the analysis of the intrinsic properties of the protein and providing invaluable insights into the structure and cross sections of well-defined (mass-selected) protein or protein complexes, a major concern with the approach is that the absence of solvent can affect the structure of the protein or complex in various ways such that the cross section obtained by IMMS may not necessarily resemble the cross section of the structure in solution. On the other hand, strong evidence from several IMMS studies [14, 15, 109, 110] has suggested protein structure is often largely retained upon transfer into the gas phase, and thus biological relevance is self-evident in the measurement. Therefore, more detailed fundamental studies are warrant to further validate the approach for its relevance to biomolecule conformation study.

The applicability of IMMS in biotherapeutic protein characterization could benefit greatly from the continued improvement in IM separation resolution. Currently, most of the commercial IMS techniques only offer a relatively low- to moderate-capability in structural resolution. The limited IMS resolution achieved by the available instrument offerings suggests that the techniques measure the average shape of an ensemble of ions of interest. Although some conclusions could still be drawn based on the distribution of drift times regarding the structural diversity of an ensemble, or the kinetics of interconverting conformational states, further enhancement in IMS resolution would certainly make the technique more acceptable to the biopharmaceutical industry and enable scientists to develop powerful new methods to meet the challenges faced by the ever evolving industry.

Acknowledgments The authors would like to thank Dr. St. John Skilton for critical review of the manuscript.

References

1. Chen G, Warrack BM, Goodenough AK, Wei H, Wang-Iverson DB, Tymiak AA (2011) Characterization of protein therapeutics by mass spectrometry: recent developments and future directions. Drug Discovery Today 16:58–63
2. Hernandez H, Robinson CV (2007) Determining the stoichiometry and interactions of macromolecular assemblies from mass spectrometry. Nat Protoc 2:715–726
3. Heck A (2008) Native mass spectrometry: a bridge between interactomics and structural biology. Nat Methods 5:927–933
4. Yin S, Loo JA (2009) Mass spectrometry detection and characterization of non-covalent protein complexes. Methods Mol Biol 492:273–282
5. Kaltashov IA, Abzalimov RR (2008) Do ionic charges in ESI MS provide useful information on macromolecular structure? J Am Soc Mass Spectrom 19:1239–1246
6. Damen CWN, Chen W, Chakraborty AB, Oosterhout MV, Mazzeo JR, Gebler JC, Schellens JHM, Rosing H, Beijnen JH (2009) Electrospray ionization quadrupole ion-mobility time-of-flight mass spectrometry as a tool to distinguish the lot-to-lot heterogeneity in n-glycosylation profile of the therapeutic monoclonal antibody Trastuzumab. J Am Soc Mass Spectrom 20:2021–2033

7. Bagal D, Zhang H, Schnier PD (2008) Gas-phase proton-transfer chemistry coupled with TOF mass spectrometry and ion mobility-MS for the facile analysis of poly(ethylene glycols) and pegylated polypeptide conjugates. Anal Chem 80:2408–2418
8. Engen JR (2009) Analysis of protein conformation and dynamics by hydrogen/deuterium exchange MS. Anal Chem 81:7870–7875
9. Hambly DM, Gross ML (2007) Chemical footprinting of proteins: •OH reactions. In: Gross ML, Caprioli R (eds) The encyclopedia of mass spectrometry, vol 6. Elsevier, Amsterdam
10. Hill HH, Siems WF, St. Louis RH, McMinn DG (1990) Ion mobility spectrometry. Anal Chem 62:A1201–A1209
11. Chen YH, Hill HH, Wittmer DP (1996) Thermal effects on electrospray ionization ion mobility spectrometry. Int J Mass Spectrom Ion Process 154:1–2
12. Sanders TM, Forrest SR (1989) Small particle-size distributions from mobility measurements. J Appl Phys 66:3317–3323
13. Whitby KT, Clark WE (1966) Electronic aerosol particle counting and size distribution measuring system for 0.015 to 1 mu size range. Tellus 18:573–582
14. Ruotolo BT, Giles K, Campuzano I, Sandercock AM, Bateman RH, Robinson CV (2005) Evidence for macromolecular protein rings in the absence of bulk water. Science 310:1658–1661
15. Liu L, Bagal D, Kitova EN, Schnier PD, Klassen JS (2009) Hydrophobic protein-ligand interactions preserved in the gas phase. J Am Chem Soc 131:15980–15981
16. Kaddis CS, Loo JA (2007) Native protein MS and ion mobility: large flying proteins with ESI. Anal Chem 79:1718–1784
17. Eiceman GA, Karpas Z (2005) Ion mobility spectrometry (2nd ed). CRC Press, Boca Raton, FL
18. Langevin P (1903) L'Ionistion des Gaz. Ann de Chim Phys 28:289–384
19. Langevin P (1905) Une Formule Fondamentale de Théorie Cinétique. Ann de Chim et de Phys 5:245–288
20. Van de Graaff RJ (1929) Mobility of ions in gases. Nature 124:10–11
21. Cravath AM (1929) The rate of formation of negative ions by electron attachment. Phys Rev 33:605–613
22. Bradbury NE, Nielsen RA (1936) Absolute values of the electron mobility in hydrogen. Phys Rev 49:388–393
23. Lovelock JE (1981) The Electron-capture detector-a personal odyssey. In: Zlatkis A, Poole CF (ed) Electron Capture, Elsevier, Amsterdam
24. McDaniel EW, Martin DW, Barnes WS (1962) Drift-tube mass spectrometer for studies of low-energy ion-molecule reactions. Rev Sci Instrum 33:2–7
25. Kebarle P, Hogg AM (1965) Mass-spectrometric study of ions at near atmospheric pressures. I. The Ionic Polymerization of Ethylene. J Chem Phys 42:668–674
26. McDaniel EW (1964) Collisional phenomena in ionized gases. Wiley, New York
27. Cohen MJ, Karasek FW (1970) Plasma chromatography-a new dimension for gas chromatography and mass spectrometry. J Chromatogr Sci 8:330–337
28. Shelimov KB, Clemmer DE, Hudgins RR, Jarrold MF (1997) Protein structure in vacuo: gas-phase conformations of BPTI and cytochrome C. J Am Chem Soc 119:2240–2248
29. Badman ER, Hoaglund-Hyzer CS, Clemmer DE (2001) Monitoring structural changes of proteins in an ion trap over ~10–200 ms: unfolding transitions in cytochrome c ions. Anal Chem 73:6000–6007
30. Valentine SJ, Clemmer DE (2002) Temperature-dependent H/D exchange of compact and elongated cytochrome c ions in the gas phase. J Am Soc Mass Spectrom 13:506–517
31. Gidden J, Ferzoco A, Baker ES, Bowers MT (2004) Duplex formation and the onset of helicity in poly d(cg)n oligonucleotides in a solvent-free environment. J Am Chem Soc 126:15132–15140
32. Mesleh MF, Hunter JM, Shvartsburg AA, Schatz GC, Jarrold MF (1996) Structural information from ion mobility measurements: effects of the long-range potential. J Phys Chem 100:16082–16086
33. Wyttenbach T, von Helden G, Batka JJ, Carlat D, Bowers MT (1997) Effect of the long-range potential on ion mobility measurements. J Am Chem Soc 8:275–282
34. Bluhm BK, Gillig KJ, Russell DH (2000) Development of a Fourier-transform ion cyclotron mass spectrometer-ion mobility spectrometer. Rev Sci Instrum 71:4078–4086

35. Lawrence AH, Barbour RJ, Sutcliffe R (1991) Identification of wood species by ion mobility spectrometry. Anal Chem 63:1217–1221
36. Valentine SJ, Clemmer DE (1997) H/D exchange levels of shape-resolved cytochrome c conformers in the gas phase. J Am Chem Soc 119:3558–3566
37. Creaser CS, Benyezzar M, Griffiths JR, Stygall JW (2000) A tandem ion trap/ion mobility spectrometer. Anal Chem 72:2724–2729
38. Clowers BH, Hill HH (2005) Mass analysis of mobility-selected ion populations using dual gate, ion mobility, quadrupole ion trap mass spectrometry. Anal Chem 77:5877–5885
39. Hoaglund CS, Valentine SJ, Sporleder CR, Reilly JP, Clemmer DE (1998) Three-dimensional ion mobility TOFMS analysis of electrosprayed biomolecules. Anal Chem 70:2236–2242
40. Mason EA, McDaniel EW (1988) Transport properties of ions in gases. Wiley, New York
41. Collins DC, Lee ML (2002) Developments in ion mobility spectrometry-mass spectrometry. Anal Bioanal Chem 372:66–73
42. Wu C, Siems WF, Asbury GR, Hill HH (1998) Electrospray ionization high-resolution ion mobility spectrometry: mass spectrometry. Anal Chem 70:4929–4938
43. Giles K, Pringle S, Worthington KR, Little D, Wildgoose JL, Bateman RH (2004) Applications of a travelling wave-based radio-frequency only stacked ring ion guide. Rapid Commun Mass Spectrom 18:2401–2414
44. Pringle SD, Kevin G, Wildgoose JL, Williams JP, Slade SE, Konstantinos T, Bateman RH, Bowers MT, Scrivens JH (2007) An investigation of the mobility separation of some peptide and protein ions using a new hybrid quadrupole/traveling wave IMS/oa-ToF instrument. Int J Mass Spectrom 261:1–12
45. Giles K, Williams JP, Campuzano I (2011) Enhancements in travelling wave ion mobility resolution. Rapid Commun Mass Spectrom 25:1559–1566
46. Shvartsburg AA, Smith RD (2008) Fundamentals of traveling wave ion mobility spectrometry. Anal Chem 80:9689–9699
47. Smith DP, Knapman TW, Campuzano I, Malham RW, Berryman JT, Radford SE, Ashcroft AE (2009) Deciphering drift time measurements from travelling wave ion mobility spectrometry-mass spectrometry. Euro J Mass Spectrom 15:113–130
48. Knapmana TW, Berrymana JT, Campuzano I, Harris SA, Ashcroft AE (2010) Considerations in experimental and theoretical collision cross-section measurements of small molecules using travelling wave ion mobility spectrometry-mass spectrometry. Int J Mass Spectrom 298:17–23
49. Guevremont R (2004) High-field asymmetric waveform ion mobility spectrometry: a new tool for mass spectrometry. J Chrom A 1058:3–19
50. Buryakov IA, Krylov EV, Nazarov EG, Rasulev UK (1993) A new method of separation of multi-atomic ions by mobility at atmospheric pressure using a high-frequency amplitude-asymmetric strong electric field. Int J Mass Spectrom Ion Processes 128:143–148
51. Purves RW, Guevremont R, Day S, Pipich CW, Matyjaszczyk MS (1998) Mass spectrometric characterization of a high-field asymmetric waveform ion mobility spectrometer. Rev Sci Instrum 69:4094–4105
52. Kolakowski BM, Mester Z (2007) Review of applications of high-field asymmetric waveform ion mobility spectrometry (FAIMS) and differential mobility spectrometry (DMS). Analyst 132:842–864
53. Barnes WS, Martin DW (1961) Mass spectrographic identification of the ion observed in hydrogen mobility experiments. Phys Rev Lett 6:110–111
54. McAfee KBJ, Edelson D (1963) Identification and mobility of ions in a townsend discharge by time-resolved mass spectrometry. Proc Phys Soc 81:382–384 (London)
55. Wyttenbach T, von Helden G, Bowers MT (1996) Gas-phase conformation of biological molecules: Bradykinin. J Am Chem Soc 118:8355–8364
56. Clemmer DE, Hudgins RR, Jarrold MF (1995) Naked protein conformations: cytochrome c in the gas phase. J Am Chem Soc 117:10141–10142
57. Valentine SJ, Anderson JG, Ellington AD, Clemmer DE (1997) Disulfide-intact and -reduced lysozyme in the gas phase: conformations and pathways of folding and unfolding. J Phys Chem B 101:3891–3900

58. Barrera NP, Di Bartolo N, Booth PJ, Robinson CV (2008) Micelles protect membrane complexes from solution to vacuum. Science 321:243–246
59. Zhou M, Sandercock AM, Fraser CS, Ridlova G, Stephens E, Schenauer MR, Yokoi-Fong T, Barsky D, Leary JA, Hershey JW, Doudna JA, Robinson CV (2008) Mass spectrometry reveals modularity and a complete subunit interaction map of the eukaryotic translation factor eIF3. Proc Natl Acad Sci USA 105:18139–18144
60. Barrera NP, Isaacson SC, Zhou M, Bavro VN, Welch A, Schaedler TA, Seeger MA, Miguel RN, Korkhov VM, van Veen HW, Venter H, Walmsley AR, Tate CG, Robinson CV (2009) Mass spectrometry of membrane transporters reveals subunit stoichiometry and interactions. Nat Methods 6:585–587
61. Hoaglund CS, Valentine SJ, Clemmer DE (1997) An ion trap interface for ESI-ion mobility experiments. Anal Chem 69:4156–4161
62. Albritton DL, Miller TM, Martin DW, McDaniel EW (1968) Mobilities of mass-identified ions in hydrogen. Phys Rev 171:94–102
63. Kanu AB, Dwivedi P, Tam M, Matz L, Hill HH (2008) Ion mobility–mass spectrometry. J Mass Spectrom 43:1–22
64. Kemper PR, Dupuis NF, Bowers MT (2009) A new higher resolution ion mobility mass spectrometer. Int J Mass Spectrom 287:46–57
65. Myung S, Lee YJ, Moon MH, Taraszka J, Sowell R, Koeniger S, Hilderbrand AE, Valentine SJ, Cherbas L, Cherbas P, Kaufmann TC, Miller DF, Mechref Y, Novotny MV, Ewing MA, Sporleder CR, Clemmer DE (2003) Development of high-sensitivity ion trap ion mobility spectrometry time-of-flight techniques: a high-throughput nano-LC-IMS-TOF separation of peptides arising from a Drosophila protein extract. Anal Chem 75:5137–5145
66. Wyttenbach T, Kemper PR, Bowers MT (2001) Design of a new electrospray ion mobility mass spectrometer. Int J Mass Spectrom 212:13–23
67. McLean JA, Ruotolo BT, Gillig KJ, Russell DH (2005) Ion mobility-mass spectrometry: a new paradigm for proteomics. Int J Mass Spectrom 240:301–305
68. Gillig KJ, Ruotolo BT, Stone EG, Russell DH (2004) An electrostatic ion guide for ion mobility-mass spectrometry. Int J Mass Spectrom 239:43–49
69. Woods AS, Ugarov M, Egan T, Koomen J, Gillig KJ, Fuhrer K, Gonin M, Schultz JA (2004) Lipid/peptide/nucleotide separation with MALDI-Ion mobility-TOF MS. Anal Chem 76:2187–2195
70. Tang K, Shvartsburg AA, Lee HN, Prior DC, Buschbach MA, Li F, Tolmachev AV, Anderson GA, Smith RD (2005) High-sensitivity ion mobility spectrometry/mass spectrometry using electrodynamic ion funnel interfaces. Anal Chem 77:3330–3339
71. Merenbloom SI, Koeniger SL, Valentine SJ, Plasencia MD, Clemmer DE (2006) IMS–IMS and IMS–IMS–IMS/MS for separating peptide and protein fragment ions. Anal Chem 78:2802–2809
72. Wilkins CL, Trimpin S (eds) (2011) Ion mobility spectrometry-mass spectrometry: theory and applications. CRC Press, New York
73. Wilm M, Mann M (1996) Analytical properties of the nano-electrospray ion source. Anal Chem 68:1–8
74. Sobott F, Hernandez H, McCammon MG, McCammon MG, Tito MA, Robinson CV (2002) A tandem mass spectrometer for improved transmission and analysis of large macromolecular assemblies. Anal Chem 74:1402–1407
75. Campuzano I, Giles K (2011) Nanospray ion mobility mass spectrometry of selected high mass species. In: Steven AT, Weil RJ (ed) Nanoproteomics: methods and protocols, Methods in Molecular Biology, vol 790 (Part 2). Springer Science, pp 57–70
76. Bagal D, Valliere-Douglass JF, Balland A, Schnier PD (2010) Resolving disulfide structural isoforms of IgG2 monoclonal antibodies by ion mobility mass spectrometry. Anal Chem 82:6751–6755
77. Atmanene C, Wagner-Rousset E, Malissard M, Chol B, Robert A, Corvaia N, Dorsselaer AV, Beck A, Sanglier-Cianferani S (2009) Extending mass spectrometry contribution to therapeutic

monoclonal antibody lead optimization: characterization of immune complexes using non-covalent ESI-MS. Anal Chem 81:6364–6373
78. Braun A, Kwee L, Labow MA, Alsenz J (1997) Protein aggregates seem to play a key role among the parameters influencing the antigenicity of interferon Alpha (IFN-α) in normal and transgenic mice. Pharm Res 14:1472–1487
79. Hermeling S, Aranha L, Damen JMA, Slijper M, Schellekens H, Crommelin DJA, Jiskoot W (2005) Structural characterization and immunogenicity in wild-type and immune tolerant mice of degraded recombinant human interferon Alpha2b. Pharm Res 22:1997–2006
80. Chen W, Chakraborty A, Skilton S, Berger S, Mazzeo J (2010) Characterizing biotherapeutic protein 3D structures by electrospray ion-mobility mass spectrometry: biological significance and comparison with X-ray crystallography and NMR measurements. Paper presented at the 58th ASMS Conference on Mass Spectrometry and Allied Topics, Salt Lake City, UT, USA, 23-27 May 2010
81. Diress A, Lorbetskie B, Larocque L, Li X, Alteen M, Isbrucker R, Girard M (2010) Study of aggregation, denaturation and reduction of interferon alpha-2 products by size-exclusion high-performance liquid chromatography with fluorescence detection and biological assays. J Chromatogr A 1217:3297–3306
82. Wang W, Nema S (2010) Protein aggregation-pathways and influencing factors. Int J Pharm 390:89–99
83. Wang W (1999) Instability, stabilization, and formulation of liquid protein pharmaceuticals. Int J Pharm 185:129–188
84. Jones AJS (1993) Analysis of polypeptides and proteins. Adv Drug Deliv Rev 10:29–90
85. Ahrer K, Buchacher A, Iberer G, Jungbauer A (2004) Detection of aggregate formation during production of human immunoglobulin G by means of light scattering. J Chromatogr A 1043:41–46
86. Stulik K, Pacakova V, Ticha M (2003) Some potentialities and drawbacks of contemporary size exclusion chromatography. J Biochem Biophys Methods 56:1–13
87. Arakawa T, Philo JS, Ejima D, Tsumoto K, Arisaka F (2007) Aggregation analysis of therapeutic proteins, part 2: analytical ultracentrifugation and dynamic light scattering. Bioprocess Int 5:36–47
88. Arakawa T, Philo JS, Haruna Sato DE, Tsumoto K (2007) Aggregation analysis of therapeutic proteins, part 3: principles and optimization of field-flow fractionation (FFF). Bioprocess Int 5:52–70
89. Pease LF, Elliott JT, Tsai D, Zachariah MR, Tarlov MJ (2008) Determination of protein aggregation with differential mobility analysis: application to IgG antibody. Biotechnol Bioeng 101:1214–1222
90. Kaufman SL, Skogen JW, Dorman FD, Zarrin F, Lewis KC (1996) Macromolecule analysis based on electrophoretic mobility in air: globular proteins. Anal Chem 68:1895–1904
91. Bacher G, Szymanski WW (2001) Charge-reduced nanoelectrospray ionization combined with differential mobility analysis of peptides, proteins, glycoproteins, non-covalent protein complexes and viruses. J Mass Spectrom 36:1038–1052
92. Adrien B, Patel R, Nowak C, Lucka A (2009) Ion mobility spectrometry for determining binding, bridging and the formation of non-covalent complexes, Presented at the 57th ASMS Conference on Mass Spectrometry and Allied Topi*cs*, Philadelphia, PA, USA, May 31–Jun 4
93. Robinson NE, Robinson AB (2004) Molecular clocks: deamidation of asparaginyl and glutaminyl residues in peptides and proteins. Althouse Press, Cave Junction, OR
94. Alfaro JF, Gillies LA, Sun HG, Dai S, Zang T, Klaene JJ, Kim BJ, Lowenson JD, Clarke SG, Karger BL, Zhou ZS (2008) Chemo-enzymatic detection of protein isoaspartate using protein isoaspartate methyltransferase and hydrazine trapping. Anal Chem 80:3882–3889
95. Cournoyer JJ, Lin C, Bowman MJ, O'Connor PB (2007) Quantitating the relative abundance of isoaspartyl residues in deamidated proteins by electron capture dissociation. J Am Soc Mass Spectrom 18:48–56
96. Chazin WJ, Koerdel J, Thulin E, Hofmann T, Drakenberg T, Forsen S (1989) Identification of an isoaspartyl linkage formed upon deamidation of bovine calbindin D9 k and structural characterization by 2D proton NMR. Biochemistry 28:8646–8653

97. Wang W, Meeler AR, Bergerud LT, Hesselberg M, Byrne M, Wu Z (2012) Quantification and characterization of antibody deamidation by peptide mapping with mass spectrometry. Int J Mass Spectrom 312:107–113
98. Motoie R, Fujii N, Tsunoda S, Nagata K, Shimo-oka T, Kinouchi T, Fujii N, Saito T, Ono K (2009) Localization of D-beta-aspartyl residue-containing proteins in various tissues. Int J Mol Sci 10:1999–2009
99. Schurter BT, Aswad DW (2000) Analysis of isoaspartate in peptides and proteins without the use of radioisotopes. Anal Biochem 282:227–231
100. Clowers B, Dwivedi P, Steiner W, Hill H, Bendiak B (2005) Separation of sodiated isobaric disaccharides and trisaccharides using electrospray ionization-atmospheric pressure ion mobility-time of flight mass spectrometry. J Am Soc Mass Spectrom 16:660–669
101. Gabryelski W, Froese KL (2003) Rapid and sensitive differentiation of anomers, linkage, and position isomers of disaccharides using high-field asymmetric waveform ion mobility spectrometry (FAIMS). J Am Soc Mass Spectrom 14:265–277
102. Yamagaki T, Sato A (2009) Isomeric oligosaccharides analyses using negative-ion electrospray ionization ion mobility spectrometry combined with collision-induced dissociation MS/MS. Anal Sci 25:985–988
103. Fenn LS, McLean JA (2011) Structural resolution of carbohydrate positional and structural isomers based on gas-phase ion mobility-mass spectrometry. Phys Chem Chem Phys 13:2196–2205
104. Chen W, Yu Y, Chakraborty A, Shion H, Skilton S (2011) Differentiating carbohydrate positional and structural isomers by ion mobility mass spectrometry. Paper presented at the 59th ASMS Conference on Mass Spectrometry and Allied Topics, Denver, CO, USA
105. Williams JP, Grabenauer M, Holland RJ, Carpenter CJ, Wormald MR, Giles K, Harvey DJ, Bateman RH, Scrivens JH, Bowers MT (2010) Characterization of simple isomeric oligosaccharides and the rapid separation of glycan mixtures by ion mobility mass spectrometry. Int J Mass Spectrom 298:119–127
106. Plasencia MD, Isailovic D, Merenbloom SI, Mechref Y, Clemmer DE (2008) Resolving and assigning N-linked glycan structural isomers from ovalbumin by IMS-MS. J Am Soc Mass Spectrom 19:1706–1715
107. Wormald MR, Petrescu AJ, Pao YL, Glithero A, Elliott T, Dwek RA (2002) Conformational studies of oligosaccharides and glycopeptides: the complementarity of NMR, X-ray crystallography and molecular modeling. Chem Rev 102:371–386
108. Harvey DJ, Sobott F, Crispin M, Wrobel A, Bonomelli C, Vasiljevic S, Scanlan CN, Scarff CA, Thalassinos K, Scrivens JH (2011) Ion mobility mass spectrometry for extracting spectra of n-glycans directly from incubation mixtures following glycan release: application to glycans from engineered glycoforms of intact, folded HIV gp120. J Am Soc Mass Spectrom 22:568–581
109. Wyttenbach T, Bowers MT (2011) Structural stability from solution to the gas phase: native solution structure of ubiquitin survives analysis in a solvent-free ion mobility-mass spectrometry environment. J Phys Chem B 115:12266–12275
110. Hopper JTS, Oldham NJ (2009) Collision induced unfolding of protein ions in the gas phase studied by ion mobility-mass spectrometry: the effect of ligand binding on conformational stability. J Am Soc Mass Spectrom 20:1851–1858
111. Bagal D, Valliere-Douglass JF, Balland A, Schnier PD (2010) Resolving disulfide structural isoforms of IgG2 monoclonal antibodies by ion mobility mass spectrometry Ion mobility mass spectrometry. Anal Chem 82:6751–6755
112. Atmanene C, Wagner-Rousset E, Malissard M, Chol B, Robert A, Corvaia N, Dorsselaer AV, Beck A, Sanglier-Cianferani S (2009) Extending mass spectrometry contribution to therapeutic monoclonal antibody lead optimization: characterization of immune complexes using noncovalent ESI–MS. Anal Chem 81:6364–6373
113. Williams JP, Grabenauer M, Holland RJ, Carpenter CJ, Wormald MR, Giles K, Harvey DJ, Bateman RH, Scrivens JH, Bowers MT (2010) Characterization of simple isomeric oligosaccharides and the rapid separation of glycanGlycan mixtures by ion mobility mass spectrometry Ion mobility mass spectrometry. Int J Mass Spectrom 298:119–127

Index

G. Chen (ed.), *Characterization of Protein Therapeutics using Mass Spectrometry*,
DOI: 10.1007/978-1-4419-7862-2, © Springer Science+Business Media New York 2013

Printed by Publishers' Graphics LLC
ESRO130521.15.16.1